Electrical Power Systems Quality

Roger C. Dugan

Mark F. McGranaghan

Surya Santoso

H. Wayne Beaty

Third Edition

New York Chicago San Francisco
Lisbon London Madrid Mexico City
Milan New Delhi San Juan
Seoul Singapore Sydney Toronto

1 2 3 4 5 6 7 8 9 0   DOC/DOC   1 8 7 6 5 4 3 2

ISBN 978-0-07-176155-0
MHID 0-07-176155-1

Sponsoring Editor	Acquisitions Coordinator	Proofreader
Joy Evangeline Bramble	Molly T. Wyand	Helen Mules
Editing Supervisor	**Project Manager**	**Art Director, Cover**
Stephen M. Smith	Aloysius Raj,	Jeff Weeks
Production Supervisor	Newgen Publishing and Data Services	**Composition**
Pamela A. Pelton	**Copy Editor**	Newgen Publishing and Data Services
	Sharma Shaktheeswari	

Printed and bound by RR Donnelley.

This book is printed on acid-free paper.

Contents

Foreword

Alex McEachern*

Almost three decades ago—back in the days when the only person with a cell phone and a pager was Dick Tracy in the comic pages—I stumbled into my first power quality problem. A minicomputer (for those of you too young to remember, a minicomputer was a refrigerator-sized device with the computing horsepower of a cheap calculator and the storage capacity of a couple of floppy disks) crashed at about 3:00 P.M. every day.

Everybody was sure it was a power problem. But nobody knew what to do.

A salesman convinced the owner of the company, whose name, oddly enough, was Jerry Lee Lewis—not the famous one—that a transient suppressor would fix any power problem. At great expense, one was installed. The minicomputer kept crashing. And nobody knew what to do.

If only we had this book back then!

Back in those days, power quality was mysterious; most of the people who claimed to understand it were interested in selling something, and few engineers were in a position to understand what appeared to be rare, random events on those dangerous power lines.

Things changed over the next couple of decades. A huge increase in sensitive loads raised the visibility of the power quality problem. Graphic recording instruments, starting with François Martzloff's automatically photographed oscilloscopes and continuing with my own digital PowerScopes, laid out what was going on for every engineer to see.

*Alex McEachern is one of the pioneers in the field of power quality monitoring and analysis. Having spent two decades studying the power quality problem, he has now decided to spend the next two decades fixing it. You can track his progress at http://www.Alex.McEachern.com.

The power quality industry developed, with independent contributions from several segments: the power conditioning industry, the electric utility power quality programs, the instrument manufacturers, and the standard-setting organizations like the IEEE and the IEC. The amount of information available to an engineer exploded. There was great stuff available, but you had to look hard for it, and you had to look in a lot of places.

The book you hold in your hands is what we have needed for years: a single, authoritative source containing all the knowledge about power quality that has accumulated in so many different places. Messrs. Dugan, McGranaghan, Santoso, and Beaty—experts all, and good people too—have performed a valuable service by gathering and organizing this information for us. I'll keep this book nearby on my bookshelf, and pull it down every time I have another question about power quality.

By the way, that power quality problem, three decades ago, was eventually solved. As the most junior engineer around, I was assigned to sit in a folding chair in front of the minicomputer every day at 3:00 p.m. and watch what happened. What happened was this: The company owner came by to visit his big investment on his daily rounds. He smoked a cigar. And, back at the minicomputer factory, the disk drive's air filter had been inadvertently omitted. The smoke particles were enough to crash the computer. Like many power quality problems, it wasn't.

Acknowledgments

This book would not have been possible without the support of our employers and colleagues. Nearly everyone in the Knoxville, Tennessee, office of Electrotek Concepts, Inc., has contributed something to this book. Some contributions are acknowledged, but there were many other contributions that cannot be acknowledged specifically. We thank all the contributors for their help. We also thank our friends in the power industry who have graciously provided us with equipment photographs and other material for inclusion in the book.

The first three of us would like to thank Wayne Beaty for making our words read better and for dealing with the publisher. As anyone who has tried this knows, it is very difficult to write a book while working full time in a demanding job. Wayne removed much of the burden from us, which is greatly appreciated.

Finally, this third edition, as well as the second edition, of the book includes the name of a new author, Surya Santoso. There is a good reason for this. Dr. Santoso was the driving force behind the expansions and updates in many of the chapters, and the other authors are very grateful for his untiring efforts.

Roger C. Dugan
Mark F. McGranaghan
Surya Santoso
H. Wayne Beaty

About the Authors

Roger C. Dugan is a Senior Technical Executive with the Electric Power Research Institute. He has a BSEE degree from Ohio University and an ME degree in electric power engineering from Rensselaer Polytechnic Institute. Mr. Dugan has more than 40 years' experience in electric power quality and distribution system analysis. He was elected a Fellow of the Institute of Electrical and Electronics Engineers (IEEE) for his work in harmonics and transients.

Mark F. McGranaghan is Vice President of the Power Delivery and Utilization Sector of the Electric Power Research Institute. He has BSEE and MSEE degrees from the University of Toledo and an MBA from the University of Pittsburgh. Mr. McGranaghan was vice chairman of the IEEE SCC 22 Power Quality Standards Coordinating Committee and a technical advisor for the IEC SC77A (International Power Quality Standards). He has taught seminars and workshops around the world and has worked with electric utilities and end users to solve power quality problems.

Dr. Surya Santoso is an Associate Professor of Electrical Engineering at the University of Texas. He has MSE and Ph.D. degrees from the University of Texas. Dr. Santoso's research interests are in electric power quality and wind power systems. His contributions include pioneering and advancing the engineering science of root-cause analysis of electric power quality disturbance phenomena, and the time-domain modeling and simulation of electromagnetic transients and wind power systems for integration studies. While working in industry, Dr. Santoso was a principal investigator for more than 20 industrial and utility studies in power quality and wind power integration. He is a senior member of the IEEE.

H. Wayne Beaty is former Senior Editor of *Electrical World* magazine and former Managing Editor of *Electric Light & Power* magazine. He is editor of McGraw-Hill's *Standard Handbook for Electrical Engineers* and *Handbook of Electric Power Calculations*. Mr. Beaty also is editor of Alexander Publications' *An Introduction to Electric Power Distribution*. He has a BSEE degree from the University of Houston and has 55 years' experience in the electric utility industry and publishing industry.

CHAPTER 1

Introduction

Both electric utilities and end users of electric power are becoming increasingly concerned about the quality of electric power. The term *power quality* has become one of the most prolific buzzwords in the power industry since the late 1980s. It is an umbrella concept for a multitude of individual types of power system disturbances. The issues that fall under this umbrella are not necessarily new. What is new is that engineers are now attempting to deal with these issues using a system approach rather than handling them as individual problems.

There are four major reasons for the increased concern:

1. Newer-generation load equipment, with microprocessor-based controls and power electronic devices, is more sensitive to power quality variations than was equipment used in the past.

2. The increasing emphasis on overall power system efficiency has resulted in continued growth in the application of devices such as high-efficiency, adjustable-speed motor drives, and shunt capacitors for power factor correction to reduce losses. This is resulting in increasing harmonic levels on power systems and has many people concerned about the future impact on system capabilities.

3. End users have an increased awareness of power quality issues. Utility customers are becoming better informed about such issues as interruptions, sags, and switching transients and are challenging the utilities to improve the quality of power delivered.

4. Many things are now interconnected in a network. Integrated processes mean that the failure of any component has much more important consequences.

The common thread running through all these reasons for increased concern about the quality of electric power is the continued push for increasing productivity for all utility customers.

1

Manufacturers want faster, more productive, and more efficient machinery. Utilities encourage this effort because it helps their customers become more profitable and also helps defer large investments in substations and generation by using more efficient load equipment. Interestingly, the equipment installed to increase the productivity is also often the equipment that suffers the most from common power disruptions. And the equipment is sometimes the source of additional power quality problems. When entire processes are automated, the efficient operation of machines and their controls becomes increasingly dependent on quality power.

Since the first edition of this book was published, there have been some developments that have had an impact on power quality:

1. Throughout the world, many governments have revised their laws regulating electric utilities with the intent of achieving more cost-competitive sources of electric energy. Deregulation of utilities has complicated the power quality problem. In many geographic areas there is no longer tightly coordinated control of the power from generation through end-use load. While regulatory agencies can change the laws regarding the flow of money, the physical laws of power flow cannot be altered. In order to avoid deterioration of the quality of power supplied to customers, regulators are going to have to expand their thinking beyond traditional reliability indices and address the need for power quality reporting and incentives for the transmission and distribution companies.

2. There has been a substantial increase of interest in distributed generation (DG), that is, generation of power dispersed throughout the power system. There are a number of important power quality issues that must be addressed as part of the overall interconnection evaluation for DG. Therefore, we have added a chapter on DG.

3. The globalization of industry has heightened awareness of deficiencies in power quality around the world. Companies building factories in new areas are suddenly faced with unanticipated problems with the electricity supply due to weaker systems or a different climate. There have been several efforts to benchmark power quality in one part of the world against other areas.

4. Indices have been developed to help benchmark the various aspects of power quality. Regulatory agencies have become involved in performance-based rate-making (PBR), which addresses a particular aspect, reliability, which is associated with interruptions. Some customers have established contracts with utilities for meeting a certain quality of power delivery. We have added a new chapter on this subject.

1.1   What Is Power Quality?

There can be completely different definitions for power quality, depending on one's frame of reference. For example, a utility may define power quality as reliability and show statistics demonstrating that its system is 99.98 percent reliable. Criteria established by regulatory agencies are usually in this vein. A manufacturer of load equipment may define power quality as those characteristics of the power supply that enable the equipment to work properly. These characteristics can be very different for different criteria.

Power quality is ultimately a consumer-driven issue, and the end user's point of reference takes precedence. Therefore, the following definition of a power quality problem is used in this book:

> Any power problem manifested in voltage, current, or frequency deviations that results in failure or misoperation of customer equipment.

There are many misunderstandings regarding the causes of power quality problems. The charts in Fig. 1.1 show the results of one survey conducted by the Georgia Power Company in which both utility personnel and customers were polled about what causes power quality problems. While surveys of other market sectors might indicate different splits between the categories, these charts clearly illustrate one common theme that arises repeatedly in such surveys: The utility's and customer's perspectives are often much different. While both tend to blame about two-thirds of the events on natural phenomena (e.g., lightning), customers, much more frequently than utility personnel, think that the utility is at fault.

When there is a power problem with a piece of equipment, end users may be quick to complain to the utility of an "outage" or "glitch" that has caused the problem. However, the utility records may indicate no abnormal events on the feed to the customer. We have investigated a case where the end-use equipment was knocked off-line 30 times in 9 months, but there were only five operations on the utility substation breaker. It must be realized that there are many events resulting in end-user problems that never show up in the utility statistics. One example is capacitor switching, which is quite common and normal on the utility system, but can cause transient overvoltages that disrupt manufacturing machinery. Another example is a momentary fault elsewhere in the system that causes the voltage to sag briefly at the location of the customer in question. This might cause an adjustable-speed drive or a distributed generator to trip off, but the utility will have no indication that anything was amiss on the feeder unless it has a power quality monitor installed.

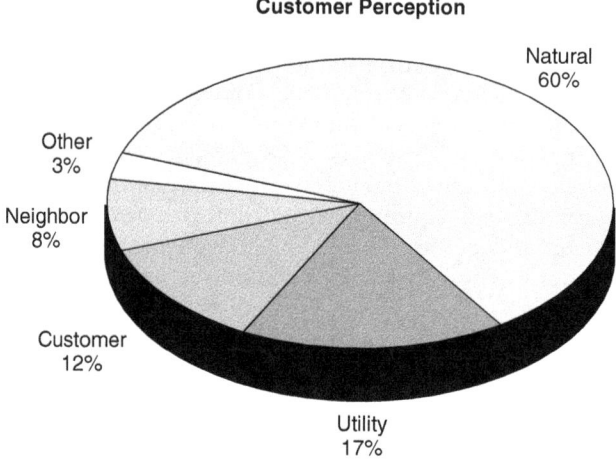

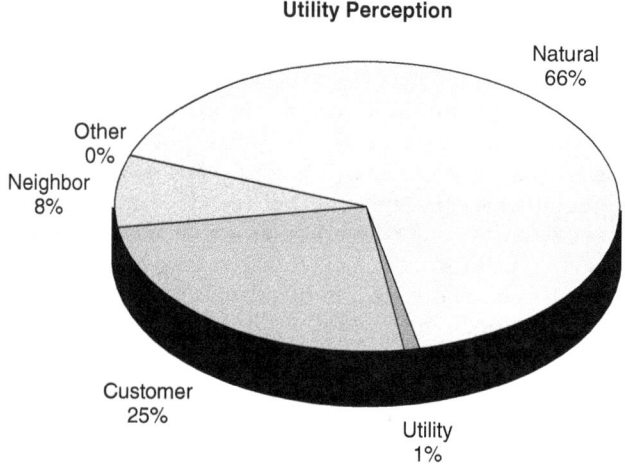

FIGURE 1.1   Results of a survey on the causes of power quality problems. (*Courtesy of Georgia Power Co.*)

In addition to real power quality problems, there are also perceived power quality problems that may actually be related to hardware, software, or control system malfunctions. Electronic components can degrade over time due to repeated transient voltages and eventually fail due to a relatively low magnitude event. Thus, it is sometimes difficult to associate a failure with a specific cause. It is becoming more common that designers of control software for microprocessor-based equipment have an incomplete knowledge of how power systems operate and do not anticipate all types of malfunction events. Thus, a device can misbehave because of a deficiency in the embedded software. This is particularly common with early versions of new computer-controlled load equipment. One of the main objectives of this book is to educate utilities, end users, and equipment suppliers alike to reduce the frequency of malfunctions caused by software deficiencies.

In response to this growing concern for power quality, electric utilities have programs that help them respond to customer concerns. The philosophy of these programs ranges from reactive, where the utility responds to customer complaints, to proactive, where the utility is involved in educating the customer and promoting services that can help develop solutions to power quality problems. The regulatory issues facing utilities may play an important role in how their programs are structured. Since power quality problems often involve interactions between the supply system and the customer facility and equipment, regulators should make sure that distribution companies have incentives to work with customers and help customers solve these problems.

The economics involved in solving a power quality problem must also be included in the analysis. It is not always economical to eliminate power quality variations on the supply side. In many cases, the optimal solution to a problem may involve making a particular piece of sensitive equipment less sensitive to power quality variations. The level of power quality required is that level which will result in proper operation of the equipment at a particular facility.

Power quality, like quality in other goods and services, is difficult to quantify. There is no single accepted definition of quality power. There are standards for voltage and other technical criteria that may be measured, but the ultimate measure of power quality is determined by the performance and productivity of end-user equipment. If the electric power is inadequate for those needs, then the "quality" is lacking.

Perhaps nothing has been more symbolic of a mismatch in the power delivery system and consumer technology than the "blinking clock" phenomenon. Clock designers created the blinking display of a digital clock to warn of possible incorrect time after loss of power and inadvertently created one of the first power quality monitors. It has made the homeowner aware that there are numerous minor disturbances occurring throughout the power delivery system that may have no ill effects other than to be detected by a clock. Many appliances now have a built-in clock, so the average household may have about a dozen clocks that must be reset when there is a brief interruption. Older-technology motor-driven clocks would simply lose a few seconds during minor disturbances and then promptly come back into synchronism.

1.2   Power Quality = Voltage Quality

The common term for describing the subject of this book is *power quality*; however, it is actually the quality of the voltage that is being addressed in most cases. Technically, in engineering terms, power is the rate of energy delivery and is proportional to the product of the voltage and current. It would be difficult to define the quality of this quantity in any meaningful manner. The power supply system can only control the quality of the voltage; it has no control over the currents that particular loads might draw. Therefore, the standards in the power quality area are devoted to maintaining the supply voltage within certain limits.

AC power systems are designed to operate at a sinusoidal voltage of a given frequency [typically 50 or 60 hertz (Hz)] and magnitude. Any significant deviation in the waveform magnitude, frequency, or purity is a potential power quality problem.

Of course, there is always a close relationship between voltage and current in any practical power system. Although the generators may provide a near-perfect sine-wave voltage, the current passing through the impedance of the system can cause a variety of disturbances to the voltage. For example,

1. The current resulting from a short circuit causes the voltage to sag or disappear completely, as the case may be.

2. Currents from lightning strokes passing through the power system cause high-impulse voltages that frequently flash over insulation and lead to other phenomena, such as short circuits.

3. Distorted currents from harmonic-producing loads also distort the voltage as they pass through the system impedance. Thus a distorted voltage is presented to other end users.

Therefore, while it is the voltage with which we are ultimately concerned, we must also address phenomena in the current to understand the basis of many power quality problems.

1.3   Why Are We Concerned about Power Quality?

The ultimate reason that we are interested in power quality is economic value. There are economic impacts on utilities, their customers, and suppliers of load equipment.

The quality of power can have a direct economic impact on many industrial consumers. There has recently been a great emphasis on revitalizing industry with more automation and more modern equipment. This usually means electronically controlled, energy-efficient equipment that is often much more sensitive to deviations in the supply voltage than were its electromechanical predecessors.

Thus, like the blinking clock in residences, industrial customers are now more acutely aware of minor disturbances in the power system. There is big money associated with these disturbances. It is not uncommon for a single, commonplace, momentary utility breaker operation to result in a $10,000 loss to an average-sized industrial concern by shutting down a production line that requires 4 hours to restart. In the semiconductor manufacturing industry, the economic impacts associated with equipment sensitivity to momentary voltage sags resulted in the development of a standard for equipment ride-through (SEMI Standard F-47, *Specification for Semiconductor Process Equipment Voltage Sag Immunity*).

The electric utility is concerned about power quality issues as well. Meeting customer expectations and maintaining customer confidence are strong motivators. With today's movement toward deregulation and competition between utilities, they are more important than ever. The loss of a disgruntled customer to a competing power supplier can have a very significant impact financially on a utility.

Besides the obvious financial impacts on both utilities and industrial customers, there are numerous indirect and intangible costs associated with power quality problems. Residential customers typically do not suffer direct financial loss or the inability to earn income as a result of most power quality problems, but they can be a potent force when they perceive that the utility is providing poor service. Home computer usage is now commonplace, and many people use the Internet daily. Users become more sensitive to interruptions when they are reliant on this technology. The sheer number of complaints requires utilities to provide staffing to handle them. Also, public interest groups frequently intervene with public service commissions, requiring the utilities to expend financial resources on lawyers, consultants, studies, and the like to counter the intervention. While all this is certainly not the result of power quality problems, a reputation for providing poor quality service does not help matters.

Load equipment suppliers generally find themselves in a very competitive market with most customers buying on lowest cost. Thus, there is a general disincentive to add features to the equipment to withstand common disturbances unless the customer specifies otherwise. Many manufacturers are also unaware of the types of disturbances that can occur on power systems. The primary responsibility for correcting inadequacies in load equipment ultimately lies with the end user who must purchase and operate it. Specifications must include power performance criteria. Since many end users are also unaware of the pitfalls, one useful service that utilities can provide is dissemination of information on power quality and the requirements of load equipment to properly operate in the real world. For instance, the SEMI F-47 standard previously referenced was developed through joint task forces consisting of semiconductor industry and utility engineers working together.

1.4   The Power Quality Evaluation Procedure

Power quality problems encompass a wide range of different phenomena, as described in Chap. 2. Each of these phenomena may have a variety of different causes and different solutions that can be used to improve the power quality and equipment performance. However, it is useful to look at the general steps that are associated with investigating many of these problems, especially if the steps can involve interaction between the utility supply system and the customer facility. Figure 1.2 gives some general steps that are often required in a power quality investigation, along with the major considerations that must be addressed at each step.

The general procedure must also consider whether the evaluation involves an existing power quality problem or one that could result from a new design or from proposed changes to the system. Measurements will play an important role for almost any power quality concern. This is the primary method of characterizing the problem or the existing system that is being evaluated. When performing the measurements, it is important to record impacts of

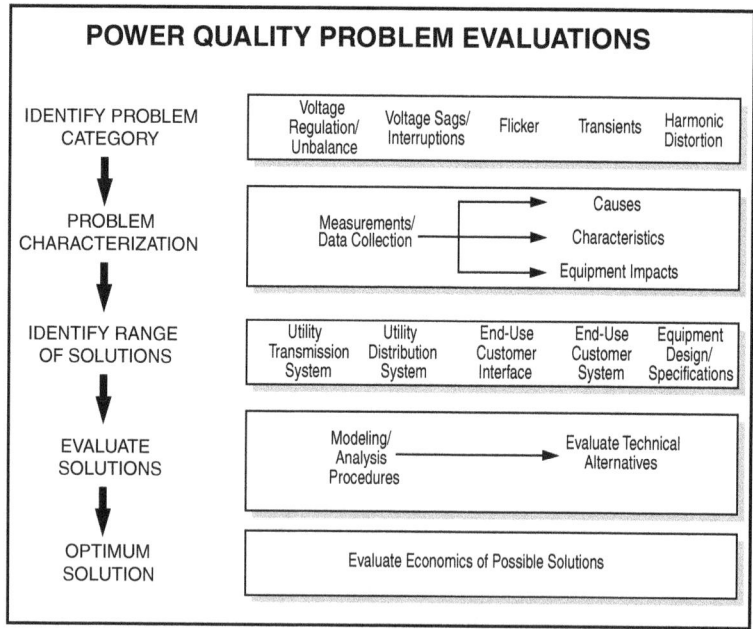

FIGURE 1.2   Basic steps involved in a power quality evaluation.

the power quality variations at the same time so that problems can be correlated with possible causes.

Solutions need to be evaluated using a system perspective, and both the economics and the technical limitations must be considered. Possible solutions are identified at all levels of the system from utility supply to the end-use equipment being affected. Solutions that are not technically viable get thrown out, and the rest of the alternatives are compared on an economic basis. The optimum solution will depend on the type of problem, the number of end users being impacted, and the possible solutions.

The overall procedure is introduced here to provide a framework for the more detailed technical information and procedures that are described in each chapter of this book. The relative role of simulations and measurements for evaluating power quality problems is described separately for each type of power quality phenomenon. The available solutions and the economics of these solutions are also addressed in the individual chapters.

1.5 Who Should Use This Book

Power quality issues frequently cross the energy meter boundary between the utility and the end user. Therefore, this book addresses issues of interest to both utility engineers and industrial engineers and technicians. Every attempt has been made to provide a balanced approach to the presentation of the problems and solutions.

The book should also be of interest to designers of manufacturing equipment, computers, appliances, and other load equipment. It will help designers learn about the environment in which their equipment must operate and the peculiar difficulties their customers might have when trying to operate their equipment. Hopefully, this book will serve as common ground on which these three entities—utility, customer, and equipment supplier—can meet to resolve problems.

This book is intended to serve both as a reference book and a textbook for utility distribution engineers and key technical personnel with industrial end users. Parts of the book are tutorial in nature for the newcomer to power quality and power systems, while other parts are very technical, intended strictly as reference for the experienced practitioner.

1.6 Overview of the Contents

The chapters of the book are organized as follows:

Chapter 2 provides background material on the different types of power quality phenomena and describes standard terms and definitions for power quality phenomena.

Chapters 3 through 7 are the heart of the book, describing four major classes of power quality variations in detail: sags and interruptions, transients, harmonics, and long-duration voltage variations. The material on harmonics has again been expanded from the second edition. Chapter 5 describes the basic harmonic phenomena, while Chap. 6 concentrates on methods for dealing with harmonic distortion.

Chapter 8 describes techniques for benchmarking power quality and how to apply power quality standards. Important standards dealing with power quality issues, primarily developed by the International Electrotechnical Commission (IEC) and the Institute for Electrical and Electronics Engineers (IEEE), are described and referenced in the chapters where they are applicable. Chapter 8 provides an overview of the overall power quality standards structure where these standards are headed. Chapter 9 addresses the subject of distributed generation (DG) interconnected to the distribution system. This chapter has been updated with some of the latest concerns over power quality disturbances due to interconnected DG. This chapter discusses the relationship between DG and power quality.

Chapter 10 provides a concise summary of key wiring and grounding problems and gives some general guidance on identifying and correcting them. Many power quality problems experienced by end users are the result of inadequate wiring or incorrect installations. However, the emphasis of this book is on power quality phenomena that can be addressed analytically and affect both sides of the meter. This chapter is included to give power quality engineers a basic understanding of the principles with respect to power quality issues.

Finally, Chap. 11 provides a guide for site surveys and power quality monitoring. There continue to be major advances in power quality monitoring technology. Power quality monitoring capability is appearing in many different types of equipment as we move toward the "Smart Grid."

CHAPTER 2

Terms and Definitions

2.1 Need for a Consistent Vocabulary

The term *power quality* is applied to a wide variety of electromagnetic phenomena on the power system. The increasing application of electronic equipment and distributed generation has heightened the interest in power quality in recent years, and this has been accompanied by the development of a special terminology to describe the phenomena. Unfortunately, this terminology has not been consistent across different segments of the industry. This has caused a considerable amount of confusion as both vendors and end users have struggled to understand why electrical equipment is not working as expected. Likewise, it is confusing to wade through the vendor jargon and differentiate between a myriad of proposed solutions.

Many ambiguous words have been used that have multiple or unclear meanings. For example, *surge* is used to describe a wide variety of disturbances that cause equipment failures or mis-operation. A *surge suppressor* can suppress some of these but will have absolutely no effect on others. Terms like *glitch* and *blink* that have no technical meaning at all have crept into the vocabulary. Unscrupulous marketers take advantage of the ignorance of the general public, selling overpriced gadgets with near-miraculous claims for improving the power quality. Of course, all these come with a money-back guarantee. Readers can protect themselves by obtaining a better understanding of power quality vocabulary and insisting on technical explanations of how a gadget works. Our basic rule: If they won't tell you what is in the box and how it works, don't buy it!

This chapter describes a consistent terminology that can be used to describe power quality variations. We also explain why some commonly used terminology is inappropriate in power quality discussions.

2.2 General Classes of Power Quality Problems

The terminology presented here reflects recent U.S. and international efforts to standardize definitions of power quality terms. The IEEE Standards Coordinating Committee 22 (IEEE SCC22) has led the main effort in the United States to coordinate power quality standards. It has the responsibilities across several societies of the IEEE, principally the Industry Applications Society and the Power Engineering Society. It coordinates with international efforts through liaisons with the IEC and the Conseil International des Grands Réseaux Électriques à Haute Tension (CIGRE; in English, International Conference on Large High-Voltage Electric Systems).

The IEC classifies electromagnetic phenomena into the groups shown in Table 2.1.[1] We will be primarily concerned with the first four classes in this book.

Conducted low-frequency phenomena
Harmonics, interharmonics
Signal systems (power line carrier)
Voltage fluctuations (flicker)
Voltage dips and interruptions
Voltage imbalance (unbalance)
Power frequency variations
Induced low-frequency voltages
DC in ac networks
Radiated low-frequency phenomena
Magnetic fields
Electric fields
Conducted high-frequency phenomena
Induced continuous-wave (CW) voltages or currents
Unidirectional transients
Oscillatory transients
Radiated high-frequency phenomena
Magnetic fields
Electric fields
Electromagnetic fields
Continuous waves
Transients
Electrostatic discharge phenomena (ESD)
Nuclear electromagnetic pulse (NEMP)

TABLE 2.1 Principal Phenomena Causing Electromagnetic Disturbances as Classified by the IEC

U.S. power industry efforts to develop recommended practices for monitoring electric power quality have added a few terms to the IEC terminology.[2] *Sag* is used as a synonym to the IEC term *dip*. The category *short-duration variations* is used to refer to *voltage dips* and *short interruptions*. The term *swell* is introduced as an inverse to sag (dip). The category *long-duration variation* has been added to deal with American National Standards Institute (ANSI) C84.1 limits. The category *noise* has been added to deal with broadband conducted phenomena. The category *waveform distortion* is used as a container category for the IEC *harmonics, interharmonics,* and *dc in ac networks* phenomena as well as an additional phenomenon from IEEE Standard 519–1992, *Recommended Practices and Requirements for Harmonic Control in Electrical Power Systems,* called *notching.*

Table 2.2 shows the categorization of electromagnetic phenomena used for the power quality community. The phenomena listed in the table can be described further by listing appropriate attributes. For steady-state phenomena, the following attributes can be used[1]:

- Amplitude
- Frequency
- Spectrum
- Modulation
- Source impedance
- Notch depth
- Notch area

For non-steady-state phenomena, other attributes may be required[1]:

- Rate of rise
- Amplitude
- Duration
- Spectrum
- Frequency
- Rate of occurrence
- Energy potential
- Source impedance

Table 2.2 provides information regarding typical spectral content, duration, and magnitude where appropriate for each category of electromagnetic phenomena.[1,4,5] The categories of the table, when used with the attributes previously mentioned, provide a means to clearly describe an electromagnetic disturbance. The categories and

Categories	Typical Spectral Content	Typical Duration	Typical Voltage
1.0 Transients Impulsive			
1.1.1 Nanosecond	5-ns rise	<50 ns	
1.1.2 Microsecond	1-µs rise	50 ns–1 ms	
1.1.3 Millisecond	0.1-ms rise	>1 ms	
1.2 Oscillatory			
1.2.1 Low frequency	<5 kHz	0.3–50 ms	0–4 pu
1.2.2 Medium frequency	5–500 kHz	20 µs	0–8 pu
1.2.3 High frequency	0.5–5 MHz	5 µs	0–4 pu
2.0 Short-duration variations			
2.1 Instantaneous			
2.1.1 Interruption		0.5–30 cycles	<0.1 pu
2.1.2 Sag (dip)		0.5–30 cycles	0.1–0.9 pu
2.1.3 Swell		0.5–30 cycles	1.1–1.8 pu
2.2 Momentary			
2.2.1 Interruption		30 cycles–3 s	<0.1 pu
2.2.2 Sag (dip)		30 cycles–3 s	0.1–0.9 pu
2.2.3 Swell		30 cycles–3 s	1.1–1.4 pu
2.3 Temporary			
2.3.1 Interruption		3 s–1 min	<0.1 pu
2.3.2 Sag (dip)		3 s–1 min	0.1–0.9 pu
2.3.3 Swell		3 s–1 min	1.1–1.2 pu
3.0 Long-duration variations			
3.1 Interruption, sustained		>1 min	0.0 pu
3.2 Undervoltages		>1 min	0.8–0.9 pu
3.3 Overvoltages		>1 min	1.1–1.2 pu
4.0 Voltage unbalance			
5.1 DC offset		Steady state	0–0.1%
5.2 Harmonics	0–100th harmonic	Steady state	0–20%
5.3 Interharmonics	0–6 kHz	Steady state	0–2%
5.4 Notching		Steady state	
5.5 Noise	Broadband	Steady state	0–1%
6.0 Voltage fluctuations	<25 Hz	Intermittent	0.1–7%
			0.2–2 Pst
7.0 Power frequency variations		<10 s	

NOTE: s = second, ns = nanosecond, µs = microsecond, ms = millisecond, kHz = kilohertz, MHz = megahertz, min = minute, pu = per unit.

TABLE 2.2   Categories and Characteristics of Power System Electromagnetic Phenomena

their descriptions are important in order to be able to classify measurement results and to describe electromagnetic phenomena which can cause power quality problems.

2.3   Transients

The term *transients* has long been used in the analysis of power system variations to denote an event that is undesirable and momentary in nature. The notion of a damped oscillatory transient due to an *RLC* network is probably what most power engineers think of when they hear the word transient.

Other definitions in common use are broad in scope and simply state that a transient is "that part of the change in a variable that disappears during transition from one steady state operating condition to another."[8] Unfortunately, this definition could be used to describe just about anything unusual that happens on the power system.

Another word in common usage that is often considered synonymous with transient is *surge*. A utility engineer may think of a surge as the transient resulting from a lightning stroke for which a surge arrester is used for protection. End users frequently use the word indiscriminantly to describe anything unusual that might be observed on the power supply ranging from sags to swells to interruptions. Because there are many potential ambiguities with this word in the power quality field, we will generally avoid using it unless we have specifically defined what it refers to.

Broadly speaking, transients can be classified into two categories, *impulsive* and *oscillatory*. These terms reflect the wave shape of a current or voltage transient. We will describe these two categories in more detail.

2.3.1   Impulsive Transient

An *impulsive transient* is a sudden, non–power frequency change in the steady-state condition of voltage, current, or both that is unidirectional in polarity (primarily either positive or negative).

Impulsive transients are normally characterized by their rise and decay times, which can also be revealed by their spectral content. For example, a 1.2×50-μs 2000-volt (V) impulsive transient nominally rises from zero to its peak value of 2000 V in 1.2 μs and then decays to half its peak value in 50 μs. The most common cause of impulsive transients is lightning. Figure 2.1 illustrates a typical current impulsive transient caused by lightning.

Because of the high frequencies involved, the shape of impulsive transients can be changed quickly by circuit components and may have significantly different characteristics when viewed from different parts of the power system. They are generally not conducted

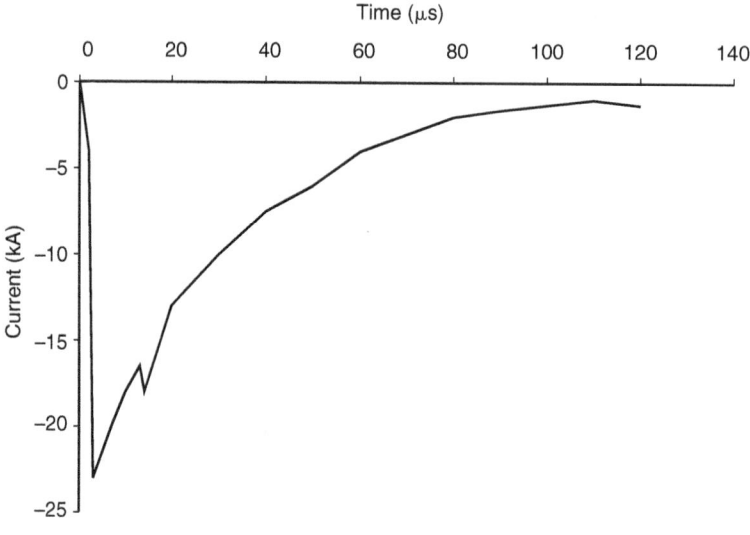

FigURE **2.1** Lightning stroke current impulsive transient.

far from the source of where they enter the power system, although they may, in some cases, be conducted for quite some distance along utility lines. Impulsive transients can excite the natural frequency of power system circuits and produce oscillatory transients.

2.3.2 Oscillatory Transient

An oscillatory transient is a sudden, non–power frequency change in the steady-state condition of voltage, current, or both that includes both positive and negative polarity values.

An oscillatory transient consists of a voltage or current whose instantaneous value changes polarity rapidly. It is described by its spectral content (predominate frequency), duration, and magnitude. The spectral content subclasses defined in Table 2.2 are high, medium, and low frequency. The frequency ranges for these classifications are chosen to coincide with common types of power system oscillatory transient phenomena.

Oscillatory transients with a primary frequency component greater than 500 kHz and a typical duration measured in microseconds (or several cycles of the principal frequency) are considered *high-frequency transients*. These transients are often the result of a local system response to an impulsive transient.

A transient with a primary frequency component between 5 and 500 kHz with duration measured in the tens of microseconds (or several cycles of the principal frequency) is termed a *medium-frequency transient*.

Back-to-back capacitor energization results in oscillatory transient currents in the tens of kilohertz as illustrated in Fig. 2.2. Cable switching results in oscillatory voltage transients in the same frequency range. Medium-frequency transients can also be the result of a system response to an impulsive transient.

A transient with a primary frequency component less than 5 kHz, and a duration from 0.3 to 50 ms, is considered a *low-frequency transient.* This category of phenomena is frequently encountered on utility subtransmission and distribution systems and is caused by many types of events. The most frequent is capacitor bank energization, which typically results in an oscillatory voltage transient with a primary frequency between 300 and 900 Hz. The peak magnitude can approach 2.0 per unit (pu), but is typically 1.3 to 1.5 pu with a duration of between 0.5 and 3 cycles depending on the system damping (Fig. 2.3).

Oscillatory transients with principal frequencies less than 300 Hz can also be found on the distribution system. These are generally associated with ferroresonance and transformer energization (Fig. 2.4). Transients involving series capacitors could also fall into this category. They occur when the system responds by resonating with low-frequency components in the transformer inrush current (second and third harmonic) or when unusual conditions result in ferroresonance.

It is also possible to categorize transients (and other disturbances) according to their *mode.* Basically, a transient in a three-phase system with a separate neutral conductor can be either *common mode* or *normal mode,* depending on whether it appears between line or neutral and ground, or between line and neutral.

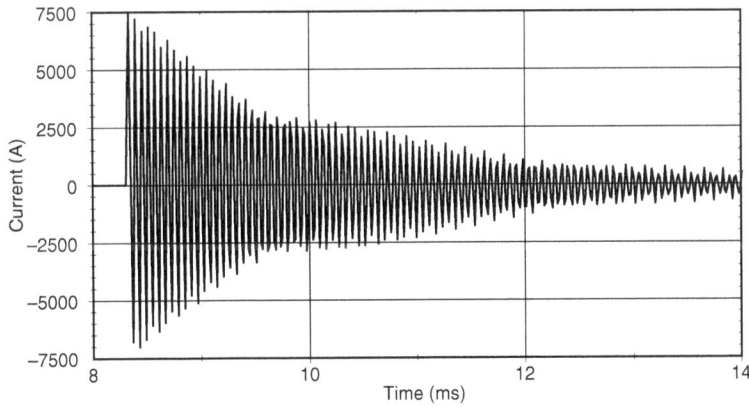

FIGURE 2.2 Oscillatory transient current caused by back-to-back capacitor switching.

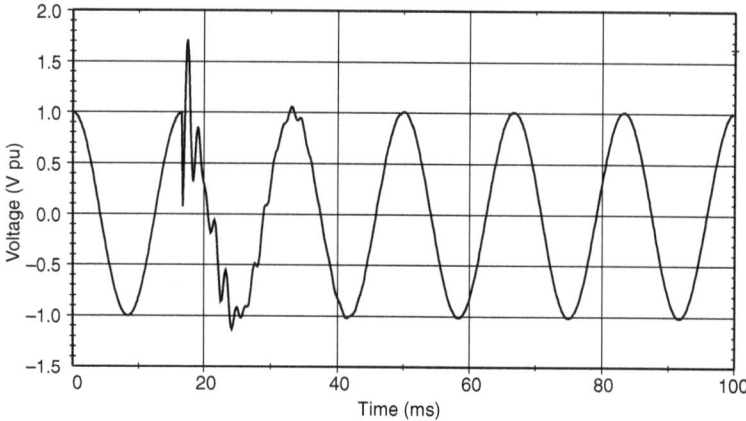

Figure 2.3   Low-frequency oscillatory transient caused by capacitor bank energization. 34.5-kV bus voltage.

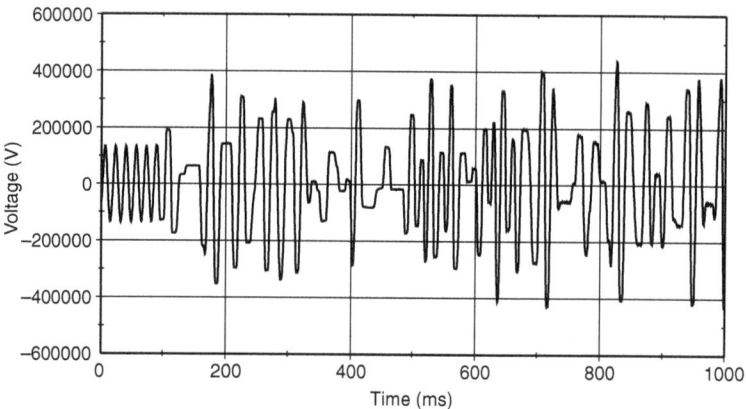

Figure 2.4   Low-frequency oscillatory transient caused by ferroresonance of an unloaded transformer.

2.4   Long-Duration Voltage Variations

Long-duration variations encompass root-mean-square (rms) deviations at power frequencies for longer than 1 min. ANSI C84.1 specifies the steady-state voltage tolerances expected on a power system. A voltage variation is considered to be long duration when the ANSI limits are exceeded for greater than 1 min.

Long-duration variations can be either *overvoltages* or *undervoltages*. Overvoltages and undervoltages generally are not the result of system faults, but are caused by load variations on the system and system switching operations. Such variations are typically displayed as plots of rms voltage versus time.

2.4.1 Overvoltage

An *overvoltage* is an increase in the rms ac voltage greater than 110 percent at the power frequency for a duration longer than 1 min.

Overvoltages are usually the result of load switching (e.g., switching off a large load or energizing a capacitor bank). The overvoltages result because either the system is too weak for the desired voltage regulation or voltage controls are inadequate. Incorrect tap settings on transformers can also result in system overvoltages.

2.4.2 Undervoltage

An *undervoltage* is a decrease in the rms ac voltage to less than 90 percent at the power frequency for a duration longer than 1 min.

Undervoltages are the result of switching events that are the opposite of the events that cause overvoltages. A load switching on or a capacitor bank switching off can cause an undervoltage until voltage regulation equipment on the system can bring the voltage back to within tolerances. Overloaded circuits can result in undervoltages also.

The term *brownout* is often used to describe sustained periods of undervoltage initiated as a specific utility dispatch strategy to reduce power demand. Because there is no formal definition for brownout and it is not as clear as the term undervoltage when trying to characterize a disturbance, the term brownout should be avoided.

2.4.3 Sustained Interruptions

When the supply voltage has been zero for a period of time in excess of 1 min, the long-duration voltage variation is considered a *sustained interruption*. Voltage interruptions longer than 1 min are often permanent and require human intervention to repair the system for restoration. The term sustained interruption refers to specific power system phenomena and, in general, has no relation to the usage of the term *outage*. Utilities use outage or interruption to describe phenomena of similar nature for reliability reporting purposes. However, this causes confusion for end users who think of an outage as any interruption of power that shuts down a process. This could be as little as one-half of a cycle. *Outage*, as defined in IEEE Standard 100,[8] does not refer to a specific phenomenon, but rather to the state of a component in a system that has failed to function as expected. Also, use of the term *interruption* in the context of power quality monitoring has no relation to reliability or other continuity of service statistics. Thus, this term has been defined to be more specific regarding the absence of voltage for long periods.

2.5 Short-Duration Voltage Variations

This category encompasses the IEC category of *voltage dips and short interruptions*. Each type of variation can be designated as *instantaneous*, *momentary*, or *temporary*, depending on its duration as defined in Table 2.2.

Short-duration voltage variations are caused by fault conditions, the energization of large loads which require high starting currents, or intermittent loose connections in power wiring. Depending on the fault location and the system conditions, the fault can cause either temporary voltage drops (*sags*), voltage rises (*swells*), or a complete loss of voltage (*interruptions*). The fault condition can be close to or remote from the point of interest. In either case, the impact on the voltage during the actual fault condition is of the short-duration variation until protective devices operate to clear the fault.

2.5.1 Interruption

An *interruption* occurs when the supply voltage or load current decreases to less than 0.1 pu for a period of time not exceeding 1 min.

Interruptions can be the result of power system faults, equipment failures, and control malfunctions. The interruptions are measured by their duration since the voltage magnitude is always less than 10 percent of nominal. The duration of an interruption due to a fault on the utility system is determined by the operating time of utility protective devices. Instantaneous reclosing generally will limit the interruption caused by a nonpermanent fault to less than 30 cycles. Delayed reclosing of the protective device may cause a momentary or temporary interruption. The duration of an interruption due to equipment malfunctions or loose connections can be irregular.

Some interruptions may be preceded by a voltage sag when these interruptions are due to faults on the source system. The voltage sag occurs between the time a fault initiates and the protective device operates. Figure 2.5 shows such a momentary interruption during which voltage on one phase sags to about 20 percent for about 3 cycles and then drops to zero for about 1.8 s until the recloser closes back in.

2.5.2 Sags (Dips)

A *sag* is a decrease to between 0.1 and 0.9 pu in rms voltage or current at the power frequency for durations from 0.5 cycle to 1 min.

The power quality community has used the term *sag* for many years to describe a short-duration voltage decrease. Although the

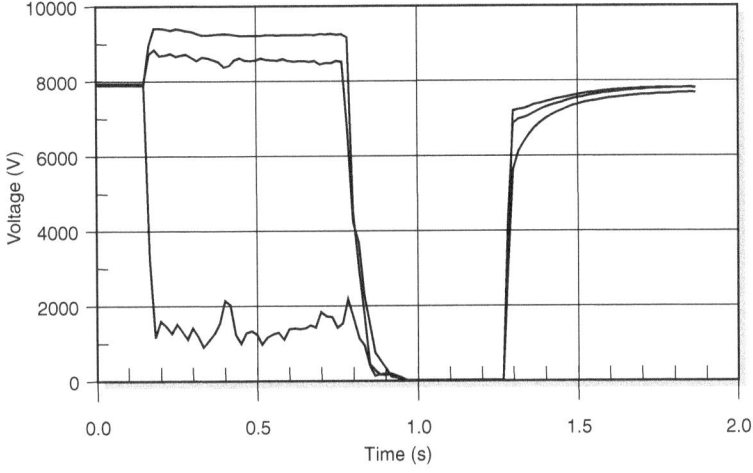

FIGURE 2.5   Three-phase rms voltages for a momentary interruption due to a fault and subsequent recloser operation.

term has not been formally defined, it has been increasingly accepted and used by utilities, manufacturers, and end users. The IEC definition for this phenomenon is *dip*. The two terms are considered interchangeable, with *sag* being the preferred synonym in the U.S. power quality community.

Terminology used to describe the magnitude of a voltage sag is often confusing. A "20 percent sag" can refer to a sag which results in a voltage of 0.8 or 0.2 pu. The preferred terminology would be one that leaves no doubt as to the resulting voltage level: "a sag to 0.8 pu" or "a sag whose magnitude was 20 percent." When not specified otherwise, a 20 percent sag will be considered an event during which the rms voltage decreased by 20 percent to 0.8 pu. The nominal, or base, voltage level should also be specified.

Voltage sags are usually associated with system faults but can also be caused by energization of heavy loads or starting of large motors. Figure 2.6 shows a typical voltage sag that can be associated with a single-line-to-ground (SLG) fault on another feeder from the same substation. An 80 percent sag exists for about 3 cycles until the substation breaker is able to interrupt the fault current. Typical fault clearing times range from 3 to 30 cycles, depending on the fault current magnitude and the type of overcurrent protection.

Figure 2.7 illustrates the effect of a large motor starting. An induction motor will draw 6 to 10 times its full load current during start-up. If the current magnitude is large relative to the available fault current in the system at that point, the resulting voltage sag can be significant. In this case, the voltage sags immediately to 80 percent

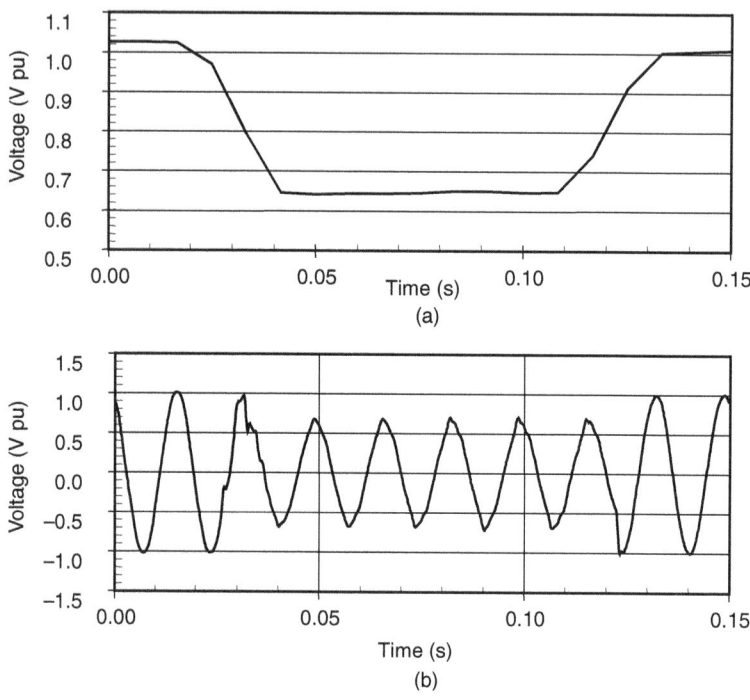

FIGURE 2.6 Voltage sag caused by an SLG fault. (a) RMS waveform for voltage sag event. (b) Voltage sag waveform.

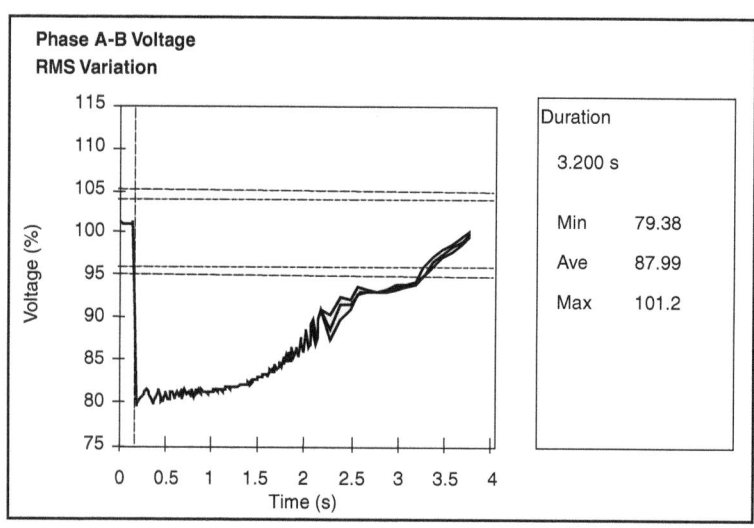

FIGURE 2.7 Temporary voltage sag caused by motor starting.

and then gradually returns to normal in about 3 s. Note the difference in time frame between this and sags due to utility system faults.

Until recent efforts, the duration of sag events has not been clearly defined. Typical sag duration is defined in some publications as ranging from 2 ms (about one-tenth of a cycle) to a couple of minutes. Undervoltages that last less than one-half cycle cannot be characterized effectively by a change in the rms value of the fundamental frequency value. Therefore, these events are considered *transients*. Undervoltages that last longer than 1 min can typically be controlled by voltage regulation equipment and may be associated with causes other than system faults. Therefore, these are classified as long-duration variations.

Sag durations are subdivided here into three categories— instantaneous, momentary, and temporary—which coincide with the three categories of interruptions and swells. These durations are intended to correspond to typical utility protective device operation times as well as duration divisions recommended by international technical organizations.[5]

2.5.3   Swells

A *swell* is defined as an increase to between 1.1 and 1.8 pu in rms voltage or current at the power frequency for durations from 0.5 cycle to 1 min.

As with sags, swells are usually associated with system fault conditions, but they are not as common as voltage sags. One way that a swell can occur is from the temporary voltage rise on the unfaulted phases during an SLG fault. Figure 2.8 illustrates a voltage swell

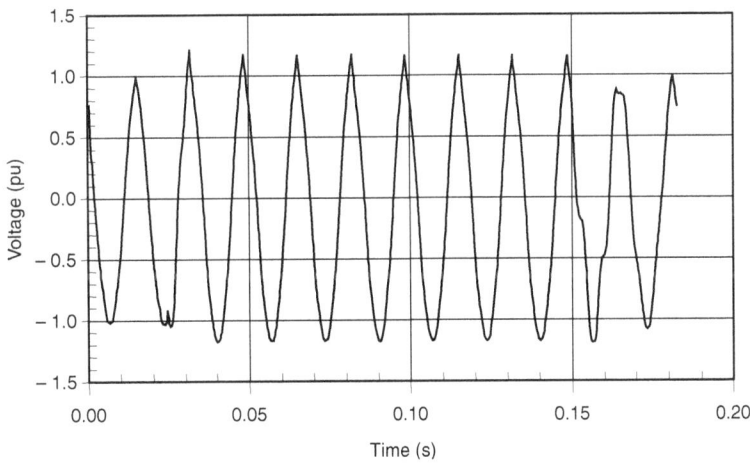

FIGURE 2.8   Instantaneous voltage swell caused by an SLG fault.

caused by an SLG fault. Swells can also be caused by switching off a large load or energizing a large capacitor bank.

Swells are characterized by their magnitude (rms value) and duration. The severity of a voltage swell during a fault condition is a function of the fault location, system impedance, and grounding. On an ungrounded system, with an infinite zero-sequence impedance, the line-to-ground voltages on the ungrounded phases will be 1.73 pu during an SLG fault condition. Close to the substation on a grounded system, there will be little or no voltage rise on the unfaulted phases because the substation transformer is usually connected delta-wye, providing a low-impedance zero-sequence path for the fault current. Faults at different points along four-wire, multigrounded feeders will have varying degrees of voltage swells on the unfaulted phases. A 15 percent swell, like that shown in Fig. 2.8, is common on U.S. utility feeders.

The term *momentary overvoltage* is used by many writers as a synonym for the term *swell*.

2.6 Voltage Imbalance

Voltage imbalance (also called *voltage unbalance*) is sometimes defined as the maximum deviation from the average of the three-phase voltages or currents, divided by the average of the three-phase voltages or currents, expressed in percent.

Imbalance is more rigorously defined in the standards[6,8,11,12] using symmetrical components. The ratio of either the negative- or

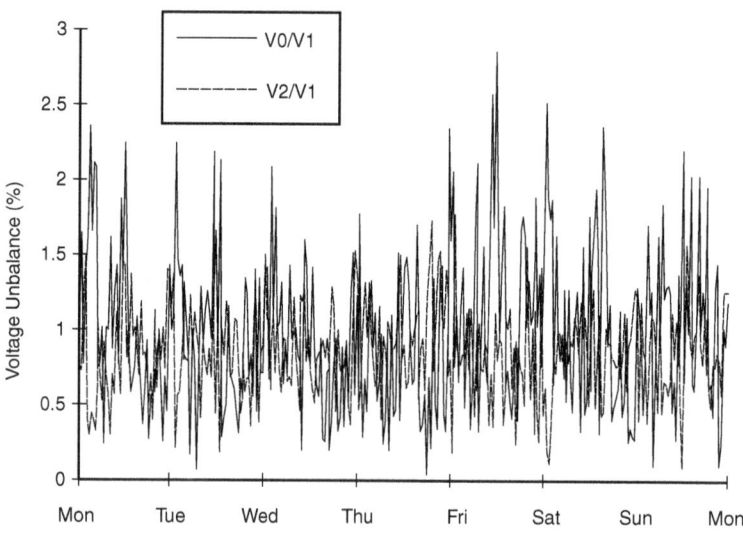

FIGURE 2.9 Voltage unbalance trend for a residential feeder.

zero-sequence component to the positive-sequence component can be used to specify the percent unbalance. Recent standards[11] specify that the negative-sequence method be used. Figure 2.9 shows an example of these two ratios for a 1-week trend of imbalance on a residential feeder.

The primary source of voltage unbalances of less than 2 percent is single-phase loads on a three-phase circuit. Voltage unbalance can also be the result of blown fuses in one phase of a three-phase capacitor bank. Severe voltage unbalance (greater than 5 percent) can result from single-phasing conditions.

2.7 Waveform Distortion

Waveform distortion is defined as a steady-state deviation from an ideal sine wave of power frequency principally characterized by the spectral content of the deviation.

There are five primary types of waveform distortion:

- DC offset
- Harmonics
- Interharmonics
- Notching
- Noise

DC Offset

The presence of a dc voltage or current in an ac power system is termed *dc offset.* This can occur as the result of a geomagnetic disturbance or asymmetry of electronic power converters. Incandescent light bulb life extenders, for example, may consist of diodes that reduce the rms voltage supplied to the light bulb by half-wave rectification. Direct current in ac networks can have a detrimental effect by biasing transformer cores so they saturate in normal operation. This causes additional heating and loss of transformer life. Direct current may also cause the electrolytic erosion of grounding electrodes and other connectors.

Harmonics

Harmonics are sinusoidal voltages or currents having frequencies that are integer multiples of the frequency at which the supply system is designed to operate (termed the *fundamental* frequency; usually 50 or 60 Hz).[6] Periodically distorted waveforms can be decomposed into a sum of the fundamental frequency and the harmonics. Harmonic distortion originates in the nonlinear characteristics of devices and loads on the power system.

Harmonic distortion levels are described by the complete harmonic spectrum with magnitudes and phase angles of each individual harmonic component. It is also common to use a single quantity, the *total harmonic distortion* (THD), as a measure of the effective value of harmonic distortion. Figure 2.10 illustrates the waveform and harmonic spectrum for a typical adjustable-speed-drive (ASD) input current. Current distortion levels can be characterized by a THD value, as previously described, but this can often be misleading. For example, many adjustable-speed drives will exhibit high THD values for the input current when they are operating at very light loads. This is not necessarily a significant concern because the *magnitude* of harmonic current is low, even though its relative distortion is high.

To handle this concern for characterizing harmonic currents in a consistent fashion, IEEE Standard 519–1992 defines another term, the *total demand distortion* (TDD). This term is the same as the THD except that the distortion is expressed as a percent of some rated load current rather than as a percent of the fundamental current magnitude at the instant of measurement. IEEE Standard 519–1992 provides guidelines

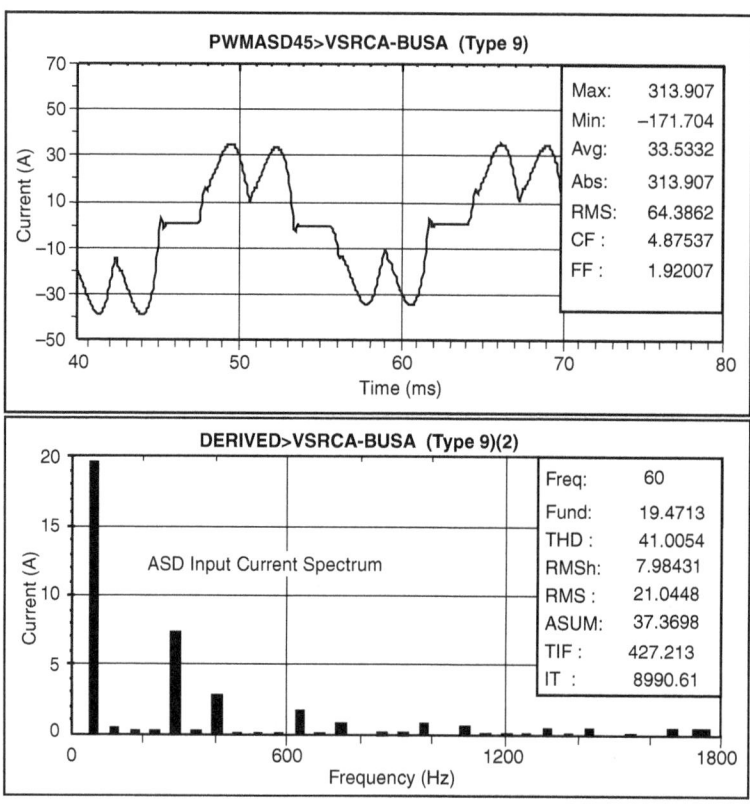

FIGURE 2.10   Current waveform and harmonic spectrum for an ASD input current.

for harmonic current and voltage distortion levels on distribution and transmission circuits.

Interharmonics

Voltages or currents having frequency components that are not integer multiples of the frequency at which the supply system is designed to operate (e.g., 50 or 60 Hz) are called *interharmonics*. They can appear as discrete frequencies or as a wideband spectrum.

Interharmonics can be found in networks of all voltage classes. The main sources of interharmonic waveform distortion are static frequency converters, cycloconverters, induction furnaces, and arcing devices. Power line carrier signals can also be considered as interharmonics.

Since the first edition of this book, considerable work has been done on this subject. There is now a better understanding of the origins and effects of interharmonic distortion. It is generally the result of frequency conversion and is often not constant; it varies with load. Such interharmonic currents can excite quite severe resonances on the power system as the varying interharmonic frequency becomes coincident with natural frequencies of the system. They have been shown to affect power-line-carrier signaling and induce visual flicker in fluorescent and other arc lighting as well as in computer display devices.

Notching

Notching is a periodic voltage disturbance caused by the normal operation of power electronic devices when current is commutated from one phase to another.

Since notching occurs continuously, it can be characterized through the harmonic spectrum of the affected voltage. However, it is generally treated as a special case. The frequency components associated with notching can be quite high and may not be readily characterized with measurement equipment normally used for harmonic analysis.

Figure 2.11 shows an example of voltage notching from a three-phase converter that produces continuous dc current. The notches occur when the current commutates from one phase to another. During this period, there is a momentary short circuit between two phases, pulling the voltage as close to zero as permitted by system impedances.

Noise

Noise is defined as unwanted electrical signals with broadband spectral content lower than 200 kHz superimposed upon the power system voltage or current in phase conductors, or found on neutral conductors or signal lines.

Noise in power systems can be caused by power electronic devices, control circuits, arcing equipment, loads with solid-state rectifiers, and switching power supplies. Noise problems are often exacerbated by improper grounding that fails to conduct noise away

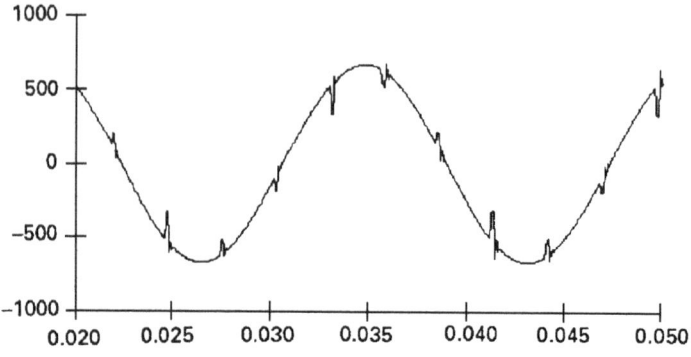

FIGURE 2.11   Example of voltage notching caused by a three-phase converter.

from the power system. Basically, noise consists of any unwanted distortion of the power signal that cannot be classified as harmonic distortion or transients. Noise disturbs electronic devices such as microcomputer and programmable controllers. The problem can be mitigated by using filters, isolation transformers, and line conditioners.

2.8   Voltage Fluctuation

Voltage fluctuations are systematic variations of the voltage envelope or a series of random voltage changes, the magnitude of which does not normally exceed the voltage ranges specified by ANSI C84.1 of 0.9 to 1.1 pu.

IEC 61000-2-1 defines various types of voltage fluctuations. We will restrict our discussion here to IEC 61000-2-1 Type (d) voltage fluctuations, which are characterized as a series of random or continuous voltage fluctuations.

Loads that can exhibit continuous, rapid variations in the load current magnitude can cause voltage variations that are often referred to as flicker. The term *flicker* is derived from the impact of the voltage fluctuation on lamps such that they are perceived by the human eye to flicker. To be technically correct, voltage fluctuation is an electromagnetic phenomenon while flicker is an undesirable result of the voltage fluctuation in some loads. However, the two terms are often linked together in standards. Therefore, we will also use the common term *voltage flicker* to describe such voltage fluctuations.

An example of a voltage waveform which produces flicker is shown in Fig. 2.12. This is caused by an arc furnace, one of the most common causes of voltage fluctuations on utility transmission and distribution systems. The flicker signal is defined by its rms magnitude expressed as a percent of the fundamental. Voltage flicker

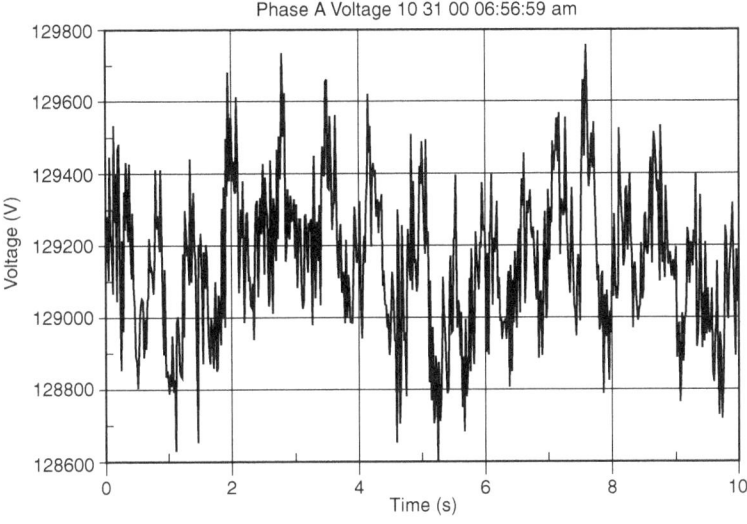

Phase A Voltage 10 31 00 06:56:59 am

FIGURE 2.12   Example of voltage fluctuations caused by arc furnace operation.

is measured with respect to the sensitivity of the human eye. Typically, magnitudes as low as 0.5 percent can result in perceptible lamp flicker if the frequencies are in the range of 6 to 8 Hz.

IEC 61000–4-15 defines the methodology and specifications of instrumentation for measuring flicker. The IEEE Voltage Flicker Working Group has adopted this standard as amended for 60-Hz power systems for use in North America. This standard devises a simple means of describing the potential for visible light flicker through voltage measurements. The measurement method simulates the lamp/eye/brain transfer function and produces a fundamental metric called short-term flicker sensation (Pst). This value is normalized to 1.0 to represent the level of voltage fluctuations sufficient to cause noticeable flicker to 50 percent of a sample observing group. Another measure called long-term flicker sensation (Plt) is often used for the purpose of verifying compliance with compatibility levels established by standards bodies and used in utility power contracts. This value is a longer-term average of Pst samples.

Figure 2.13 illustrates a trend of Pst measurements taken at a 161-kV substation bus serving an arc furnace load. Pst samples are normally reported at 10-min intervals. A statistical evaluation process defined in the measurement standard processes instantaneous flicker measurements to produce the Pst value. The Plt value is produced every 2 h from the Pst values.

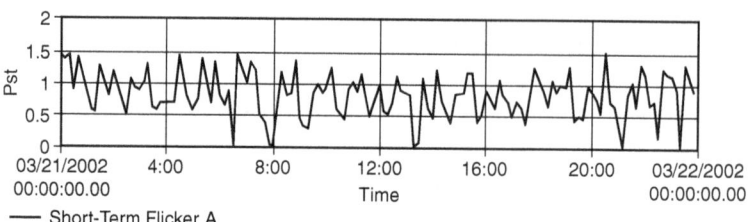

FIGURE 2.13 Flicker (Pst) at 161-kV substation bus measured according to IEC Standard 61000-4-15. (*Courtesy of Dranetz-BMI/Electrotek Concepts.*)

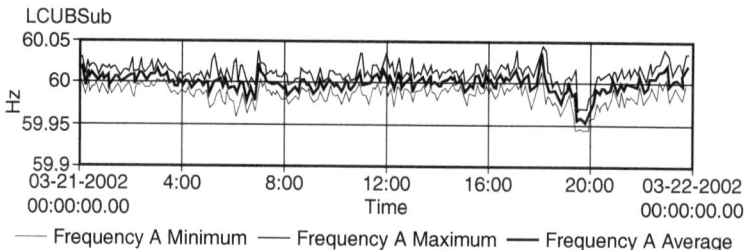

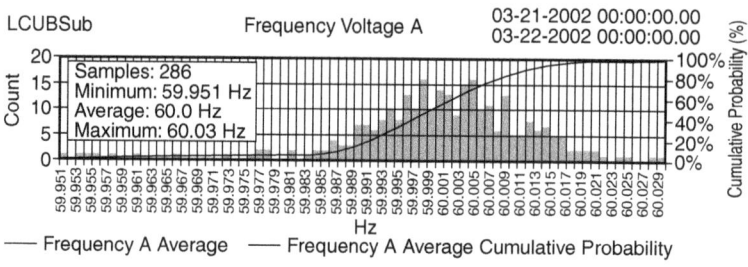

FIGURE 2.14 Power frequency trend and statistical distribution at 13-kV substation bus. (*Courtesy of Dranetz-BMI/Electrotek Concepts.*)

2.9 Power Frequency Variations

Power frequency variations are defined as the deviation of the power system fundamental frequency from it specified nominal value (e.g., 50 or 60 Hz).

The power system frequency is directly related to the rotational speed of the generators supplying the system. There are slight variations in frequency as the dynamic balance between load and generation changes. The size of the frequency shift and its duration depend on the load characteristics and the response of the generation control system to load changes. Figure 2.14 illustrates frequency variations for a 24-h period on a typical 13-kV substation bus.

Frequency variations that go outside of accepted limits for normal steady-state operation of the power system can be caused by faults on the bulk power transmission system, a large block of load being disconnected, or a large source of generation going off-line.

On modern interconnected power systems, significant frequency variations are rare. Frequency variations of consequence are much more likely to occur for loads that are supplied by a generator isolated from the utility system. In such cases, governor response to abrupt load changes may not be adequate to regulate within the narrow bandwidth required by frequency-sensitive equipment.

Voltage notching can sometimes be mistaken for frequency deviation. The notches may come sufficiently close to zero to cause errors in instruments and control systems that rely on zero crossings to derive frequency or time.

2.10 Power Quality Terms

So that you will be better able to understand the material in this book, we have included the definitions of many common power quality terms that are relevant to the material in this book. For the most part, these definitions coincide with current industry efforts to define power quality terms.[2] We have also included other terms relevant to the material in this book.

active filter Any of a number of sophisticated power electronic devices for eliminating harmonic distortion. See *passive filter*.

CBEMA curve A set of curves representing the withstand capabilities of computers in terms of the magnitude and duration of the voltage disturbance. Developed by the Computer Business Equipment Manufacturers Association (CBEMA), it had become the de facto standard for measuring the performance of all types of equipment and power systems and is commonly referred to by this name.[9] CBEMA has been replaced by the Information Technology Industry Council (ITI), and a new curve has been developed that is commonly referred to as the ITI curve. See *ITI curve*.

common mode voltage The noise voltage that appears equally from current-carrying conductor to ground.[2]

coupling A circuit element, or elements, or a network that may be considered common to the input mesh and the output mesh and through which energy may be transferred from one to another.[8]

crest factor A value reported by many power quality monitoring instruments representing the ratio of the crest value of the measured waveform to the root mean square (rms) of the waveform. For example, the crest factor of a sinusoidal wave is 1.414.

critical load Devices and equipment whose failure to operate satisfactorily jeopardizes the health or safety of personnel, and/or results in loss of function, financial loss, or damage to property deemed critical by the user.

current distortion Distortion in the ac line current. See *distortion.*

differential mode voltage The voltage between any two of a specified set of active conductors.

dip See *sag.*

distortion Any deviation from the normal sine wave for an ac quantity.

distributed generation (DG) Generation dispersed throughout the power system as opposed to large, central station power plants. In the context used in this book, DG typically refers to units less than 10 megawatts (MW) in size that are interconnected with the distribution system rather than the transmission system.

dropout A loss of equipment operation (discrete data signals) due to noise, sag, or interruption.

dropout voltage The voltage at which a device will release to its deenergized position (for this document, the voltage at which a device fails to operate).

electromagnetic compatibility The ability of a device, equipment, or system to function satisfactorily in its electromagnetic environment without introducing intolerable electromagnetic disturbances to anything in that environment.[2,3]

equipment grounding conductor The conductor used to connect the non–current carrying parts of conduits, raceways, and equipment enclosures to the grounded conductor (neutral) and the grounding electrode at the service equipment (main panel) or secondary of a separately derived system (e.g., isolation transformer). See National Fire Protection Association (NFPA) 70–1993, Section 100.[7]

failure mode The effect by which failure is observed.[8]

fast tripping Refers to the common utility protective relaying practice in which the circuit breaker or line recloser operates faster than a fuse can blow. Also called fuse saving. Effective for clearing transient faults without a sustained interruption, but is somewhat controversial because industrial loads are subjected to a momentary or temporary interruption.

fault Generally refers to a short circuit on the power system.

fault, transient   A short circuit on the power system usually induced by lightning, tree branches, or animals, which can be cleared by momentarily interrupting the current.

ferroresonance   An irregular, often chaotic type of resonance that involves the nonlinear characteristic of iron-core (ferrous) inductors. It is nearly always undesirable when it occurs in the power delivery system, but it is exploited in technologies such as constant-voltage transformers to improve the power quality.

flicker   An impression of unsteadiness of visual sensation induced by a light stimulus whose luminance or spectral distribution fluctuates with time.[2]

frequency deviation   An increase or decrease in the power frequency. The duration of a frequency deviation can be from several cycles to several hours.

frequency response   In power quality usage, generally refers to the variation of impedance of the system, or a metering transducer, as a function of frequency.

fundamental (component)   The component of order 1 (50 to 60 Hz) of the Fourier series of a periodic quantity.[2]

ground   A conducting connection, whether intentional or accidental, by which an electric circuit or electrical equipment is connected to the earth, or to some conducting body of relatively large extent that serves in place of the earth. Note: It is used for establishing and maintaining the potential of the earth (or of the conducting body) or approximately that potential, on conductors connected to it, and for conducting ground currents to and from earth (or the conducting body).[8]

ground electrode   A conductor or group of conductors in intimate contact with the earth for the purpose of providing a connection with the ground.[7]

ground grid   A system of interconnected bare conductors arranged in a pattern over a specified area and on or buried below the surface of the earth. The primary purpose of the ground grid is to provide safety for workers by limiting potential differences within its perimeter to safe levels in case of high currents that could flow if the circuit being worked became energized for any reason or if an adjacent energized circuit faulted. Metallic surface mats and gratings are sometimes utilized for the same purpose.[8] This is not necessarily the same as a signal reference grid.

ground loop   A potentially detrimental loop formed when two or more points in an electrical system that are nominally at ground potential are connected by a conducting path such that either or both points are not at the same ground potential.[8]

ground window   The area through which all grounding conductors, including metallic raceways, enter a specific area. It is often used in communications systems through which the building grounding system is connected to an area that would otherwise have no grounding connection.

harmonic (component)   A component of order greater than 1 of the Fourier series of a periodic quantity.[2]

harmonic content   The quantity obtained by subtracting the fundamental component from an alternating quantity.

harmonic distortion   Periodic distortion of the sine wave. See *distortion* and *total harmonic distortion (THD)*.

harmonic filter   On power systems, a device for filtering one or more harmonics from the power system. Most are passive combinations of inductance, capacitance, and resistance. Newer technologies include active filters that can also address reactive power needs.

harmonic number   The integral number given by the ratio of the frequency of a harmonic to the fundamental frequency.[2]

harmonic resonance   A condition in which the power system is resonating near one of the major harmonics being produced by nonlinear elements in the system, thus exacerbating the harmonic distortion.

impulse   A pulse that, for a given application, approximates a unit pulse or a Dirac function.[2] When used in relation to monitoring power quality, it is preferable to use the term impulsive transient in place of impulse.

impulsive transient   A sudden, nonpower frequency change in the steady-state condition of voltage or current that is unidirectional in polarity (primarily either positive or negative).

instantaneous   When used to quantify the duration of a short-duration variation as a modifier, this term refers to a time range from one-half cycle to 30 cycles of the power frequency.

instantaneous reclosing   A term commonly applied to reclosing of a utility breaker as quickly as possible after an interrupting fault current. Typical times are 18 to 30 cycles.

interharmonic (component)   A frequency component of a periodic quantity that is not an integer multiple of the frequency at which the supply system is designed to operate (e.g., 50 or 60 Hz).

interruption, momentary (electrical power systems) An interruption of a duration limited to the period required to restore service by automatic or supervisory-controlled switching operations or by manual switching at locations where an operator is immediately available. Note: Such switching operations must be completed in a specified time not to exceed 5 min.

interruption, momentary (power quality monitoring) A type of short-duration variation. The complete loss of voltage (<0.1 pu) on one or more phase conductors for a time period between 30 cycles and 3 s.

interruption, sustained (electrical power systems) Any interruption not classified as a momentary interruption.

interruption, sustained (power quality) A type of long-duration variation. The complete loss of voltage (<0.1 pu) on one or more phase conductors for a time greater than 1 min.

interruption, temporary A type of short-duration variation. The complete loss of voltage (<0.1 pu) on one or more phase conductors for a time period between 3 s and 1 min.

inverter A power electronic device that converts direct current to alternating current of either power frequency or a frequency required by an industrial process. Common inverters today employ pulse-width modulation to create the desired frequency with minimal harmonic distortion.

islanding Refers to a condition in which distributed generation is isolated on a portion of the load served by the utility power system. It is usually an undesirable situation, although there are situations where controlled islands can improve the system reliability.

isolated ground An insulated equipment grounding conductor run in the same conduit or raceway as the supply conductors. This conductor is insulated from the metallic raceway and all ground points throughout its length. It originates at an isolated ground-type receptacle or equipment input terminal block and terminates at the point where neutral and ground are bonded at the power source. See NFPA 70–1993, Section 250–74, Exception #4 and Section 250–75, Exception.[7]

isolation Separation of one section of a system from undesired influences of other sections.

ITI curve A set of curves published by the Information Technology Industry Council (ITI) representing the withstand capabilities of computers connected to 120-V power systems in terms of the magnitude and duration of the voltage disturbance.[10] The ITI curve replaces the curves originally developed by the ITI's predecessor organization, the Computer Business Equipment Manufacturers Association (CBEMA).[9] See *CBEMA curve*.

linear load   An electrical load device that, in steady-state operation, presents an essentially constant load impedance to the power source throughout the cycle of applied voltage.

long-duration variation   A variation of the rms value of the voltage from nominal voltage for a time greater than 1 min. Usually further described using a modifier indicating the magnitude of a voltage variation (e.g., undervoltage, overvoltage, or voltage interruption).

low-side surges   A term coined by distribution transformer designers to describe the current surge that appears to be injected into the transformer secondary terminals during a lightning strike to grounded conductors in the vicinity.

momentary   When used to quantify the duration of a short-duration variation as a modifier, refers to a time range at the power frequency from 30 cycles to 3 s.

noise   Unwanted electrical signals that produce undesirable effects in the circuits of the control systems in which they occur.[8] (For this document, "control systems" is intended to include sensitive electronic equipment in total or in part.)

nominal voltage (Vn)   A nominal value assigned to a circuit or system for the purpose of conveniently designating its voltage class (as 208/120, 480/277, 600).[6]

nonlinear load   Electrical load that draws current discontinuously or whose impedance varies throughout the cycle of the input ac voltage waveform.

normal mode voltage   A voltage that appears between or among active circuit conductors.

notch   A switching (or other) disturbance of the normal power voltage waveform, lasting less than a half-cycle, which is initially of opposite polarity than the waveform and is thus subtracted from the normal waveform in terms of the peak value of the disturbance voltage. This includes complete loss of voltage for up to a half-cycle.

oscillatory transient   A sudden, nonpower frequency change in the steady-state condition of voltage or current that includes both positive- or negative-polarity value.

overvoltage   When used to describe a specific type of long-duration variation, refers to a voltage having a value of at least 10 percent above the nominal voltage for a period of time greater than 1 min.

passive filter A combination of inductors, capacitors, and resistors designed to eliminate one or more harmonics. The most common variety is simply an inductor in series with a shunt capacitor, which short-circuits the major distorting harmonic component from the system.

phase shift The displacement in time of one voltage waveform relative to other voltage waveform(s).

power factor, displacement The power factor of the fundamental frequency components of the voltage and current waveforms.

power factor (true) The ratio of active power (watts) to apparent power (voltamperes).

Plt The long-term flicker severity level as defined by IEC 61000-4-15, based on an observation period of 2 h.

Pst The short-term flicker severity level as defined by IEC 61000-4-15, based on an observation period of 10 min. A Pst value greater than 1.0 corresponds to the level of irritability for 50 percent of the persons subjected to the measured flicker.

pulse An abrupt variation of short duration of a physical quantity followed by a rapid return to the initial value.

pulse-width modulation (PWM) A common technique used in inverters to create an ac waveform by controlling the electronic switch to produce varying-width pulses. Minimizes power frequency harmonic distortion in some applications, but care must be taken to properly filter out the switching frequencies, which are commonly 3 to 6 kHz.

reclosing The common utility practice used on overhead lines of closing the breaker within a short time after clearing a fault, taking advantage of the fact that most faults are transient, or temporary.

recovery time The time interval needed for the output voltage or current to return to a value within the regulation specification after a step load or line change.[8] Also may indicate the time interval required to bring a system back to its operating condition after an interruption or dropout.

recovery voltage The voltage that occurs across the terminals of a pole of a circuit-interrupting device upon interruption of the current.[8]

rectifier A power electronic device for converting alternating current to direct current.

resonance A condition in which the natural frequencies of the inductances and capacitances in the power system are excited and sustained by disturbing phenomena. This can result in excessive voltages and currents. Waveform distortion, whether harmonic or nonharmonic, is probably the most frequent excitation source. Also, various short-circuit and open-circuit faults can result in resonant conditions.

safety ground See *equipment grounding conductor.*

sag A decrease to between 0.1 and 0.9 pu in rms voltage or current at the power frequency for durations of 0.5 cycle to 1 min.

shield As normally applied to instrumentation cables, refers to a conductive sheath (usually metallic) applied, over the insulation of a conductor or conductors, for the purpose of providing means to reduce coupling between the conductors so shielded and other conductors that may be susceptible to, or which may be generating, unwanted electrostatic or electromagnetic fields (noise).

shielding Shielding is the use of a conducting and/or ferromagnetic barrier between a potentially disturbing noise source and sensitive circuitry. Shields are used to protect cables (data and power) and electronic circuits. They may be in the form of metal barriers, enclosures, or wrappings around source circuits and receiving circuits.

shielding (of utility lines) The construction of a grounded conductor or tower over the lines to intercept lightning strokes in an attempt to keep the lightning currents out of the power system.

short-duration variation A variation of the rms value of the voltage from nominal voltage for a time greater than one-half cycle of the power frequency but less than or equal to 1 min. Usually further described using a modifier indicating the magnitude of a voltage variation (e.g., sag, swell, or interruption) and possibly a modifier indicating the duration of the variation (e.g., instantaneous, momentary, or temporary).

signal reference grid (or plane) A system of conductive paths among interconnected equipment, which reduces noise-induced voltages to levels that minimize improper operation. Common configurations include grids and planes.

sustained When used to quantify the duration of a voltage interruption, refers to the time frame associated with a long-duration variation (i.e., greater than 1 min).

swell A temporary increase in the rms value of the voltage of more than 10 percent of the nominal voltage, at the power frequency, for durations from 0.5 cycle to 1 min.

sympathetic tripping When a circuit breaker on an unfaulted feeder section trips unnecessarily due to backfeed into a fault elsewhere. Most commonly occurs when sensitive ground fault relaying is employed.

synchronous closing Generally used in reference to closing all three poles of a capacitor switch in synchronism with the power system to minimize transients.

temporary When used to quantify the duration of a short-duration variation as a modifier, refers to a time range from 3 s to 1 min.

total demand distortion (TDD) The ratio of the root mean square of the harmonic current to the rms value of the rated or maximum demand fundamental current, expressed as a percent.

total disturbance level The level of a given electromagnetic disturbance caused by the superposition of the emission of all pieces of equipment in a given system.[2]

total harmonic distortion (THD) The ratio of the root mean square of the harmonic content to the rms value of the fundamental quantity, expressed as a percent of the fundamental.[8]

transient Pertaining to or designating a phenomenon or a quantity that varies between two consecutive steady states during a time interval that is short compared to the time scale of interest. A transient can be a unidirectional impulse of either polarity or a damped oscillatory wave with the first peak occurring in either polarity.[2]

triplen harmonics A term frequently used to refer to the odd multiples of the third harmonic, which deserve special attention because of their natural tendency to be zero sequence.

undervoltage When used to describe a specific type of long-duration variation, refers to a measured voltage having a value at least 10 percent below the nominal voltage for a period of time greater than 1 min. In other contexts, such as distributed generation protection, the time frame of interest would be measured in cycles or seconds.

voltage change A variation of the root mean square or peak value of a voltage between two consecutive levels sustained for definite but unspecified durations.[6]

voltage dip See *sag*.

voltage distortion Distortion of the ac line voltage. See *distortion*.

voltage fluctuation A series of voltage changes or a cyclical variation of the voltage envelope.[6]

voltage imbalance (unbalance)   A condition in which the three-phase voltages differ in amplitude or are displaced from their normal 120 degree phase relationship or both. Frequently expressed as the ratio of the negative-sequence or zero-sequence voltage to the positive-sequence voltage, in percent.

voltage interruption   Disappearance of the supply voltage on one or more phases. Usually qualified by an additional term indicating the duration of the interruption (e.g., momentary, temporary, or sustained).

voltage magnification   The magnification of capacitor switching oscillatory transient voltage on the primary side by capacitors on the secondary side of a transformer.

voltage regulation   The degree of control or stability of the rms voltage at the load. Often specified in relation to other parameters, such as input-voltage changes, load changes, or temperature changes.

waveform distortion   A steady-state deviation from an ideal sine wave of power frequency principally characterized by the spectral content of the deviation.

2.11   Ambiguous Terms

Much of the history of the power quality movement has been marked by a fair amount of hype as a number of equipment vendors have jockeyed for position in the marketplace. This book attempts to apply a strong engineering interpretation of all areas of power quality and remove the hype and mystery. Marketers have created many colorful phrases to entice potential customers to buy. Unfortunately, many of these terms are ambiguous and cannot be used for technical definitions.

The following words are commonly used but have a variety of meanings or no meaning at all. For example, what is a "power surge"? This term has probably been used at some time to describe each of the disturbance phenomena described in this book. Is there really a surge in the power? Or is it only the voltage? Power is related to the product of voltage and current. Normally, voltage is the quantity causing the observed disturbance and the resulting power will not necessarily be directly proportional to the voltage. The solution will generally be to correct or limit the voltage as opposed to doing something to address the power. Therefore, the following terms are meaningless in terms of describing an event and determining a solution:

Blackout   Glitch

Blink   Outage

Brownout	Interruption
Bump	Power surge
Clean ground	Raw power
Clean power	Spike
Dirty ground	Surge
Dirty power	Wink

The unqualified use of these words for describing power quality phenomena is discouraged. Try to use the standard terms where possible, or qualify nonstandard terms with appropriate explanation.

2.12   CBEMA and ITI Curves

One of the most frequently employed displays of data to represent the power quality is the so-called CBEMA curve. A portion of the curve adapted from IEEE Standard 446[9] that we typically use in our analysis of power quality monitoring results is shown in Fig. 2.15. This curve was originally developed by CBEMA to describe the tolerance of mainframe computer equipment to the magnitude and duration of voltage variations on the power system. While many modern computers have greater tolerance than this, the curve has become a standard design target for sensitive equipment to be applied on the power system and a common format for reporting power quality variation data.

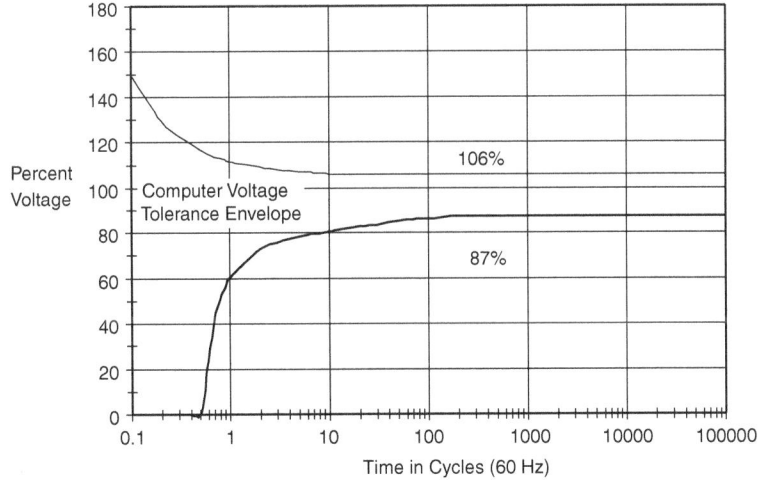

FIGURE 2.15   A portion of the CBEMA curve commonly used as a design target for equipment and a format for reporting power quality variation data.

The axes represent magnitude and duration of the event. Points below the envelope are presumed to cause the load to drop out due to lack of energy. Points above the envelope are presumed to cause other malfunctions such as insulation failure, overvoltage trip, and overexcitation. The upper curve is actually defined down to 0.001 cycle where it has a value of about 375 percent voltage. We typically employ the curve only from 0.1 cycle and higher due to limitations in power quality monitoring instruments and differences in opinion over defining the magnitude values in the subcycle time frame.

The CBEMA organization has been replaced by ITI,[10] and a modified curve has been developed that specifically applies to common 120-V computer equipment (see Fig. 2.16). The concept is similar to the CBEMA curve. Although developed for 120-V computer equipment, the curve has been applied to general power quality evaluation like its predecessor curve.

Both curves are used as a reference in this book to define the withstand capability of various loads and devices for protection from

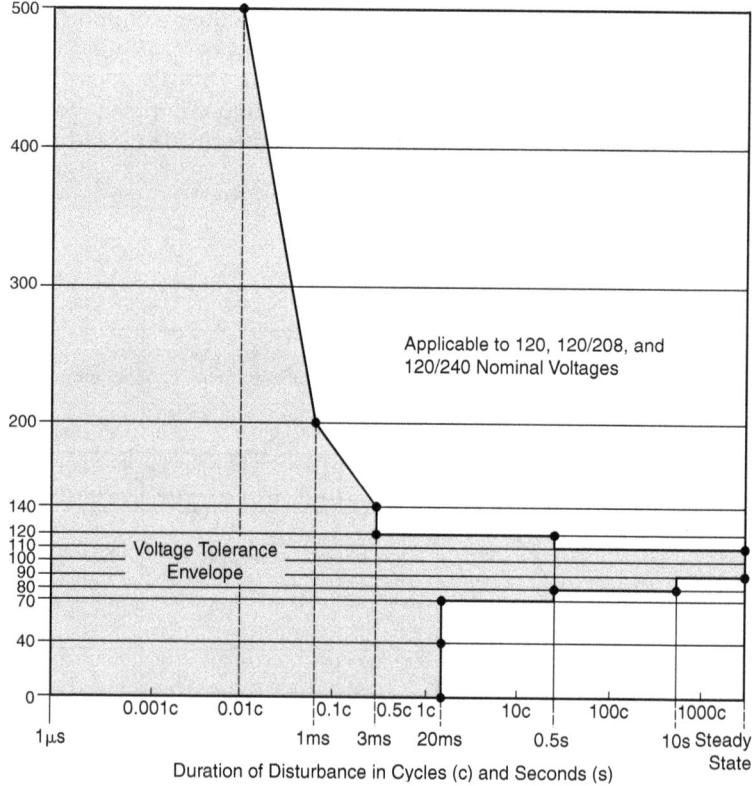

Figure 2.16 ITI curve for susceptibility of 120-V computer equipment.

power quality variations. For display of large quantities of power quality monitoring data, we frequently add a third axis to the plot to denote the number of events within a certain predefined cell of magnitude and duration. If restricted to just the two-dimensional views shown in Fig. 2.16, the plot tends to turn into a solid mass of points over time, which is not useful.

2.13 References

1. TC77WG6 (Secretary) 110-R5, *Draft Classification of Electromagnetic Environments*, January 1991.
2. IEEE Standard 1159–1995, *Recommended Practice on Monitoring Electric Power*.
3. IEC 50 (161), *International Electrotechnical Vocabulary*, chap. 161: "Electromagnetic Compatibility," 1989.
4. UIE-DWG-3-92-G, *Guide to Quality of Electrical Supply for Industrial Installations— Part 1: General Introduction to Electromagnetic Compatibility (EMC), Types of Disturbances and Relevant Standards*. Advance UIE Edition. "Disturbances" Working Group GT 2.
5. UIE-DWG-2-92-D, *UIE Guide to Measurements of Voltage Dips and Short Interruptions Occurring in Industrial Installations*.
6. IEC 61000-2-1(1990–05), "Description of the Environment—Electromagnetic Environment for Low Frequency Conducted Disturbances and Signaling in Public Power Supply Systems," *Electromagnetic Compatibility (EMC)—Part 2, Environment*, Section 1, 1990.
7. ANSI/NFPA 70–1993, *National Electrical Code*.
8. IEEE Standard 100–1992, *IEEE Standard Dictionary of Electrical and Electronic Terms*.
9. IEEE Standard 446–1987, *IEEE Recommended Practice for Emergency and Standby Power Systems for Industrial and Commercial Applications* (IEEE Orange Book).
10. Information Technology Industry Council (ITI), 1250 Eye Street NW, Suite 200, Washington, D.C. (http://www.itic.org).
11. IEC 61000-4-30 77A/356/CDV, *Power Quality Measurement Methods*, 2001.
12. IEC 61000-4-15, *Flicker Meter—Functional and Design Specifications*, 1997.

CHAPTER 3

Voltage Sags and Interruptions

Voltage sags and interruptions are related power quality problems. Both are usually the result of faults in the power system and switching actions to isolate the faulted sections. They are characterized by rms voltage variations outside the normal operating range of voltages.

A *voltage sag* is a short-duration (typically 0.5 to 30 cycles) reduction in rms voltage caused by faults on the power system and the starting of large loads, such as motors. Momentary interruptions (typically no more than 2 to 5 s) cause a complete loss of voltage and are a common result of the actions taken by utilities to clear transient faults on their systems. Sustained interruptions of longer than 1 min are generally due to permanent faults.

Utilities have been faced with rising numbers of complaints about the quality of power due to sags and interruptions. There are a number of reasons for this, with the most important being that customers in all sectors (residential, commercial, and industrial) have more sensitive loads. The influx of digital computers and other types of electronic controls is at the heart of the problem. Computer controls tend to lose their memory, and the processes that are being controlled also tend to be more complex and, therefore, take much more time to restart. Industries are relying more on automated equipment to achieve maximum productivity to remain competitive. Thus, an interruption has considerable economic impact. [1,3]

3.1 Utility Distribution System Designs

In describing voltage sag and interruption phenomena, it is important to examine the key characteristics of utility distribution circuits. Common distribution systems in North America are of four-wire, multi-grounded-neutral and three-wire, delta uni-grounded designs. The European distribution systems consist of medium- and low-voltage circuits. The medium-voltage circuit has three-wire design with a resonant grounding at the wye neutral point, while the low-voltage circuit is similar to the North American's four-wire multi-grounded neutral design. Elsewhere around the world, distribution systems are mixtures of North American and European designs.

3.1.1 Four-Wire Multi-Grounded Neutral Systems

A four-wire multi-grounded neutral system illustrated in Fig. 3.1 is the most common type of distribution system design in North America. The main three-phase feeders are constructed with four wires—the three phase wires and the neutral wire. The neutral is connected to ground (grounded) every few poles and at locations where distribution transformers or other equipment are connected. Most residential and small commercial customers are served from single-phase transformers. Many of the single-phase loads are served from single-phase laterals off the main three-phase feeder trunk. The primary sides of the single-phase service transformers are connected line-to-neutral. If the feeder has capacitor banks, they would be typically connected in grounded-wye, as shown in Fig. 3.1, although there are numerous exceptions. Likewise, if voltage regulators (not shown) are present, they would typically be connected in grounded-wye.

Single-phase laterals are very common since they are relatively inexpensive to build. The lateral would consist of one phase wire, fully insulated for line-to-neutral voltage, and one neutral wire that is grounded at regular intervals. The neutral wire need not be insulated and the single-phase pole lines are often built without crossarms. Thus, the line can be constructed at low cost.

Because the common service voltage is low (120 V), the primary distribution lines have to be brought close to the point of use. This is in contrast to the European-style system that is discussed later in Sec. 3.1.3.

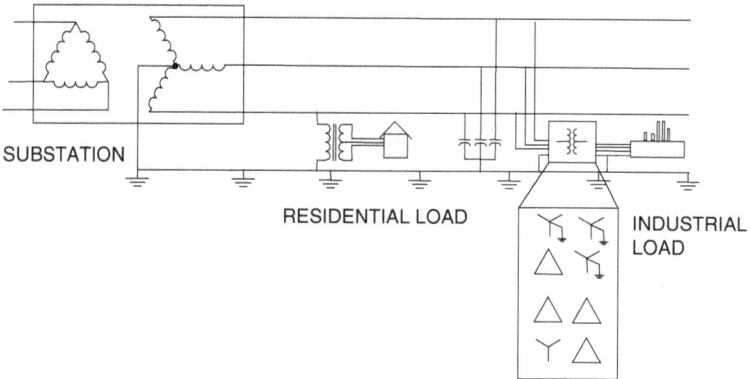

FIGURE 3.1 Four-wire multi-grounded neutral distribution system.

Industrial and larger commercial loads have three-phase service. A variety of transformer connections are used. Today the most common connections are grounded-wye/grounded-wye, followed by delta/grounded-wye. Other connections may also be used as shown in Fig. 3.1. This type of system also permits three-phase loads to be served from only two primary phases by employing an open-wye/open-delta connection. The effectively grounded neutral makes it possible to do this and other similar tricks of the trade to provide service at lower cost.

In most areas, surge arresters are used at each distribution transformer and cable riser pole. The four-wire multi-grounded neutral system makes it possible to use arresters rated for line-to-neutral voltage, achieving significant cost savings. This is the predominant practice of North American utilities.

One common exception to the four-wire multi-grounded neutral system is found on the West Coast. Ungrounded systems are commonly employed for voltages up through the 15-kV class. The neutral at the substation is grounded, but only three phase wires are carried along the feeders. Single-phase transformers are connected phase-to-phase similar to three-wire delta systems (see Sec. 3.1.2). Sometimes, these ungrounded systems become effectively grounded due to extensive use of direct-buried underground cable. The neutrals (shields) of the cables are necessarily grounded to achieve uniform dielectric stress on the cable insulation. However, the transformer connections would not necessarily rely on this ground to carry power.

3.1.2 Three-Wire Delta Systems

Many older systems were three-wire delta systems similar to that shown in Fig. 3.2. Single-phase service transformers are connected line-to-line. In areas where surge arresters are used, which includes

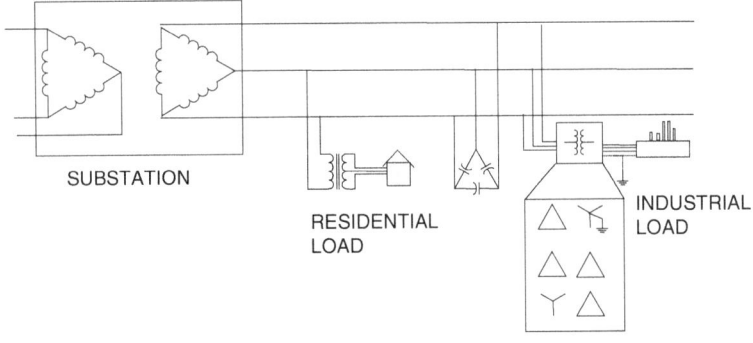

FIGURE 3.2 Three-wire delta distribution system.

most areas, two would be required for each transformer. Capacitor banks and voltage regulator banks, if present, would be typically connected in delta. Single-phase laterals would consist of two of the phases with full insulation required for each phase.

There are fewer options for the three-phase service transformer connection. Grounded-wye primary connections are not used on these systems, although they may be used on the low-voltage side.

As higher primary distribution voltages were adopted throughout North America, there were many economies to be realized by switching from three-wire delta systems to four-wire multi-grounded neutral systems. This includes systems on the West Coast for voltages above the 15-kV class. Reduced insulation levels can be employed in transformers and other line equipment, resulting in considerable savings. There are also efficiencies with respect to overcurrent protection, such as fusing, that permit more compact and less costly designs to be used.

3.1.3   European-Style Distribution Systems

European-style distribution systems have some notable differences (Fig. 3.3). In North America, "distribution" almost always refers to primary distribution, ranging from 2.4 to 34.5 kV. This corresponds to the European medium-voltage distribution system. The European-style design also makes extensive use of low-voltage distribution, which is generally 400 V line-to-line. The term "distribution substation" generally refers to medium/low-voltage transformers, which Americans would typically call distribution transformers. These would nearly always be three-phase transformers supplying perhaps as many as 100 residences. This is possible because of the

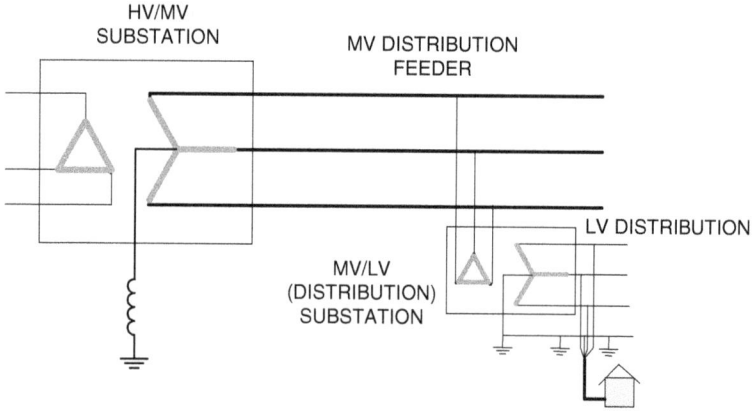

Figure 3.3   Basic European-style distribution (medium-voltage) system.

higher utilization voltage. While a typical North American distribution feeder may have hundreds of distribution transformers, the comparable European-style design might have nearly an order of magnitude fewer—all three-phase and connected delta/grounded-wye.

The low-voltage distribution system is designed similarly to the four-wire multi-grounded neutral system that was described previously. In many nations, all customers receive three-phase service simply by tapping the 400 V lines. Lamps and household appliances would be connected line-to-neutral, operating at approximately 230 V.

European-style distribution systems are much more uniformly designed than North American systems. Nearly all transformers are connected delta/grounded-wye as shown. High/medium-voltage substation transformers will typically be grounded through an inductance. This would have the purpose of either simply limiting the ground-fault current or of attempting to extinguish the ground-fault current by resonant tuning with the shunt capacitance of the feeder. This is intended to virtually eliminate voltage sags and interruptions due to the common single-line-to-ground (SLG) fault.

Power factor correction capacitors have historically not been common on medium-voltage distribution. This may be changing in some areas because of needs to increase system capacities and efficiencies. Power factor correction would be expected to take place at the low-voltage distribution level.

3.1.4 Radial Distribution Configuration/Structure

Around the world, there is a mixture of North American and European-style designs, depending on which country had the greatest influence on the development of electric power systems in a particular region. Regardless of basic style, most primary distribution systems throughout the world are operated radially for economic reasons. The basic overcurrent protection scheme is illustrated in Fig. 3.4.

Engineers will often refer to radial distribution systems as "looped" or "network" systems, but the tie switch between two adjacent feeders is normally open. Thus, it is radial while in operation except for brief periods where the utility may perform closed transition switching during feeder reconfiguration for maintenance purposes.

The radial distribution system is connected to one main source of power at a time (there may be alternate sources in case of emergency). Moving out onto the feeders from the substation, there will be a number of overcurrent protection devices in series. There is at least a feeder breaker, but there may also be other automatic fault interrupters, such as the line reclosers shown, farther down the feeder. The lowest

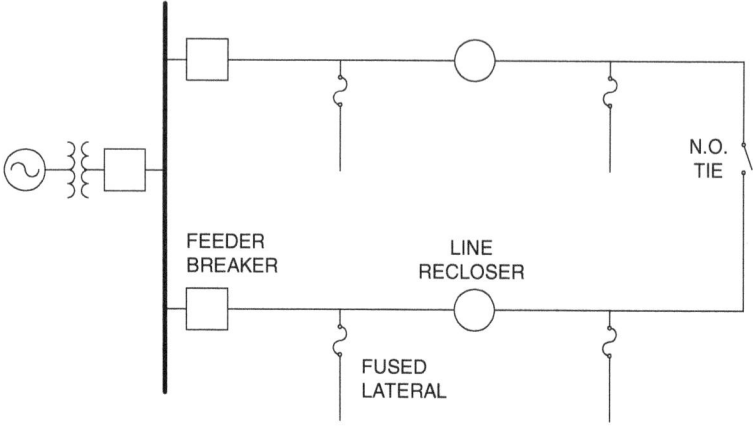

FIGURE **3.4**   Radial distribution system overcurrent protection design.

level of overcurrent protection is generally provided by a fuse. This may be a lateral fuse as shown or the fuse in some piece of equipment such as a transformer. These devices predominantly use simple overcurrent measurements to determine if there is a fault. Some will also sense direction to prevent nuisance tripping on faults out of their protective zones. Breakers are nearly always three-phase devices. Reclosers may be either three-phase or single-phase interrupters. Fuses are always single-phase devices.

Because fuses are at the lowest level, they dictate the basic timing of all overcurrent devices upstream. Proper coordination for permanent faults is generally to start with the fuse time–current characteristic (TCC) and set each successive upline relay in the automatic devices a little slower. An exception is when fuse saving is employed. Then the breaker or recloser will operate faster than the fuse once or twice in an attempt to save the fuse for temporary faults such as a lightning strike or a tree brushing against the line. More details of overcurrent coordinations on utility distribution systems are described in Sec. 3.8.

3.2   Sources of Sags and Interruptions

Voltage sags and interruptions are generally caused by faults (short circuits) on the utility system.[4] Consider a customer that is supplied from the feeder supplied by circuit breaker 1 on the diagram shown in Fig. 3.5. If there is a fault on the same feeder, the customer will experience a voltage sag during the fault followed by an interruption when the breaker opens to clear the fault. If the fault is temporary in nature, a reclosing operation on the breaker should be successful and

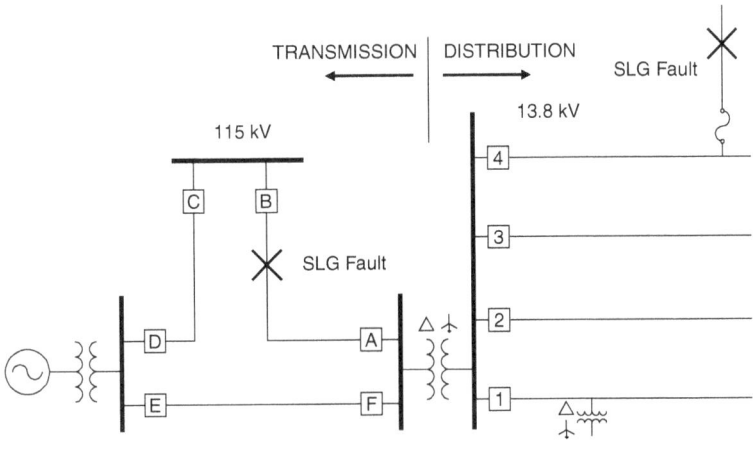

Figure 3.5 Fault locations on the utility power system.

the interruption will only be temporary. It will usually require about 5 or 6 cycles for the breaker to operate, during which time a voltage sag occurs. The breaker will remain open for typically a minimum of 12 cycles up to 5 s depending on utility reclosing practices. Sensitive equipment will almost surely trip during this interruption.

A much more common event would be a fault on one of the other feeders from the substation, i.e., a fault on a parallel feeder, or a fault somewhere on the transmission system (see the fault locations shown in Fig. 3.5). In either of these cases, the customer will experience a voltage sag during the period that the fault is actually on the system. As soon as breakers open to clear the fault, normal voltage will be restored at the customer.

Note that to clear the fault shown on the transmission system, both breakers A and B must operate. Transmission breakers will typically clear a fault in 5 or 6 cycles. In this case there are two lines supplying the distribution substation and only one has a fault. Therefore, customers supplied from the substation should expect to see only a sag and not an interruption. The distribution fault on feeder 4 may be cleared either by the lateral fuse or the breaker, depending on the utility's fuse-saving practice.

Any of these fault locations can cause equipment to misoperate in customer facilities. The relative importance of faults on the transmission system and the distribution system will depend on the specific characteristics of the systems (underground versus overhead distribution, lightning flash densities, overhead exposure, etc.) and the sensitivity of the equipment to voltage sags. Figure 3.6 shows an example of the breakdown of the events that caused equipment misoperation for one industrial customer. Note that faults on the

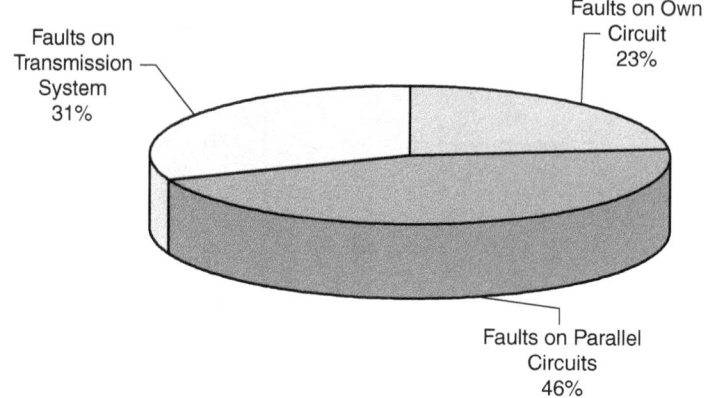

Faults on Transmission System 31%

Faults on Own Circuit 23%

Faults on Parallel Circuits 46%

FIGURE 3.6   Example of fault locations that caused misoperation of sensitive production equipment at an industrial facility (the example system had multiple overhead distribution feeders and an extensive overhead transmission system supplying the substation).

customer feeder only accounted for 23 percent of the events that resulted in equipment misoperation. This illustrates the importance of understanding the voltage sag performance of the system and the equipment sensitivity to these events.

Figures 3.7 and 3.8 show an interesting utility fault event recorded for an Electric Power Research Institute research project[4,11] by 8010 PQNode* instruments at two locations in the power system. The top chart in each of the figures is the rms voltage variation with time, and the bottom chart is the first 175 ms of the actual waveform. Figure 3.7 shows the characteristic measured at a customer location on an unfaulted part of the feeder. Figure 3.8 shows the momentary interruption (actually two separate interruptions) observed downline from the fault. The interrupting device in this case was a line recloser that was able to interrupt the fault very quickly in about 2.5 cycles. This device can have a variety of settings. In this case, it was set for two fast operations and two delayed operations. Figure 3.7 shows only the brief sag to 65 percent voltage for the first fast operation. There was an identical sag for the second operation. While this is very brief sag that is virtually unnoticeable by observing lighting blinks, many industrial processes would have shut down.

Figure 3.8 clearly shows the voltage sag prior to fault-clearing and the subsequent two fast recloser operations. The reclose time (the time the recloser was open) was a little more than 2 s, a very common time for a utility line recloser. Apparently, the fault—perhaps, a tree branch—was not cleared completely by the first operation, forcing a second. The system was restored after the second operation.

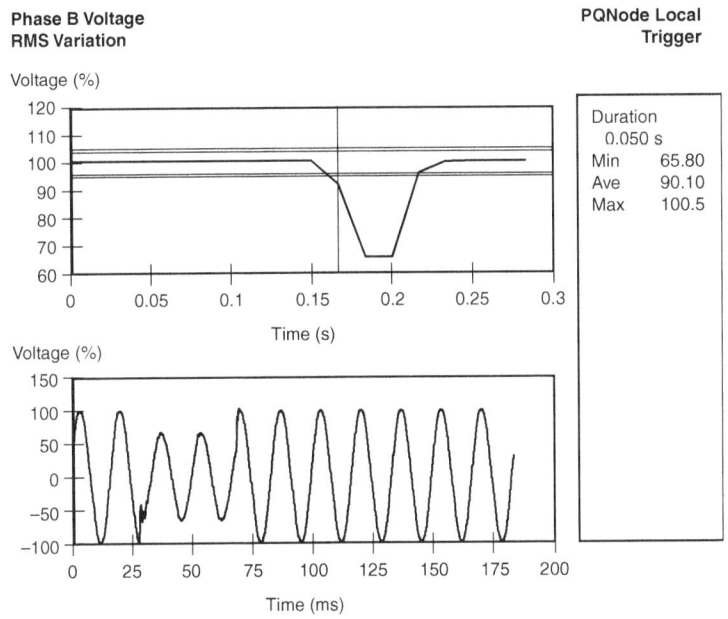

FIGURE 3.7   Voltage sag due to a short-circuit fault on a parallel utility feeder.

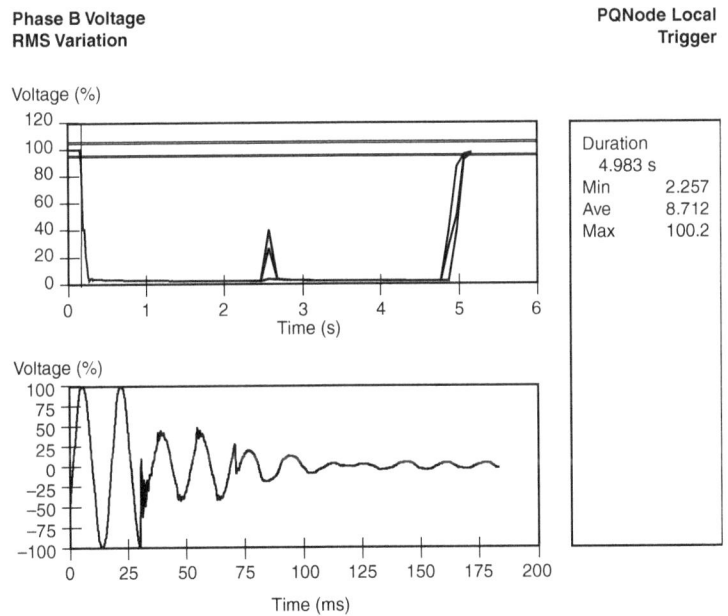

FIGURE 3.8   Utility short-circuit fault event with two fast trip operations of utility line recloser.[11]

There are a few things to note about this typical event that will tie in with other material in this book:

1. The voltage did not go to zero during the fault as is often assumed in textbook examples. There are few examples of the textbook case in real life.

2. The line recloser detected the fault and operated very quickly. There is a common misconception that fault interruption is slower on the distribution system than on the transmission system. While it can be slower, it can also be faster.

3. Since the voltage did not collapse to zero during the fault, induction machines will continue to have excitation and continue to feed the fault. This can be an especially important consideration for distributed generation (Chap. 9).

3.3   Estimating Voltage Sag Performance

It is important to understand the expected voltage sag performance of the supply system so that facilities can be designed and equipment specifications developed to assure the optimum operation of production facilities. The following is a general procedure for working with industrial customers to assure compatibility between the supply system characteristics and the facility operation:

1. Determine the number and characteristics of voltage sags that result from transmission system faults.

2. Determine the number and characteristics of voltage sags that result from distribution system faults (for facilities that are supplied from distribution systems).

3. Determine the equipment sensitivity to voltage sags. This will determine the actual performance of the production process based on voltage sag performance calculated in steps 1 and 2.

4. Evaluate the economics of different solutions that could improve the performance, either on the supply system (fewer voltage sags) or within the customer facility (better immunity).

The steps in this procedure are discussed in more detail throughout this chapter.

3.3.1   Area of Vulnerability

The concept of an *area of vulnerability* has been developed to help evaluate the likelihood of sensitive equipment being subjected to voltage lower than its *minimum voltage sag ride-through capability*.[5] The latter term is defined as the minimum voltage magnitude a piece of equipment can withstand or tolerate without misoperation or failure. This is also known as the equipment voltage sag immunity or susceptibility limit. An area of vulnerability is determined by the total circuit miles of exposure to faults that can cause voltage magnitudes at an end-user facility to drop below the equipment minimum voltage sag ride-through capability. Figure 3.9 shows an example of an area of vulnerability diagram for motor contactor and adjustable-speed-drive loads at an end-user facility served from the distribution system. The loads will be subject to faults on both the transmission system and the distribution system. The actual number of voltage sags that a facility can expect is determined by combining the area of vulnerability with the expected fault performance for this portion of the power system. The expected fault performance is usually determined from historical data.

3.3.2 Equipment Sensitivity to Voltage Sags

Equipment within an end-user facility may have different sensitivity to voltage sags.[8] Equipment sensitivity to voltage sags is very dependent on the specific load type, control settings, and applications. Consequently, it is often difficult to identify which characteristics of a given voltage sag are most likely to cause equipment to misoperate. The most commonly used characteristics are the duration and magnitude of the sag. Other less commonly used characteristics include phase shift and unbalance, missing voltage, three-phase voltage unbalance during the sag event, and the point-in-the-wave at which the sag initiates and terminates.

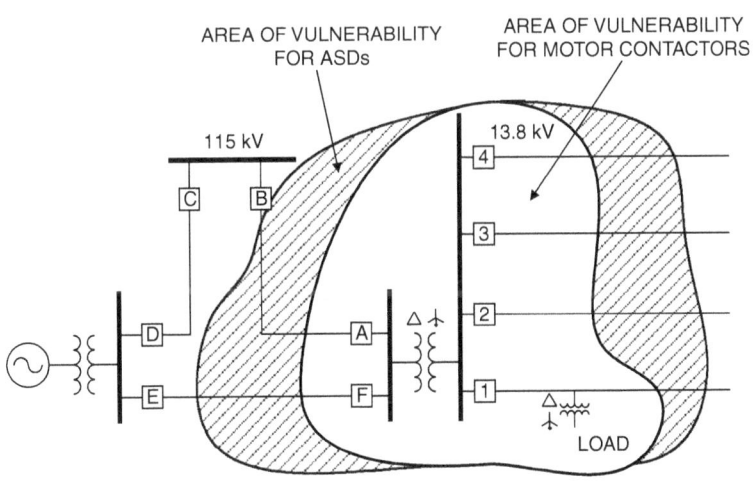

FIGURE 3.9   Illustration of an area of vulnerability.

Generally, equipment sensitivity to voltage sags can be divided into three categories:

- *Equipment sensitive to only the magnitude of a voltage sag.* This group includes devices such as undervoltage relays, process controls, motor drive controls,[6] and many types of automated machines (e.g., semiconductor manufacturing equipment). Devices in this group are sensitive to the minimum (or maximum) voltage magnitude experienced during a sag (or swell). The duration of the disturbance is usually of secondary importance for these devices.

- *Equipment sensitive to both the magnitude and duration of a voltage sag.* This group includes virtually all equipment that uses electronic power supplies. Such equipment misoperates or fails when the power supply output voltage drops below specified values. Thus, the important characteristic for this type of equipment is the duration that the rms voltage is below a specified threshold at which the equipment trips.

- *Equipment sensitive to characteristics other than magnitude and duration.* Some devices are affected by other sag characteristics such as the phase unbalance during the sag event, the point-in-the-wave at which the sag is initiated, or any transient oscillations occurring during the disturbance. These characteristics are more subtle than magnitude and duration, and their impacts are much more difficult to generalize. As a result, the rms variation performance indices defined here are focused on the more common magnitude and duration characteristics.

For end users with sensitive processes, the voltage sag ride-through capability is usually the most important characteristic to consider. These loads can generally be impacted by very short duration events, and virtually all voltage sag conditions last at least 4 or 5 cycles (unless the fault is cleared by a current-limiting fuse). Thus, one of the most common methods to quantify equipment susceptibility to voltage sags is using a magnitude-duration plot as shown in Fig. 3.10. It shows the voltage sag magnitude that will cause equipment to misoperate as a function of the sag duration.

The curve labeled CBEMA represents typical equipment sensitivity characteristics. The curve was developed by the CBEMA and was adopted in IEEE 446 (Orange Book). Since the association reorganized in 1994 and was subsequently renamed the Information Technology Industry Council (ITI), the CBEMA curve was also updated and renamed the ITI curve. Typical loads will likely trip off when the voltage is below the CBEMA, or ITI, curve.

The curve labeled ASD represents an example ASD voltage sag ride-through capability for a device that is very sensitive to voltage

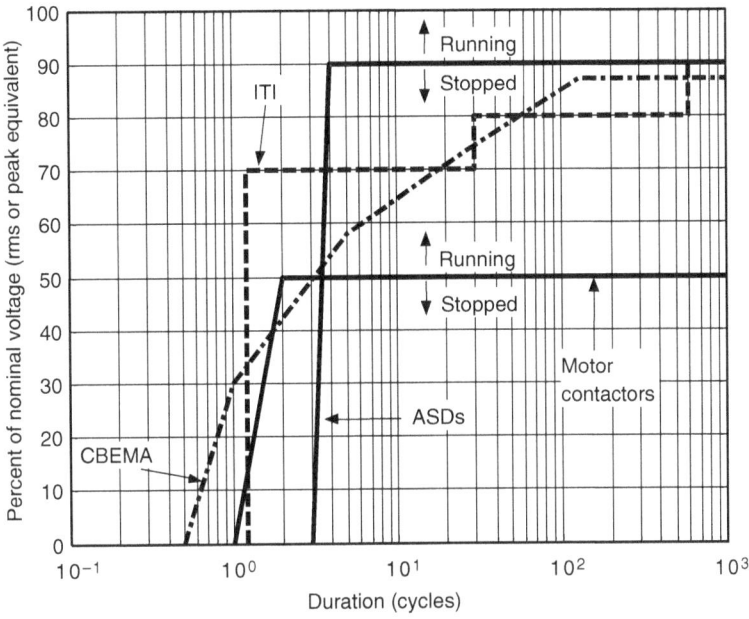

FIGURE **3.10** Typical equipment voltage sag ride-through capability curves.

sags. It trips for sags below 0.9 pu that last for only 4 cycles. The contactor curve represents typical contactor sag ride-through characteristics. It trips for voltage sags below 0.5 pu that last for more than 1 cycle.

The area of vulnerability for motor contactors shown in Fig. 3.9 indicates that faults within this area will cause the end-user voltage to drop below 0.5 pu. Motor contactors having a minimum voltage sag ride-through capability of 0.5 pu would have tripped out when a fault causing a voltage sag with duration of more than 1 cycle occurs within the area of vulnerability. However, faults outside this area will not cause the voltage to drop below 0.5 pu. The same discussion applies to the area of vulnerability for ASD loads. The less sensitive the equipment, the smaller the area of vulnerability will be (and the fewer times sags will cause the equipment to misoperate).

3.3.3 Transmission System Sag Performance Evaluation

The voltage sag performance for a given customer facility will depend on whether the customer is supplied from the transmission system or from the distribution system. For a customer supplied from the transmission system, the voltage sag performance will depend on only the transmission system fault performance. On the other hand, for a customer supplied from the distribution system, the voltage sag performance will depend on the fault performance on both the transmission and distribution systems.

This section discusses procedures to estimate the transmission system contribution to the overall voltage sag performance at a facility. Section 3.3.4 focuses on the distribution system contribution to the overall voltage sag performance.

Transmission line faults and the subsequent opening of the protective devices rarely cause an interruption for any customer because of the interconnected nature of most modern-day transmission networks. These faults do, however, cause voltage sags. Depending on the equipment sensitivity, the unit may trip off, resulting in substantial monetary losses. The ability to estimate the expected voltage sags at an end-user location is therefore very important.

Most utilities have detailed short-circuit models of the interconnected transmission system available for programs such as ASPEN* One Liner (Fig. 3.11). These programs can calculate the voltage throughout the system resulting from faults around the system. Many of them can also apply faults at locations along the transmission lines to help calculate the area of vulnerability at a specific location.

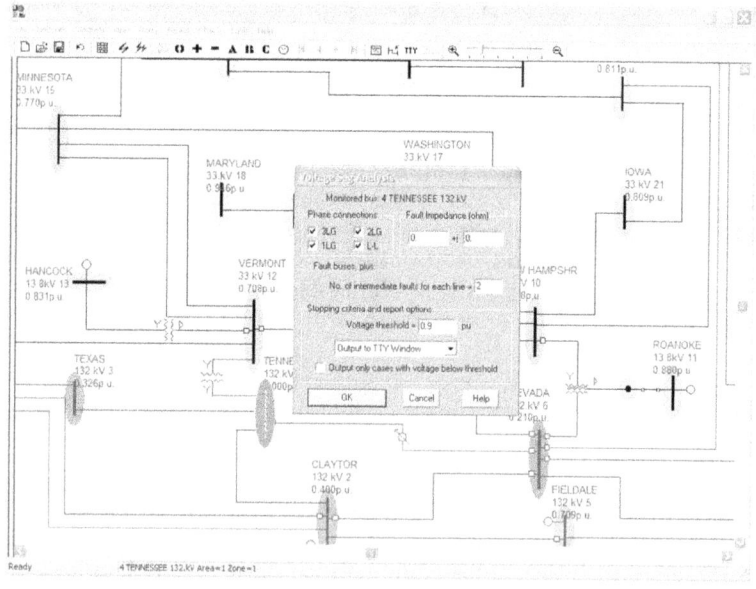

FIGURE 3.11   Example of modeling the transmission system in a short-circuit program for calculation of the area of vulnerability.

The area of vulnerability describes all the fault locations that can cause equipment to misoperate. The type of fault must also be considered in this analysis. SLG faults will not result in the same voltage sag at the customer equipment as a three-phase fault. The characteristics at the end-use equipment also depend on how the voltages are changed by transformer connections and how the equipment is connected, i.e., phase-to-ground or phase-to-phase. Table 3.1 summarizes voltages at the customer transformer secondary for a SLG fault at the primary.

The relationships in Table 3.1 illustrate the fact that a SLG fault on the primary of a delta–wye grounded transformer does not result in zero voltage on any of the phase-to-ground or phase-to-phase voltages on the secondary of the transformer. The magnitude of the

Transformer connection (pri/sec)	Phase-to-phase $V_{ab} V_{bc} V_{ca}$			Phase-to-neutral $V_{an} V_{bn} V_{cn}$			Phasor diagram
	0.58	1.00	0.58	0.00	1.00	1.00	
	0.58	1.00	0.58	0.33	0.88	0.88	
	0.33	0.88	0.88	----	----	----	
	0.33	0.88	0.88	0.58	0.58	1.00	

TABLE 3.1 Transformer Secondary Voltages with a SLG Fault on the Primary

lowest secondary voltage depends on how the equipment is connected:

- Equipment connected line-to-line would experience a minimum voltage of 33 percent.
- Equipment connected line-to-neutral would experience a minimum voltage of 58 percent.

This illustrates the importance of both transformer connections and the equipment connections in determining the actual voltage that equipment will experience during a fault on the supply system.

Math Bollen[16] developed the concept of voltage sag "types" to describe the different voltage sag characteristics that can be experienced at the end-user level for different fault conditions and system configurations. The five types that can commonly be experienced are illustrated in Fig. 3.12. These fault types can be used to conveniently summarize the expected performance at a customer location for different types of faults on the supply system.

Example of Evaluation

Table 3.2 is an example of an area of vulnerability for a customer supplied from a transmission system. The table only shows voltage sags below 90 percent at a customer bus (the rightmost column) due to SLG faults along a transmission line between "From Bus" and "To Bus" shown in columns two and three.

The data in Table 3.2 groups the area of vulnerability based on three transmission voltages, 115 kV, 230 kV, and 525 kV. The first row provides the following information. The total circuit miles of exposure for Tall Timber—Cedar Bluff 115 kV line resulting in voltage sags below 90 percent at a customer bus is 3.3 miles. The lowest voltage sags caused by SLG faults anywhere along the 3.3-mile stretch

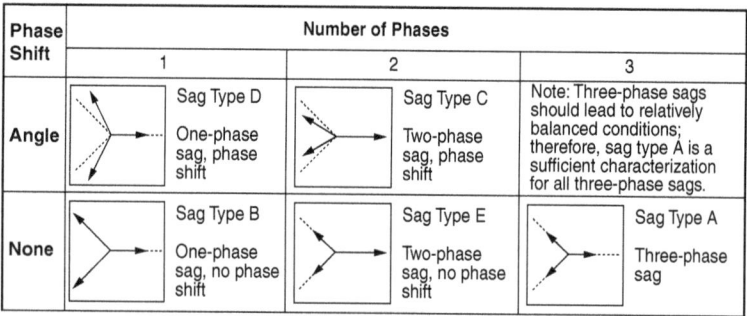

Phase Shift	Number of Phases		
	1	2	3
Angle	Sag Type D One-phase sag, phase shift	Sag Type C Two-phase sag, phase shift	Note: Three-phase sags should lead to relatively balanced conditions; therefore, sag type A is a sufficient characterization for all three-phase sags.
None	Sag Type B One-phase sag, no phase shift	Sag Type E Two-phase sag, no phase shift	Sag Type A Three-phase sag

Figure 3.12   Voltage sag types at end-use equipment that result from different types of faults and transformer connections.

Voltage (kV)	From bus	To bus	Length (miles)	SLG fault performance (faults/year)	Phase voltage at end user bus (per unit)
115	Tall Timber	Cedar Bluff	3.3	0.59	0.516
115	Cedar Bluff	So Peters	8.52	1.54	0.598
115	Sledge	So Peters	1.89	0.34	0.746
115	Patricks	No Peters	0.1	0.02	0.804
115	Sledge	Kingston Pk	10.2	1.84	0.826
115	Sledge	Gallager View	14.38	2.59	0.829
115	Patricks	Woodview	5.98	1.08	0.851
115	Sledge	Deep Eddy	10.1	1.82	0.859
115	Sledge	Enfield	17.56	3.17	0.862
230	Cedar Bluff	Gallager View	10	0.99	0.612
230	Sledge	Cedar Bluff	10.66	1.06	0.722
230	Manor	Gallager View	7	0.69	0.765
230	Airport	Gallager View	0.15	0.01	0.813
230	Red River	Duval	1.98	0.20	0.825
230	McNeil	Boulder	0.96	0.10	0.838
230	Glass Mt	Boulder	0.84	0.08	0.841
230	Crossland	Boulder	0.96	0.10	0.841
230	Weller	Boulder	1.09	0.11	0.842
230	Tincup	Boulder	0.92	0.09	0.842
230	Swan	Boulder	0.84	0.08	0.843
230	Ashton	Vista	6	0.59	0.843
230	Topridge	Spring H	12.2	1.21	0.847
230	Airport	Topridge	20.57	2.04	0.850
230	Vista	Topridge	10	0.99	0.851
230	Woodview	Gallager View	26.24	2.60	0.853
230	Buckner	Spring H	0.31	0.03	0.857
525	ManorV	Windy Ridge	2.45	0.05	0.747
525	Vista	Giacomo	18.33	0.37	0.776
525	Vista	Twin Ck	17.84	0.36	0.776
525	TwinCk	Topridge	8.22	0.16	0.791
525	Vista	Lion	39.38	0.79	0.792
525	Bent Bow	Boulder	0.62	0.01	0.799
525	Grand Falls	Boulder	0.72	0.01	0.799
525	Westwood	Boulder	18.94	0.38	0.818
525	Westwood	Boulder	18.96	0.38	0.823
525	Lion	Duval CAPS	64.82	1.30	0.825
525	Star	Twin Ck	23.79	0.48	0.842
525	Windy Ridge	Checker	68.97	1.38	0.843
525	Giacomo	Lime Ck	42.35	0.85	0.853

TABLE 3.2   Calculating Expected Sag Performance at a Specific Customer Site for a Given Voltage Level

of line is 0.516 per unit (pu).The SLG fault performance for this line is 0.59 faults/year or 1 faults for every 1.7 years.

The expected 90 percent voltage sag performance due to 115-kV line exposure to SLG faults is the sum of individual SLG fault performance for each 115-kV line. The expected performance would be 0.59 + 1.54 + ⋯ + 1.82 + 3.17 = 12.99 events/year. This means that the customer bus would expect 12.99 voltage sag events with magnitudes of 0.9 pu and below annually. If a 80 percent voltage sag performance is desired, the customer should expect to see 0.59 + 1.54 + 0.34 = 2.47 events/year.

Figure 3.13 shows the expected voltage sag performance at a customer bus for sensitivity thresholds between 0.5 and 0.9 pu. The performance is given for each transmission voltage level as well as the entire transmission system. The expected voltage sag performance for a given sensitivity level can be quickly determined from the figure. Note that, expected voltage sag performance in the above examples have been computed using the historical fault performance of individual lines.

Individual historical fault performance data may be difficult to obtain, however, the average fault performance for each voltage level is usually available. Assuming that the average fault performance for 115-kV lines is 17.5 faults/100 miles/year, the expected 90 percent

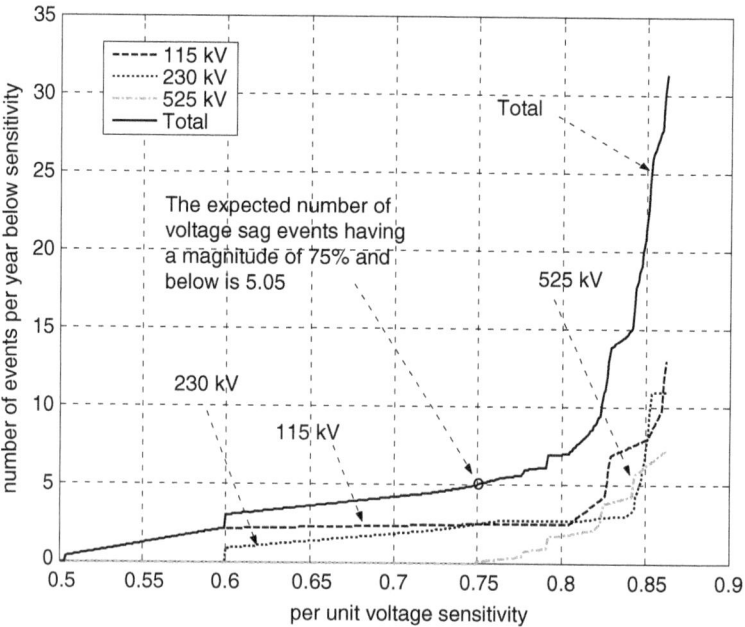

FIGURE 3.13   Estimated voltage sag performance at customer equipment due to transmission system faults.

Location:	Customer bus		
Performance of voltage sags for a threshold of 90%			
Sags due to SLG faults only			
Voltage level (kV)	**Exposure (miles)**	**Fault performance (faults/100 miles/year)**	**Expected sag (sag/year)**
115	72.03	17.5	12.60
230	110.72	10.0	11.07
525	325.39	2.0	6.51
Total			30.18

TABLE 3.3 Calculating Expected Sag Performance at Specific Customer Site for a Given Voltage Level

voltage sag performance would be $(3.3 + 8.52 + \ldots + 10.1 + 17.56) \times 17.5/100 = 12.60$ events/yr. The expected voltage sag performance for other voltage levels can be computed similarly as shown in Table 3.3.

3.3.4 Utility Distribution System Sag Performance Evaluation

Customers that are supplied at distribution voltage levels are impacted by faults on both the transmission system and the distribution system. The analysis at the distribution level must also include momentary interruptions caused by the operation of protective devices to clear the faults.[7] These interruptions will most likely trip out sensitive equipment. The example presented in this section illustrates data requirements and computation procedures for evaluating the expected voltage sag and momentary interruption performance. The overall voltage sag performance at an end-user facility is the total of the expected voltage sag performance from the transmission and distribution systems.

Figure 3.14 shows a typical distribution system with multiple feeders and fused branches, and protective devices. The utility protection scheme plays an important role in the voltage sag and momentary interruption performance. The critical information needed to compute voltage sag performance can be summarized as follows:

- Number of feeders supplied from the substation.
- Average feeder length.
- Average feeder reactance.
- Short-circuit equivalent reactance at the substation.
- Feeder reactors, if any.
- Average feeder fault performance which includes three-phase-line-to-ground (3LG) faults and single-line-to-ground

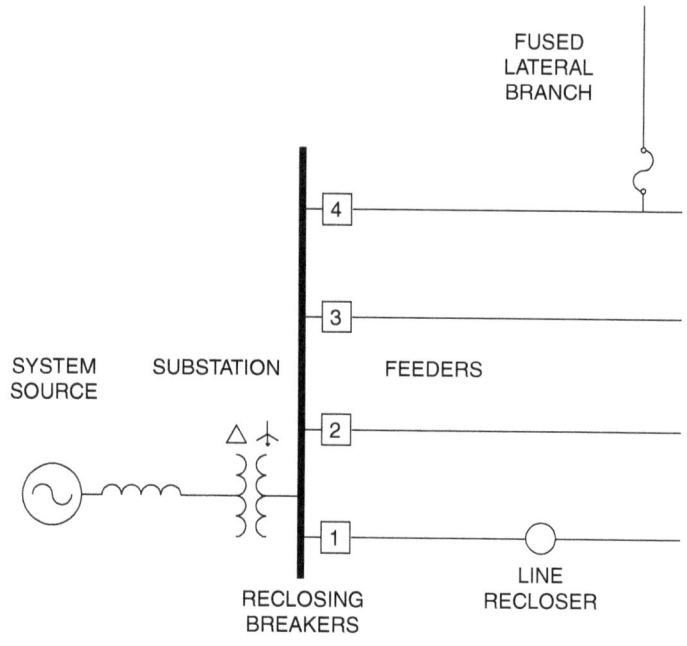

FIGURE 3.14 Typical distribution system illustrating protection devices.

(SLG) faults in faults per mile per month. The feeder performance data may be available from protection logs. However, data for faults that are cleared by downline fuses or downline protective devices may be difficult to obtain and this information may have to be estimated.

There are two possible locations for faults on the distribution systems, i.e., on the same feeder and on parallel feeders. An area of vulnerability defining the total circuit miles of fault exposures that can cause voltage sags below equipment sag ride-through capability at a specific customer needs to be defined. The computation of the expected voltage sag performance can be performed as follows:

Faults on Parallel Feeders

Voltage experienced at the end-user facility following a fault on parallel feeders can be estimated by calculating the expected voltage magnitude at the substation. The voltage magnitude at the substation is impacted by the fault impedance and location, the configuration of the power system, and the system protection scheme. Figure 3.15 illustrates the effect of the distance between the substation and the fault locations for 3LG and SLG faults on a radial distribution system. The SLG fault curve shows the A–B phase bus voltage on the

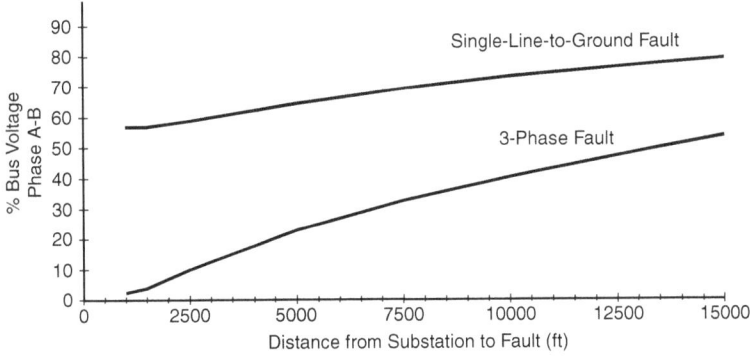

FIGURE 3.15 Example of voltage sag magnitude at an end-user location as a function of the fault location along a parallel feeder circuit.

secondary of a delta–wye–grounded step-down transformer, with an A phase-to-ground fault on the primary. The actual voltage at the end-user location can be computed by converting the substation voltage using Table 3.3. The voltage sag performance for a specific sensitive equipment having the minimum ride-through voltage of v_s can be computed as follows:

$$E_{\text{parallel}}(v_s) = N_1 \times E_{p1} + N_3 \times E_{p3}$$

where N_1 and N_3 are the fault performance data for SLG and 3LG faults in faults per miles per month, and E_{p1} and E_{p3} are the total circuit miles of exposure to SLG and 3LG faults on parallel feeders that result in voltage sags below the minimum ride-through voltage v_s at the end-user location.

Faults on the Same Feeder

In this step the expected voltage sag magnitude at the end-user location is computed as a function of fault location on the same feeder. Note that, however, the computation is performed only for fault locations that will result in a sag but will not result in a momentary interruption, which will be computed separately. Examples of such fault locations include faults beyond a downline recloser or a branched fuse that is coordinated to clear before the substation recloser. The voltage sag performance for specific sensitive equipment with ride-through voltage v_s is computed as follows:

$$E_{\text{same}}(v_s) = N_1 \times E_{s1} + N_3 \times E_{s3}$$

where E_{s1} and E_{s3} are the total circuit miles of exposure to SLG and 3LG on the same feeders that result in voltage sags below v_s at the end-user location.

The total expected voltage sag performance for the minimum ride-through voltage v_s would be the sum of expected voltage sag performance on the parallel and the same feeders, i.e., $E_{parallel}(v_s)$ + $E_{same}(v_s)$. The total expected sag performance can be computed for other voltage thresholds, which then can be plotted to produce a plot similar to ones in Fig. 3.13.

The expected interruption performance at the specified location can be determined by the length of exposure that will cause a breaker or other protective device in series with the customer facility to operate. For example, if the protection is designed to operate the substation breaker for any fault on the feeder, then this length is the total exposure length. The expected number of interruptions can be computed as follows:

$$E_{int} = L_{int} \times (N_1 + N_3)$$

where L_{int} is the total circuit miles of exposure to SLG and 3LG that results in interruptions at an end-user facility.

3.4 Fundamental Principles of Protection

Several things can be done by the utility, end user, and equipment manufacturer to reduce the number and severity of voltage sags and to reduce the sensitivity of equipment to voltage sags. Figure 3.16

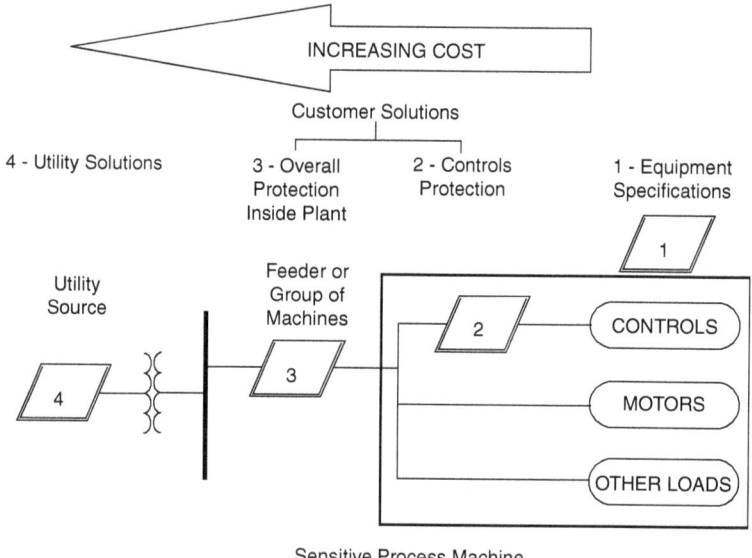

FIGURE 3.16 Approaches for voltage sag ride-through.

illustrates voltage sag solution alternatives and their relative costs. As this chart indicates, it is generally less costly to tackle the problem at its lowest level, close to the load. The best answer is to incorporate ride-through capability into the equipment specifications themselves. This essentially means keeping problem equipment out of the plant, or at least identifying ahead of time power conditioning requirements. Several ideas, outlined here, could easily be incorporated into any company's equipment procurement specifications to help alleviate problems associated with voltage sags:

1. Equipment manufacturers should have voltage sag ride-through capability curves (similar to the ones shown previously) available to their customers so that an initial evaluation of the equipment can be performed. Customers should begin to demand that these types of curves be made available so that they can properly evaluate equipment.

2. The company procuring new equipment should establish a procedure that rates the importance of the equipment. If the equipment is critical in nature, the company must make sure that adequate ride-through capability is included when the equipment is purchased. If the equipment is not important or does not cause major disruptions in manufacturing or jeopardize plant and personnel safety, voltage sag protection may not be justified.

3. Equipment should at least be able to ride through voltage sags with a minimum voltage of 70 percent (ITI curve). The relative probability of experiencing a voltage sag to 70 percent or less of nominal is much less than experiencing a sag to 90 percent or less of nominal. A more ideal ride-through capability for short-duration voltage sags would be 50 percent, as specified by the semiconductor industry in Standard SEMI F-47.[17]

As we entertain solutions at higher levels of available power, the solutions generally become more costly. If the required ride-through cannot be obtained at the specification stage, it may be possible to apply an uninterruptible power supply (UPS) system or some other type of power conditioning to the machine control. This is applicable when the machines themselves can withstand the sag or interruption, but the controls would automatically shut them down.

At level 3 in Fig. 3.16, some sort of backup power supply with the capability to support the load for a brief period is required. Level 4 represents alterations made to the utility power system to significantly reduce the number of sags and interruptions.

3.5   Solutions at the End-User Level

Solutions to improve the reliability and performance of a process or facility can be applied at many different levels. The different technologies available should be evaluated based on the specific requirements of the process to determine the optimum solution for improving the overall voltage sag performance. As illustrated in Fig. 3.16, the solutions can be discussed at the following different levels of application:

1. *Protection for small loads [e.g., less than 5 kilovoltamperes (kVA)].* This usually involves protection for equipment controls or small, individual machines. Many times, these are single-phase loads that need to be protected.

2. *Protection for individual equipment or groups of equipment up to about 300 kVA.* This usually represents applying power conditioning technologies within the facility for protection of critical equipment that can be grouped together conveniently. Since usually not all the loads in a facility need protection, this can be a very economical method of dealing with the critical loads, especially if the need for protection of these loads is addressed at the facility design stage.

3. *Protection for large groups of loads or whole facilities at the low-voltage level.* Sometimes such a large portion of the facility is critical or needs protection that it is reasonable to consider protecting large groups of loads at a convenient location (usually the service entrance). New technologies are available for consideration when large groups of loads need protection.

4. *Protection at the medium-voltage level or on the supply system.* If the whole facility needs protection or improved power quality, solutions at the medium-voltage level can be considered.

The size ranges in these categories are quite arbitrary, and many of the technologies can be applied over a wider range of sizes. The following sections describe the major technologies available and the levels where they can be applied.

3.5.1   Ferroresonant Transformers

Ferroresonant transformers, also called constant-voltage transformers (CVTs), can handle most voltage sag conditions. (See Fig. 3.17.) CVTs are especially attractive for constant, low-power loads. Variable loads, especially with high inrush currents, present more of a problem for CVTs because of the tuned circuit on the

Figure 3.17 Examples of commercially available constant-voltage transformers (CVTs) (www.sola-hevi-duty.com).

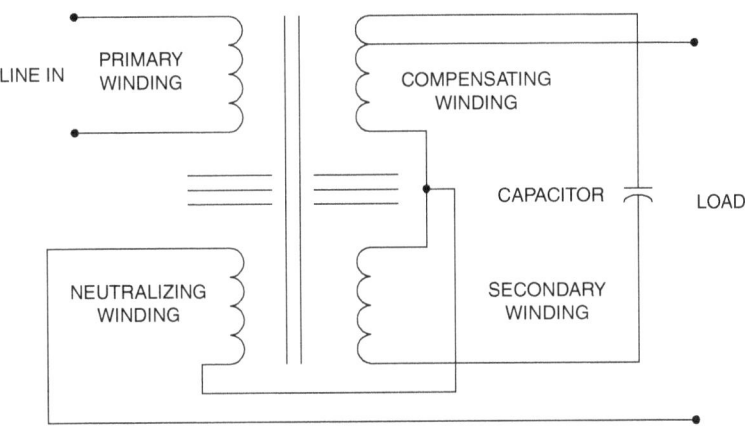

Figure 3.18 Schematic of ferroresonant constant-voltage transformer.

output. Ferroresonant transformers are basically 1:1 transformers which are excited high on their saturation curves, thereby providing an output voltage which is not significantly affected by input voltage variations. A typical ferroresonant transformer schematic circuit diagram is shown in Fig. 3.18.

Figure 3.19 shows the voltage sag ride-through improvement of a process controller fed from a 120-VA ferroresonant transformer. With

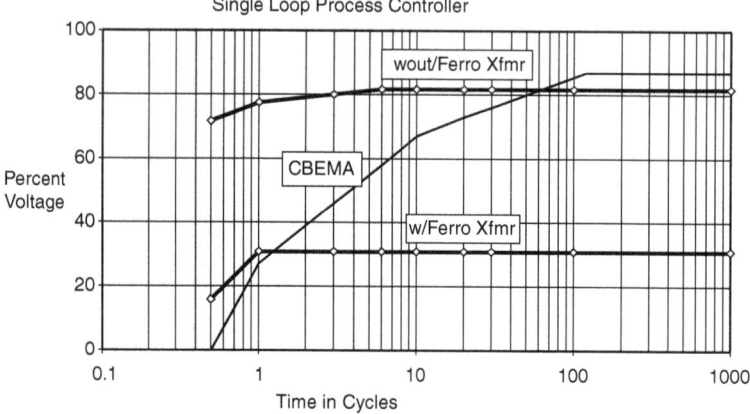

FIGURE 3.19   Voltage sag improvement with ferroresonant transformer.

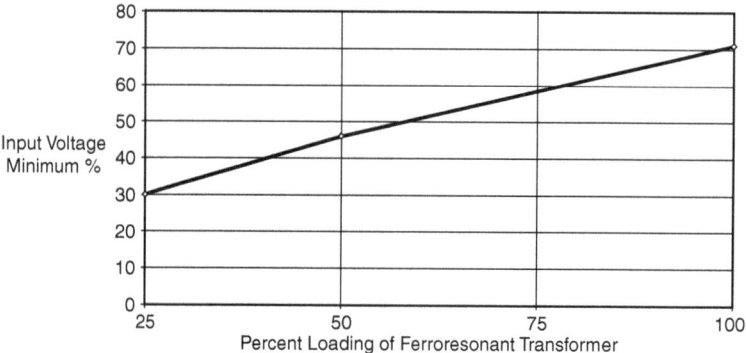

FIGURE 3.20   Voltage sag versus ferroresonant transformer loading.

the CVT, the process controller can ride through a voltage sag down to 30 percent of nominal, as opposed to 82 percent without one. Notice how the ride-through capability is held constant at a certain level. The reason for this is the small power requirement of the process controller, only 15 VA.

Ferroresonant transformers should be sized significantly larger than the load. Figure 3.20 shows the allowable voltage sag as a percentage of nominal voltage (that will result in at least 90 percent voltage on the CVT output) versus ferroresonant transformer loading, as specified by one manufacturer. At 25 percent of loading, the allowable voltage sag is 30 percent of nominal, which means that the CVT will output over 90 percent normal voltage as long as the input voltage is above 30 percent. This is important since the plant voltage rarely falls below 30 percent of nominal during voltage sag conditions. As the loading is increased, the corresponding ride-through capability is reduced, and when the ferroresonant transformer is overloaded (e.g., 150 percent loading), the voltage will collapse to zero.

3.5.2 Magnetic Synthesizers

Magnetic synthesizers use a similar operating principle to CVTs except they are three-phase devices and take advantage of the three-phase magnetics to provide improved voltage sag support and regulation for three-phase loads. They are applicable over a size range from about 15 to 200 kVA and are typically applied for process loads of larger computer systems where voltage sags or steady-state voltage variations are important issues. A block diagram of the process is shown in Fig. 3.21.

Energy transfer and line isolation are accomplished through the use of nonlinear chokes. This eliminates problems such as line noise. The ac output waveforms are built by combining distinct voltage pulses from saturated transformers. The waveform energy is stored in the saturated transformers and capacitors as current and voltage. This energy storage enables the output of a clean waveform with little harmonic distortion. Finally, three-phase power is supplied through a zigzag transformer. Figure 3.22 shows a magnetic synthesizer's voltage sag ride-through capability as compared to the CBEMA curve, as specified by one manufacturer.[*]

3.5.3 Active Series Compensators

Advances in power electronic technologies and new topologies for these devices have resulted in new options for providing voltage sag ride-through support to critical loads. One of the important new

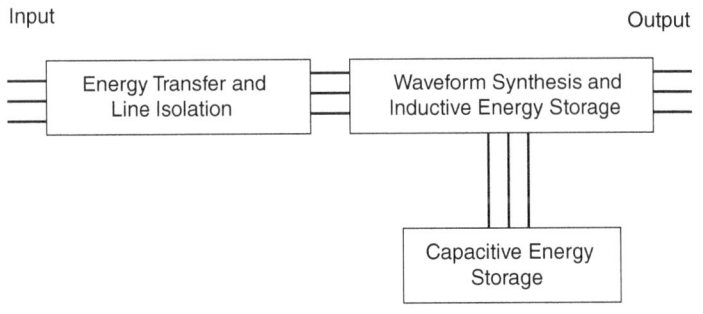

FIGURE 3.21 Block diagram of magnetic synthesizer.

---

[*] Liebert Corporation.

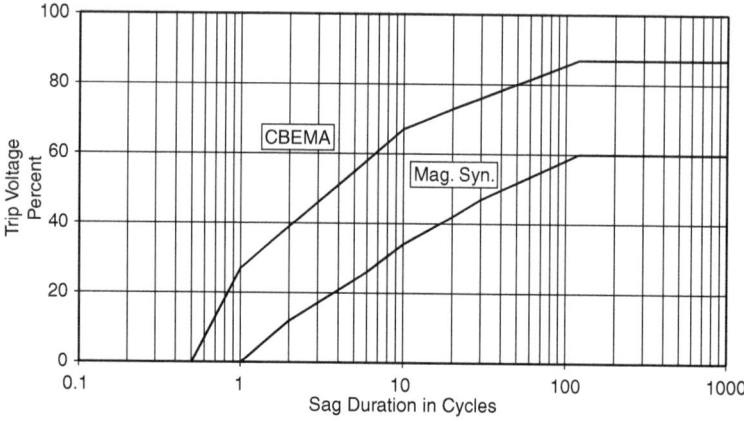

FIGURE 3.22   Magnetic synthesizer voltage sag ride-through capability.

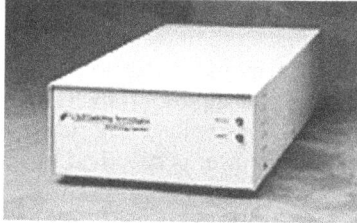

FIGURE 3.23   Example of active series compensator for single-phase loads up to about 5 kVA (www. softswitch.com).

options is a device that can boost the voltage by injecting a voltage in series with the remaining voltage during a voltage sag condition. These are referred to as *active series compensation devices.* They are available in size ranges from small single-phase devices (1 to 5 kVA) to very large devices that can be applied on the medium-voltage systems (2 MVA and larger). Figure 3.23 is an example of a small single-phase compensator that can be used to provide ride-through support for single-phase loads.

A one-line diagram illustrating the power electronics that are used to achieve the compensation is shown in Fig. 3.24. When a disturbance to the input voltage is detected, a fast switch opens and the power is supplied through the series-connected electronics. This circuit adds or subtracts a voltage signal to the input voltage so that the output voltage remains within a specified tolerance during the disturbance. The switch is very fast so that the disturbance seen by the load is less than a quarter cycle in duration. This is fast enough to avoid problems with almost all sensitive loads. The circuit can

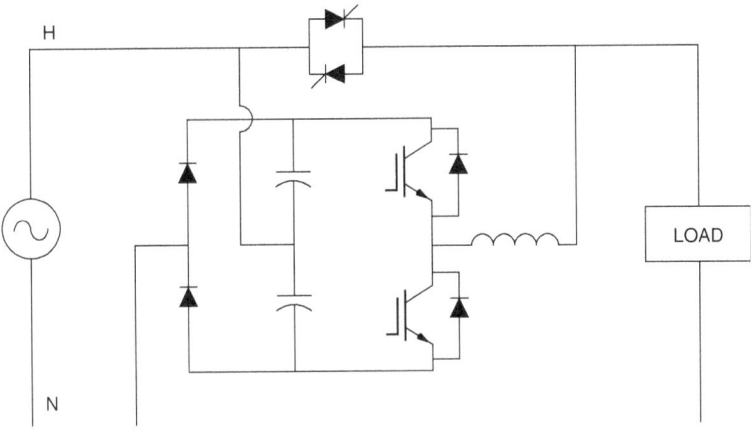

FIGURE 3.24   Topology illustrating the operation of the active series compensator.

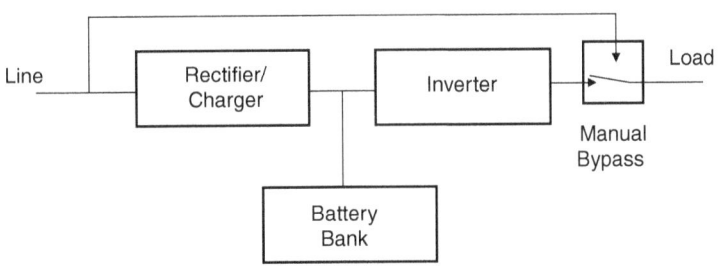

FIGURE 3.25   On-line UPS.

provide voltage boosting of about 50 percent, which is sufficient for almost all voltage sag conditions.

3.5.4   On-Line UPS

Figure 3.25 shows a typical configuration of an on-line UPS. In this design, the load is always fed through the UPS. The incoming ac power is rectified into dc power, which charges a bank of batteries. This dc power is then inverted back into ac power, to feed the load. If the incoming ac power fails, the inverter is fed from the batteries and continues to supply the load. In addition to providing ride-through for power outages, an on-line UPS provides very high isolation of the critical load from all power line disturbances. However, the on-line operation increases the losses and may be unnecessary for protection of many loads.

3.5.5 Standby UPS

A standby power supply (Fig. 3.26) is sometimes termed *off-line UPS* since the normal line power is used to power the equipment until a disturbance is detected and a switch transfers the load to the battery-backed inverter. The transfer time from the normal source to the battery-backed inverter is important. The CBEMA curve shows that 8 ms is the lower limit on interruption through for power-conscious manufacturers. Therefore a transfer time of 4 ms would ensure continuity of operation for the critical load. A standby power supply does not typically provide any transient protection or voltage regulation as does an on-line UPS. This is the most common configuration for commodity UPS units available at retail stores for protection of small computer loads.

UPS specifications include kilovoltampere capacity, dynamic and static voltage regulation, harmonic distortion of the input current and output voltage, surge protection, and noise attenuation. The specifications should indicate, or the supplier should furnish, the test conditions under which the specifications are valid.

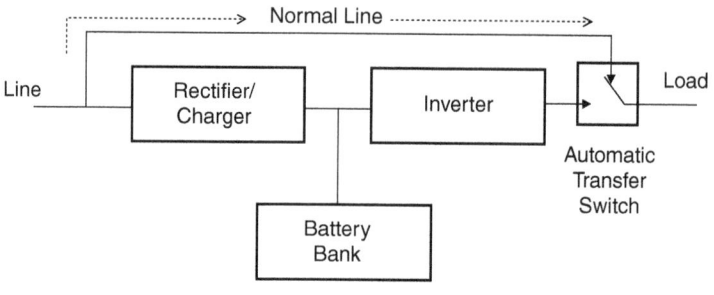

Figure 3.26 Standby UPS.

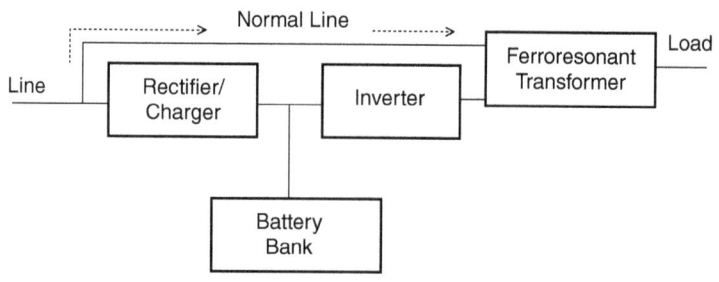

Figure 3.27 Hybrid UPS.

3.5.6   Hybrid UPS

Similar in design to the standby UPS, the hybrid UPS (Fig. 3.27) utilizes a voltage regulator on the UPS output to provide regulation to the load and momentary ride-through when the transfer from normal to UPS supply is made.

3.5.7   Motor-Generator Sets

Motor-generator (M-G) sets come in a wide variety of sizes and configurations. This is a mature technology that is still useful for isolating critical loads from sags and interruptions on the power system. The concept is very simple, as illustrated in Fig. 3.28. A motor powered by the line drives a generator that powers the load. Flywheels on the same shaft provide greater inertia to increase ride-through time. When the line suffers a disturbance, the inertia of the machines and the flywheels maintain the power supply for several seconds. This arrangement may also be used to separate sensitive loads from other classes of disturbances such as harmonic distortion and switching transients.

While simple in concept, M-G sets have disadvantages for some types of loads:

1. There are losses associated with the machines, although they are not necessarily larger than those in other technologies described here.

2. Noise and maintenance may be issues with some installations.

3. The frequency and voltage drop during interruptions as the machine slows. This may not work well with some loads.

Another type of M-G set uses a special synchronous generator called a written-pole motor that can produce a constant 60-Hz frequency as the machine slows. It is able to supply a constant output

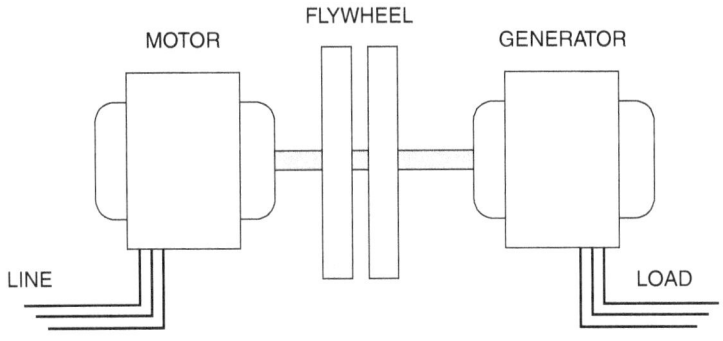

FIGURE 3.28   Block diagram of typical M-G set with flywheel.

by continually changing the polarity of the rotor's field poles. Thus, each revolution can have a different number of poles than the last one. Constant output is maintained as long as the rotor is spinning at speeds between 3150 and 3600 revolutions per minute (rpm). Flywheel inertia allows the generator rotor to keep rotating at speeds above 3150 rpm once power shuts off. The rotor weight typically generates enough inertia to keep it spinning fast enough to produce 60 Hz for 15 s under full load.

Another means of compensating for the frequency and voltage drop while energy is being extracted is to rectify the output of the generator and feed it back into an inverter. This allows more energy to be extracted, but also introduces losses and cost.

3.5.8   Flywheel Energy Storage Systems

Motor-generator sets are only one means to exploit the energy stored in flywheels. A modern flywheel energy system uses high-speed flywheels and power electronics to achieve sag and interruption ride-through from 10 s to 2 min. Figure 3.29 shows an example of a

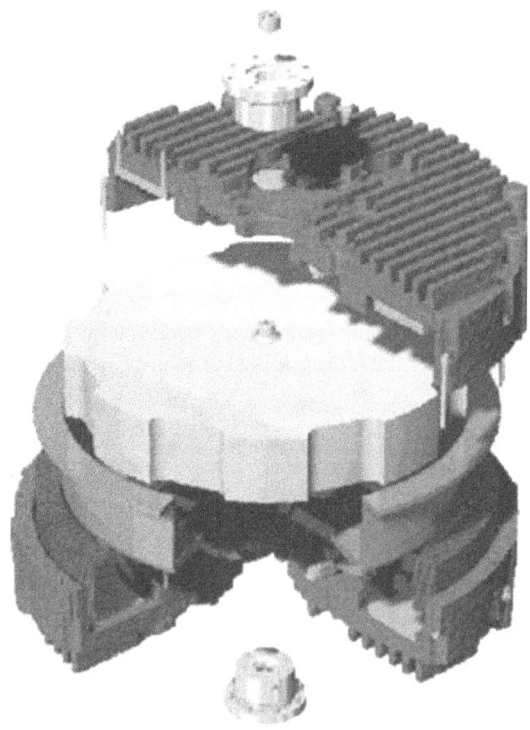

FIGURE 3.29   Cutaway view of an integrated motor, generator, and flywheel used for energy storage systems. (*Courtesy of Active Power, Inc.*)

flywheel used in energy storage systems. While M-G sets typically operate in the open and are subject to aerodynamic friction losses, these flywheels operate in a vacuum and employ magnetic bearings to substantially reduce standby losses. Designs with steel rotors may spin at approximately 10,000 rpm, while those with composite rotors may spin at much higher speeds. Since the amount of energy stored is proportional to the square of the speed, a great amount of energy can be stored in a small space.

The rotor serves as a one-piece storage device, motor, and generator. To store energy, the rotor is spun up to speed as a motor. When energy is needed, the rotor and armature act as a generator. As the rotor slows when energy is extracted, the control system automatically increases the field to compensate for the decreased voltage. The high-speed flywheel energy storage module would be used in place of the battery in any of the UPS concepts previously presented.

3.5.9 Superconducting Magnetic Energy Storage (SMES) Devices

An SMES device can be used to alleviate voltage sags and brief interruptions.[2] The energy storage in an SMES-based system is provided by the electric energy stored in the current flowing in a superconducting magnet. Since the coil is lossless, the energy can be released almost instantaneously. Through voltage regulator and inverter banks, this energy can be injected into the protected electrical system in less than 1 cycle to compensate for the missing voltage during a voltage sag event.

The SMES-based system has several advantages over battery-based UPS systems:

1. SMES-based systems have a much smaller footprint than batteries for the same energy storage and power delivery capability.[13]

2. The stored energy can be delivered to the protected system more quickly.

3. The SMES system has virtually unlimited discharge and charge duty cycles. The discharge and recharge cycles can be performed thousands of times without any degradation to the superconducting magnet.

The recharge cycle is typically less than 90 s from full discharge.

Figure 3.30 shows the functional block diagram of a common system. It consists of a superconducting magnet, voltage regulators, capacitor banks, a dc-to-dc converter, dc breakers, inverter modules, sensing and control equipment, and a series-injection transformer. The superconducting magnet is constructed of a niobium titanium

(NbTi) conductor and is cooled to approximately 4.2 kelvin (K) by liquid helium. The cryogenic refrigeration system is based on a two-stage recondenser. The magnet electrical leads use high-temperature superconductor (HTS) connections to the voltage regulator and controls. The magnet might typically store about 3 megajoules (MJ).

In the example system shown, energy released from the SMES passes through a current-to-voltage converter to charge a 14-microfarad (mF) dc capacitor bank to 2500 Vdc. The voltage regulator keeps the dc voltage at its nominal value and also provides protection control to the SMES. The dc-to-dc converter reduces the dc voltage down to 750 Vdc. The inverter subsystem module consists of six single-phase inverter bridges. Two IGBT inverter bridges rated 450 amperes (A) rms are paralleled in each phase to provide a total rating of 900 A per phase. The switching scheme for the inverter is based on the pulse-width modulation (PWM) approach where the carrier signal is a sine-triangle with a frequency of 4 kHz.[15]

A typical SMES system can protect loads of up to 8 MVA for voltage sags as low as 0.25 pu. It can provide up to 10 s of voltage sag ride-through depending on load size. Figure 3.31 shows an example where the grid voltage experiences a voltage sag of 0.6 pu for approximately 7 cycles. The voltage at the protected load remains virtually unchanged at its pre-fault value.

3.5.10 Static Transfer Switches and Fast Transfer Switches

There are a number of alternatives for protection of an entire facility that may be sensitive to voltage sags. These include dynamic voltage

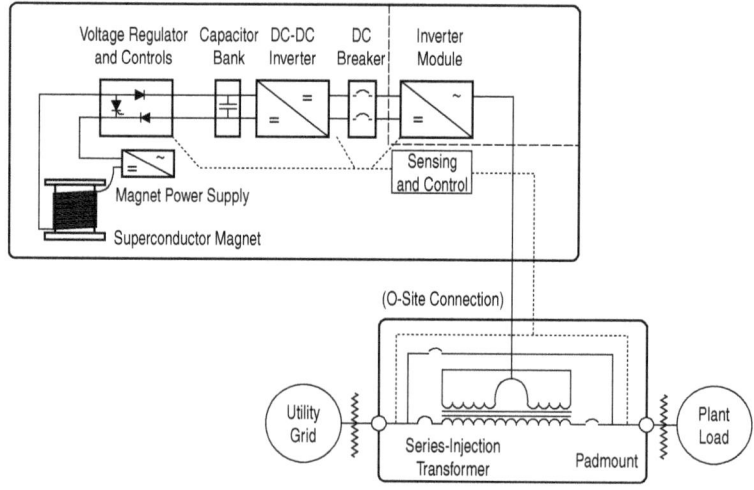

FIGURE 3.30 Typical power quality–voltage regulator (PQ-VR) functional block diagram. (*Courtesy of American Superconductor, Inc.*)

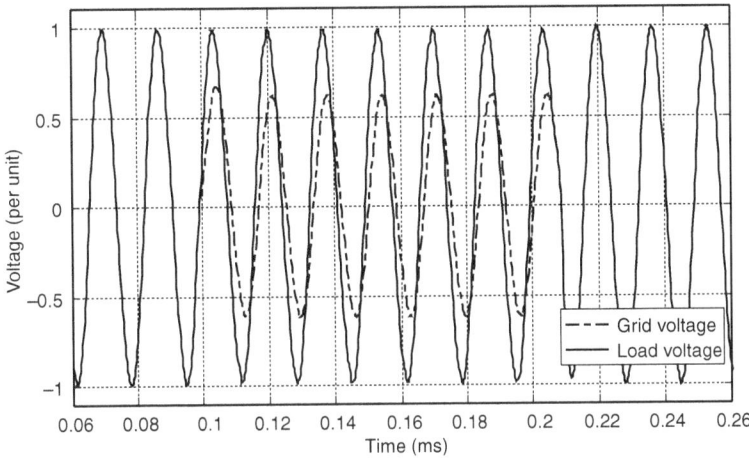

FIGURE **3.31** SMES-based system providing ride-through during voltage sag event.

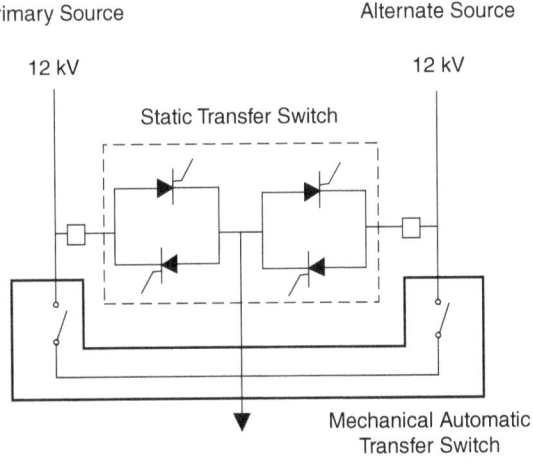

FIGURE **3.32** Configuration of a static transfer switch used to switch between a primary supply and a backup supply in the event of a disturbance. The controls would switch back to the primary supply after normal power is restored.

restorers (DVRs) and UPS systems that use technology similar to the systems described previously but applied at the medium-voltage level. Another alternative that can be applied at either the low-voltage level or the medium-voltage level is the automatic transfer switch.

Automatic transfer switches can be of various technologies, ranging from conventional breakers to static switches. Conventional transfer switches will switch from the primary supply to a backup supply in seconds. Fast transfer switches that use vacuum breaker

Figure 3.33   Example of a static transfer switch application at medium voltage.

technology are available that can transfer in about 2 electrical cycles. This can be fast enough to protect many sensitive loads. Static switches use power electronic switches to accomplish the transfer within about a quarter of an electrical cycle. The transfer switch configuration is shown in Fig. 3.32. An example medium-voltage installation is shown in Fig. 3.33.

The most important consideration in the effectiveness of a transfer switch for protection of sensitive loads is that it requires two independent supplies to the facility. For instance, if both supplies come from the same substation bus, then they will both be exposed to the same voltage sags when there is a fault condition somewhere in the supply system. If a significant percentage of the events affecting the facility are caused by faults on the transmission system, the fast transfer switch might have little benefit for protection of the equipment in the facility.

3.6   Evaluating the Economics of Different Ride-Through Alternatives

The economic evaluation procedure to find the best option for improving voltage sag performance consists of the following steps:

1. Characterize the system power quality performance.

2. Estimate the costs associated with the power quality variations.

3. Characterize the solution alternatives in terms of costs and effectiveness.

4. Perform the comparative economic analysis.

We have already presented the methodology for characterizing the expected voltage sag performance, and we have outlined the major technologies that can be used to improve the performance of the facility. Now, we will focus on evaluating the economics of the different options.

3.6.1 Estimating the Costs for the Voltage Sag Events

The costs associated with sag events can vary significantly from nearly zero to several million dollars per event. The cost will vary not only among different industry types and individual facilities but also with market conditions. Higher costs are typically experienced if the end product is in short supply and there is limited ability to make up for the lost production. Not all costs are easily quantified or truly reflect the urgency of avoiding the consequences of a voltage sag event.

The cost of a power quality disturbance can be captured primarily through three major categories:

- Product-related losses, such as loss of product and materials, lost production capacity, disposal charges, and increased inventory requirements.

- Labor-related losses, such as idled employees, overtime, cleanup, and repair.

- Ancillary costs such as damaged equipment, lost opportunity cost, and penalties due to shipping delays.

Focusing on these three categories will facilitate the development of a detailed list of all costs and savings associated with a power quality disturbance. One can also refer to appendix A of IEEE Std. 1346–1998 for a more detailed explanation of the factors to be considered in determining the cost of power quality disturbances.

Costs will typically vary with the severity (both magnitude and duration) of the power quality disturbance. This relationship can often be defined by a matrix of weighting factors. The weighting factors are developed using the cost of a momentary interruption as the base.

Usually, a momentary interruption will cause a disruption to any load or process that is not specifically protected with some type of energy storage technology. Voltage sags and other power quality variations will always have an impact that is some portion of this total shutdown.

If a voltage sag to 40 percent causes 80 percent of the economic impact that a momentary interruption causes, then the weighting factor for a 40 percent sag would be 0.8. Similarly, if a sag to 75 percent only results in 10 percent of the costs that an interruption causes, then the weighting factor is 0.1.

After the weighting factors are applied to an event, the costs of the event are expressed in per unit of the cost of a momentary interruption. The weighted events can then be summed and the total is the total cost of all the events expressed in the number of equivalent momentary interruptions.

Table 3.4 provides an example of weighting factors that were used for one investigation. The weighting factors can be further expanded to differentiate between sags that affect all three phases and sags that only affect one or two phases. Table 3.5 combines the weighting factors with expected performance to determine a total annual cost associated with voltage sags and interruptions. The cost is 16.9 times the cost of an interruption. If an interruption costs $40,000, the total costs associated with voltage sags and interruptions would be $676,000 per year (see Chap. 8 for alternative costing methods).

Category of event	Weighting for economic analysis
Interruption	1.0
Sag with minimum voltage below 50%	0.8
Sag with minimum voltage between 50% and 70%	0.4
Sag with minimum voltage between 70% and 90%	0.1

TABLE 3.4 Example of Weighting Factors for Different Voltage Sag Magnitudes

Category of event	Weighting for economic analysis	Number of events per year	Total equivalent interruptions
Interruption	1	5	5
Sag with minimum voltage below 50%	0.8	3	2.4
Sag with minimum voltage between 50% and 70%	0.4	15	6
Sag with minimum voltage between 70% and 90%	0.1	35	3.5
Total			16.9

TABLE 3.5 Example of Combining the Weighting Factors with Expected Voltage Sag Performance to Determine the Total Costs of Power Quality Variations

3.6.2   Characterizing the Cost and Effectiveness for Solution Alternatives

Each solution technology needs to be characterized in terms of cost and effectiveness. In broad terms the solution cost should include initial procurement and installation expenses, operating and maintenance expenses, and any disposal and/or salvage value considerations. A thorough evaluation would include less obvious costs such as real estate or space-related expenses and tax considerations. The cost of the extra space requirements can be incorporated as a space rental charge and included with other annual operating expenses. Tax considerations may have several components, and the net benefit or cost can also be included with other annual operating expenses. Table 3.6 provides an example of initial costs and annual operating costs for some general technologies used to improve performance for voltage sags and interruptions. These costs are provided for use in the example and should not be considered indicative of any particular product.

Besides the costs, the solution effectiveness of each alternative needs to be quantified in terms of the performance improvement that can be achieved. Solution effectiveness, like power quality costs, will typically vary with the severity of the power quality

Alternative category	Typical cost	Operating and maintenance costs (% of initial costs per year)
Controls protection (<5 kVA)		
CVTs	$1000/kVA	10
UPS	$500/kVA	25
Dynamic sag corrector	$250/kVA	5
Machine protection (10–300 kVA)		
UPS	$500/kVA	15
Flywheel	$500/kVA	7
Dynamic sag corrector	$200/kVA	5
Facility protection (2–10 MVA)		
UPS	$500/kVA	15
Flywheel	$500/kVA	5
DVR (50% voltage boost)	$300/kVA	5
Static switch (10 MVA)	$600,000	5
Fast transfer switch (10 MVA)	$150,000	5

TABLE 3.6   Example Costs for Different Types of Power Quality Improvement Technologies

disturbance. This relationship can be defined by a matrix of "% sags avoided" values. Table 3.7 illustrates this concept for the example technologies from Table 3.6 as they might apply to a typical industrial application.

3.6.3   Performing Comparative Economic Analysis

The process of comparing the different alternatives for improving performance involves determining the total annual cost for each alternative, including both the costs associated with the voltage sags (remember that the solutions do not typically eliminate these costs completely) and the annualized costs of implementing the solution. The objective is to minimize these annual costs (power quality costs + solution costs).

Comparing the different power quality solution alternatives in terms of their total annual costs (annual power quality costs + annual power quality solution costs) identifies those solution(s) with lower costs that warrant more detailed investigations. The do-nothing solution is generally included in the comparative analysis and is typically identified as the base case. The do-nothing solution has a

		Sags, %		
	Interruption, %	Minimum voltage below 50%	Minimum voltage between 50% and 70%	Minimum voltage between 70% and 90%
CVT (controls)	0	20	70	100
Dynamic sag corrector/DVR	0	20	90	100
Flywheel ride-through technologies	70	100	100	100
UPS (battery ride-through technologies)	100	100	100	100
Static switch	100	80	70	50
Fast transfer switch	80	70	60	40

*The entries in this table represent the percentage of voltage sags or interruptions in each category that are corrected to levels that will no longer cause equipment impacts in the facility.

TABLE 3.7   Effectiveness of the Power Quality Improvement Options for a Particular Example Case.*

zero annual power quality solution cost but has the highest annual power quality costs.

Many of the costs (power quality and operation and maintenance) are by their nature annual costs. The costs associated with purchasing and installing various solution technologies are one-time up-front costs that can be annualized using an appropriate interest rate and assumed lifetime or evaluation period.

Figure 3.34 gives an example of this type of analysis for a typical industrial facility. The facility has a total load of 5 MW, but only about 2 MW of load needs to be protected to avoid production disruptions. The voltage sag performance was given in Table 3.5. The costs for an interruption are $40,000 per event, and the costs for voltage sags are based on the weighting factors given previously. The six options given in Table 3.7 are analyzed, and the annual costs are presented. The annualized costs are calculated based on a 15-year life and an interest rate of 10 percent.

It is interesting to note that all the options reduce the total annual costs (in other words, any of these options would have a net benefit to the facility with the assumed interest rate and lifetime when compared to the existing conditions). It is also interesting that the best solution in this case involves applying equipment on the utility side (fast transfer switch). However, this has a major assumption that a backup feeder would be available and that there would be no charge

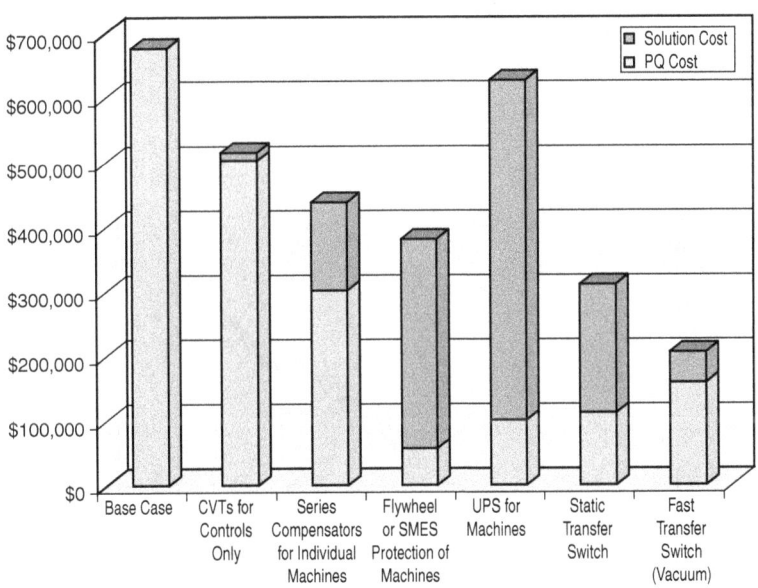

Figure 3.34 Example of comparing solution alternatives with the base case using total annualized costs.

from the utility for providing a connection to this backup feeder except for the equipment and operating costs.

More commonly, the solution would be implemented in the facility and either a dynamic sag corrector or flywheel-based standby power supply might make sense for protecting the 2 MW of sensitive loads. In this case, protecting just the controls with CVTs does not provide the best solution because the machines themselves are sensitive to voltage sags.

3.7 Motor-Starting Sags

Motors have the undesirable effect of drawing several times their full load current while starting. This large current will, by flowing through system impedances, cause a voltage sag which may dim lights, cause contactors to drop out, and disrupt sensitive equipment. The situation is made worse by an extremely poor starting displacement factor—usually in the range of 15 to 30 percent.

The time required for the motor to accelerate to rated speed increases with the magnitude of the sag, and an excessive sag may prevent the motor from starting successfully. Motor starting sags can persist for many seconds, as illustrated in Fig. 3.35.

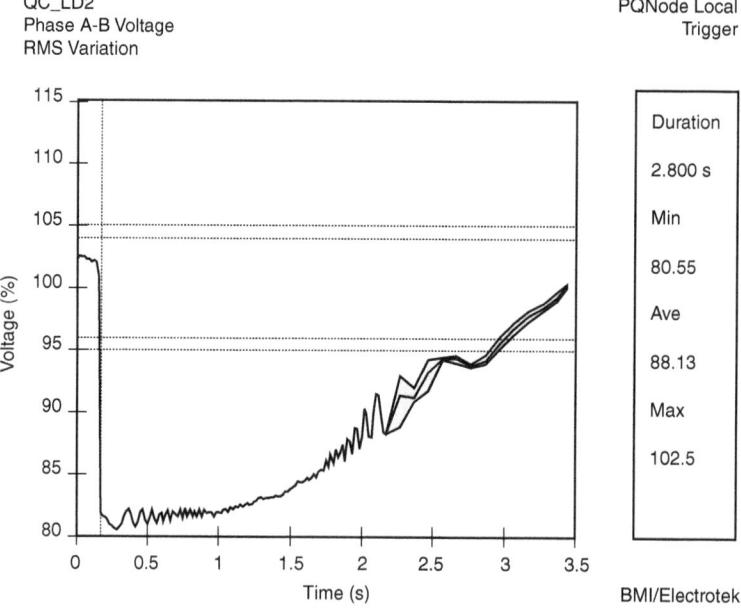

FIGURE 3.35 Typical motor-starting voltage sag.

3.7.1   Motor-Starting Methods

Energizing the motor in a single step (*full-voltage starting*) provides low cost and allows the most rapid acceleration. It is the preferred method unless the resulting voltage sag or mechanical stress is excessive.

Autotransformer starters have two autotransformers connected in open delta. Taps provide a motor voltage of 80, 65, or 50 percent of system voltage during start-up. Line current and starting torque vary with the square of the voltage applied to the motor, so the 50 percent tap will deliver only 25 percent of the full-voltage starting current and torque. The lowest tap which will supply the required starting torque is selected.

Resistance and reactance starters initially insert an impedance in series with the motor. After a time delay, this impedance is shorted out. Starting resistors may be shorted out over several steps; starting reactors are shorted out in a single step. Line current and starting torque vary directly with the voltage applied to the motor, so for a given starting voltage, these starters draw more current from the line than with autotransformer starters, but provide higher starting torque. Reactors are typically provided with 50, 45, and 37.5 percent taps.

Part-winding starters are attractive for use with dual-rated motors (220/440 V or 230/460 V). The stator of a dual-rated motor consists of two windings connected in parallel at the lower voltage rating, or in series at the higher voltage rating. When operated with a part-winding starter at the lower voltage rating, only one winding is energized initially, limiting starting current and starting torque to 50 percent of the values seen when both windings are energized simultaneously.

Delta–wye starters connect the stator in wye for starting and then, after a time delay, reconnect the windings in delta. The wye connection reduces the starting voltage to 57 percent of the system line-line voltage; starting current and starting torque are reduced to 33 percent of their values for full-voltage start.

3.7.2   Estimating the Sag Severity during Full-Voltage Starting

As shown in Fig. 3.35, starting an induction motor results in a steep dip in voltage, followed by a gradual recovery. If full-voltage starting is used, the sag voltage, in pu of nominal system voltage, is

$$V_{Min}(pu) = \frac{V(pu) \cdot kVA_{SC}}{kVA_{LR} + kVA_{SC}},$$

where
$V(pu)$ = actual system voltage, in pu of nominal
kVA_{LR} = motor locked rotor kVA
kVA_{SC} = system short-circuit kVA at motor

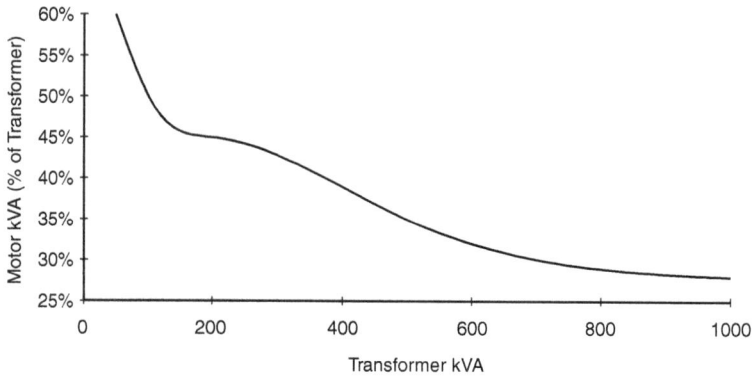

FIGURE 3.36   Typical motor versus transformer size for full-voltage starting sags of 90 percent.

Figure 3.36 illustrates the results of this computation for sag to 90 percent of nominal voltage, using typical system impedances and motor characteristics.

If the result is above the minimum allowable steady-state voltage for the affected equipment, then the full-voltage starting is acceptable. If not, then the sag magnitude versus duration characteristic must be compared to the voltage tolerance envelope of the affected equipment. The required calculations are fairly complicated and best left to a motor-starting or general transient analysis computer program. The following data will be required for the simulation:

- Parameter values for the standard induction motor equivalent circuit: R_1, X_1, R_2, X_2, and X_M.
- Number of motor poles and rated rpm (or slip).
- WK^2 (inertia constant) values for the motor and the motor load.
- Torque versus speed characteristic for the motor load.

3.8   Utility System Fault-Clearing Issues

Utility feeder design and fault-clearing practices have a great influence on the voltage sag and interruption performance at a distribution-connected load.[12] Ways to improve the performance will now be explored.

Utilities have two basic options to continue to reduce the number and severity of faults on their system:

1. Prevent faults.
2. Modify fault-clearing practices.

Utilities derive important benefits from activities that prevent faults. These activities not only result in improved customer satisfaction, but prevent costly damage to power system equipment. Fault prevention activities include tree trimming, adding line arresters, insulator washing, and adding animal guards. Insulation on utility lines cannot be expected to withstand all lightning strokes. However, any line that shows a high susceptibility to lightning-induced faults should be investigated. On transmission lines, shielding can be analyzed for its effectiveness in reducing direct lightning strokes. Tower footing resistance is an important factor in backflashovers from static wire to a phase wire. If the tower footing resistance is high, the surge energy from a lightning stroke will not be absorbed by the ground as quickly. On distribution feeders, shielding may also be an option as is placing arresters along the line frequently. Of course, one of the main problems with overhead distribution feeders is that storms blow tree limbs into the lines. In areas where the vegetation grows quickly, it is a formidable task to keep trees properly trimmed.

Improved fault-clearing practices may include adding line reclosers, eliminating fast tripping, adding loop schemes, and modifying feeder design. These practices may reduce the number and/or duration of momentary interruptions and voltage sags, but utility system faults can never be eliminated completely.

3.8.1   Overcurrent Coordination Principles

It is important to understand the operation of the utility system during fault conditions. There are certain physical limitations to interrupting the fault current and restoring power. This places certain minimum requirements on loads that are expected to survive such events without disruption. There are also some things that can be done better on the utility system to improve the power quality than on the load side. Therefore, we will address the issues relevant to utility fault clearing with both the end user (or load equipment designer) and the utility engineer in mind.

There are two fundamental types of faults on power systems:

1. *Transient (temporary) faults.* These are faults due to such things as overhead line flashovers that result in no permanent damage to the system insulation. Power can be restored as soon as the fault arc is extinguished. Automatic switchgear can do this within a few seconds. Some transient faults are self-clearing.

2. *Permanent faults.* These are faults due to physical damage to some element of the insulation system that requires intervention by a line crew to repair. The impact on the end user is an outage that lasts from several minutes to a few hours.

The chief objective of the utility system fault-clearing process, besides personnel safety, is to limit the damage to the distribution system. Therefore, the detection of faults and the clearing of the fault current must be done with the maximum possible speed without resulting in false operations for normal transient events.

The two greatest concerns for damage are typically

1. Arcing damage to conductors and bushings.

2. Through-fault damage to substation transformers, where the windings become displaced by excessive forces, resulting in a major failure.

A radial distribution system is designed so that only one fault interrupter must operate to clear a fault. For permanent faults, that same device, or another, operates to *sectionalize* the feeder. That is, the faulted section is isolated so that power may be restored to the rest of the loads served from the sound sections. Orchestrating this process is referred to as the *coordination* of the overcurrent protection devices. While this is simple in concept, some of the behaviors of the devices involved can be quite complex. What is remarkable about this is that nearly all of the process is performed automatically by autonomous devices employing only local intelligence.

Overcurrent protection devices appear in series along a feeder. For permanent fault coordination, the devices operate progressively slower as one moves from the ends of the feeders toward the substation. This helps ensure the proper sectionalizing of the feeder so that only the faulted section is isolated. However, this principle is often violated for temporary faults, particularly if fuse saving is employed. The typical hierarchy of overcurrent protection devices on a feeder is

1. *Feeder breaker in the substation.* This is a circuit breaker capable of interrupting typically 40 kA of current and controlled by separate relays. When the available fault current is less than 20 kA, it is common to find reclosers used in this application.

2. *Line reclosers mounted on poles at midfeeder.* The simplest are self-contained with hydraulically operated timing, interrupting, and reclosing mechanisms. Others have separate electronic controls.

3. *Fuses on many lateral taps off the main feeder.*

The power quality issues relating to the placement and operation of these devices are now explored.

3.8.2 Fuses

The most basic overcurrent protective element on the system is a fuse. Fuses are relatively inexpensive and maintenance-free. For those reasons, they are generally used in large numbers on most utility distribution systems to protect individual transformers and feeder branches (sometimes called laterals or lateral branches). Figure 3.37 shows a typical overhead line fused cutout. The fundamental purpose of fuses is to operate on permanent faults and isolate (sectionalize) the faulted section from the sound portion of the feeder. They are positioned so that the smallest practical section of the feeder is disturbed.

Fuses detect overcurrent by melting the fuse element, which generally is made of a metal such as tin or silver. This initiates some sort of arcing action that will lead to the interruption of the current. There are two basic kinds of fuse technologies used in power systems:

1. Expulsion fuses (as in Fig. 3.37).

2. Current-limiting fuses (see Sec. 3.8.12).

The essential difference between the two is the way the arc is quenched. This also gives the fuses different power quality characteristics. An expulsion fuse creates an arc inside a tube with an ablative coating. This creates high-pressure gases that expel the arc plasma and fuse remnants out the bottom of the cutout, often with a loud report similar to a firearm. This cools the arc such that it will not reignite after the alternating current naturally goes through zero. This can be as short as one-half cycle for high currents to several cycles for low fault currents. This determines the duration of the voltage sag observed at loads. An expulsion fuse is considerably less expensive than a current-limiting fuse.

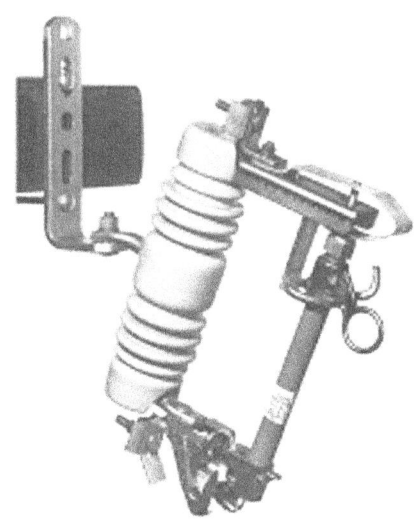

FIGURE 3.37 Typical utility fused cutout with expulsion fuse. (*Courtesy of Cooper Power Systems.*)

A current-limiting fuse dissipates the energy in the arc in a closed environment, typically by melting a special sand within an insulating tube. This process actually quenches the arc very quickly, forcing the current to zero before that would naturally occur. This can have some beneficial impacts on the voltage sag characteristics (see Fig. 3.52).

Because it is based on a piece of metal that must accumulate heat until it reaches its melting temperature, it takes a fuse different amounts of time to operate at different levels of fault current. The time decreases as the current level increases, giving a fuse its distinctive inverse TCC, as shown in Fig. 3.38. To achieve full-range coordination with fuses, all other overcurrent protective devices in the distribution system must adopt this same basic shape. The fuse TCC is typically given as a band between two curves as shown. The leftmost edge represents the minimum melting time, while the rightmost edge represents the maximum clearing time for different current levels.

Some aspects of coordinating with the fuse characteristic relevant to power quality are as follows:

1. If the utility employs fuse saving on temporary faults, the coordinating fault interrupter must have a TCC to the left of the minimum melting curve.

2. For a permanent fault, the coordinating device must have a TCC to the right of the clearing curve to allow the fuse to

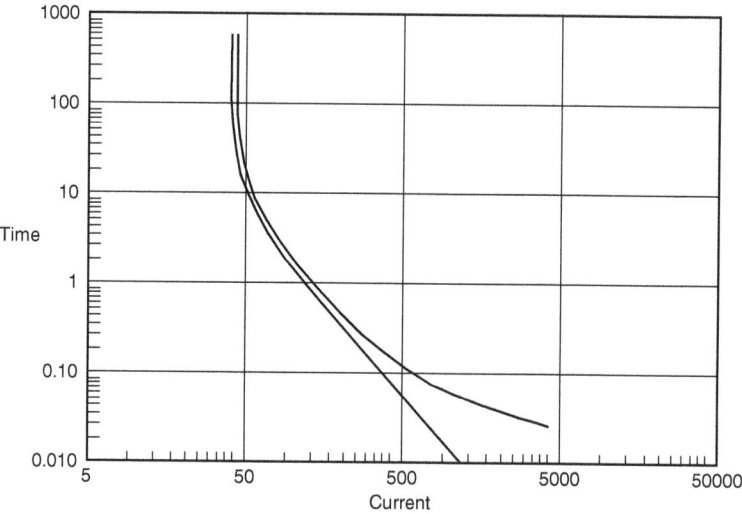

FIGURE 3.38   The inverse TCC of a fuse that dictates the shape of the characteristic of all other devices for series overcurrent coordination.

melt and clear first. Otherwise, many other customers will be interrupted.

3. Repeated fault currents, inrush currents from reclosing, and lightning stroke currents can damage the fuse element, generally shifting the TCC to the left. This will result in inadvertent interruptions of customers downline from the fuse.

4. For high current values with operating time less than 0.1 s, it is difficult to guarantee that an upline mechanical fault interrupter will be able to save the fuse.

3.8.3 Reclosing

Because most faults on overhead lines are transient, the power can be successfully restored within several cycles after the current is interrupted. Thus, most automatic circuit breakers are designed to reclose 2 or 3 times, if needed, in rapid succession. The multiple operations are designed to permit various sectionalizing schemes to operate and to give some more persistent transient faults a second chance to clear. There are special circuit breakers for utility distribution systems called, appropriately, *reclosers*, that were designed to perform the fault interruption and reclosing function particularly well. The majority of faults will be cleared on the first operation. Figure 3.39 shows a typical single-phase recloser and Figs. 3.40 and 3.41 show two different three-phase designs in common usage. These devices are generally pole-mounted on overhead utility lines, although a pad-mounted version also exists. The oil-insulated designs are the most common, but sulfur hexafluoride (SF6)–insulated and encapsulated solid dielectric designs are also popular.

These devices can be found in numerous places along distribution feeders and sometimes in substations. They are typically applied at the head of sections subjected to numerous temporary faults. However, they may be applied nearly anywhere a convenient, low-cost primary-side circuit breaker is needed.

Because they are designed for fuse-saving (fast tripping) applications, reclosers are some of the fastest mechanical fault interrupters employed on the utility system. While they are typically rated for no faster than 3 to 6 cycles, many examples of interruptions as short as 1.5 cycles have been observed with power quality monitors. This can be beneficial to limiting sag durations. Where fast tripping is not employed, the recloser control will commonly delay operation to more than 6 cycles to allow time for downline fuses to clear.

Reclosing is quite prevalent in North American utility systems. Utilities in regions of low lightning incidence may reclose only once because they assume that the majority of their faults will be permanent. In lightning-prone regions, it is common to attempt to

Figure 3.39 Typical pole-mounted, oil-insulated single-phase line recloser. (*Courtesy of Cooper Power Systems.*)

clear the fault as many as 4 times. Figure 3.42 illustrates the two most common sequences in use on four-shot reclosers:

1. One fast operation, three delayed
2. Two fast, two delayed

See Sec. 3.8.4 for a more detailed explanation of fast and delayed operations. Reclosers tend to have uniform reclose intervals between operations. The original hydraulic reclosers were limited to about 1 to 2 s, and this setting has been retained by many utilities although modern electronically controlled reclosers can be set for any value. It is common for the first reclose interval on some types of reclosers to be set for *instantaneous reclose,* which will result in closure in 12 to 30 cycles (0.2 to 0.5 s). This is done to reduce the time of the interruption and improve the power quality. However, there are some conflicts created by this, such as with DG disconnecting times (see Sec. 9.5.2).

FIGURE 3.40   Typical standard three-phase oil-insulated line recloser with vacuum interrupters. (*Courtesy of Cooper Power Systems.*)

FIGURE 3.41   Newer three-phase line recloser with vacuum interrupters encapsulated in solid dielectric insulation. (*Courtesy of Cooper Power Systems.*)

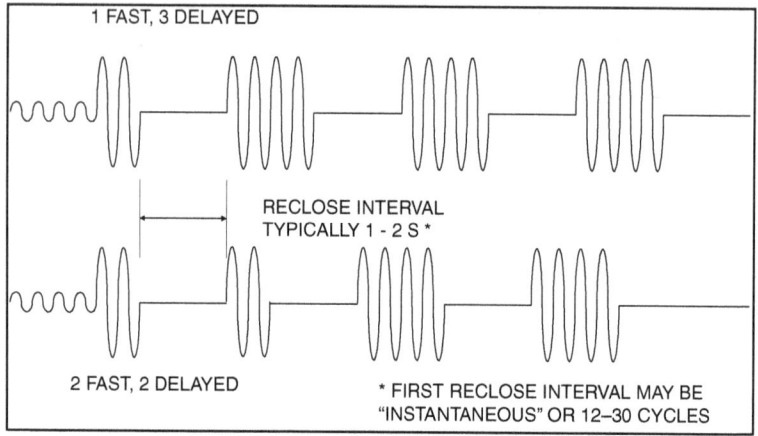

FIGURE **3.42** Common reclosing sequences for line reclosers in use in the United States.

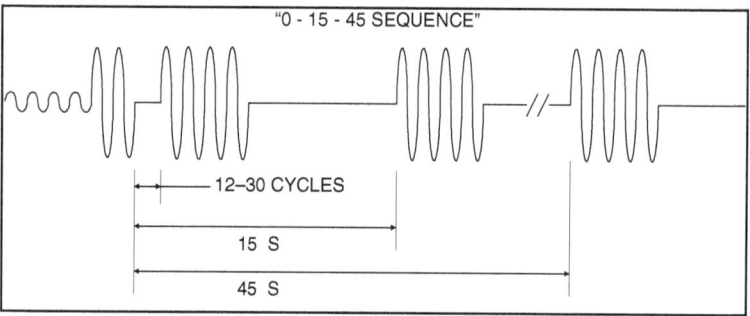

FIGURE **3.43** A common reclosing sequence for substation breakers in the United States.

Substation circuit breakers often have a different style of reclosing sequence as shown in Fig. 3.43. This stems from a different evolution of relaying technology. Reclosing times are counted from the first tripping signal of the first operation. Thus, the common "0–15–45" operating sequence recloses essentially as fast as possible on the first operation, with approximately 15- and 30-s intervals between the next two operations.

Although the terminology may differ, modern breakers and reclosers can both be set to have the same operating sequences to meet load power quality requirements. Utilities generally choose one technology over the other based on cost or construction standards.

It is generally fruitless to have automatic reclosers on distribution circuits with underground cable unless a significant portion of the system is overhead and exposed to trees or lightning.

3.8.4 Fuse Saving

Ideally, utility engineers would like to avoid blowing fuses needlessly on transient faults because a line crew must be dispatched to change them. Line reclosers were designed specifically to help save fuses. Substation circuit breakers can use instantaneous ground relaying to accomplish the same thing. The basic idea is to have the mechanical circuit-interrupting device operate very quickly on the first operation so that it clears before any fuses downline from it have a chance to melt. When the device closes back in, power is fully restored in the majority of the cases and no human intervention is required. The only inconvenience to the customer is a slight blink. This is called the *fast* operation of the device, or the *instantaneous trip.*

If the fault is still there upon reclosing, there are two options in common usage:

1. *Switch to a slow, or delayed, tripping characteristic.* This is frequently the only option for substation circuit breakers; they will operate only one time on the instantaneous trip. This philosophy assumes that the fault is now permanent and switching to a delayed operation will give a downline fuse time to operate and clear the fault by isolating the faulted section.

2. *Try a second fast operation.* This philosophy is used where experience has shown a significant percentage of transient faults need two chances to clear while saving the fuses. Some line constructions and voltage levels have a greater likelihood that a lightning-induced arc may reignite and need a second chance to clear. Also, a certain percentage of tree faults will burn free if given a second shot.

Many utilities have abandoned fuse saving in selected areas due to complaints about power quality. The fast, or instantaneous, trip is eliminated so that breakers and reclosers have only time-delayed operations (see Sec. 3.8.7).

3.8.5 Reclosers with Pulse Closing Technology

A recent innovation in recloser technology holds the promise of improving the power quality compared to standard reclosing practices. Automatic reclosing has been very effective in clearing temporary faults which account for about 50 percent to 80 percent of faults on distribution circuits with overhead lines. The typical reclosing practice clears the majority of faults on the first operation and restores service to the affected line. Unfortunately, a persistent temporary or permanent fault forces a second and, possibly, up to four total operations that reclose into a fault (see Fig. 3.44). End users upstream and downstream from the recloser will be subjected to repeated voltage sags and attempts to restore service that fail. The fault I^2t let-through energy causes excessive thermal, mechanical, and electrical stresses on power apparatus. Since the duration of the second and subsequent operations is generally determined by the delayed TCC curve, equipment that carries the fault current is subjected to even higher let-through energy stresses than the first operation.

Reclosers with pulse-closing technology, developed by S&C Electric Company, attempt to eliminate the pitfalls described above and thereby improve the overall power quality during automatic fault-clearing operations. Instead of reclosing automatically on the second operation, the recloser sends a current pulse immediately after the first fast operation, i.e. the initial trip, to determine if the fault is still present. The interrupter contacts close briefly at a specific point-on-wave, i.e. a few degrees beyond the voltage peak. As shown in Fig. 3.44, the closing causes an asymmetrical current to flow with the lesser magnitude appearing first. Because the contacts close for less than 2 ms and immediately open, the asymmetric current is interrupted at its first zero crossing. The first half-cycle of the current serves as a current pulse. It has minimal let-through energy with a duration between 3 and 8 ms. The current pulse is used to determine if the line is faulted. The recloser will reclose after the current pulse indicates the fault arc has extinguished. If the fault is permanent, after a predetermined number of current pulses, the recloser will lock out isolating the faulty-line segment.[19]

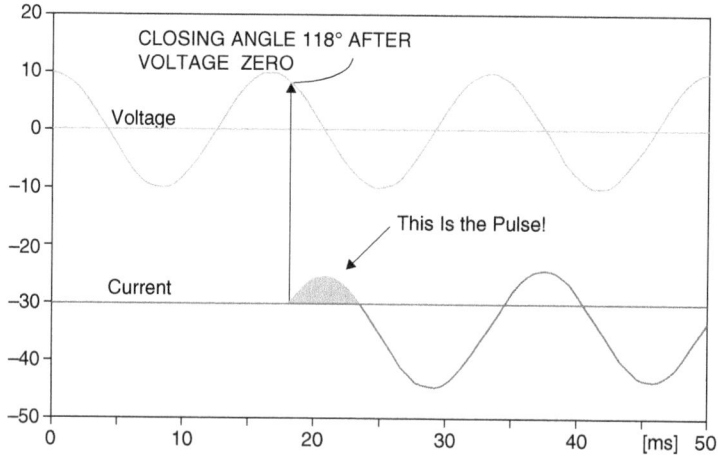

FIGURE 3.44   Interrupter contacts are closed at a specific point-on-wave to create a pulse current with minimum fault let-through energy. The pulse current is used to determine if the line segment is faulty.

Figure 3.45 shows an installation of a three-phase line recloser with vacuum interrupters and pulse-closing technology. The recloser is equipped with three current sensors and six voltage sensors, three on each side of load and source contacts. In addition to a battery backup, the control and switch mechanisms are self-powered from either source or load sides of the recloser unit. The recloser can perform single-phase tripping to reduce instantaneous and momentary interruption on two other healthy phases in case of a SLG fault.

An actual fault-clearing event was recorded by the pulse-closing interrupter sensors during a two-phase fault. Figure 3.46 shows three-phase current (I1, I2, and I3) and three-phase voltages on the source side (VX1, VX2, and VX3), and on the load side (VY1, VY2, and VY3) the interrupters. Immediately after the initial three-phase trip, current pulses are sent to the faulty lines. From the I3 current trace, it is clear that Pole 3 is pulsed first, approximately 20 and 28 cycles after the initial trip. Since the resulting current pulses are greater than the anticipated fault current threshold, Pole 3 does not reclose. It is pulsed again about 40 cycles later. The resulting current pulse is below the fault current threshold indicating fault arc has dissipated. Pole 3 then recloses. The same process is repeated for Pole 2 and 1 as shown in I2 and I1 current traces. Service is restored after all three poles reclose

Figure 3.45   A three-phase line recloser with pulse-closing technology. (Courtesy of S&C Electric Company.)

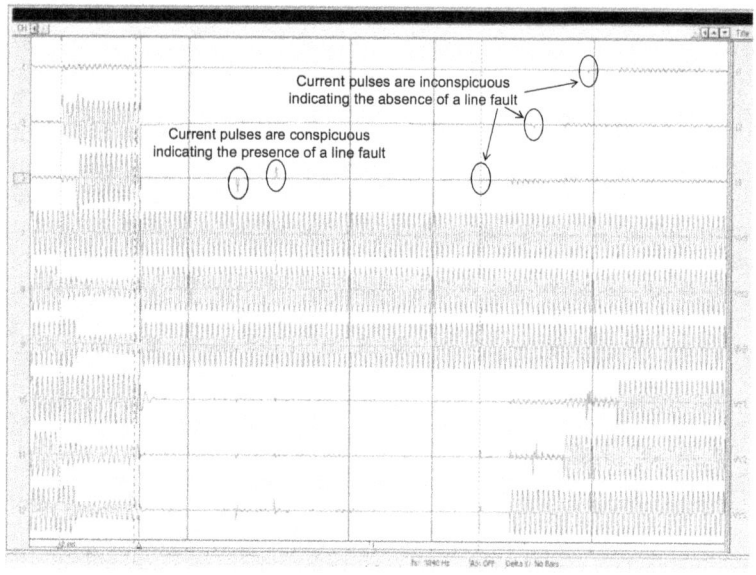

Current pulses are inconspicuous indicating the absence of a line fault

Current pulses are conspicuous indicating the presence of a line fault

FIGURE 3.46   Current and voltage waveforms during a two-phase fault clearing using a three-phase recloser equipped with pulse-closing technology. The recloser anticipates the fault condition and recloses only after the fault arc has dissipated.

successfully. During the fault-clearing process, the recloser with pulse technology anticipates the fault condition correctly thereby avoiding a reclose onto an active fault and subjecting consumers to additional power quality variations.

3.8.6   Reliability

The term *reliability* in the utility context usually refers to the amount of time end users are totally without power for an extended period of time (i.e., a sustained interruption). Definitions of what constitutes a sustained interruption vary among utilities from 1 to 5 min. This is what many utilities refer to as an "outage." Current power quality standards efforts are leaning toward calling any interruption of power for longer than 1 min a sustained interruption (see Chap. 2). In any case, reliability is affected by the permanent faults on the system that must be repaired before service can be restored.

Of course, many industrial end users have a different view of what constitutes reliability because even momentary interruptions for transient faults can knock their processes off-line and require several hours to get back into production. There is a movement to extend the traditional reliability indices to include momentary interruptions as well.

The traditional reliability indices for utility distribution systems are defined as follows:[8]

SAIFI: System average interruption frequency index

$$\text{SAIFI} = \frac{(\text{no. of customers interrupted})(\text{no. of interruptions})}{\text{total no. of customers}}$$

SAIDI: System average interruption duration index

$$\text{SAIDI} = \frac{\Sigma(\text{no. of customers affected})(\text{duration of outage})}{\text{total no. of customers}}$$

CAIFI: Customer average interruption frequency index

$$\text{CAIFI} = \frac{\text{total no. of customer interruptions}}{\text{no. of customers affected}}$$

CAIDI: Customer average interruption duration index

$$\text{CAIDI} = \frac{\Sigma\,\text{customer interruption durations}}{\text{total no. customer interruptions}}$$

ASAI: Average system availability index

$$\text{ASAI} = \frac{\text{customer hours service availability}}{\text{customer hours service demand}}$$

where customer hours service demand = 8760 for an entire year. Typical target values for these indices are

Index	Target
SAIFI	1.0
SAIDI	1.0–1.5 h
CAIDI	1.0–1.5 h
ASAI	0.99983

These are simply design targets, and actual values can, of course, vary significantly from this. Burke[9] reports the results of a survey in which the average SAIFI was 1.18, SAIDI was 76.93 min, CAIDI was 76.93 min, and ASAI was 0.999375. We have experience with utilities whose SAIFI is usually around 0.5 and SAIDI is between 2.0 and 3.0 h. This means that the fault rate was lower than typical, at least for the bulk of the customers, but the time to repair the faults was longer. This could be common for feeders with mixed urban and rural sections. The faults are more common in the rural sections, but fewer customers are affected and it takes longer to find and repair faults.

3.8.7 Impact of Eliminating Fuse Saving

One of the more common ways of dealing with complaints about momentary interruptions is to disable the fast-tripping, or fuse-saving, feature of the substation breaker or recloser. This avoids interrupting the entire feeder for a fault on a tap. This has been a very effective way many utilities have found to deal with *complaints* about the quality of the power. It simply minimizes the number of people inconvenienced by any single event. The penalty is that customers on the affected fused tap will suffer a sustained interruption until the fuse can be replaced, even for a transient fault. There is also an additional cost to the utility to make the trouble call, and it can have an adverse impact on the reliability indices by which some utilities are graded.

In a Utility Power Quality Practices survey conducted in 1991 by Cooper Power Systems for EPRI Project RP3098–1, 40 percent of participating utilities indicated that they have responded to customer complaints by removing fast tripping. Sixty percent of participating investor-owned utilities (IOUs) but only 30 percent of participating public power utilities (largely rural electric cooperatives) indicated that they followed this practice. This may validate a widely held belief that customer sensitivity to momentary interruptions is much greater in urban areas than in rural areas. Since the time of this survey, our experience would indicate this trend is continuing, if not accelerating.

This solution to power quality complaints does not sit well with many utility engineers. They would prefer the optimal technical and economical solution, which would make use of the fast-trip capability of breakers and reclosers. This not only saves operating costs, but it improves the reliability indices by which utility performance is measured. Momentary interruptions have traditionally not been reported in these indices, but only the permanent outages. However, when we consider the economic impact of both the end user and the utility (i.e., a value-based analysis), the utility costs can be swamped by the costs to industrial end users.[11]

If the utility has been in the practice of fuse saving, there will generally be some additional costs to remove fast tripping. For example, the fused cutouts along the main feed may have to be changed for better coordination. In some cases, additional lateral fuses will have to be added so that the main feeder is better protected from faults on branches. Considering engineering time, estimates for the cost of instituting this may be from $20,000 to $40,000 per feeder. Additional operating costs to change fuses that would not have blown otherwise may be as high as $2000 per year.

While these costs may seem high, they can appear relatively small if we compare them to the costs of an end user such as a plastic bag manufacturer who can sell all the output of the plant. A single breaker operation can cost $3000 to $10,000 in lost production and extra labor charges. Thus, it is economical in the global, or value-based, sense to remove fast tripping if at least three to five interruptions (momentary and sustained combined) are eliminated each year.

The impact on the reliability indices is highly dependent on the structure of the feeder and what other sectionalizing is done. The impact can be negligible if the critical industrial load is close to the substation and the rest of the feeder can be isolated with a line recloser that does employ fast tripping. The farther out the feeder one goes with no fuse saving, the greater the impact on the reliability indices. Therefore, it is advantageous to limit the area of vulnerability to as small an area as possible and to feed sensitive customers with a high economic value of service as close to substations as possible. See Sec. 3.8.8 for more details.

Removing fast tripping will not eliminate all events that cause problems for industrial users. It will only eliminate most of the momentary interruptions. However, it will do nothing for voltage sags due to faults on the transmission system, other feeders, or even on fused laterals. These events can account for one-half to two-thirds of the events that disrupt industrial processes. As a rule of thumb, removing fast tripping will eliminate about one-third of the industrial process disruptions in areas where lightning-induced faults are a problem. Of course, this figure will depend on the types of processes being served by the feeder.

A particular problem is when there are faults close to the substation on other feeders, or even the same feeder, but on fused taps. This causes a deep sag on all feeders connected to the bus. Two approaches that have been proposed to deal with this are to (1) install reactors on each line coming from the substation bus to limit the maximum bus sag to about 60 percent[12] and (2) install current-limiting fuses on all branch laterals near the substation so that sags are very brief (see Sec. 3.8.12).

Residential end users may be quite vocal about the number of interruptions they experience, but, in most cases, there is little direct economic impact for a momentary interruption. Perhaps, the biggest nuisance is resetting the dozen or so digital clocks found in households. In fact, there may be more cases of adverse economic impact if fast tripping were eliminated. For example, homes with sump pumps may suffer more cases of flooded basements if they suffer sustained interruptions because their lateral fuse blew for a temporary fault during a thunderstorm. Some utilities have taken another approach with the residential complaint problem by employing *instantaneous reclosing* on residential feeders while retaining the fast tripping. By getting the reclose interval down to 12 to 20 cycles, the momentary interruption is so brief that the majority of digital clocks seem to be able to ride through it. This would not be fast enough to help with larger industrial loads. However, there is mounting anecdotal evidence that many more modern loads such as adjustable-speed drives are now able to ride through these brief interruptions. Instantaneous reclosing is not always possible, particularly, if distributed generation is served from the feeder.

One unintended consequence of eliminating fast tripping may be more substation transformer failures. Reclosers operate 4 times by default before locking out. On some designs, dialing out the fast trips will simply result in four delayed operations for permanent faults. There is almost never a reason for four delayed operations. This subjects the substation transformer to unnecessary through-fault current events, which will shorten the life of the transformer. The number of delayed operations should be decreased to two if there are only fuses downline to coordinate with. If there are other mechanical interrupters or sectionalizers, three operations may be needed.

3.8.8   Increased Sectionalizing

The typical utility primary distribution feeder in the United States is a radial feed from the substation breaker. In its simplest form, it consists of a main three-phase feeder with fused one-phase and three-phase taps as shown in Fig. 3.47.

The first step in sectionalizing the feeder further to improve overall reliability is to add a line recloser as shown in Fig. 3.48. If only traditional reliability is of concern, one might place the recloser halfway down the feeder or at the half-load point. For power quality

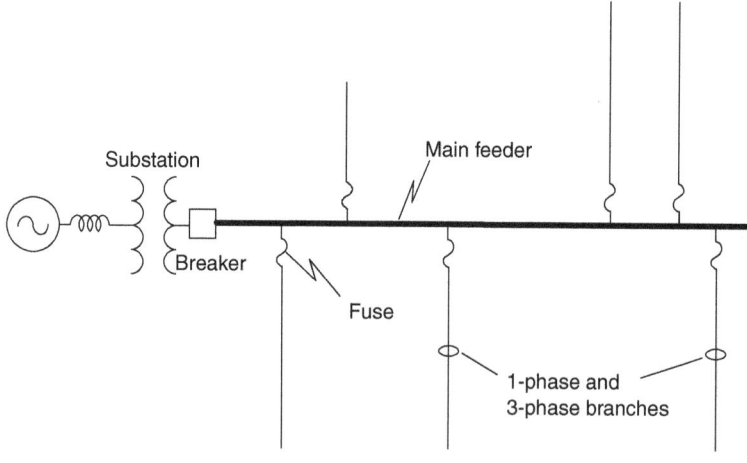

FIGURE 3.47 Typical main line feeder construction with fused taps.

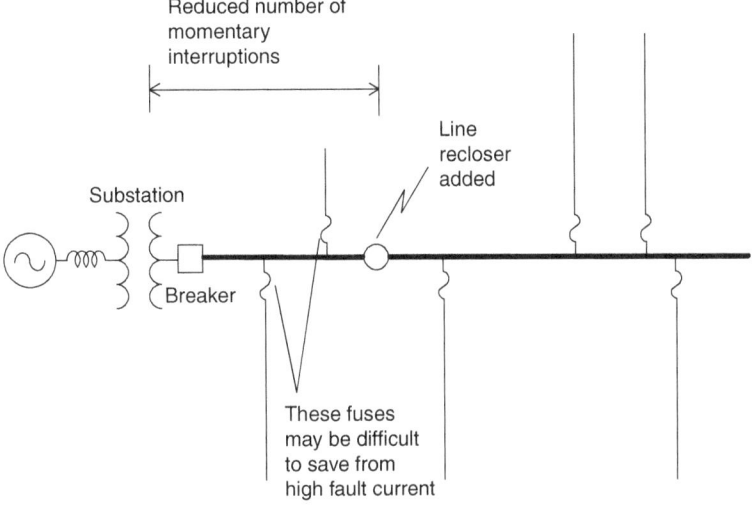

FIGURE 3.48 Adding a line recloser to the main feeder as the first step in sectionalizing.

concerns, it might be better for the recloser to be located closer to the substation, depending on the location of critical loads. One possible criterion is to place it at the first point where the fault current has dropped to where one can nearly always guarantee coordination with the fuses on fast tripping. Another would be to place the recloser just downline from the bulk of the critical loads that are likely to complain about momentary interruptions.

With this concept, the fast tripping can be removed from the substation breaker while only sacrificing fuse saving on a small portion of the feeder. As pointed out previously, it is often difficult to achieve fuse saving near the substation anyway. A special effort is made to keep the first section of the main feeder free of transient faults. This would include more frequent maintenance such as tree trimming and insulator washing. Also, extraordinary measures can be taken to prevent lightning flashover, for example, line shielding or the application of line arresters at least every two or three spans.

The question about how much the reliability is compromised by eliminating the fast tripping is often raised. We performed a reliability analysis on a number of feeders to study this. One feeder used in the study was a single main feeder concept like that in Fig. 3.47, except that the single-phase laterals were uniformly spaced down the feeder. We used the urban feeder described by Warren[10] as the prototype. We'll refer to this as feeder 1. It is a uniform, 8-mile feeder with identical fused taps every 0.25 mile and a total load of 6400 kVA.

While this may not be a realistic feeder, it is a good feeder for study so that the general trends of certain actions can be determined. We assumed values of 0.1 faults/year/mile on the main feeder and 0.25 faults/year/mile on the fused taps, with 80 percent of the faults being transient. A uniform repair time of 3 h was assumed for permanent faults.

We first looked at the base case (case 1 in Table 3.8) assuming that the utility was employing fuse saving and that 100 percent of the fuses could be saved on transient faults. For case 2, the fast tripping of the substation breaker was disabled and it was assumed that none of the tap fuses could be saved. Finally, for case 3, we placed a three-phase recloser 1 mile from the substation and assumed that all fuses downline were saved. The resulting SAIFI and SAIDI reliability indices are shown in Table 3.8.

Typical target values for both SAIFI and SAIDI (in hours) in an urban environment are 1.0. While none of these cases are particularly bad, it is apparent that removing fast tripping has a very significant negative effect on the reliability indices (compare case 2 with case 1). The SAIFI increases by about 60 percent. This example involves a very regular, well-sectionalized feeder with a fuse on every tap, where a blown fuse takes out less than 3 percent of the customers. For other feeder structures, the effect can be more pronounced (see the following discussion on feeder 2), but this serves to illustrate the point that the reliability can be expected to deteriorate when fast tripping is eliminated.

The SAIDI increases only slightly. The largest change is in the number of fuse operations, which increased by a factor of 5. Thus, the utility can expect considerably more trouble calls during stormy weather.

If we would subsequently add a line recloser as described for case 3, the reliability indices and number of fuse operations return to almost the same values as the base case. In fact, the reliability indices are slightly better because of the increased sectionalizing in the line, although there are more nuisance fuse blowings in the first section than in case 1. Thus, if we also place a line recloser past the majority of the critical loads, eliminating fast tripping at the substation will

Case	SAIFI	SAIDI (h)	Annual fuse operations
1	0.184	0.551	1.2
2	0.299	0.666	6.0
3	0.182	0.516	1.88

TABLE **3.8**   Reliability Indices Computed for Feeder 1

Case	SAIFI	SAIDI (h)
1	0.43	1.28
2	1.51	2.37
3	0.47	1.29

TABLE **3.9**   Reliability Indices for Feeder 2

probably not have a significant negative impact on overall reliability. Of course, this assumes that the more critical loads are close to the substation.

We studied the same three cases for another feeder, which we will call feeder 2. This feeder is, perhaps, more typical of mixed urban and rural feeders in much of the United States. Space does not allow a complete description of the topology. The main difference from feeder 1 is that the feeder structure is more random and the sectionalizing is much more coarse with far fewer lateral fuses. The fault rate was assumed to be the same as for feeder 1. The SAIFI and SAIDI for the three cases for feeder 2 are shown in Table 3.9. The number of fuse blowings has not been computed.

For feeder 2, many more customers are inconvenienced by each fuse blowing. Thus, the SAIFI jumps by more than a factor of 3 when fast tripping is removed. This emphasizes the need for good sectionalizing of the feeder to keep the impact on reliability at a minimum. As with feeder 1, the case 3 reliability indices return to nearly the same values as case 1.

What about the power quality? Those customers in the first section of line are going to see much improved power quality as well as improved reliability. In our study of feeder 1, the average number of interruptions, both momentary and sustained, dropped from 15 per year to a little more than 1 per year. This is a dramatic improvement! Unfortunately, the number of interruptions for the remainder of the customers—downline from the recloser—remain unchanged. What can be done about this?

The first temptation is to add another line recloser farther down on the main feeder. The customers served from the portion of the feeder between the reclosers will see an improvement. If we place the second recloser 4 miles downline on our uniform 8-mile feeder example, the average annual interruption rate drops to about 8.3. However, again, the customers at the end will see less improvement on the number of interruptions.

One can continue placing additional line reclosers in series on the main feeder and larger branch feeders to achieve even greater sectionalizing while still retaining desirable practices like fuse saving. In this way, the portion of the feeder disturbed by a fault decreases. This will generally improve the reliability (with diminishing returns) but may not have much effect on the perceived power quality.

The actions that have the most effect on the number of interruptions on the portion of the feeder that is downline from the recloser are

1. Reduce the fault rate by tree trimming, line arresters, animal guards, or other fault prevention techniques.

2. Provide more parallel paths into the service area.

3. Do not trip phases that are not involved in the fault (see Sec. 3.8.11).

There are at least two options for providing additional parallel paths:

1. Build more conventional feeders from the substation.

2. Use more three-phase branches from the main feeder to serve the load.

The first approach is fairly straightforward: Simply build a new feeder from the substation out. This could certainly improve the reliability and power quality by simply reducing the number of customers inconvenienced by each interruption, but this may not be an economical alternative. It also may not achieve as great of an improvement in the interruption rate as some of the approaches associated with the second option. Let's investigate further the second idea: using more three-phase branches off the main feeder, which has the potential of being less costly in most cases.

There are two concepts being put forward. The first involves coming out a short distance from the substation and dividing the feeder into two or three subfeeders. This could typically cut the number of interruptions by almost one-half or two-thirds, respectively, when compared to serving the same customers with a single, long main feeder. The point at which this branch occurs is a little beyond the point where it becomes practical to save lateral fuses on temporary faults. A three-phase recloser is placed in each branch near this point. It would be wise to separate the reclosers by some distance of line to reduce the chances of sympathetic tripping, where a recloser on the unfaulted branch trips as a result of the transient currents related to the fault. Figure 3.49 depicts how this principle might be put into practice on an existing feeder with minimal rebuilding, assuming the existence of three-phase feeders of sufficient conductor size in the locations indicated.

The second proposal, as depicted in Fig. 3.50, is to first build a highly reliable main feeder that extends a significant distance into the service area. Very few loads are actually served directly off this

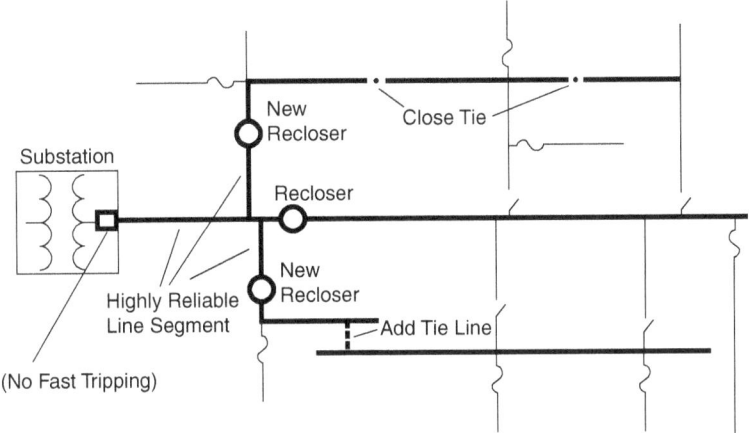

FIGURE 3.49 Reconfiguring a feeder with parallel subfeeders to reduce the average number of interruptions to all customers.

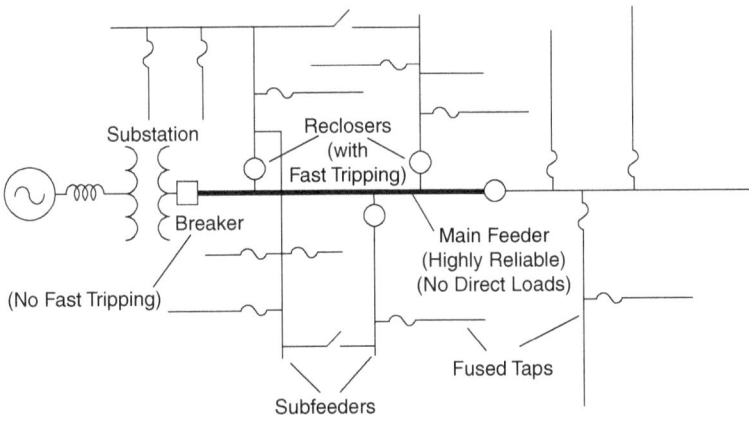

FIGURE 3.50 Designing a feeder with multiple three-phase subfeeds off a highly reliable main feeder.

main feeder. Instead, the loads are served off three-phase branch feeders that are tapped off the main feeder periodically. A three-phase line recloser is used at the head of each branch feeder. Of course, there is no fast tripping at the substation to limit momentary interruptions on the main feeder. Special efforts will be made to prevent faults on this part of the feeder. Essentially, the main feeder becomes an extension of the substation bus that is permitted by design to have a few more faults over its lifetime than the bus. And the branch feeders are analogous to having separate feeds to each part of the service area directly from the substation, but, hopefully, with considerably less cost.

Whether either of these ideas is suitable for a particular utility is dependent on many factors including terrain, load density, load distribution, and past construction practices. These ideas are presented here simply as alternatives to consider for achieving overall lower average interruption rates than is possible by stacking fault interrupters in series. Although these practices may not become widespread, they may be very useful for dealing with specific difficult power quality complaint problems stemming from excessive interruptions.

3.8.9 Midline or Tap Reclosers

Despite responding to complaints by removing fast tripping, about 40 percent of the utilities surveyed indicated they were interested in adding *more* line reclosers to improve customers' perceptions of power quality. This would accomplish greater sectionalizing of the feeder and, perhaps, permit the use of fuse-saving practices on the bulk of the feeder again. This practice is very effective if the whole feeder is being interrupted for faults that are largely constrained to a particular region. By putting the recloser farther out on the feeder, it will attempt to clear the fault first so that the number of customers inconvenienced by a blink is reduced. If it is also necessary to eliminate fast tripping on the substation breaker, only a smaller portion of the feeder nearer the substation is threatened with the possibility of having a fuse blow on a transient fault, as explained previously. This is not much different than the normal case because of the difficulty in preventing fuse blowing in the high fault current regions near the substation anyway.

A few utilities have actually done the opposite to this and removed line reclosers in response to complaints about momentary interruptions. Perhaps, a section of the feeder ran through heavily wooded areas causing frequent operations of the recloser, or the device was responding to high ground currents due to harmonics or a load imbalance, causing false trips. Whatever the reason, this is an unusual practice and is counter to the direction most utilities seem to be taking. The main question at this point does not seem to be about whether more line reclosers are needed but about how to go about applying them to achieve the dual goal of increased power quality and reliability of service.

3.8.10 Instantaneous Reclosing

Instantaneous reclosing is the practice of reclosing within 12 to 30 cycles after interrupting the fault, generally only on the first operation. This has been a standard feature of breakers and reclosers for some time, and some utilities use it as standard practice, particularly on substation breakers. However, the practice has never been universally accepted. Many utilities reclose no faster than 2 s (the standard reclosing interval on a hydraulic recloser) and some wait even longer.

After it was observed that many digital clocks and even some motor-driven loads can successfully ride through a 12- to 30-cycle interruption, some utilities began to experiment with using instantaneous reclosing while retaining the fast tripping to save fuses. One utility trying this on 12-kV feeders reported that there was no significant increase in the number of breaker and recloser operations and that the number of complaints had diminished.[11] Therefore, it is something that other utilities might consider, with the caution that the same experience may not be achieved at higher voltage levels and with certain line designs.

Instantaneous reclosing has had a bad reputation with some utility engineers. One risk is that there is insufficient time for the arc products to disperse and the fault will not clear. Some utilities have had this experience with higher distribution voltage levels and particular line constructions. When this happens, substation transformers are subjected to repeated through-faults unnecessarily. This could result in increased failures of the transformers. However, if there is no indication that instantaneous reclosing is causing increased breaker operations, it should be safe to use it.

Another concern is that very high torques will be generated in rotating machines upon reclosing. This is a particular issue with distributed generation because 12 to 30 cycles may not be sufficient time to guarantee that the generator's protective relaying will detect a problem on the utility side and be off-line (see Chap. 9). Reclosing intervals on feeders with DG should be at least 1 to 2 s so that there is less chance the utility will reclose into the DG out of synchronism. Some utilities allow 5 s. One way the utility can help prevent such an occurrence is to use a common recloser accessory that blocks reclosing when there is voltage present on the load side. This may add significant cost if suitable potential transformers are not already installed.

3.8.11 Single-Phase Tripping

Most of the three-phase breakers and reclosers on the utility distribution system operate all three phases simultaneously. One approach that has been suggested to minimize the exposure of customers to momentary outages is to trip only the faulted phase or phases. Because many of the loads are single phase, this would automatically reduce the exposure by two-thirds for most faults. The main problem with this is that it is possible to damage some three-phase loads if they are single-phased for a substantial length of time. Thus, it is generally considered to be undesirable to use single-phase reclosers on three-phase branches with significant three-phase loads. Of course, this is done quite commonly when only one-phase loads are being served.

This problem is solved by a three-phase breaker, or recloser, that is capable of operating each phase independently until it is determined that the fault is permanent. Then, to prevent single-phasing of three-phase loads, all three phases are opened if the fault is permanent and the interrupter locks out. Such devices are available from distribution equipment suppliers (see Fig. 3.41).

3.8.12 Current-Limiting Fuses

Current-limiting fuses are often used in electrical equipment where the fault current is very high and an internal fault could result in a catastrophic failure. Since they are more expensive than conventional expulsion links, their application is generally limited to locations where the fault current is in excess of 2000 to 3000 A. Figure 3.51 shows examples of current-limiting fuses. There are various designs, but the basic configuration is that of a thin ribbon element or wire wound around a form and encased in a sealed insulating tube filled

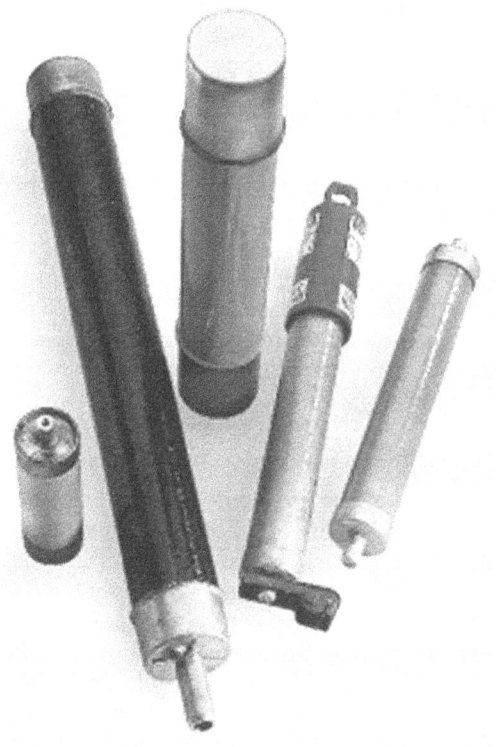

Figure 3.51 Various types of current-limiting fuses used in utility applications. (*Courtesy of Cooper Power Systems.*)

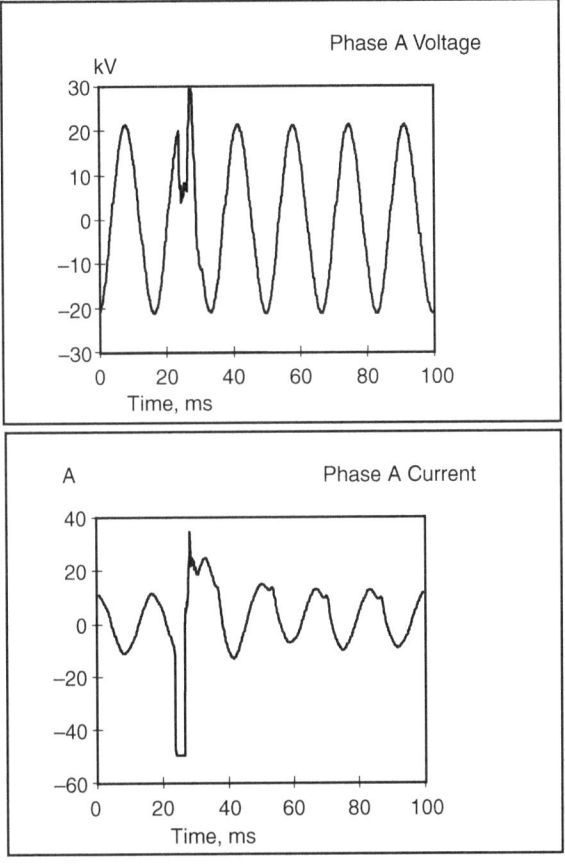

FIGURE 3.52 Typical current-limiting fuse operation showing brief sag followed by peak arc voltage when the fuse clears.

with a special sand. The tube is constructed of stout material such as a fiberglass-epoxy resin composite to withstand the pressures during the interruption process without rupturing. The element melts in many places simultaneously and, with the aid of the melting sand, very quickly builds up a voltage drop that opposes the flow of current. The current is forced to zero in about one-fourth of a cycle.

The main purpose of current-limiting fuses is to prevent damage due to excessive fault current. They have the beneficial side effect with respect to power quality that the voltage sag resulting from the fault is very brief. Figure 3.52 shows typical voltage and current waveforms from a current-limiting fuse operation during a SLG fault. The voltage drops immediately to about 40 percent because of the fault, but shortly recovers and overshoots to about 140 percent as the peak arc voltage develops in the fuse to cut off current flow. Note that the current waveform is clipped by the instrumentation. The voltage sag is so short that not many industrial processes will be adversely affected. Therefore, one proposed practice is to install current-limiting fuses on each lateral branch in the high fault current region near the substation to reduce the number of sags that affect industrial processes.

When current-limiting fuses were first installed on utility systems in great numbers, there was the fear that the peak arc voltage transient, which exceeds system voltage, would cause damage to arresters and to insulation in the system. This has not proven to be a significant problem. The overvoltage is on the same order as capacitor-switching transient overvoltages, which occur several times a day on most utility systems without serious consequences.

3.8.13   Adaptive Relaying

Adaptive relaying is the practice of changing the relaying characteristics of the overcurrent protective device to suit the present system conditions.

One relevant thing that is currently being done with adaptive relaying is the enabling and disabling of fast tripping of breakers in response to weather conditions. This is generally done through a radio or telecommunications link to the utility control center. It could also be done with local devices that have the ability to detect the presence of nearby lightning or rain. If a storm is approaching, fast tripping is enabled to save the fuses from the anticipated high incidence of temporary faults. End users are more understanding and less likely to complain about interruptions during storms.

At other times, fast tripping is disabled and the fuses are allowed to blow. This does not mean there will be no temporary faults without storms. Animals can climb electrical structures at any time and cause a fault. Vegetation growth may be sufficient to cause faults. However, the public is generally much less understanding about an interruption on a clear day.

3.8.14   Ignoring Third-Harmonic Currents

The level of third-harmonic currents has been increasing due to the increase in the numbers of computers and other types of electronic loads on the system. The residual current (sum of the three-phase currents) on many feeders contains as much third harmonic as it does fundamental frequency. A common case is to find each of the phase currents to be moderately distorted with a THD of 7 to 8 percent, consisting primarily of the third harmonic. The third-harmonic currents sum directly in the neutral so that the third harmonic is 20 to 25 percent of the phase current, which is often as large, or larger, than the fundamental frequency current in the neutral (see Chaps. 5 and 6).

Because the third-harmonic current is predominantly zero-sequence, it affects the ground-fault relaying. There have been incidents where there have been false trips and lockout due to excessive harmonic currents in the ground-relaying circuit. At least one of the events we have investigated has been correlated with capacitor switching where it is suspected that the third-harmonic current was amplified somewhat due to resonance. There may be many more events that we have not heard about, and it is expected that the problem will only get worse in the future.

The simplest solution is to raise the ground-fault pickup level when operating procedures will allow. Unfortunately, this makes fault detection less sensitive, which defeats the purpose of having ground relaying, and some utilities are restrained by standards from raising the ground trip level. It has been observed that if the third harmonic could be filtered out, it might be possible to set the ground relaying to be *more* sensitive. The third-harmonic current is almost entirely a function of load and is not a component of fault current. When a fault occurs, the current seen by the relaying is predominantly sinusoidal. Therefore, it is not necessary for the relaying to be able to monitor the third harmonic for fault detection.

The first relays were electromagnetic devices that basically responded to the effective (rms) value of the current. Thus, for years, it has been common practice to design electronic relays to duplicate that response and digital relays have also generally included the significant lower harmonics. In retrospect, it would have been better if the third harmonic would have been ignored for ground-fault relays.

There is still a valid reason for monitoring the third harmonic in phase relaying because phase relaying is used to detect overload as well as faults. Overload evaluation is generally an rms function.

3.8.15   Utility Fault Prevention

One sure way to eliminate complaints about utility fault-clearing operations is to eliminate faults altogether. Of course, there will always be some faults, but there are many things that can be done to dramatically reduce the incidence of faults.[18]

Overhead Line Maintenance

Tree trimming.   This is one of the more effective methods of reducing the number of faults on overhead lines. It is a necessity, although the public may complain about the environmental and aesthetic impact.

Insulator washing.   Like tree trimming in wooded regions, insulator washing is necessary in coastal and dusty regions. Otherwise, there will be numerous insulator flashovers for even a mild rainstorm without lightning.

Shield wires.   Shield wires for lightning are common for utility transmission systems. They are generally not applied on distribution feeders except where lines have an unusually high incidence of lightning strikes. Some utilities construct their feeders with the neutral on top, perhaps even extending the pole, to provide shielding. No shielding is perfect.

Improving pole grounds.   Several utilities have reported doing this to improve the power quality with respect to faults. However, we are not certain of all the reasons for doing this. Perhaps, it makes the faults easier to detect. If shielding is employed, this will reduce the backflashover rate. If not, it would not seem that this would provide any benefit with respect to lightning unless combined with line arrester applications (see Line Arresters).

Modified conductor spacing.   Employing a different line spacing can sometimes increase the withstand to flashover or the susceptibility to getting trees in the line.

Tree wire (insulated/covered conductor).   In areas where tree trimming is not practical, insulated or covered conductor can reduce the likelihood of tree-induced faults.

UD Cables
Fault prevention techniques in underground distribution (UD) cables are generally related to preserving the insulation against voltage surges. The insulation degrades significantly as it ages, requiring increasing efforts to keep the cable sound. This generally involves arrester protection schemes to divert lightning surges coming from the overhead system, although there are some efforts to restore insulation levels through injecting fluids into the cable.

Since nearly all cable faults are permanent, the power quality issue is more one of finding the fault location quickly so that the cable can be manually sectionalized and repaired. Fault location devices available for that purpose are addressed in Sec. 3.8.16.

Line Arresters
To prevent overhead line faults, one must either raise the insulation level of the line, prevent lightning from striking the line, or prevent the voltage from exceeding the insulation level. The third idea is becoming more popular with improving surge arrester designs. To accomplish this, surge arresters are placed every two or three poles along the feeder as well as on distribution transformers. Some utilities place them on all three phases, while other utilities place them only on the phase most likely to be struck by lightning. To support some of the recent ideas about improving power quality, or providing custom power with super-reliable main feeders, it will be necessary to put arresters on every phase of every pole.

Presently, applying line arresters in addition to the normal arrester at transformer locations is done only on line sections with a history of numerous lightning-induced faults. But some utilities have claimed that applying line arresters is not only more effective than shielding, but it is more economical.[14]

Some sections of urban and suburban feeders will naturally approach the goal of an arrester every two or three poles because the density of load requires the installation of a distribution transformer at least that frequently. Each transformer will normally have a primary arrester in lightning-prone regions. A word of caution: the failure rate of aging arresters may negate the benefits.

3.8.16   Fault Locating

Finding faults quickly is an important aspect of reliability and the quality of power.

Faulted Circuit Indicators

Finding cable faults is often quite a challenge. The cables are underground, and it is generally impossible to see the fault, although occasionally there will be a physical display. To expedite locating the fault, many utilities use "faulted circuit indicators," or simply "fault indicators," to locate the faulted section more quickly. These are devices that flip a target indicator when the current exceeds a particular level. The idea is to put one at each pad-mount transformer; the last one showing a target will be located just before the faulted section.

There are two main schools of thought on the selection of ratings of faulted circuit indicators. The more traditional school says to choose a rating that is 2 to 3 times the maximum expected load on the cable. This results in a fairly sensitive fault detection capability.

The opposing school says that this is too sensitive and is the reason that many fault indicators give a false indication. A false indication delays the location of the fault and contributes to degraded reliability and power quality. The reason given for the false indication is that the energy stored in the cable generates sufficient current to trip the indicator when the fault occurs. Thus, a few indicators downline from the fault may also show the fault. The solution to this problem is to apply the indicator with a rating based on the maximum fault current available rather than on the maximum load current. This is based on the assumption that most cable faults quickly develop into bolted faults. Therefore, the rating is selected allowing for a margin of 10 to 20 percent.

Another issue impacting the use of fault indicators is DG. With multiple sources on the feeder capable of supplying fault current, there will be an increase in false indications. In some cases, it is likely that all the fault indicators between the generator locations and the fault will be tripped. It will be a challenge to find new technologies that work adequately in this environment. This is just one example of the subtle impacts on utility practice resulting from sufficient DG penetration to significantly alter fault currents.

Fault indicators must be reset before the next fault event. Some must be reset manually, while others have one of a number of techniques for detecting, or assuming, the restoration of power and resetting automatically. Some of the techniques include test point reset, low-voltage reset, current reset, electrostatic reset, and time reset.

Locating Cable Faults without Fault Indicators

Without fault indicators, the utility must rely on more manual techniques for finding the location of a fault. There are a large number of different types of fault-locating techniques and a detailed description of each is beyond the scope of this report. Some of the general classes of methods follow.

Thumping. This is a common practice with numerous minor variations. The basic technique is to place a dc voltage on the cable that is sufficient to cause the fault to be reestablished and then try to detect by sight, sound, or feel the physical display from the fault. One common way to do this is with a capacitor bank that can store enough energy to generate a sufficiently loud noise. Those standing on the ground on top of the fault can feel and hear the "thump" from the discharge. Some combine this with cable radar techniques to confirm estimates of distance. Many are concerned with the potential damage to the sound portion of the cable due to thumping techniques.

Cable radar and other pulse methods. These techniques make use of traveling-wave theory to produce estimates of the distance to the fault. The wave velocity on the cable is known. Therefore, if an impulse is injected into the cable, the time for the reflection to return will be proportional to the length of the cable to the fault. An open circuit will reflect the voltage wave back positively while a short circuit will reflect it back negatively. The impulse current will do the opposite. If the routing of the cable is known, the fault location can be found simply by measuring along the route. It can be confirmed and fine-tuned by thumping the cable. On some systems, there are several taps off the cable. The distance to the fault is only part of the story; one has to determine which branch it is on. This can be a very difficult problem that is still a major obstacle to rapidly locating a cable fault.

Tone. A tone system injects a high-frequency signal on the cable, and the route of the cable can be followed by a special receiver. This technique is sometimes used to trace the cable route while it is energized, but is also useful for fault location because the tone will disappear beyond the fault location.

Fault chasing with a fuse. The cable is manually sectionalized, and then each section is reenergized until a fuse blows. The faulted section is determined by the process of elimination or by observing the physical display from the fault. Because of the element of danger and the possibility of damaging cable components, some utilities strongly discourage this practice. Others require the use of small current-limiting fuses, which minimize the amount of energy permitted into the fault. This can be an expensive and time-consuming procedure that some consider to be the least effective of fault-locating methods and one that should be used only as a last resort. This also subjects end users to nuisance voltage sags.

3.9 Fault Locating Using Voltage and Current Measurements

With the advances in microprocessor-based relays and power quality monitors, fault locating is increasingly performed using voltage and current measurements. These measurements are taken while the fault is in the circuit, and are used to estimate the apparent impedance to the fault location seen from the relay or the power quality monitor. A number of impedance-based methods have been demonstrated to be robust and effective in accurately estimating fault locations particularly in overhead distribution circuits. Impedance-based methods are often employed in conjunction with or instead of fault indicators described in the earlier section. When arc voltage during the fault is taken into account in computing the apparent impedance, it is possible to estimate locations of incipient faults with reasonable accuracy. This method is especially useful in preventing actual short-circuit faults from materializing as the utility would have sufficient time to inspect and mitigate the problems. Fault locating with only current measurements is also possible, although the accuracy may be compromised slightly compared to that when both voltage and current measurements are available. This current-only method uses a fault current profile to estimate fault location. Fault locating methods described above are presented below along with their applications.

3.9.1 Impedance-Based Fault Location Methods

Impedance-based methods are commonly used to locate faults on distribution systems because of their simplicity and ease of implementation. The main inputs for these methods are the voltage and current measurements recorded by a power quality monitor or a relay. They also require the positive- and zero-sequence z_1 and z_0 line impedance values (in Ω pu length) while estimating fault location. The output of these methods is in terms of distance to fault location.

The loop reactance method,[20] the positive-sequence reactance method,[21] and the Takagi method[21,22] are three commonly used impedance-based methods. Each of these methods is briefly reviewed below.

A simple two-bus distribution system is shown in Fig. 3.53. The source impedance is represented by Z_S. There is a power quality monitor (PQM) at Bus B_1. It continuously measures the voltage V and current I_{tot} at this bus. The load impedance Z_{load} which is present at Bus B_2, draws current I_{load}. The distance between Bus B_1 and Bus B_2 is d expressed in miles. Further, Z_1 and Z_0 (in Ω) are the positive- and zero-sequence impedances of the distribution line between buses B_1 and B_2. The positive- and zero-sequence line impedances (in Ω per-unit length) are z_1 and z_0, respectively. It is to be noted that $Z_1 = d \times z_1$ and $Z_0 = d \times z_0$.

A SLG fault occurs at Bus B_2. The total current I_{tot} recorded by the power quality monitor is the sum of the fault current I_F and the load current I_{load}. The loop reactance, positive-sequence reactance and Takagi methods attempt to estimate the distance to fault location d using the measured voltage V, the measured current I_{tot} and the positive- and zero-sequence z_1 and z_0 line impedances.

a. Loop Reactance Method

The method computes the loop reactance to the fault location X_t, using the following expression:

$$X_t = \text{imag}\left(\frac{V}{3I_{tot0}}\right)$$

where V and I_{tot0} are measured voltage and zero-sequence component of the measured current, respectively. It is to be noted that the loop reactance method does not require values of positive- and zero-sequence z_1 and z_0 line impedances while estimating reactance to

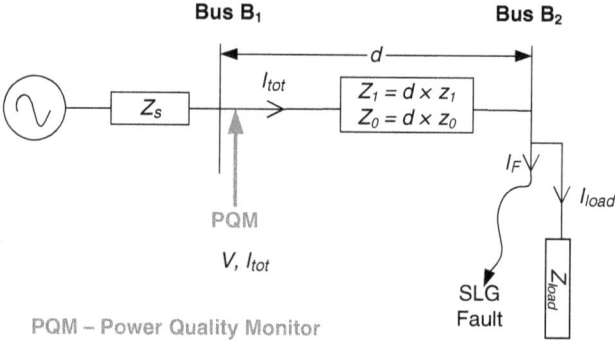

PQM – Power Quality Monitor

FIGURE 3.53   Single-line diagram for a simple distribution system.

fault location. The distance to fault location d can be computed as follows:

$$d = \frac{X_t}{x_t}$$

where, x_t is the loop reactance in Ω per-unit length and is defined as:

$$x_t = \text{imag}\left(\frac{2z_1 + z_0}{3}\right)$$

This method performs well when used to conduct fault location on systems serving light loads.

b. Positive-Sequence Reactance Method
The positive-sequence reactance method estimates the positive-sequence reactance to fault location X_1 as follows:

$$X_1 = \text{imag}(Z_1) = \text{imag}\left(\frac{V}{I_S}\right)$$

Current I_S is given as: $I_S = I_{tot} + kI_{tot0}$. The factor k is computed using values of positive- and zero-sequence z_1 and z_0 line impedances,

$$k = \frac{z_0 - z_1}{z_1}.$$

This method requires values of z_1 and z_0 line impedances in addition to voltage and current measurements, while estimating reactance to fault location. The distance-to-fault d is obtained as follows:

$$d = \frac{X_1}{x_1}$$

where x_1 = positive-sequence reactance (in Ω per-unit length) = imag(z_1). This method works with reasonable accuracy on light-load as well as heavy-load systems.

c. Takagi Method
The Takagi method uses the following expression to compute the distance to fault location:

$$d = \frac{\text{imag}(VI_{sup}^*)}{\text{imag}(z_1 I_S I_{sup}^*)}$$

where I_S is defined as $I_S = I_{tot} + kI_{tot0}$ while I_{sup} is the superposition current (the asterisk denotes complex conjugate of I_{sup}). It is given by

$I_{sup} = I_{tot} - I_{pre\text{-}flt}$ where $I_{pre\text{-}flt}$ is the pre-fault load current. Based on this expression, the Takagi method requires pre-fault load current data, z_1 and z_0 line impedance values as well as voltage and current measurements, while estimating fault location. The performance of this method is comparable to that of the positive-sequence reactance method.

Error Sources for the Impedance-Based Methods

There are various factors which cause fault location estimates of the impedance-based methods to be erroneous. All impedance-based methods use fault voltage and current measurements to estimate fault location. Inadequacies in instrument transformers can lead to errors in measurement of currents and voltages. These erroneous measurements have an adverse impact on the accuracy of the three methods. Further, each method uses positive- and zero-sequence z_1 and z_0 line impedance values while computing distance-to-fault. The values of z_1 and z_0 depend on various parameters like sizes of phase and neutral conductors, distances between conductors of different phases and earth resistivity at the particular location. If any of these parameters is modeled inaccurately, it leads to errors in computation of z_1 and z_0. Incorrect z_1 and z_0 values in turn introduce errors in the fault location estimates of the impedance-based methods.

The simple distribution feeder shown in Fig. 3.53 has a pair of buses and a single section of distribution lines between the two buses. However in a practical feeder, there are multiple buses and numerous sections of single- as well as three-phase distribution lines. There is a wide variation in conductor sizes for different sections. The distances between the conductors may not be the same for all sections of the feeder. As a result, the values of z_1 and z_0 are different for various sections of the feeder. Such a feeder is called a *non-homogeneous feeder*. The impedance-based methods are derived based on the assumption that the distribution feeder has the same z_1 and z_0 values for all sections of distribution lines. While applying these methods on non-homogeneous feeders, values of z_1 and z_0 corresponding to the longest section of distribution lines, are used in these methods. This simplification may cause errors in estimation of fault location. Lastly, most of the impedance-based methods either neglect load currents or use simplified load models while estimating fault location. As the load in a practical system does not conform to the oversimplified models used in the impedance-based methods, estimation accuracy can be adversely affected.

Sample Application of Impedance-Based Methods to a Real-World Fault Event

The single-line diagram for a utility circuit is shown in Fig. 3.54. The rated voltage at the substation is 23.9 kV. An SEL-351S relay is present at the substation for line protection. The relay samples the three-phase line currents and the line-to-neutral voltages at a rate of 16 samples per cycle. The main line is protected by surge arresters, at about every 1000 feet.

On June 16, at 8:45 pm, several industrial customers complained about a momentary outage on this circuit. An investigation by the utility engineers into the cause of this interruption found one of the surge arresters severely damaged. A dead squirrel was found at the bottom of the pole as shown in Fig. 3.55. The arrester did not have a squirrel guard. As the faulted equipment was identified in this case, the utility is aware of the exact fault distance from the substation. The faulted surge arrester is located on a three-phase lateral 5.33 miles from the substation. The fault location is shown in Fig. 3.54.

The voltage and current waveforms for this momentary SLG fault on phase C were recorded by the relay and are shown in Fig. 3.56. In addition, the relay fault event log is shown in Fig. 3.57.

Figure 3.56 shows that the pre-fault load current is about 200 A. The maximum fault current magnitude shown in the relay fault log is 1.907 kA. The most commonly occurring conductor type in the circuit has a positive-sequence impedance $z_1 = 0.1680 + j\,0.6218\ \Omega/\text{mile}$ and a zero-sequence impedance $z_0 = 0.4372 + j\,1.9305\ \Omega/\text{mile}$. These line impedances are used in the three impedance-based methods along with the fault voltage and current measurements. The fault location estimates are shown in Table 3.10.

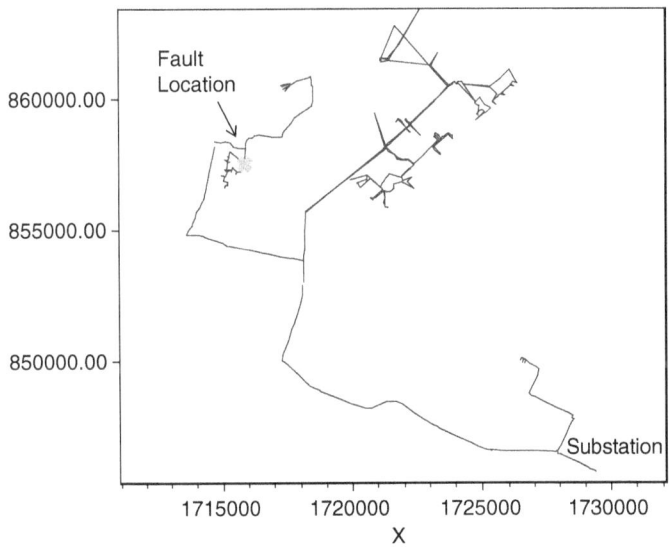

Figure 3.54 Utility circuit diagram.

FIGURE **3.55** Faulted arrester in utility circuit.

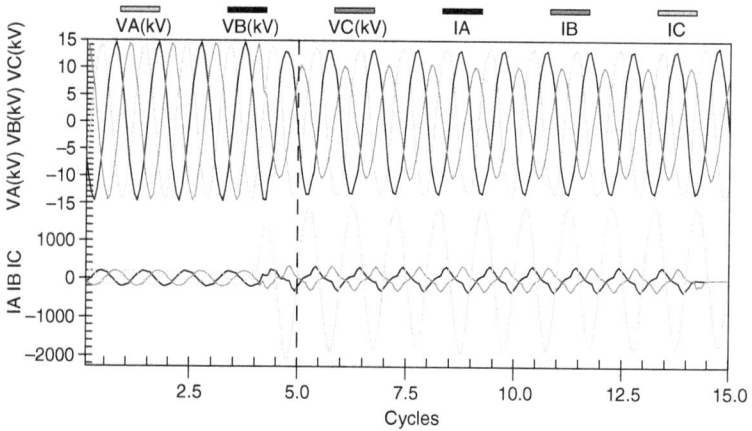

FIGURE **3.56** Fault voltage and current waveforms for a Phase C to ground on utility circuit.

It can be seen from that the fault location estimate obtained using the loop reactance method is the most accurate. It differs from the actual fault location by 0.3 miles. The positive-sequence reactance and Takagi methods give nearly equal fault location estimates. The distance-to-fault estimates of these two methods are within 0.5 to 0.6 miles of the actual fault location.

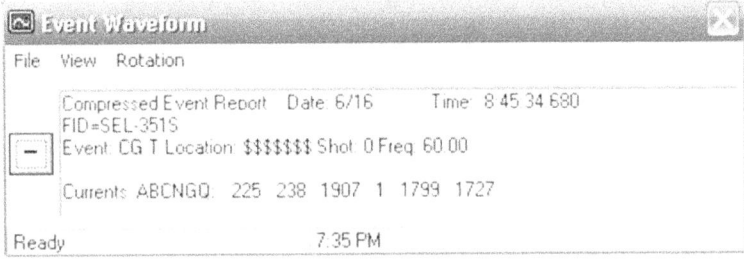

Event Waveform   ☒

File   View   Rotation

Compressed Event Report   Date: 6/16   Time: 8:45:34.680
FID=SEL-351S
Event: CG T Location: $$$$$$$ Shot: 0 Freq: 60.00

Currents ABCNGQ   225   238   1907   1   1799   1727

Ready   7:35 PM

FIGURE 3.57   Relay fault log.

Actual fault location	5.33 mile
Fault location analysis	
Method used	**Fault location estimate**
Loop reactance	5.08 mile
Positive-sequence reactance	4.78 mile
Takagi	4.82 mile

TABLE 3.10   Reactance to fault estimates for events shown in Fig. 3.56

3.9.2   Locating Incipient Faults

Cable failure is a gradual process characterized by precursor self-clearing or incipient faults which eventually result in a permanent fault.[23] Such a phenomenon is very common in a cable splice following moisture penetration into the splice that results in the insulation break down. An electric arc is produced and it evaporates the moisture, creating high pressure vapors which in turn extinguish the arc, making such faults self-clearing. A typical self-clearing fault leading to a permanent fault in an underground cable is shown in Figure 3.58.

The unique characteristics of self-clearing faults[24] are explained with this example as follows.

1. Fault duration is less than one cycle, generally ¼ to ½ cycle. Self-clearing events shown in Fig. 3.58a and Fig. 3.58b represent half-cycle blips in phase B taking place on the same day nearly an hour-and-a-half apart.

2. Fault generally starts near the peak of the voltage waveform. In Fig. 3.558a and Fig. 3.58b. the fault occurs at the peak of the positive and negative voltage cycle, respectively.

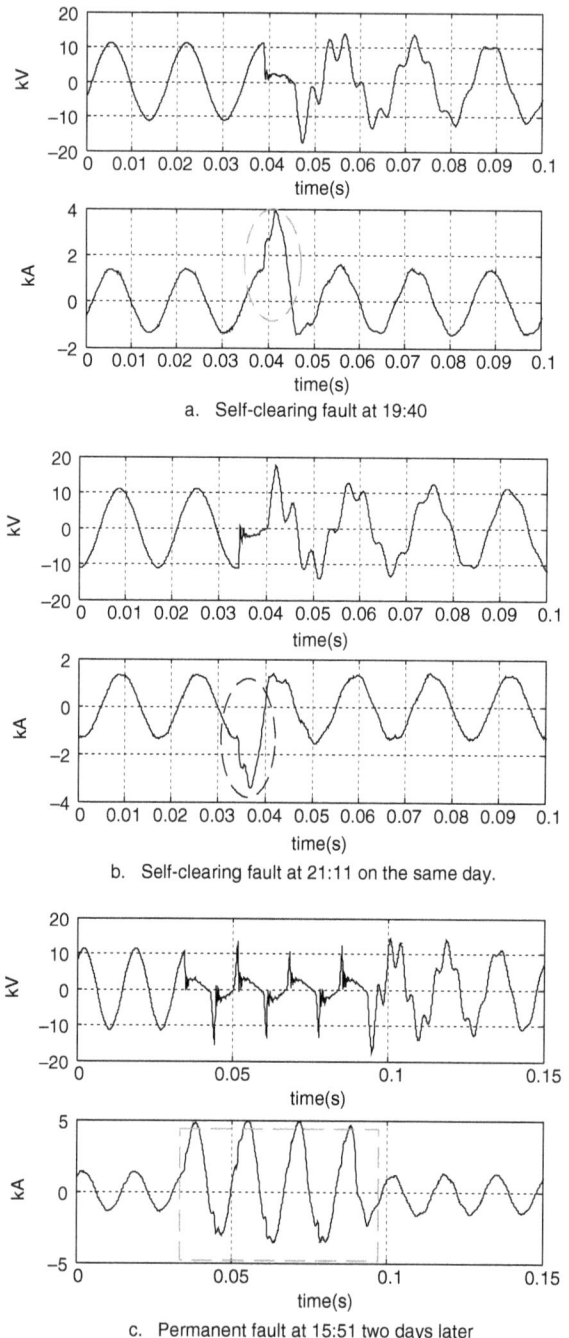

a. Self-clearing fault at 19:40

b. Self-clearing fault at 21:11 on the same day.

c. Permanent fault at 15:51 two days later

FIGURE 3.58 Self-clearing faults leading to permanent fault in underground cable captured at the same monitor.

3. No overcurrent protective device operates because a relay generally needs more than ½ cycles to detect a fault.

4. They are generally precursors to permanent faults on the same phase. The permanent fault in phase B is shown in Fig. 3.58c.

5. Frequency of incipient fault occurrence increases over time. There might be one or two such events initially, but their frequency can increase rapidly before they are about to turn permanent.

As described in the previous paragraph, arc voltage is an important characteristic of incipient cable faults. An arc is a self-sustained electrical discharge caused by short-circuits on the power system. It exhibits a low voltage drop and is capable of sustaining large currents. The arc voltage remains constant over a wide range of currents and arc lengths. Hence, the arc resistance is a nonlinear function of the voltage. Generally, it is preferred to measure the arc in terms of the voltage rather than the resistance.[25] Arc voltage is strongly dependent on the inherent physical properties of the faulted equipment. Arc current waveform is predictable and can be expressed analytically once the circuit parameters are known. However, it is very difficult to obtain an analytical expression for the arc voltage because of the nonlinear arc resistance. The arc resistance, being strongly dependent on the physical properties of the faulted equipment results in a distorted arc voltage curve as shown in Fig. 3.59. In general, there is a blip at the start of the waveform (it is not an ideal square wave) because the arc cools off at the current zero decreasing the ionization rate and increasing the arc resistance. Once the temperature increases, the voltage flattens out. The presence of high odd harmonics makes the arc voltage waveform resemble a distorted square wave shape.

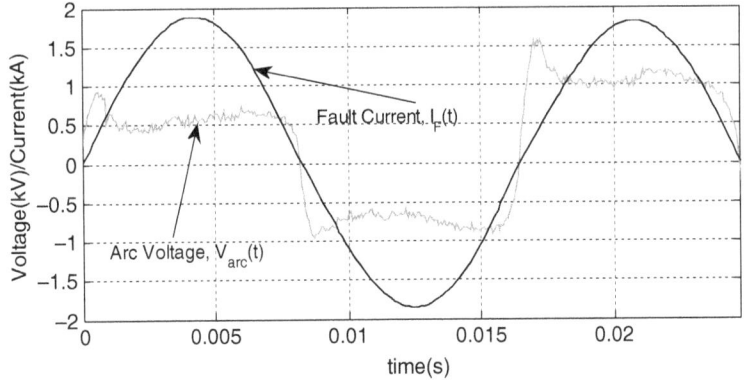

FIGURE 3.59 Typical arc voltage and current waveforms.

The following assumptions have been made to simplify the analysis of the arc voltage.

1. Arc voltage has an ideal square wave shape. This implies that the arc voltage is constant irrespective of the magnitude of the fault current.

2. Arc current and voltage are in phase.

In general, the arc voltage during the fault is a combination of a square wave shape and a random white noise. Figure 3.60 shows the voltage and current captured by a PQM during a fault immediately downstream from the monitor. It can be seen that the recorded fault phase voltage exhibits the classic square wave shape characteristics of an electric arc. Hence, the arc voltage is an important quantity that needs to be taken into account when calculating the fault location.

Impedance-based distribution fault location methods are most commonly used with voltage and current waveforms captured by PQMs.[21,26] These algorithms generally work in the phasor-domain and thus require duration of one or more cycles to provide reasonable location estimates. In addition, they do not take the fault arc voltage into consideration when estimating the fault location. Given these requirements and the short duration of incipient faults, the applicability of impedance-based algorithms is limited.

The use of an arc voltage-based fault location method was proposed in Refs. [27, 28] and a simplified algorithm is described in

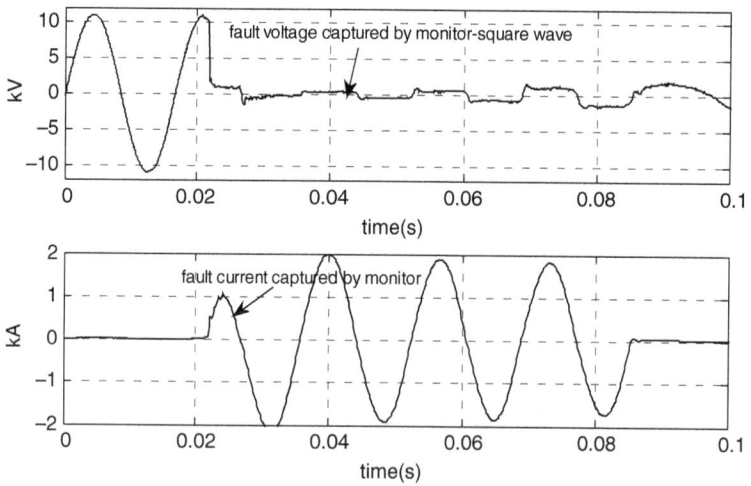

Figure 3.60 Voltage and current recorded by a monitor when fault occurs immediately downstream from it.

Refs. [29, 30]. This algorithm is applicable to SLG faults. and it estimates the arc voltage magnitude in the affected phase and the resistance and reactance to the fault. The single-line diagram for a faulted circuit can be represented as shown in Fig. 3.61.

The voltage at the PQM site in the time-domain, V_F for the faulted phase can be written as follows.

$$V_F = R \cdot I_F + L \cdot \frac{dI_F}{dt} + V_{arc} \cdot \text{sign}(I_F)$$

where
V_F is the fault phase voltage measured at PQM site
I_F is the fault current
L is the line inductance
R is the line resistance
$V_{arc,mag}$ is the amplitude of the ideal square wave arc voltage
sign $(I_F) = 1$, if $I_F > 0$ and -1 if $I_F \leq 0$

Note that R and L above are the self-resistance and inductance of the line conductor, respectively. The effect of mutual coupling is not taken into consideration. Furthermore, the self-impedance of a line is the loop impedance. In a distribution system, the faulted phase current typically contains a component of the load current, which can lead to inaccurate fault location estimates. On the contrary, the neutral current ($I_n = I_a + I_b + I_c$) during the fault purely consists of the fault current, as the balanced load current in the three phases gets canceled out. As a result, the use of neutral current in the fault location algorithm yields better estimates. Hence, neutral current recorded at the monitor is used instead of faulted phase current, I_F. The above expression is applicable at every point of the voltage and current waveform obtained from the PQMs. Common sampling rates of PQM waveforms are 128 or 256 samples per cycle, which results in an overdetermined system of equations. Such an overdetermined system of equations is solved using the non-negative least square

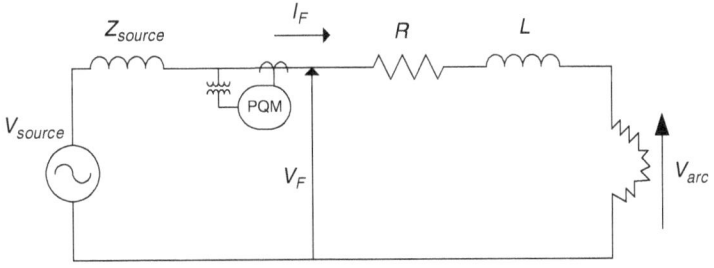

FIGURE 3.61 Faulted phase circuit during a SLG fault.

method. A sample window is defined as the number of sample points analyzed at a given time instant. After computing the unknowns at one time instant the window is shifted to the next instant and the process is repeated for the entire waveform. Fault duration is the most important factor that determines the width of the sample window. The rule of thumb for window selection can be stated as, *the window length should be less than the fault duration.* Note that the arc voltage during the fault may vary considerably due to the tendency of the arc to elongate or shrink over a period of time. Nevertheless, the resistance and reactance to the fault should be constant during a fault. Once constant values for R and L are known, the impedance from the monitoring point to the fault location can be obtained and hence the fault location.

A series of faulted voltage and current waveforms captured by the same monitor on a 13.8-kV distribution feeder were shown in Fig. 3.58. Now the arc voltage-based incipient fault location algorithm is applied to determine the location of these events in terms of the reactance to the fault from the monitor. The monitor samples the voltage at a rate of 256 samples per cycle, while the current is sampled at a rate of 128 samples per cycle. The current is up-sampled to match the voltage sampling level. Furthermore, the voltage and current waveforms are smoothed before using them in the fault-locating algorithm to yield less noise in the results. A moving average filter is used for this purpose. Based on the discussion in the previous paragraph, the window length chosen for the two incipient faults in Fig. 3.58a and Fig. 3.58b is 100 samples, i.e. between half and quarter cycles. On the other hand, for the permanent fault shown in Fig. 3.58c, the window length chosen is 256 samples, or one cycle. The reactance to fault, X_L estimates for all three events are shown in Table 3.11 and Fig. 3.62. For this case, the actual reactance to the fault is unknown. However, assuming the estimated reactance during the permanent fault as the actual value, the errors for the results obtained by the precursor events are given in .

The average absolute error in the reactance estimates obtained from the self-clearing faults is 7.37 percent or 0.0250Ω. It can be concluded that the first two self-clearing events were precursors

Event	Mean reactance estimate (Ω)	Absolute error (Ω)	Absolute error (%)
Self-clearing fault 1	0.344	0.0050	1.47
Self-clearing fault 2	0.384	0.0450	13.27
Average		0.0250	7.37
Permanent fault	0.339	N/A	N/A

TABLE **3.11**   Reactance to fault estimates for events shown in Fig. 3.62

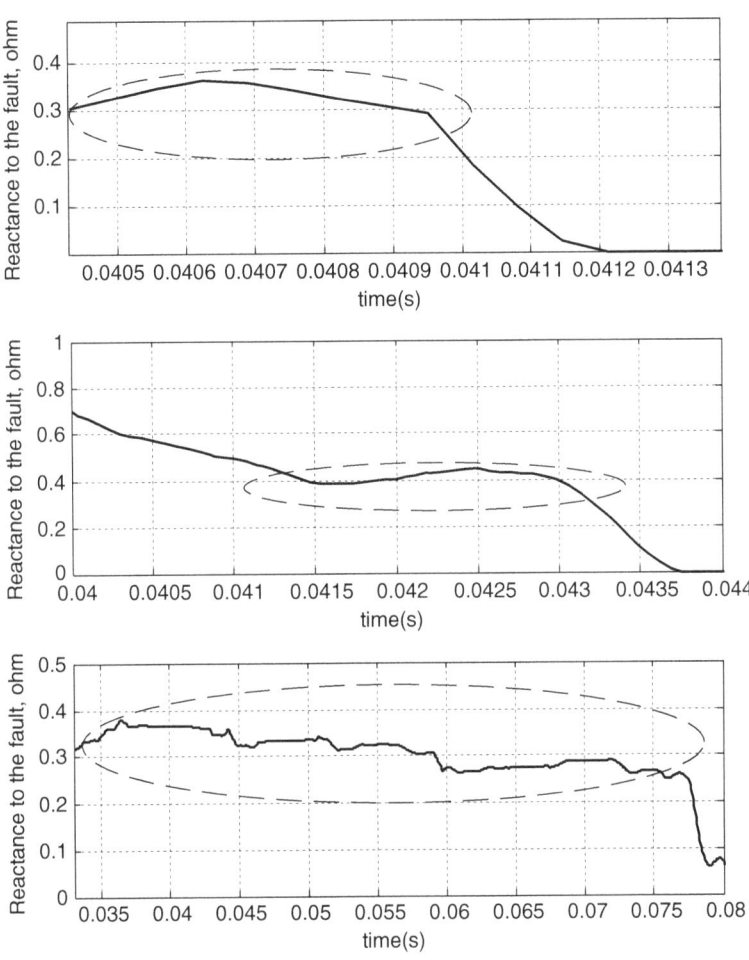

FIGURE 3.62   Reactance to fault estimates for events shown in Figure 3.58.

to the permanent fault event. In this way, the arc voltage-based fault-locating algorithm can be effectively used to detect subcycle incipient faults in underground cables before they turn permanent.

3.9.3   Fault Current Profile

Impedance-based methods like Takagi and positive-sequence method are widely used to locate faults in a distribution system because of their simplicity and ease of application.[20,21,26] Unfortunately, certain assumptions made by impedance-based methods affect the accuracy of fault location estimates. For example, as seen in Sec. 3.9.1, loop reactance method does not consider pre-fault current. Pre-fault current plays an important role in fault location estimation and for faults wherein the load current constitutes a significant percentage of the total fault current seen by the relay at the substation, neglecting the load current results in a high error in estimation.[31] Other methods like Takagi and positive-sequence method assume that the load is lumped beyond the fault point. However, when a fault occurs on the distribution feeder, loads may be present between the relay and the fault point. Hence, the lumped load assumption does not hold true and affects the accuracy of the location estimates.[31]

Another drawback of the impedance-based methods is that they require zero-sequence line impedance when computing fault location. Zero-sequence line impedance is prone to error. Moreover, utility systems are generally heterogeneous, i.e., pole configurations and line or wire sizes are not uniform. In such cases, impedance-based methods assume that the system is homogenous and use the line parameters of the most frequently occurring line configuration. This assumption of homogeneity further introduces error in the estimation.[21,26]

To minimize the errors associated with impedance-based methods, the fault current profile approach is an alternative method for locating faults in a distribution system.[20,32] It requires the availability of current waveforms measured during the occurrence of the fault. Utilities may have the circuit model of the distribution feeder in any of a number of distribution software packages. The circuit model is assumed to closely represent the actual feeder and is used to build a reference current profile of short-circuit fault current versus reactance or distance to the fault. When a fault occurs in the distribution system, the fault current magnitude is extrapolated on the current profile to get a location estimate.

The fault current profile attempts to move away from the commonly used impedance-based algorithms since they do not use line impedance values directly in its estimation. Unfortunately impedances are still present in the circuit model. However, since the circuit model is directly used in building the current profile, this approach allows for changes in line impedance and hence avoids errors in estimation introduced because of the system homogeneity assumption. Using the circuit model also takes into account the variability of loads present in the system. The approach is straightforward and does not require any computational effort. It also indicates the branch or lateral in which the fault is located in and hence significantly speeds up the fault restoration process. Progress Carolina (formerly Carolina Power & Light) has implemented this approach and Lampley reported that their fault location estimations were accurate to within 0.5 miles of the actual fault location 75 percent of the time.[20,32] For the remaining cases, the error was no more than 1 or 2 miles from the actual location.[20,32] A brief description of the method and its demonstration on a utility test case is presented in the next subsections.

Description of the Method and Application

The fault current profile approach requires only fault current magnitude (no phase angle), pre-fault current and circuit model of the distribution feeder to determine location of a fault.[20,32] Faults are placed at successive incremental distances from the relay in the circuit model. The corresponding short-circuit fault current measured by the relay is plotted against positive-sequence reactance or distance to fault from the relay location. A typical short-circuit fault current profile is an exponentially decreasing curve, since the short-circuit fault current decreases with distance from the relay.

When developing the current profile, it is always built along one particular direction from the relay to the end of the feeder. Now, it is simple to build a fault current profile for a single feeder length with no laterals and branches. However, generally distribution systems are complex, having many branches and single-phase laterals. When building the current profile for such a feeder, each branch or lateral is considered as an individual path and the current profile is built along every path from the relay monitoring location. Hence, the current profile would consist of multiple sub-plots, with each sub-plot indicating a particular path.

In the following the application of the fault current profile approach on the utility test case shown in Sec. 3.9.1 is demonstrated. The circuit model of the utility distribution feeder is available in the EPRI OpenDSS program. The SEL 351-S relay at the substation records a fault current magnitude of 1907 A and pre-fault current of 208 A per phase.

Before building the fault current profile, the pre-fault current is taken into account by matching the load level in the circuit model with the real world load current level. A load flow analysis on the circuit model indicates that the peak load current is 400 A per phase. Hence, the system is operating under half load condition. The current profile should also be built under half-load condition. Loads in the circuit model are switched off till the desired load level is achieved. From the pre-fault current recorded by the relay, no information regarding which loads have been switched off or on in the real world can be obtained. Only the load level can be determined. Hence the loads are adjusted to obtain the desired load current in the circuit model. To determine whether this inaccuracy will affect the fault location estimates, a case study was performed on the distribution feeder. A number of cases were taken in which a half-load condition was achieved by switching off loads at different buses. In each case, for a fault at a particular bus, the total current seen by the relay at the substation varied in the range of a few amperes. Therefore, this inaccuracy in terms of distance to fault estimation will not lead to significant error.

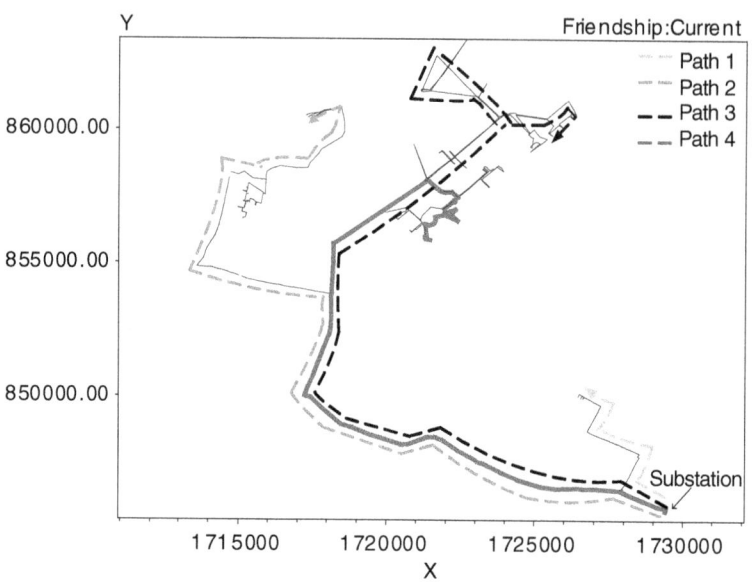

Figure 3.63 Paths for short-circuit fault current profile.

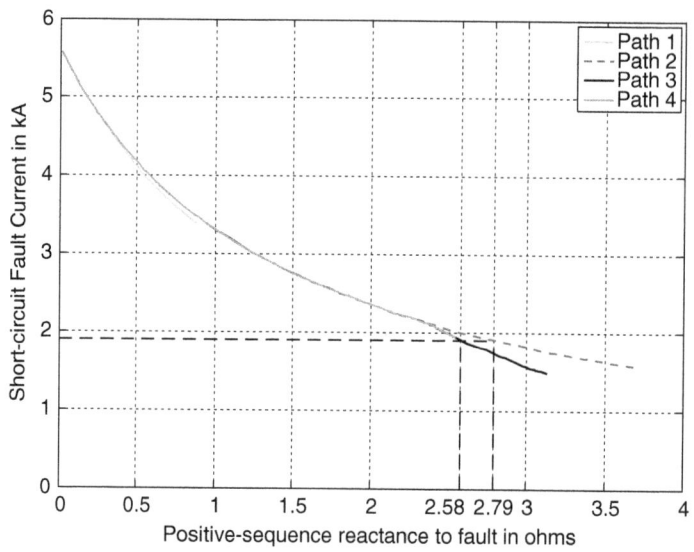

Figure 3.64 Short-circuit fault current profile.

The distribution feeder has four branches and hence four paths are identified as shown in Fig. 3.63. SLG faults are placed at every bus along the path and the developed fault current profile of short-circuit fault current versus the positive-sequence reactance to fault is shown in Fig. 3.64 . The fault current magnitude of 1907 A intersects the paths 2 and 3 at a positive-sequence reactance value of 2.58 Ω and 2.79 Ω. This corresponds to distances of 4.5 and 6.22 miles from the substation along paths 2 and 3, respectively. The utility may utilize customer outage reports or recloser operation status to determine which location has the maximum probability of a fault. In this case, the estimate of 4.54 miles on path 2 is reported as the fault location. The actual location of the fault is known to be at 5.33 miles from the substation along path 2. The error in estimation is 0.79 miles.

To determine the source of the error, a SLG fault was placed in the circuit model at the known fault location. The short-circuit fault current computed was 1705 A while the relay had recorded a fault current of 1907 A. The difference in the two currents indicates that there are discrepancies between the circuit model and the actual distribution feeder which result in an error in estimation. Hence, when implementing the current profile approach for fault location, the engineer should make every effort to ensure that the circuit model accurately represents the distribution feeder.

3.10   References

1. J. Lamoree, J. C. Smith, P. Vinett, T. Duffy, M. Klein, "The Impact of Voltage Sags on Industrial Plant Loads," *First International Conference on Power Quality, PQA '91*, Paris, France.
2. P. Vinett, R. Temple, J. Lamoree, C. De Winkel, E. Kostecki, "Application of a Superconducting Magnetic Energy Storage Device to Improve Facility Power Quality," *Proceedings of the Second International Conference on Power Quality: Enduse Applications and Perspectives, PQA '92*, Atlanta, GA, September 1992.
3. G. Beam, E. G. Dolack, C. J. Melhorn, V. Misiewicz, M. Samotyj, "Power Quality Case Studies, Voltage Sags: The Impact on the Utility and Industrial Customers," *Third International Conference on Power Quality, PQA '93*, San Diego, CA, November 1993.
4. J. Lamoree, D. Mueller, P. Vinett, W. Jones, "Voltage Sag Analysis Case Studies," *1993 IEEE I&CPS Conference*, St. Petersburg, FL.
5. M. F. McGranaghan, D. R. Mueller, M. J. Samotyj, "Voltage Sags in Industrial Systems," *IEEE Transactions on Industry Applications*, vol. 29, no. 2, March/April 1993.
6. Le Tang, J. Lamoree, M. McGranaghan, H. Mehta, "Distribution System Voltage Sags: Interaction with Motor and Drive Loads," *IEEE Transmission and Distribution Conference*, Chicago, IL, April 10–15, 1994, pp. 1–6.
7. EPRI RP 3098–1, *An Assessment of Distribution Power Quality*, Electric Power Research Institute, Palo Alto, CA.
8. *IEEE Standard Guide for Electric Power Distribution Reliability Indices*, IEEE Standard, 1366–2001.
9. J. J. Burke, *Power Distribution Engineering: Fundamentals and Applications,* Marcel Dekker, Inc., 1994.
10. C. M. Warren, "The Effect of Reducing Momentary Outages on Distribution Reliability Indices," *IEEE Transactions on Power Delivery*, July 1993, pp. 1610–1617.

11. R. C. Dugan, L. A. Ray, D. D. Sabin, et al., "Impact of Fast Tripping of Utility Breakers on Industrial Load Interruptions," *Conference Record of the 1994 IEEE/ IAS Annual Meeting,* Vol. III, Denver, October 1994, pp. 2326–2333.

12. T. Roughan and P. Freeman, "Power Quality and the Electric Utility, Reducing the Impact of Feeder Faults on Customers," *Proceedings of the Second International Conference on Power Quality: End-use Applications and Perspectives (PQA '92),* EPRI, Atlanta, GA, September 28–30, 1992.

13. J. Lamoree, Le Tang, C. De Winkel, P. Vinett, "Description of a Micro-SMES System for Protection of Critical Customer Facilities," *IEEE Transactions on Power Delivery,* April 1994, pp. 984–991.

14. R. A. Stansberry, "Protecting Distribution Circuits: Overhead Shield Wire Versus Lightning Surge Arresters," *Transmission & Distribution,* April 1991, pp. 56ff.

15. S. Santoso, R. Zavadil, D. Folts, M. F McGranaghan, T. E Grebe, "Modeling and Analysis of a 1.7 MVA SMES-based Sag Protector," *Proceedings of the 4th International Conference on Power System Transients Conference,* Rio de Janeiro, Brazil, June 24–28, 2001, pp. 115–119.

16. M. H. J. Bollen, *Understanding Power Quality Problems, Voltage Sags and Interruptions,* IEEE Press Series on Power Engineering, The Institute of Electrical and Electronics Engineers, Inc., New York, 2000.

17. SEMI Standard F-47, *Semiconductor Equipment and Materials International,* 1999.

18. IEEE Standard 1346–1998, *Recommended Practice for Evaluating Electric Power System Compatibility with Electronic Process Equipment.*

19. F. Goodman and C. McCarthy, "Applications assessment of pulse closing technology," 20th International Conference on Electricity Distribution, CIRED, Prague, 8–11 June 2009, paper 1028.

20. *"Distribution Fault Location—Filed Data and Analysis."* Electric Power Research Institute, Palo Alto, CA, Technical report 1012438, April 2006.

21. K. Zimmerman and D. Costello, "Impedance-based fault location experience," in *Rural Electric Power Conference, 2006 IEEE,* 2006, pp. 1–16.

22. T. Takagi, Y. Yamakoshi, M. Yamaura, R. Kondow and T. Matsushima, "Development of a New Type Fault Locator Using the One-Terminal Voltage and Current Data." In IEEE Transactions on Power Apparatus and Systems, Aug. 1982, Issue 8, Vols. PAS-101, pp. 2892–2898.

23. B. Clegg, *Underground Cable Fault Location,* 1st ed. McGraw-Hill, 1993.

24. L. A. Kojovic and C. W. Williams, Jr., "Sub-cycle detection of incipient cable splice faults to prevent cable damage," in *Power Engineering Society Summer Meeting,* 2000. IEEE, 2000.

25. H. Ayrton, *The Electric Arc* , 1st ed. Kessinger Publishing, 2007.

26. "IEEE Guide for Determining Fault Location on AC Transmission and Distribution Lines," *IEEE Std C37.114–2004,* 2005.

27. Z. Radojevic and J.-R. Shin, "New one terminal digital algorithm for adaptive reclosing and fault distance calculation on transmission lines," in *Power Delivery, IEEE Transactions on,* vol. 21, no. 3, pp. 1231–1237, 2006.

28. T. Short, D. Sabin, M. McGranaghan, "Using PQ monitoring and substation relays for fault location on distribution systems," in *Rural Electric Power Conference, 2007 IEEE,* May 2007, pp. B3–B3–7.

29. *"Distribution Fault Location—Prototypes, Algorithms and New Technologies."* Electric Power Research Institute, Palo Alto, CA, Technical report 1013825, March 2008.

30. *"Distribution Fault Location and Waveform Characterization."* Electric Power Research Institute, Palo Alto, CA: 2009. 1017842.

31. N. Karnik, S. Das, S. Kulkarni, S. Santoso, "Effect of Load Current on Fault Location Estimates of Impedance-based Methods," Proceedings of the IEEE PES General Meeting , Detroit, Michigan, 2011.

32. Lampley, G. C., "Fault Detection and Location on Electrical Distribution System Case Study," presented at IEEE Rural Electric Power Conference, 2002.

CHAPTER 4

Transient Overvoltages

4.1 Sources of Transient Overvoltages

There are two main sources of transient overvoltages on utility systems: capacitor switching and lightning. These are also sources of transient overvoltages as well as a myriad of other switching phenomena within end-user facilities. Some power electronic devices generate significant transients when they switch. As described in Chap. 2, transient overvoltages can be generated at high frequency (load switching and lightning), medium frequency (capacitor energizing), or low frequency.

4.1.1 Capacitor Switching

Capacitor switching is one of the most common switching events on utility systems. Capacitors are used to provide reactive power (in units of vars) to correct the power factor, which reduces losses and supports the voltage on the system. They are a very economical and generally trouble-free means of accomplishing these goals. Alternative methods such as the use of rotating machines and electronic var compensators are much more costly or have high maintenance costs. Thus, the use of capacitors on power systems is quite common and will continue to be.

One drawback to the use of capacitors is that they yield oscillatory transients when switched. Some capacitors are energized all the time (a fixed bank), while others are switched according to load levels. Various control means, including time, temperature, voltage, current, and reactive power, are used to determine when the capacitors are switched. It is common for controls to combine two or more of these functions, such as temperature with voltage override.

One of the common symptoms of power quality problems related to utility capacitor-switching overvoltages is that the problems appear at nearly the same time each day. On distribution feeders with industrial loads, capacitors are frequently switched by time clock in anticipation of an increase in load with the beginning of the working day. Common problems are adjustable-speed-drive trips and malfunctions of other electronically controlled load equipment that occur without a noticeable blinking of the lights or impact on other, more conventional loads.

Figure 4.1 shows the one-line diagram of a typical utility feeder capacitor-switching situation. When the switch is closed, a transient similar to the one in Fig. 4.2 may be observed upline from the capacitor at the monitor location. In this particular case, the capacitor switch contacts close at a point near the system voltage peak. This is a common occurrence for many types of switches because the insulation across the switch contacts tends to break down when the voltage across the switch is at a maximum value. The voltage across the capacitor at this instant is zero. Since the capacitor voltage cannot change instantaneously, the system voltage at the capacitor location is briefly pulled down to zero and rises as the capacitor begins to charge toward the system voltage. Because the power system source is inductive, the capacitor voltage overshoots and rings at the natural frequency of the system. At the monitoring location shown, the initial change in voltage will not go completely to zero because of the impedance between the observation point and the switched capacitor. However, the initial drop and subsequent ringing transient that is indicative of a capacitor-switching event will be observable to some degree.

The overshoot will generate a transient between 1.0 and 2.0 per unit (pu) depending on system damping. In this case the transient observed at the monitoring location is about 1.34 pu. Utility capacitor-switching transients are commonly in the 1.3- to 1.4-pu range but have also been observed near the theoretical maximum.

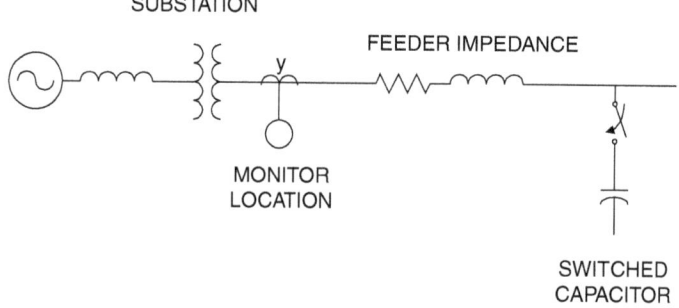

Figure 4.1 One-line diagram of a capacitor-switching operation corresponding to the waveform in Fig. 4.2.

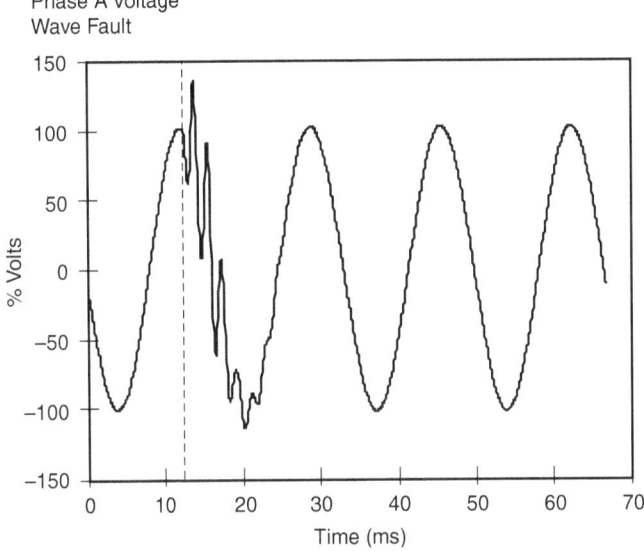

Phase A Voltage
Wave Fault

Figure 4.2   Typical utility capacitor-switching transient reaching 134 percent voltage, observed upline from the capacitor.

The transient shown in the oscillogram propagates into the local power system and will generally pass through distribution transformers into customer load facilities by nearly the amount related to the turns ratio of the transformer. If there are capacitors on the secondary system, the voltage may actually be magnified on the load side of the transformer if the natural frequencies of the systems are properly aligned (see Sec. 4.1.2). While such brief transients up to 2.0 pu are not generally damaging to the system insulation, they can often cause misoperation of electronic power conversion devices. Controllers may interpret the high voltage as a sign that there is an impending dangerous situation and subsequently disconnect the load to be safe. The transient may also interfere with the gating of thyristors.

Switching of grounded-wye transformer banks may also result in unusual transient voltages in the local grounding system due to the current surge that accompanies the energization. Figure 4.3 shows the phase current observed for the capacitor-switching incident described in the preceding text. The transient current flowing in the feeder peaks at nearly 4 times the load current.

4.1.2   Magnification of Capacitor-Switching Transients

A potential side effect of adding power factor correction capacitors at the customer location is that they may increase the impact of utility

Phase A Current
Wave Fault

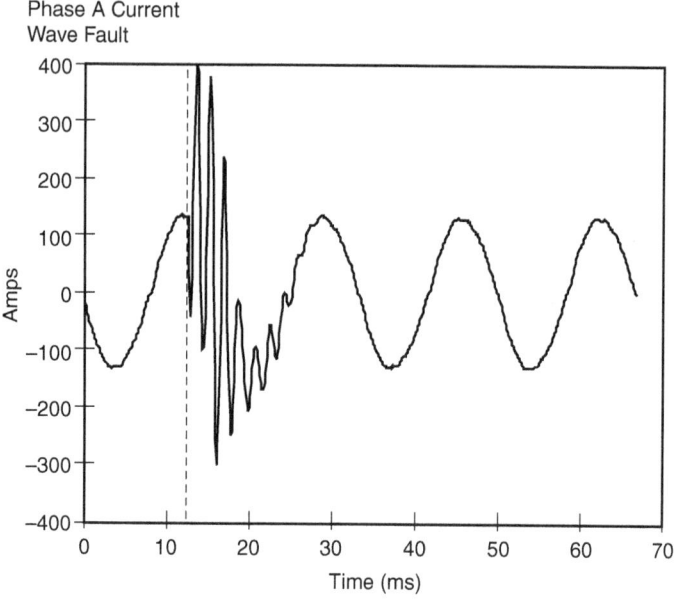

FIGURE 4.3   Feeder current associated with capacitor-switching event.

capacitor-switching transients on end-use equipment. As shown in Sec. 4.1.1, there is always a brief voltage transient of at least 1.3 to 1.4 pu when capacitor banks are switched. The transient is generally no higher than 2.0 pu on the primary distribution system, although ungrounded capacitor banks may yield somewhat higher values. Load-side capacitors can magnify this transient overvoltage at the end-user bus for certain low-voltage capacitor and step-down transformer sizes. The circuit of concern for this phenomenon is illustrated in Fig. 4.4. Transient overvoltages on the end-user side may reach as high as 3.0 to 4.0 pu on the low-voltage bus under these conditions, with potentially damaging consequences for all types of customer equipment.

Magnification of utility capacitor-switching transients at the end-user location occurs over a wide range of transformer and capacitor sizes. Resizing the customer's power factor correction capacitors or step-down transformer is therefore usually not a practical solution. One solution is to control the transient overvoltage at the utility capacitor. This is sometimes possible using synchronous closing breakers or switches with preinsertion resistors. These solutions are discussed in more detail in Sec. 4.4.2.

At the customer location, high-energy surge arresters can be applied to limit the transient voltage magnitude at the customer bus. Energy levels associated with the magnified transient will typically be

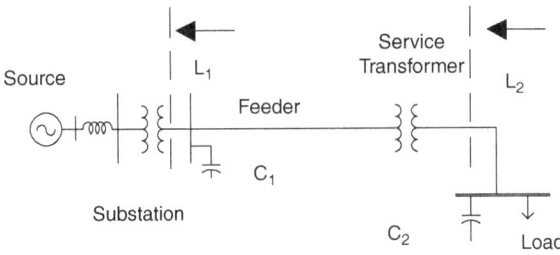

(a) Voltage magnification at customer capacitor due to energizing capacitor on utility system

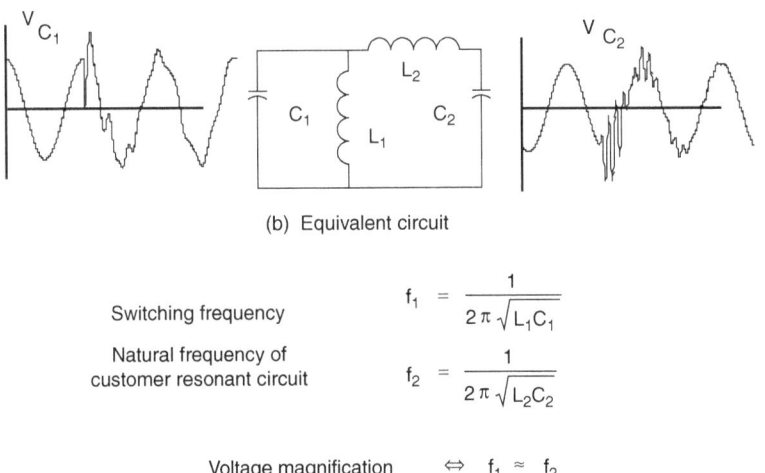

(b) Equivalent circuit

Switching frequency

$$f_1 = \frac{1}{2\pi\sqrt{L_1 C_1}}$$

Natural frequency of customer resonant circuit

$$f_2 = \frac{1}{2\pi\sqrt{L_2 C_2}}$$

Voltage magnification $\Leftrightarrow$ $f_1 \approx f_2$

FIGURE **4.4**   Voltage magnification of capacitor bank switching.

about 1 kJ. Figure 4.5 shows the expected arrester energy for a range of low-voltage capacitor sizes. Newer high-energy metal-oxide varistor (MOV) arresters for low-voltage applications can withstand 2 to 4 kJ.

It is important to note that the arresters can only limit the transient to the arrester protective level. This will typically be approximately 1.8 times the normal peak voltage (1.8 pu). This may not be sufficient to protect sensitive electronic equipment that might only have a withstand capability of 1.75 pu [1200-V peak inverse voltage (PIV) rating of many silicon-controlled rectifiers (SCRs) used in the industrial environment]. It may not be possible to improve the protective characteristics of the arresters substantially because these characteristics are limited by the physics of the metal-oxide materials. Therefore, for proper coordination, it is important to carefully evaluate the withstand capabilities of sensitive equipment used in applications where these transients can occur.

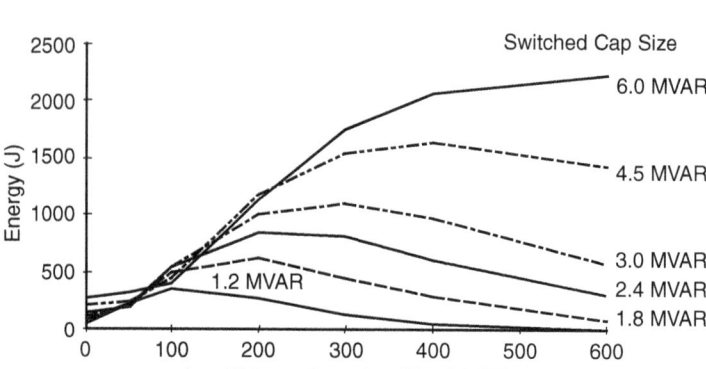

Step-Down Transformer = 1500 kVA

FIGURE 4.5   Arrester energy duty caused by magnified transient.

Another means of limiting the voltage magnification transient is to convert the end-user power factor correction banks to harmonic filters. An inductance in series with the power factor correction bank will decrease the transient voltage at the customer bus to acceptable levels. This solution has multiple benefits including providing correction for the displacement power factor, controlling harmonic distortion levels within the facility, and limiting the concern for magnified capacitor-switching transients.

In many cases, there are only a small number of load devices, such as adjustable-speed motor drives, that are adversely affected by the transient. It is frequently more economical to place line reactors in series with the drives to block the high-frequency magnification transient. A 3 percent reactor is generally effective. While offering only a small impedance to power frequency current, it offers a considerably larger impedance to the transient. Many types of drives have this protection inherently, either through an isolation transformer or through a dc bus reactance.

4.1.3   Restrikes During Capacitor Deenergizing

When a grounded-wye capacitor bank is energized using a mechanical oil-filled switch, the switching is most likely to occur at or near the system voltage peak. At this instant, the capacitor voltage is zero while the system voltage is near maximum. The largest potential difference between contacts is 1.0 pu. Due to the slow-moving nature of the mechanical switch and the large potential difference, insulation across the switch contacts tends to break down giving rise to an electric arc. It allows capacitive current to flow, thereby energizing the capacitor before the actual switch contacts close or touch. This phenomenon is known as pre-strike. It occurs in many types of switches without closing control. It is usually considered as a natural behavior of the switch and should not raise any concern because capacitor voltage prior to energizing is negligible. The resulting peak transient voltage is no worse than that when the capacitor is intentionally energized at the system voltage peak.

On the other hand, restrikes (or reignitions) during capacitor deenergizing do raise concerns. Restrikes or reignitions occur when arcs between parting contacts are re-established after initial current extinction causing unintended re-energizing of the capacitor. Because of the trapped charges following the initial extinction, significant overvoltage well above 2 pu can result. It should be noted that a successful capacitor deenergizing (i.e., no restrike) does not normally yield any transient overvoltage.

Restrikes can occur multiple times and in succession over a short time period. It can also be intermittent over a longer time span. In either circumstance, a restrike poses problems to the power system as well as to the switching device itself. A potentially damaging capacitor restrike case is illustrated in Fig. 4.6 using a time-domain simulation model.

Consider a grounded-wye capacitor being deenergized. System and capacitor voltages along with current flowing in the capacitor are shown in Fig. 4.6. Following the first current extinction, the capacitor maintains a voltage of –1.0 pu because of the trapped charges. The first restrike occurs at the next immediate system peak voltage causing system and capacitor voltages to overshoot above 2.0 pu. The capacitor is re-energized because the arc has been re-established even though contacts are parting away. Certain types of circuit breakers may be able to interrupt the high-frequency current at its first zero-crossing following the reignition. The magnitude of capacitor voltage depends on where in time the second current extinction takes places. If it happens in one of the first few zero-crossing of the inrush current, the capacitor voltage would be higher compared to if it happens in the last few zero-crossings. A second restrike can occur, if the voltage potential difference across contacts exceeds the withstand-dielectric-breakdown voltage. This second restrike has the potential of causing an even higher overvoltage transient. As the contacts are parting even further, the subsequent current extinction is usually more successful, hence resulting in a capacitor deenergizing. Should current extinction be unsuccessful, fuses protecting individual capacitor units may blow.

Although worst-case restrikes occur at the system voltage peak, restrikes can occur anywhere between zero and peak voltage depending on the switch condition. If it occurs near voltage zero, the resulting overvoltage transient would be benign. A voltage waveform of a restrike during the opening of a 34.5-kV, 9.6-Mvar substation capacitor bank is shown in Fig. 4.7. The cause of the

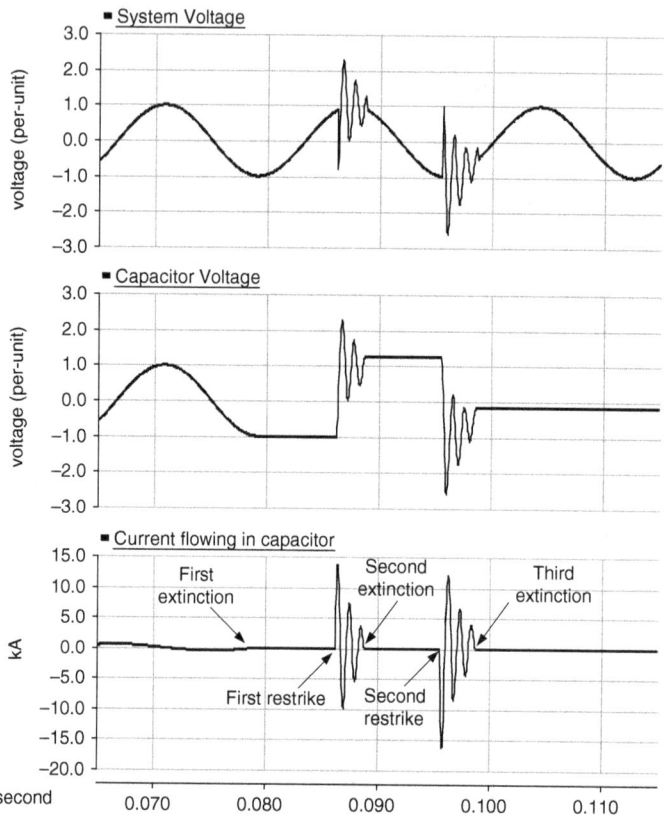

FIGURE 4.6   System and capacitor voltages along with capacitor current during a successive restrike while deenergizing the capacitor.

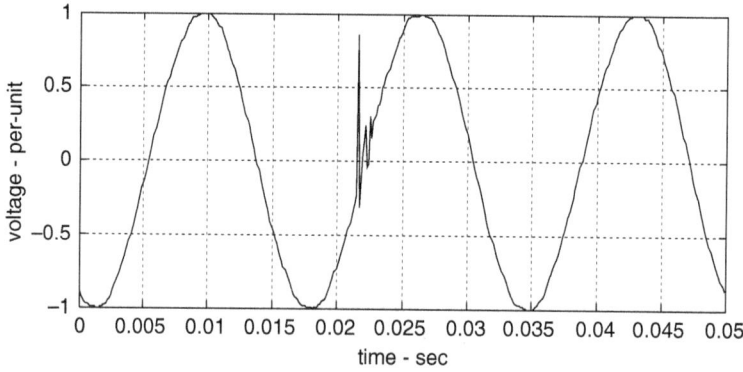

FIGURE 4.7   A voltage waveform during a restrike on opening of a 34.5-kV, 9.6-Mvar capacitor.

restrike is due to the build-up of burrs in the contacts of the circuit switcher. Power quality monitors have been used to detect the incidence of restrikes and mitigate the problem so as to avoid switch failures.

4.1.4 Lightning

Lightning is a potent source of impulsive transients. We will not devote space to the physical phenomena here because that topic is well documented in other reference books.[1-3] We will concentrate on how lightning causes transient overvoltages to appear on power systems.

Figure 4.8 illustrates some of the places where lightning can strike that results in lightning currents being conducted from the power system into loads.

The most obvious conduction path occurs during a direct strike to a phase wire, either on the primary or the secondary side of the transformer. This can generate very high overvoltages, but some analysts question whether this is the most common way that lightning surges enter load facilities and cause damage. Very similar transient overvoltages can be generated by lightning currents flowing along ground conductor paths. Note that there can be numerous paths for lightning currents to enter the grounding system. Common ones, indicated by the dotted lines in Fig. 4.8, include the primary ground, the secondary ground, and the structure of the load facilities. Note also that strikes to the primary phase are conducted to the ground circuits through the arresters on the service transformer. Thus, many more lightning impulses may be observed at loads than one might think.

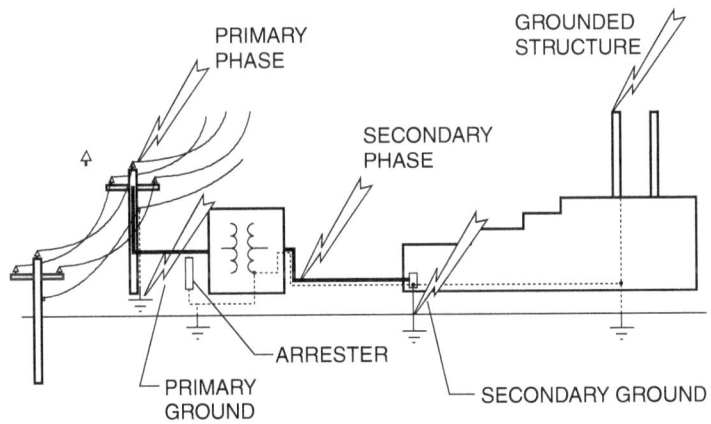

FIGURE 4.8 Lightning strike locations in which lightning impulses will be conducted into load facilities.

Keep in mind that grounds are never perfect conductors, especially for impulses. While most of the surge current may eventually be dissipated into the ground connection closest to the strike, there will be substantial surge currents flowing in other connected ground conductors in the first few microseconds of the strike.

A direct strike to a phase conductor generally causes line flashover near the strike point. Not only does this generate an impulsive transient, but it causes a fault with the accompanying voltage sags and interruptions. The lightning surge can be conducted a considerable distance along utility lines and cause multiple flashovers at pole and tower structures as it passes. The interception of the impulse from the phase wire is fairly straightforward if properly installed surge arresters are used. If the line flashes over at the location of the strike, the tail of the impulse is generally truncated. Depending on the effectiveness of the grounds along the surge current path, some of the current may find its way into load apparatus. Arresters near the strike may not survive because of the severe duty (most lightning strokes are actually many strokes in rapid-fire sequence).

Lightning does not have to actually strike a conductor to inject impulses into the power system. Lightning may simply strike near the line and induce an impulse by the collapse of the electric field. Lightning may also simply strike the ground near a facility causing the local ground reference to rise considerably. This may force currents along grounded conductors into a remote ground, possibly passing near sensitive load apparatus.

Many investigators in this field postulate that lightning surges enter loads from the utility system through the interwinding capacitance of the service transformer as shown in Fig. 4.9. The concept is that the lightning impulse is so fast that the inductance of the

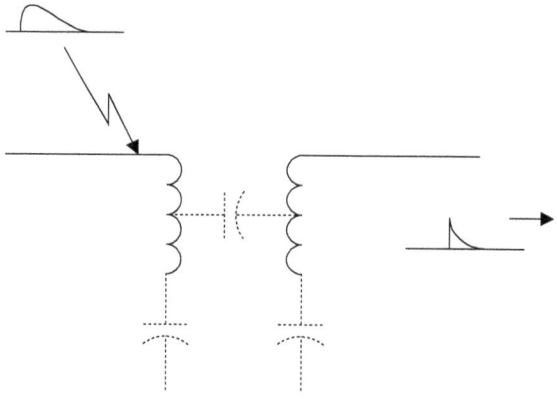

FIGURE 4.9 Coupling of impulses through the interwinding capacitance of transformers.

transformer windings blocks the first part of the wave from passing through by the turns ratio. However, the interwinding capacitance may offer a ready path for the high-frequency surge. This can permit the existence of a voltage on the secondary terminals that is much higher than what the turns ratio of the windings would suggest.

The degree to which capacitive coupling occurs is greatly dependent on the design of the transformer. Not all transformers have a straightforward high-to-low capacitance because of the way the windings are constructed. The winding-to-ground capacitance may be greater than the winding-to-winding capacitance, and more of the impulse may actually be coupled to ground than to the secondary winding. In any case, the resulting transient is a very short single impulse, or train of impulses, because the interwinding capacitance charges quickly. Arresters on the secondary winding should have no difficulty dissipating the energy in such a surge, but the rates of rise can be high. Thus, lead length becomes very important to the success of an arrester in keeping this impulse out of load equipment.

Many times, a longer impulse, which is sometimes oscillatory, is observed on the secondary when there is a strike to a utility's primary distribution system. This is likely due not to capacitive coupling through the service transformer but to conduction around the transformer through the grounding systems as shown in Fig. 4.10. This is a particular problem if the load system offers a better ground and much of the surge current flows through conductors in the load facility on its way to ground.

The chief power quality problems with lightning stroke currents entering the ground system are

1. They raise the potential of the local ground above other grounds in the vicinity by several kilovolts. Sensitive electronic equipment that is connected between two ground references, such as a computer connected to the telephone

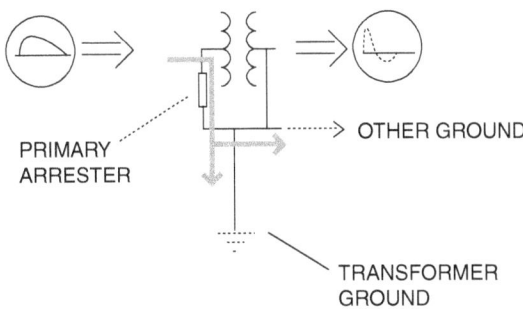

PRIMARY
ARRESTER

OTHER GROUND

TRANSFORMER
GROUND

Figure **4.10** Lightning impulse bypassing the service transformer through ground connections.

system through a modem, can fail when subjected to the lightning surge voltages.

2. They induce high voltages in phase conductors as they pass through cables on the way to a better ground.

The problems are related to the so-called low-side surge problem that is described in Sec. 4.5.3.

Ideas about lightning are changing with recent research.[10] Lightning causes more flashovers of utility lines than previously thought. Evidence is also mounting that lightning stroke current wavefronts are faster than previously thought and that multiple strikes appear to be the norm rather than the exception. Durations of some strokes may also be longer than reported by earlier researchers. These findings may help explain failures of lightning arresters that were thought to have adequate capacity to handle large lightning strokes.

4.1.5 Ferroresonance

The term *ferroresonance* refers to a special kind of resonance that involves capacitance and iron-core inductance. The most common condition in which it causes disturbances is when the magnetizing impedance of a transformer is placed in series with a system capacitor. This happens when there is an open-phase conductor. Under controlled conditions, ferroresonance can be exploited for useful purpose such as in a constant-voltage transformer (see Chap. 3).

Ferroresonance is different than resonance in linear system elements. In linear systems, resonance results in high sinusoidal voltages and currents of the resonant frequency. Linear-system resonance is the phenomenon behind the magnification of harmonics in power systems (see Chaps. 5 and 6). Ferroresonance can also result in high voltages and currents, but the resulting waveforms are usually irregular and chaotic in shape. The concept of ferroresonance can be explained in terms of linear-system resonance as follows.

Consider a simple series RLC circuit as shown in Fig. 4.11. Neglecting the resistance R for the moment, the current flowing in the circuit can be expressed as follows:

$$I = \frac{E}{j(X_L - |X_C|)}$$

where E = driving voltage
X_L = reactance of L
X_C = reactance of C

When $X_L = |X_C|$, a series-resonant circuit is formed, and the equation yields an infinitely large current that in reality would be limited by R.

An alternate solution to the series RLC circuit can be obtained by writing two equations defining the voltage across the inductor, i.e.,

$$v = jX_L I$$

$$v = E j|X_C|I$$

where v is a voltage variable. Figure 4.12 shows the graphical solution of these two equations for two different reactances, X_L and X_L'. X_L' represents the series-resonant condition. The intersection point between the capacitive and inductive lines gives the voltage across inductor E_L. The voltage across capacitor E_C is determined as shown in Fig. 4.12. At resonance, the two lines will intersect at infinitely large voltage and current since the $|X_C|$ line is parallel to the X_L' line.

Now, let us assume that the inductive element in the circuit has a nonlinear reactance characteristic like that found in transformer

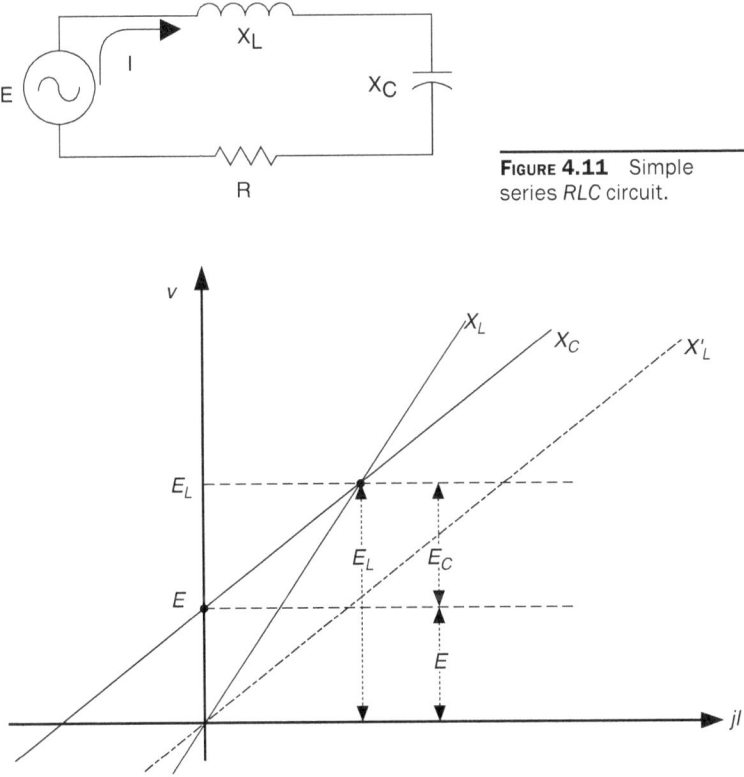

FIGURE **4.11** Simple series RLC circuit.

FIGURE **4.12** Graphical solution to the linear LC circuit.

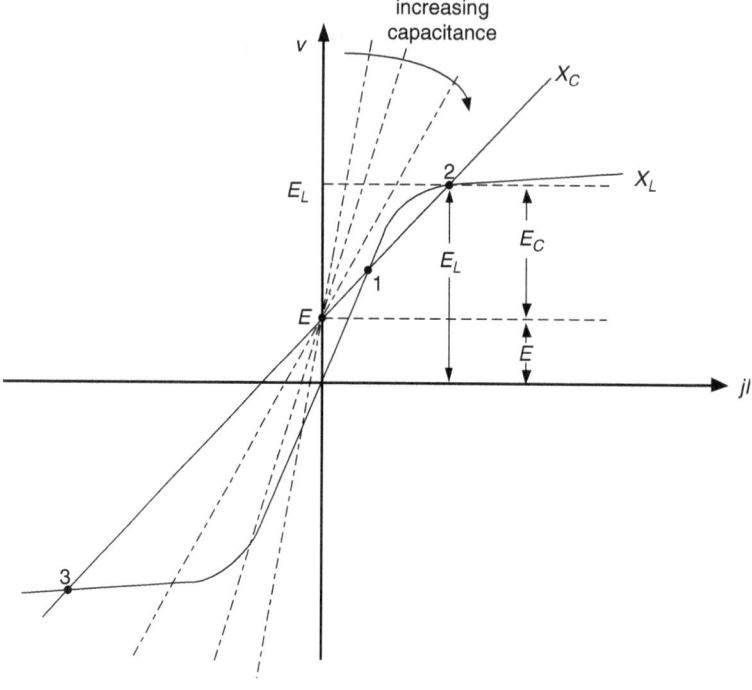

Figure 4.13   Graphical solution to the nonlinear *LC* circuit.

magnetizing reactance. Figure 4.13 illustrates the graphical solution of the equations following the methodology just presented for linear circuits. While the analogy cannot be made perfectly, the diagram is useful to help understand ferroresonance phenomena.

It is obvious that there may be as many as three intersections between the capacitive reactance line and the inductive reactance curve. Intersection 2 is an unstable solution, and this operating point gives rise to some of the chaotic behavior of ferroresonance. Intersections 1 and 3 are stable and will exist in the steady state. Intersection 3 results in high voltages and high currents.

Figures 4.14 and 4.15 show examples of ferroresonant voltages that can result from this simple series circuit. The same inductive characteristic was assumed for each case. The capacitance was varied to achieve a different operating point after an initial transient that pushes the system into resonance. The unstable case yields voltages in excess of 4.0 pu, while the stable case settles in at voltages slightly over 2.0 pu. Either condition can impose excessive duty on power system elements and load equipment.

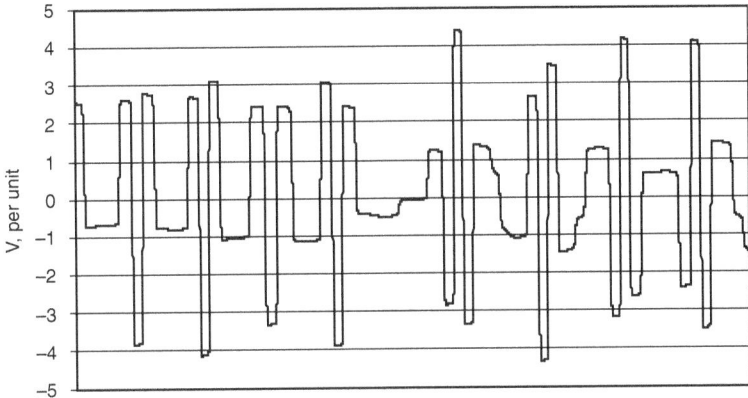

FIGURE 4.14   Example of unstable, chaotic ferroresonance voltages.

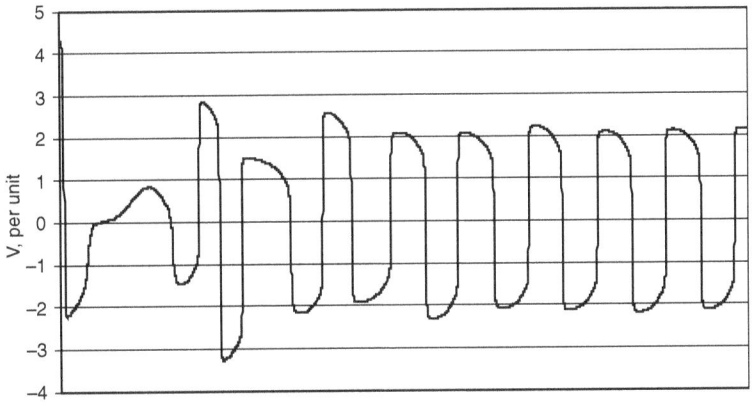

FIGURE 4.15   Example of ferroresonance voltages settling into a stable operating point (intersection 3) after an initial transient.

For a small capacitance, the $|X_C|$ line is very steep, resulting in an intersection point on the third quadrant only. This can yield a range of voltages from less than 1.0 pu to voltages like those shown in Fig. 4.15.

When C is very large, the capacitive reactance line will intersect only at points 1 and 3. One operating state is of low voltage and lagging current (intersection 1), and the other is of high voltage and leading current (intersection 3). The operating points during ferroresonance can oscillate between intersection points 1 and 3 depending on the applied voltage. Often, the resistance in the circuit prevents operation at point 3 and no high voltages will occur.

In practice, ferroresonance most commonly occurs when unloaded transformers become isolated on underground cables of a certain range of lengths. The capacitance of overhead distribution lines is generally insufficient to yield the appropriate conditions.

The minimum length of cable required to cause ferroresonance varies with the system voltage level. The capacitance of cables is nearly the same for all distribution voltage levels, varying from 40 to 100 nF per 1000 feet (ft), depending on conductor size. However, the magnetizing reactance of a 35-kV-class distribution transformer is several times higher (the curve is steeper) than a comparably sized 15-kV-class transformer. Therefore, damaging ferroresonance has been more common at the higher voltages. For delta-connected transformers, ferroresonance can occur for less than 100 ft of cable. For this reason, many utilities avoid this connection on cable-fed transformers. The grounded wye-wye transformer has become the most commonly used connection in underground systems in North America. It is more resistant, but not immune, to ferroresonance because most units use a three-legged or five-legged core design that couples the phases magnetically. It may require a minimum of several hundred feet of cable to provide enough capacitance to create a ferroresonant condition for this connection.

The most common events leading to ferroresonance are

- Manual switching of an unloaded, cable-fed, three-phase transformer where only one phase is closed (Fig. 4.16a). Ferroresonance may be noted when the first phase is closed upon energization or before the last phase is opened on deenergization.

- Manual switching of an unloaded, cable-fed, three-phase transformer where one of the phases is open (Fig. 4.16b). Again, this may happen during energization or deenergization.

- One or two riser-pole fuses may blow leaving a transformer with one or two phases open. Single-phase reclosers may also cause this condition. Today, many modern commercial loads have controls that transfer the load to backup systems when they sense this condition. Unfortunately, this leaves the transformer without any load to damp out the resonance.

It should be noted that these events do not always yield noticeable ferroresonance. Some utility personnel claim to have worked with underground cable systems for decades without seeing ferroresonance. System conditions that help increase the likelihood of ferroresonance include

- Higher distribution voltage levels, most notably 25- and 35-kV-class systems

- Switching of lightly loaded and unloaded transformers

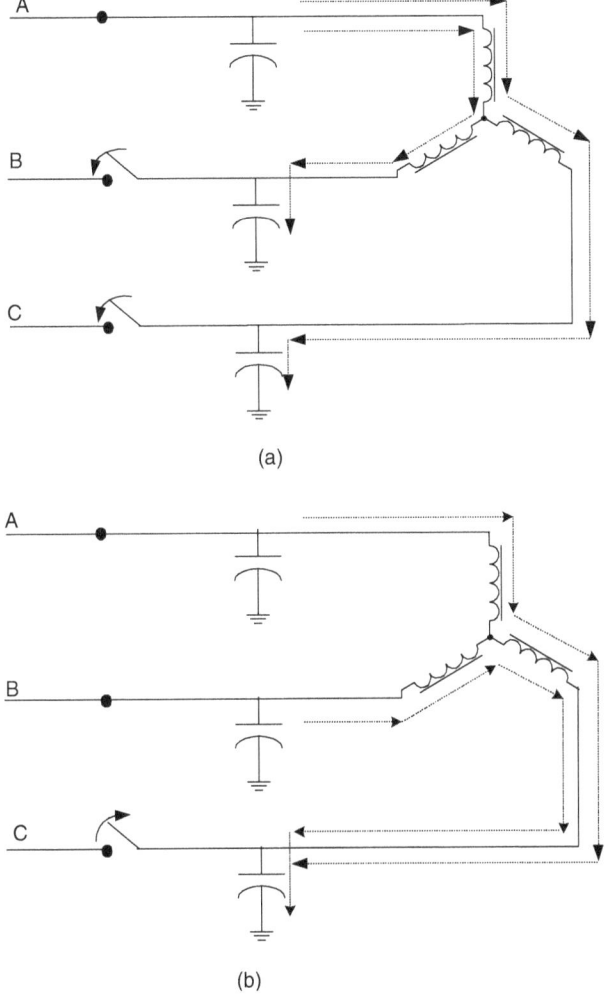

FIGURE 4.16 Common system conditions where ferroresonance may occur: (a) one phase closed, (b) one phase open.

- Ungrounded transformer primary connections
- Very lengthy underground cable circuits
- Cable damage and manual switching during construction of underground cable systems
- Weak systems, i.e., low short-circuit currents
- Low-loss transformers
- Three-phase systems with single-phase switching devices

While it is easier to cause ferroresonance at the higher voltage levels, its occurrence is possible at all distribution voltage levels. The proportion of losses, magnetizing reactance, and capacitance at lower levels may limit the effects of ferroresonance, but it can still occur.

There are several modes of ferroresonance with varying physical and electrical manifestations. Some have very high voltages and currents, while others have voltages close to normal. There may or may not be failures or other evidence of ferroresonance in the electrical components. Therefore, it may be difficult to tell if ferroresonance has occurred in many cases, unless there are witnesses or power quality measurement instruments.

Common indicators of ferroresonance are as follows.

Audible Noise

During ferroresonance, there may be an audible noise, often likened to that of a large bucket of bolts being shaken, whining, a buzzer, or an anvil chorus pounding on the transformer enclosure from within. The noise is caused by the magnetostriction of the steel core being driven into saturation. While difficult to describe in words, this noise is distinctively different and louder than the normal hum of a transformer. Most electrical system operating personnel are able to recognize it immediately upon first hearing it.

Overheating

Transformer overheating often, although not always, accompanies ferroresonance. This is especially true when the iron core is driven deep into saturation. Since the core is saturated repeatedly, the magnetic flux will find its way into parts of the transformer where the flux is not expected such as the tank wall and other metallic parts. The stray flux heating is often evidenced from the charring or bubbling of the paint on the top of the tank. This is not necessarily an indication that the unit is damaged, but damage can occur in this situation if ferroresonance has persisted sufficiently long to cause overheating of some of the larger internal connections. This may in turn damage solid insulation structures beyond repair. It should be noted that some transformers exhibiting signs of ferroresonance such as loud, chaotic noises do not show signs of appreciable heating. The design of the transformer and the ferroresonance mode determine how the transformer will respond.

High Overvoltages and Surge Arrester Failure

When overvoltages accompany ferroresonance, there could be electrical damage to both the primary and secondary circuits. Surge arresters are common casualties of the event. They are designed to intercept brief overvoltages and clamp them to an acceptable level. While they may be able to withstand several overvoltage events, there is a definite limit to their energy absorption capabilities. Low-voltage arresters in end-user facilities are more susceptible than utility arresters, and their failure is sometimes the only indication that ferroresonance has occurred.

Flicker

During ferroresonance the voltage magnitude may fluctuate wildly. End users at the secondary circuit may actually see their light bulbs flicker. Some electronic appliances may be very susceptible to such voltage excursions. Prolonged exposure can shorten the expected life of the equipment or may cause immediate failure. In facilities that transfer over to the UPS system in the event of utility-side disturbances, repeated and persistent sounding of the alarms on the UPS may occur as the voltage fluctuates.

4.1.6   Other Switching Transients

Line energization transients occur, as the term implies, when a switch is closed connecting a line to the power system. They generally involve higher-frequency content than capacitor energizing transients. The transients are a result of a combination of traveling-wave effects and the interaction of the line capacitance and the system equivalent source inductance. Traveling waves are caused by the distributed nature of the capacitance and inductance of the transmission or distribution line. Line energizing transients typically result in rather benign overvoltages at distribution voltage levels and generally do not cause any concern. It is very unusual to implement any kind of switching control for line energizing except for transmission lines operating at 345 kV and above. Line energizing transients usually die out in about 0.5 cycle.

The energization transients on distribution feeder circuits consist of a combination of line energizing transients, transformer energizing inrush characteristics, and load inrush characteristics. Figure 4.17 shows a typical case in which the monitor was located on the line side of the switch. The initial transient frequency is above 1.0 kHz and appears as a small amount of "hash" on the front of the waveform. Following the energization, the voltage displays noticeable distortion caused by the transformer inrush current that contains a number of low-order harmonic components, including the second and fourth harmonics. This is evidenced by the lack of symmetry in the voltage waveform in the few cycles recorded. This will eventually die out in nearly all cases. The first peak of the current waveform displays the basic characteristic of magnetizing inrush, which is subsequently swamped by the load inrush current.

Line energizing transients do not usually pose a problem for end-user equipment. Equipment can be protected from the high-frequency components with inductive chokes and surge protective devices if necessary. The example shown in Fig. 4.17 is relatively benign and should pose few problems. Cases with less load may exhibit much more oscillatory behavior.

Another source for overvoltages that is somewhat related to switching is the common single-line-to-ground fault. On a system with high, zero-sequence impedance, the sound phases will experience

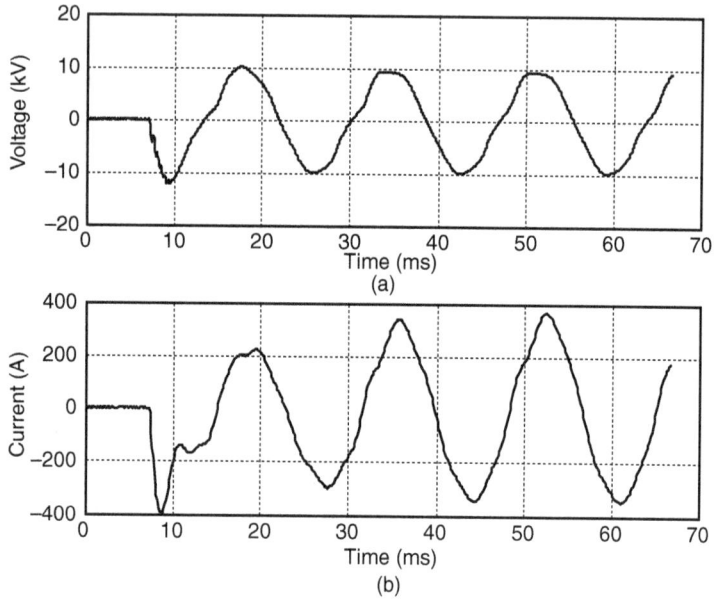

FIGURE **4.17**   Energizing a distribution feeder: (a) voltage and (b) current waveforms.

a voltage rise during the fault. The typical voltage rise on effectively grounded four-wire, multigrounded neutral systems is generally no more than 15 to 20 percent. On systems with neutral reactors that limit the fault current, for example, the voltage rise may reach 40 to 50 percent. This overvoltage is temporary and will disappear after the fault is cleared. These overvoltages are not often a problem, but there are potential problems if the fault clearing is slow:

- Some secondary arresters installed by end users attempt to clamp the voltage to as low as 110 percent voltage in the— perhaps mistaken—belief that this offers better insulation protection. Such arresters are subject to failure when conducting several cycles of power frequency current.

- Adjustable-speed-drive controls may presume a failure if the dc bus voltage goes too high and trips the machine.

- Distributed generation interconnected with the utility system will often interpret voltages in excess of 120 percent as warranting immediate disconnection (less than 10 cycles). Therefore, nuisance tripping is a likely result.

Of course, the actual impact of this overvoltage on the secondary side of the system depends heavily on the service transformer connection. While the common grounded wye-wye connection will transform the voltages directly, transformers with a delta connection will help protect the load from seeing overvoltages due to these faults.

4.2 Principles of Overvoltage Protection

The fundamental principles of overvoltage protection of load equipment are

1. Limit the voltage across sensitive insulation.
2. Divert the surge current away from the load.
3. Block the surge current from entering the load.
4. Bond grounds together at the equipment.
5. Reduce, or prevent, surge current from flowing between grounds.
6. Create a low-pass filter using limiting and blocking principles.

Figure 4.18 illustrates these principles, which are applied to protect from a lightning strike.

The main function of surge arresters and transient voltage surge suppressors (TVSSs) is to limit the voltage that can appear between two points in the circuit. This is an important concept to understand. One of the common misconceptions about varistors, and similar devices, is that they somehow are able to absorb the surge or divert it to ground independently of the rest of the system. That may be a

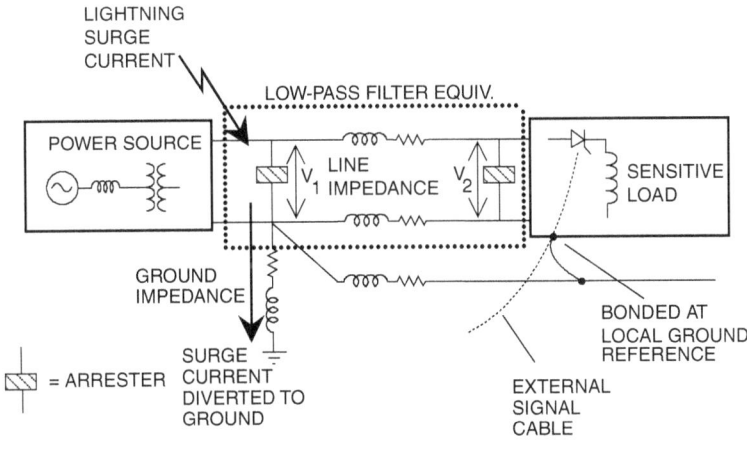

FIGURE 4.18 Demonstrating the principles of overvoltage protection.

beneficial side effect of the arrester application if there is a suitable path for the surge current to flow into, but the foremost concern in arrester application is to place the arresters directly across the sensitive insulation that is to be protected so that the voltage seen by the insulation is limited to a safe value. Surge currents, just like power currents, must obey Kirchoff's laws. They must flow in a complete circuit, and they cause a voltage drop in every conductor through which they flow.

One of the points to which arresters, or surge suppressors, are connected is frequently the local ground, but this need not be the case. Keep in mind that the local ground may not remain at zero potential during transient impulse events.

Surge suppression devices should be located as closely as possible to the critical insulation with a minimum of lead length on all terminals. While it is common to find arresters located at the main panels and subpanels, arresters applied at the point where the power line enters the load equipment are generally the most effective in protecting that particular load. In some cases, the best location is actually inside the load device. For example, many electronic controls made for service in the power system environment have protectors (MOV arresters, gaps, zener diodes, or surge capacitors) on every line that leaves the cabinet.

In Fig. 4.18 the first arrester is connected from the line to the neutral-ground bond at the service entrance. It limits the line voltage V_1 from rising too high relative to the neutral and ground voltage at the panel. When it performs its voltage-limiting action, it provides a low-impedance path for the surge current to travel onto the ground lead. Note that the ground lead and the ground connection itself have significant impedance. Therefore, the potential of the whole power system is raised with respect to that of the remote ground by the voltage drop across the ground impedance. For common values of surge currents and ground impedances, this can be several kilovolts.

One hopes, in this situation, that most of the surge energy will be discharged through the first arrester directly into ground. In that sense, the arrester becomes a surge "diverter." This is another important function related to surge arrester application. In fact, some prefer to call a surge arrester a surge diverter because its voltage-limiting action offers a low-impedance path around the load being protected. However, it can only be a diverter if there is a suitable path into which the current can be diverted. That is not always easy to achieve, and the surge current is sometimes diverted toward another critical load where it is not wanted.

In this figure, there is another possible path for the surge current—the signal cable indicated by the dotted line and bonded to the safety ground. If this is connected to another device that is referenced to ground elsewhere, there will be some amount of surge current flowing down the safety ground conductor. Damaging voltages can be impressed across the load as a result. The first arrester at the service entrance is electrically too remote to provide adequate load protection. Therefore, a second arrester is applied at the load— again, directly across the insulation to be protected. It is connected "line to neutral" so that it only protects against normal mode transients. This illustrates the principles without complicating the diagram but should be considered as the *minimum* protection one would apply to protect the load. Frequently, surge suppressors will have suppression on all lines to ground, all lines to neutral, and neutral to ground.

While lightning surge currents are seeking a remote ground reference, many transient overvoltages generated by switching will be those of a normal mode and will not seek ground. In cases where surge currents are diverted into other load circuits, arresters must be applied at each load along the path to ensure protection.

Note that the signal cable is bonded to the local ground reference at the load just before the cable enters the cabinet. It might seem that this creates an unwanted ground loop. However, it is essential to achieving protection of the load and the low-voltage signal circuits. Otherwise, the power components can rise in potential with respect to the signal circuit reference by several kilovolts. Many loads have multiple power and signal cables connected to them. Also, a load may be in an environment where it is close to another load and operators or sensitive equipment are routinely in contact with both loads. This raises the possibility that a lightning strike may raise the potential of one ground much higher than the others. This can cause a flashover across the insulation that is between the two ground references or cause physical harm to operators. Thus, all ground reference conductors (safety grounds, cable shields, cabinets, etc.) should be bonded together at the load equipment. The principle is not to prevent the local ground reference from rising in potential with the surge; with lightning, that is impossible. Rather, the principle is to tie the references together so that all power and signal cable references in the vicinity rise together.

This phenomenon is a common reason for failure of electronic devices. The situation occurs in TV receivers connected to cables, computers connected to modems, computers with widespread peripherals powered from various sources, and in manufacturing facilities with networked machines.

Since a few feet of conductor make a significant difference at lightning surge frequencies, it is sometimes necessary to create a special low-inductance, ground reference plane for sensitive electronic equipment such as mainframe computers that occupy large spaces.[4]

Efforts to block the surge current are most effective for high-frequency surge currents such as those originating with lightning strokes and capacitor-switching events. Since power frequency currents must pass through the surge suppressor with minimal additional impedance, it is difficult and expensive to build filters that are capable of discriminating between low-frequency surges and power frequency currents.

Blocking can be done relatively easily for high-frequency transients by placing an inductor, or choke, in series with the load. The high surge voltage will drop across the inductor. One must carefully consider that high voltage could damage the insulation of both the inductor and the loads. However, a line choke alone is frequently an effective means to block such high-frequency transients as line-notching transients from adjustable-speed drives.

The blocking function is frequently combined with the voltage-limiting function to form a *low-pass filter* in which there is a shunt-connected voltage-limiting device on either side of the series choke. Figure 4.18 illustrates how such a circuit naturally occurs when there are arresters on both ends of the line feeding the load. The line provides the blocking function in proportion to its length. Such a circuit has very beneficial overvoltage protection characteristics. The inductance forces the bulk of fast-rising surges into the first arrester. The second arrester then simply has to accommodate what little surge energy gets through. Such circuits are commonly built into outlet strips for computer protection.

Many surge-protection problems occur because the surge current travels between two, or more, separate connections to ground. This is a particular problem with lightning protection because lightning currents are seeking ground and basically divide according to the ratios of the impedances of the ground paths. The surge current does not even have to enter the power, or phase, conductors to cause problems. There will be a significant voltage drop along the ground conductors that will frequently appear across critical insulation. The grounds involved may be entirely within the load facility, or some of the grounds may be on the utility system.

Ideally, there would be only one ground path for lightning within a facility, but many facilities have multiple paths. For example, there may be a driven ground at the service entrance or substation transformer and a second ground at a water well that actually creates a better ground. Thus, when lightning strikes, the bulk of the surge current will tend to flow toward the well. This can impress an excessively high voltage across the pump insulation, even if the electrical system is not intentionally bonded to a second ground. When lightning strikes, the potentials can become so great that the power system insulation will flash over somewhere.

The amount of current flowing between the grounds may be reduced by improving all the intentional grounds at the service entrance and nearby on the utility system. This will normally reduce, but not eliminate entirely, the incidence of equipment failure within the facility due to lightning. However, some structures also have significant lightning exposure, and the damaging surge currents can flow back into the utility grounds. It doesn't matter which direction the currents flow; they cause the same problems. Again, the same principle applies, which is to improve the grounds for the structure to minimize the amount of current that might seek another path to ground.

When it is impractical to keep the currents from flowing between two grounds, *both* ends of any power or signal cables running between the two grounds must be protected with voltage-limiting devices to ensure adequate protection. This is common practice for both utility and end-user systems where a control cabinet is located quite some distance from the switch, or other device, being controlled.

4.3 Devices for Overvoltage Protection

4.3.1 Surge Arresters and Transient Voltage Surge Suppressors

Arresters and TVSSs protect equipment from transient overvoltages by limiting the maximum voltage, and the terms are sometimes used interchangeably. However, TVSSs are generally associated with devices used at the load equipment. A TVSS will sometimes have more surge-limiting elements than an arrester, which most commonly consists solely of MOV blocks. An arrester may have more energy-handling capability; however, the distinction between the two is blurred by common language usage.

The elements that make up these devices can be classified by two different modes of operation, *crowbar* and *clamping.*

Crowbar devices are normally open devices that conduct current during overvoltage transients. Once the device conducts, the line voltage will drop to nearly zero due to the short circuit imposed across the line. These devices are usually manufactured with a gap filled with air or a special gas. The gap arcs over when a sufficiently high overvoltage transient appears. Once the gap arcs over, usually power frequency current, or "follow current," will continue to flow in the gap until the next current zero. Thus, these devices have the disadvantage that the power frequency voltage drops to zero or to a very low value for at least one-half cycle. This will cause some loads to drop off-line unnecessarily.

Clamping devices for ac circuits are commonly nonlinear resistors (varistors) that conduct very low amounts of current until an overvoltage occurs. Then they start to conduct heavily, and their impedance drops rapidly with increasing voltage. These devices effectively conduct increasing amounts of current (and energy) to limit the voltage rise of a surge. They have an advantage over gap-type devices in that the voltage is not reduced below the conduction level when they begin to conduct the surge current. Zener diodes are also used in this application. Example characteristics of MOV arresters for load systems are shown in Figs. 4.19 and 4.20.

MOV arresters have two important ratings. The first is maximum continuous operating voltage (MCOV), which must be higher than the line voltage and will often be at least 125 percent of the system nominal voltage. The second rating is the energy dissipation rating (in joules). MOVs are available in a wide range of energy ratings. Figure 4.20 shows the typical energy-handling capability versus operating voltages.

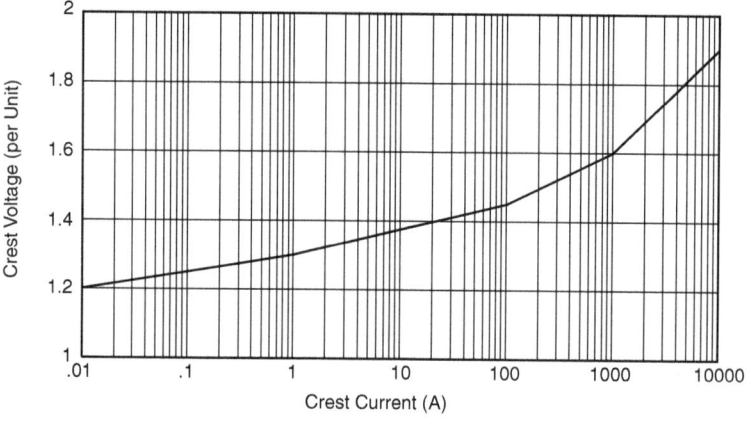

FIGURE 4.19   Crest voltage versus crest amps.

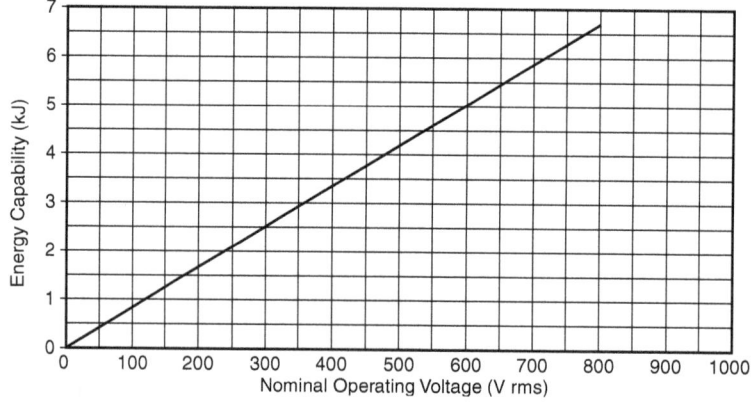

FIGURE 4.20   Energy capability versus operating voltage.

4.3.2   Isolation Transformers

Figure 4.21 shows a diagram of an isolation transformer used to attenuate high-frequency noise and transients as they attempt to pass from one side to the other. However, some common-mode and normal-mode noise can still reach the load. An electrostatic shield, as shown in Figure 4.22, is effective in eliminating common-mode noise. However, some normal-mode noise can still reach the load due to magnetic and capacitive coupling.

The chief characteristic of isolation transformers for electrically isolating the load from the system for transients is their leakage inductance. Therefore, high-frequency noise and transients are kept from reaching the load, and any load-generated noise and transients are kept from reaching the rest of the power system. Voltage notching due to power electronic switching is one example of a problem that can be limited to the load side by an isolation transformer. Capacitor-switching and lightning transients coming from the utility system

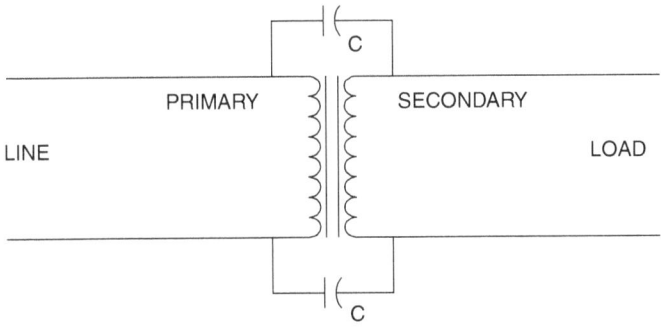

Figure 4.21   Isolation transformer.

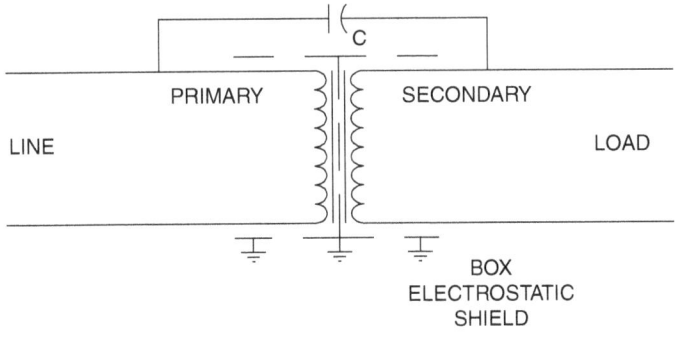

Figure 4.22   Isolation transformer with electrostatic shield.

can be attenuated, thereby preventing nuisance tripping of adjustable-speed drives and other equipment.

An additional use of isolation transformers is that they allow the user to define a new ground reference, or *separately derived system*. This new neutral-to-ground bond limits neutral-to-ground voltages at sensitive equipment.

4.3.3 Low-Pass Filters

Low-pass filters use the pi-circuit principle illustrated in Fig. 4.18 to achieve even better protection for high-frequency transients. For general usage in electric circuits, low-pass filters are composed of series inductors and parallel capacitors. This LC combination provides a low-impedance path to ground for selected resonant frequencies. In surge protection usage, voltage clamping devices are added in parallel to the capacitors. In some designs, there are no capacitors.

Figure 4.23 shows a common hybrid protector that combines two surge suppressors and a low-pass filter to provide maximum protection. It uses a gap-type protector on the front end to handle high-energy transients. The low-pass filter limits transfer of high-frequency transients. The inductor helps block high-frequency transients and forces them into the first suppressor. The capacitor limits the rate of rise, while the nonlinear resistor (MOV) clamps the voltage magnitude at the protected equipment.

Other variations on this design will employ MOVs on both sides of the filters and may have capacitors on the front end as well.

4.3.4 Low-Impedance Power Conditioners

Low-impedance power conditioners (LIPCs) are used primarily to interface with the switch-mode power supplies found in electronic equipment. LIPCs differ from isolation transformers in that these conditioners have a much lower impedance and have a filter as part of their design (Fig. 4.24). The filter is on the output side and protects against high-frequency, source-side, common-mode, and normal-mode disturbances (i.e., noise and impulses). Note the new neutral-to-ground connection that can be made on the load side because of

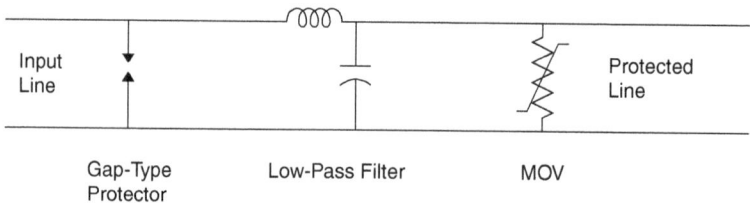

| Input Line | | Low-Pass Filter | | Protected Line |

Gap-Type Protector Low-Pass Filter MOV

FIGURE 4.23 Hybrid transient protector.

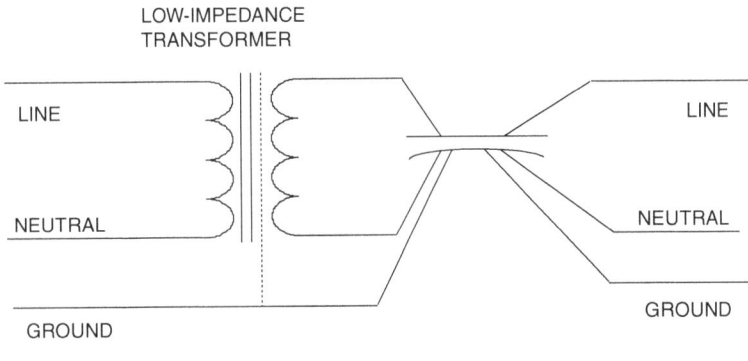

LOW-IMPEDANCE
TRANSFORMER

LINE

NEUTRAL

GROUND

LINE

NEUTRAL

GROUND

FIGURE **4.24** Low-impedance power conditioner.

the existence of an isolation transformer. However, low- to medium-frequency transients (capacitor switching) can cause problems for LIPCs: The transient can be magnified by the output filter capacitor.

4.3.5 Utility Surge Arresters

The three most common surge arrester technologies employed by utilities are depicted in Fig. 4.25. Most arresters manufactured today use a MOV as the main voltage-limiting element. The chief ingredient of a MOV is zinc oxide (ZnO), which is combined with several proprietary ingredients to achieve the necessary characteristics and durability. Older-technology arresters, of which there are still many installed on the power system, used silicon carbide (SiC) as the energy-dissipating nonlinear resistive element. The relative discharge voltages for each of these three technologies are shown in Fig. 4.26.

Originally, arresters were little more than spark gaps, which would result in a fault each time the gap sparked over. Also, the sparkover transient injected a very steep fronted voltage wave into the apparatus being protected, which was blamed for many insulation failures. The addition of an SiC nonlinear resistance in series with a spark gap corrected some of these difficulties. It allowed the spark gap to clear and reseal without causing a fault and reduced the sparkover transient to perhaps 50 percent of the total sparkover voltage (Fig. 4.26a). However, insulation failures were still blamed on this front-of-wave transient. Also, there is substantial power-follow current after sparkover, which heats the SiC material and erodes the gap structures, eventually leading to arrester failures or loss of protection.

Gaps are necessary with the SiC because an economical SiC element giving the required discharge voltage is unable to withstand continuous system operating voltage. The development of MOV technology enabled the elimination of the gaps. This technology could withstand continuous system voltage without gaps and still

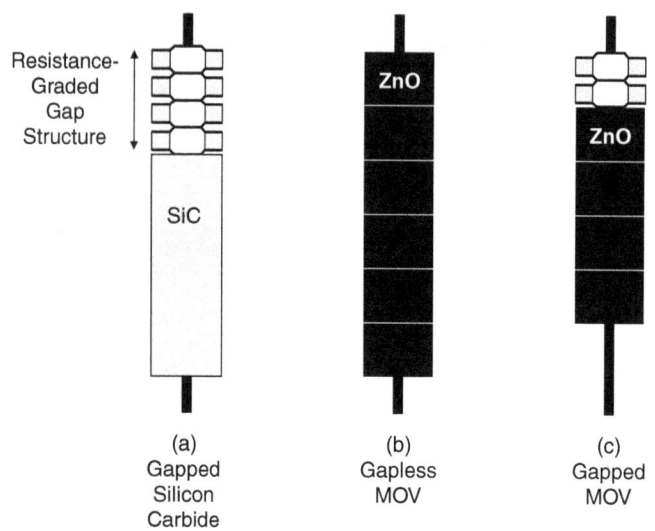

FIGURE **4.25** Three common utility surge arrester technologies.

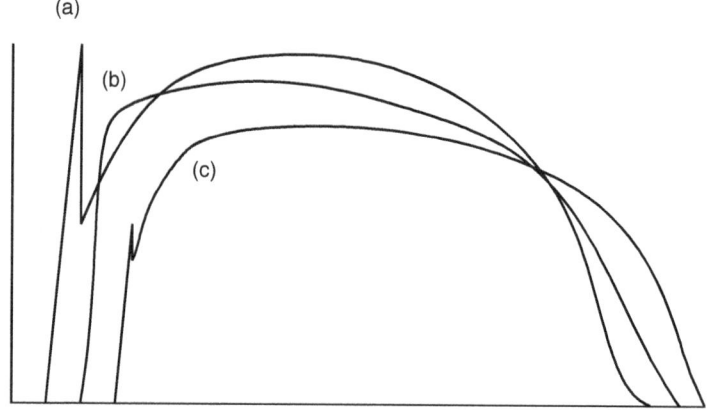

FIGURE **4.26** Comparative lightning wave discharge voltage characteristics for an 8 × 20 μs wave corresponding to the utility surge arrester technologies in Fig. 4.25.

provide a discharge voltage comparable to the SiC arresters (see Fig. 4.26*b*). By the late 1980s, SiC arrester technology was being phased out in favor of the gapless MOV technology. The gapless MOV provided a somewhat better discharge characteristic without the objectionable sparkover transient. The majority of utility distribution arresters manufactured today are of this design.

The gapped MOV technology was introduced commercially about 1990 and has gained acceptance in some applications where there is need for increased protective margins. By combining resistance-graded gaps (with SiC grading rings) and MOV blocks, this arrester technology has some very interesting, and counterintuitive, characteristics. It has a lower lightning-discharge voltage (Fig. 4.26c), but has a higher transient overvoltage withstand characteristic than a gapless MOV arrester. To achieve the required protective level for lightning, gapless MOV arresters typically begin to conduct heavily for low-frequency transients at about 1.7 pu. There are some system conditions where the switching transients will exceed this value for several cycles and cause failures. Also, applications such as aging underground cable systems demand lower lightning-discharge characteristics.

The gapped MOV technology removes about one-third of the MOV blocks and replaces them with a gap structure having a lightning sparkover approximately one-half of the old SiC technology. The smaller number of MOV blocks yields a lightning-discharge voltage typically 20 to 30 percent less than a gapless MOV arrester. Because of the capacitive and resistive interaction of the grading rings and MOV blocks, most of the front-of-wave impulse voltage of lightning transients appears across the gaps. They spark over very early into the MOV blocks, yielding a minor sparkover transient on the front.[5] For switching transients, the voltage divides by resistance ratios and most of it appears first across the MOV blocks, which hold off conduction until the gaps spark over. This enables this technology to achieve a transient overvoltage withstand of approximately 2.0 pu in typical designs. Additionally, the energy dissipated in the arrester is less than dissipated by gapless designs for the same lightning current because of the lower voltage discharge of the MOV blocks. There is no power-follow current because there is sufficient MOV capability to block the flow. This minimizes the erosion of the gaps. In several ways, this technology holds the promise of yielding a more capable and durable utility surge arrester.

Utility surge arresters are manufactured in various sizes and ratings. The three basic rating classes are designated distribution, intermediate, and station in increasing order of their energy-handling capability. Most of the arresters applied on primary distribution feeders are distribution class. Within this class, there are both small-block and heavy-duty designs. One common exception to this is that sometimes intermediate- or station-class arresters are applied at riser poles to obtain a better protective characteristic (lower discharge voltage) for the cable.

4.4   Utility Capacitor-Switching Transients

This section describes how utilities can deal with problems related to capacitor-switching transients.

4.4.1 Switching Times

Capacitor-switching transients are very common and usually not damaging. However, the timing of switching may be unfortunate for some sensitive industrial loads. For example, if the load picks up the same time each day, the utility may decide to switch the capacitors coincident with that load increase. There have been several cases where this coincides with the beginning of a work shift and the resulting transient causes several adjustable-speed drives to shut down shortly after the process starts. One simple and inexpensive solution is to determine if there is a switching time that might be more acceptable. For example, it may be possible to switch on the capacitor a few minutes before the beginning of the shift and before the load actually picks up. It may not be needed then, but probably will not hurt anything. If this cannot be worked out, other, more expensive solutions will have to be found.

4.4.2 Preinsertion Resistors

Preinsertion resistors, also known as closing resistors, can reduce the capacitor-switching transient considerably. The first peak of the transient is usually the most damaging. The idea is to insert a resistor into the circuit briefly so that the first peak is damped significantly. This is an established technology and is quite effective.

Figure 4.27 shows one example of a capacitor switch with preinsertion resistors to reduce transients. The preinsertion is accomplished by the movable contacts sliding past the resistor contacts first before mating with the main contacts. As a result, the resistor is part of the circuit for about 4 to 15 μs (or 30 to 90 percent of a 60-Hz cycle.[17] The equivalent circuit describing the action of preinsertion is illustrated in Fig. 4.28.

When a movable contact meets the resistor contact, the resistor is in the circuit (S1 is closed, while S2 is open). When the movable contact mates with the main contact, S2 contact is closed which essentially removes the preinsertion resistor from the 60 hertz (Hz) circuit.[17] The effectiveness of the resistors is dependent on capacitor size and available short-circuit current at the capacitor location. Table 4.1 shows expected maximum transient overvoltages upon energization for various conditions, both with and without the preinsertion resistors. These are the maximum values expected; average values are typically 1.3 to 1.4 pu without resistors and 1.1. to 1.2 pu with resistors.

Figure 4.29 illustrates a simulated voltage waveform corresponding to the energizing of a 115-kV, 42-Mvar capacitor using switches without preinsertion resistors. The peak transient overvoltage is 1.8 pu.

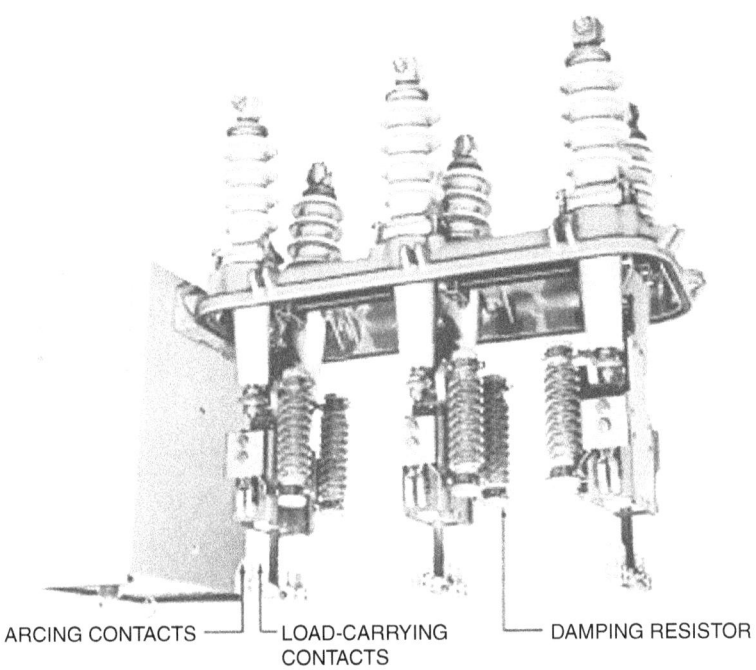

ARCING CONTACTS ——— └—LOAD-CARRYING └— DAMPING RESISTOR
CONTACTS

FIGURE 4.27 Capacitor switch with preinsertion resistors. (*Courtesy of Cooper Power systems.*)

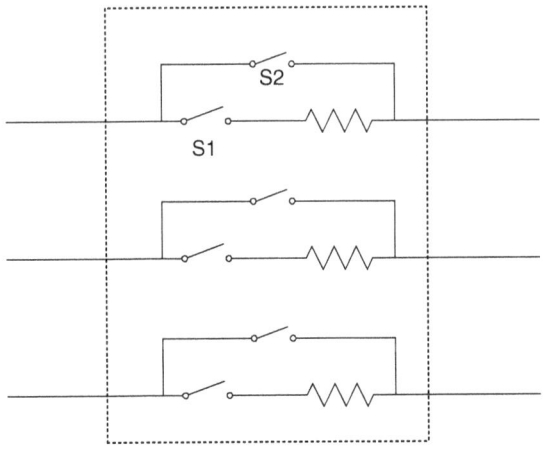

FIGURE 4.28 An equivalent circuit of switches with preinsertion resistors.

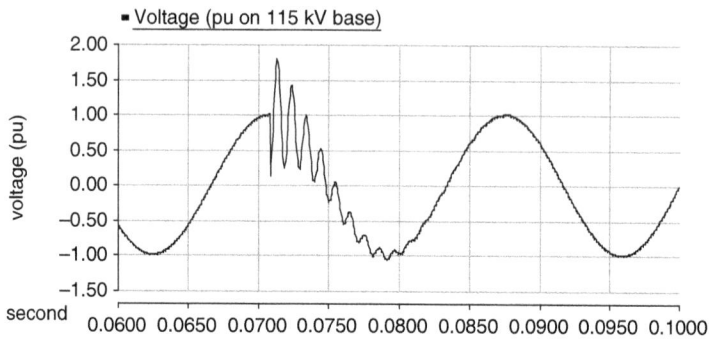

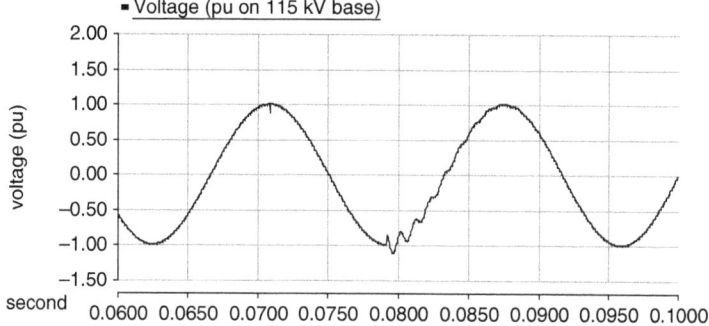

FIGURE 4.29   Simulated voltage waveforms during capacitor energizing using switches without (top) and with (bottom) preinsertion resistors.

Size, kvar	Available short circuit, kA	Without resistor (pu)	With 6.4Ω resistor (pu)
900	4	1.95	1.55
900	9	1.97	1.45
900	14	1.98	1.39
1200	4	1.94	1.50
1200	9	1.97	1.40
1200	14	1.98	1.34
1800	4	1.92	1.42
1800	9	1.96	1.33
1800	14	1.97	1.28

Courtesy Cooper Power Systems

TABLE 4.1   Peak Transient Overvoltages due to Capacitor Switching With and Without Preinsertion Resistor

A common size of preinsertion for this application is between 70 and 80 Ω.[18] Using 80-Ω preinsertion resistors and with preinsertion time of one-half of a cycle, the peak transient overvoltage is reduced to 0.13 pu. In addition to reducing peak transient overvoltages, a preinsertion resistor also reduces the capacitive inrush current during energizing significantly. This is especially useful during back-to-back energizing where inrush currents are typically very high.

Switches with preinsertion reactors have also been developed for this purpose. The inductor is helpful in limiting the higher-frequency components of the transient. In some designs, the reactors are intentionally built with high resistance so that they appear lossy to the energization transient. This helps the transient damp out quickly.

4.4.3 Synchronous Closing

Another popular strategy for reducing transients on capacitor switching is to use a synchronous closing breaker. Synchronous closing prevents transients by timing the contact closure such that the system voltage closely matches the capacitor voltage at the instant the contacts mate. This avoids the step change in voltage that normally occurs when capacitors are energized, causing the circuit to oscillate. A successful synchronous closing requires the switch to maintain a sufficient dielectric strength to withstand the system voltage until its contacts touch. Otherwise, a pre-strike would defeat the purpose of controlled closing.

For a grounded-wye capacitor, the control objective for each phase is to be energized at voltage zero. For an ungrounded-wye capacitor, the first phase is energized at voltage zero, the second phase is energized when its voltage is identical to the voltage of the first phase, and finally, the third phase is energized at voltage zero. A small timing error (± 1 ms) is acceptable as long as the peak overvoltage transient is below 1.2 pu. Figure 4.30 shows simulated voltage waveforms due to capacitor energizing. The first phase (top) is energized with switches having no closing control, while the second and third phases are with controlled closing. The timing error for the second phase is negligible resulting in minimal transient since the instantaneous system voltage matches the second-phase capacitor voltage. The timing error for the third phase is 1 ms resulting in a noticeable oscillatory transient voltage with a peak of 1.12 pu.

Actual three-phase voltage and current waveforms of a 150-Mvar, 345-kV grounded-wye capacitor are shown in Fig. 4.31. The capacitor was energized using switches equipped with synchronous closing control. The timing errors are well below 1 ms resulting in negligible oscillatory voltage transient. Since a power quality monitor is installed immediately upstream from the circuit switchers, inrush currents were clearly recorded as well (bottom). The peak inrush currents in all three phases are remarkably low (below 1 kA peak) because the instantaneous voltage at the switching instant is near

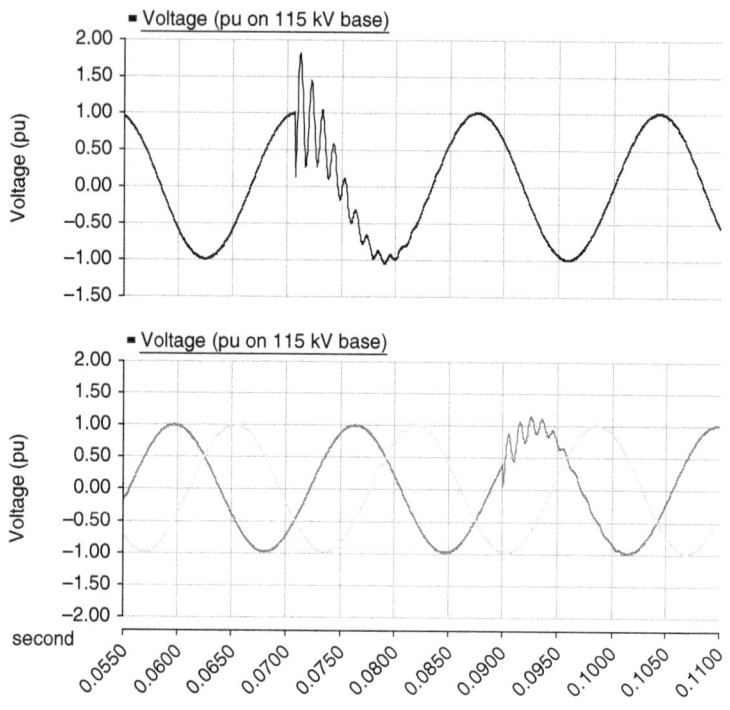

FIGURE 4.30   Simulated voltage waveforms during capacitor energizing using switches equipped without (top) and with (bottom) synchronous closing control.

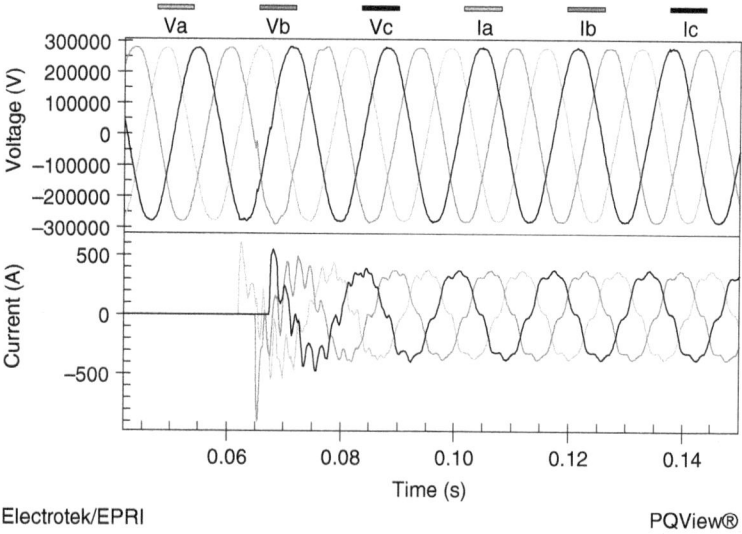

Electrotek/EPRI   PQView®

FIGURE 4.31   Actual voltage (top) and inrush current (b) waveforms during capacitor energizing using switches equipped synchronous closing control.

zero. Hence, switches with synchronous closing control reduce both oscillatory voltage transients and inrush currents during energizing significantly.

Figure 4.32 shows one example of a circuit breaker designed for this purpose. This breaker would normally be applied on the utility subtransmission or transmission system (72- and 145-kV classes). This is a three-phase SF_6 breaker that uses a specially designed operating mechanism with three independently controllable drive rods. It is capable of closing within 1 ms of voltage zero. The electronic control samples variables such as ambient temperature, control voltage, stored energy, and the time since the last operation to compensate the algorithms for the timing forecast. The actual performance of the breaker is sampled to adjust the pole timing for future operations to compensate for wear and changes in mechanical characteristics.

Figure 4.33 shows a vacuum switch made for this purpose. It is applied on 46-kV-class capacitor banks. It consists of three independent poles with separate controls. The timing for synchronous closing is determined by anticipating an upcoming voltage zero. Its success is dependent on the consistent operation of the vacuum switch. The switch reduces capacitor inrush currents by an order of magnitude and voltage transients to about 1.1 pu. A similar switch may also be used at distribution voltages.

Figure 4.34 shows one phase of a three-phase synchronous switch used for distribution capacitor banks. This particular technology uses a vacuum switch encapsulated in a solid dielectric.

Figure 4.32 Synchronous closing breaker. (*Courtesy of ABB, Inc.*)

Figure 4.33   Synchronous closing capacitor switch. (*Courtesy of Joslyn Hi-Voltage Corporation.*)

Figure 4.34   One pole of a synchronous closing switch for distribution capacitor banks. (*Courtesy of Cooper Power Systems.*)

Each of the switches described here requires a sophisticated microprocessor-based control. Understandably, a synchronous closing system is more expensive than a straightforward capacitor switch. However, it is frequently a cost-effective solution when capacitor-switching transients are disrupting end-user loads.

4.4.4 Capacitor Location

For distribution feeder banks, a switched capacitor may be too close to a sensitive load or at a location where the transient overvoltages tend to be much higher. Often, it may be possible to move the capacitor downline or to another branch of the circuit and eliminate the problem. The strategy is either to create more damping with more resistance in the circuit or to get more impedance between the capacitor and the sensitive load.

The success of this strategy will depend on a number of factors. Of course, if the capacitor is placed at a large load to supply reactive power specifically for that load, moving the bank may not be an option. Then, techniques for soft switching or switching at noncritical times must be explored. Besides utility-side solutions, one should also explore load-side solutions. In some cases, it will be more cost-effective to harden load equipment against capacitor-switching transients by the application of line chokes, TVSSs, etc.

4.5 Utility System Lightning Protection

Many power quality problems stem from lightning. Not only can the high-voltage impulses damage load equipment, but the temporary fault that follows a lightning strike to the line causes voltage sags and interruptions. Here are some strategies for utilities to use to decrease the impact of lightning.

4.5.1 Shielding

One of the strategies open to utilities for lines that are particularly susceptible to lightning strikes is to shield the line by installing a grounded neutral wire over the phase wires. This will intercept most lightning strokes before they strike the phase wires. This can help, but will not necessarily prevent line flashovers because of the possibility of backflashovers.

Shielding overhead utility lines is common at transmission voltage levels and in substations, but is not common on distribution lines because of the added cost of taller poles and the lower benefit due to lower flashover levels of the lines. On distribution circuits, the grounded neutral wire is typically installed underneath the phase conductors to facilitate the connection of line-to-neutral connected equipment such as transformers and capacitors.

Shielding is not quite as simple as adding a wire and grounding it every few poles. When lightning strikes the shield wire, the voltages at

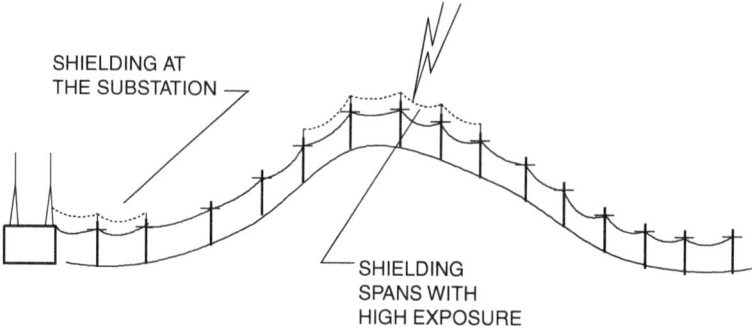

SHIELDING AT
THE SUBSTATION

SHIELDING
SPANS WITH
HIGH EXPOSURE

FIGURE 4.35 Shielding a portion of a distribution feeder to reduce the incidence of temporary lightning-induced faults.

the top of the pole will still be extremely high and could cause backflashovers to the line. This will result in a temporary fault. To minimize this possibility, the path of the ground lead down the pole must be carefully chosen to maintain adequate clearance with the phase conductors. Also, the grounding resistance plays an important role in the magnitude of the voltage and must be maintained as low as possible.

However, when it becomes obvious that a particular section of feeder is being struck frequently, it may be justifiable to retrofit that section with a shield wire to reduce the number of transient faults and to maintain a higher level of power quality. Figure 4.35 illustrates this concept. It is not uncommon for a few spans near the substation to be shielded. The substation is generally shielded anyway, and this helps prevent high-current faults close to the substation that can damage the substation transformer and breakers. It is also common near substations for distribution lines to be underbuilt on transmission or subtransmission structures. Since the transmission is shielded, this provides shielding for the distribution as well, provided adequate clearance can be maintained for the ground lead. This is not always an easy task.

Another section of the feeder may crest a ridge giving it unusual exposure to lightning. Shielding in that area may be an effective way of reducing lightning-induced faults. Poles in the affected section may have to be extended to accommodate the shield wire and considerable effort put into improving the grounds. This increases the cost of this solution. It is possible that line arresters would be a more economical and effective option for many applications.

4.5.2 Line Arresters

Another strategy for lines that are struck frequently is to apply arresters periodically along the phase wires. Normally, lines flash over first at the pole insulators. Therefore, preventing insulator flashover will reduce the interruption and sag rate significantly.

Stansberry[6] argues that this is more economical than shielding and results in fewer line flashovers. Neither shielding nor line arresters will prevent all flashovers from lightning. The aim is to significantly reduce flashovers in particular trouble spots.

As shown in Fig. 4.36, the arresters bleed off some of the stroke current as it passes along the line. The amount that an individual arrester bleeds off will depend on the grounding resistance. The idea is to space the arresters sufficiently close to prevent the voltage at unprotected poles in the middle from exceeding the basic impulse insulation level (BIL) of the line insulators. This usually requires an arrester at every second or third pole. In the case of a feeder supplying a highly critical load, or a feeder with high ground resistance, it may be necessary to place arresters at every pole. A transients study of different configurations will show what is required.

Some utilities place line arresters only on the top phase when one phase is mounted higher than the others. In other geometries, it will be necessary to put arresters on all three phases to achieve a consistent reduction in flashovers.

Figure 4.37 shows a typical utility arrester that is used for overhead line protection applications. This model consists of MOV blocks

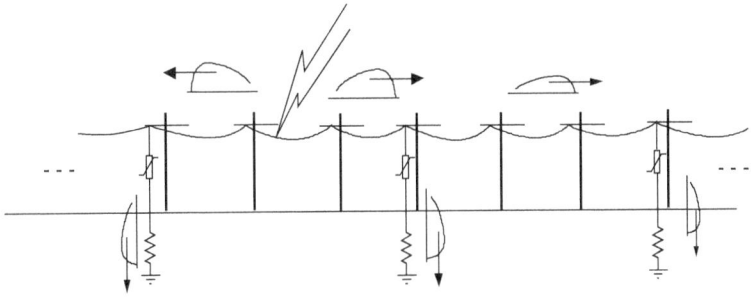

FIGURE 4.36 Periodically spaced line arresters help prevent flashovers.

FIGURE 4.37 Typical polymer-housed utility distribution arrester for overhead line applications. (*Courtesy of Cooper Power Systems.*)

encapsulated in a polymer housing that is resistant to sunlight and other natural elements. Older-technology models used porcelain housings like that shown on the primary side of the transformer in Fig. 4.39.

There are already sufficient arresters on many lines in densely populated areas in North America to achieve sufficient line protection. These arresters are on the distribution transformers, which are installed close together and in sufficient numbers in these areas to help protect the lines from flashover.

4.5.3   Low-Side Surges

Some utility and end-user problems with lightning impulses are closely related. One of the most significant ones is called the "low-side surge" problem by many utility engineers.[7] The name was coined by distribution transformer designers because it appears from the transformer's perspective that a current surge is suddenly injected into the low-voltage side terminals. Utilities have not applied secondary arresters at low-voltage levels in great numbers. From the customer's point of view it appears to be an impulse coming from the utility and is likely to be termed a *secondary surge*.

Both problems are actually different side effects of the same surge phenomenon—lightning current flowing from either the utility side or the customer side along the service cable neutral. Figure 4.38 shows one possible scenario. Lightning strikes the primary line, and the current is discharged through the primary arrester to the pole ground lead. This lead is also connected to the X2 bushing of the transformer at the top of the pole. Thus, some of the current will flow toward the load ground. The amount of current into the load ground is primarily

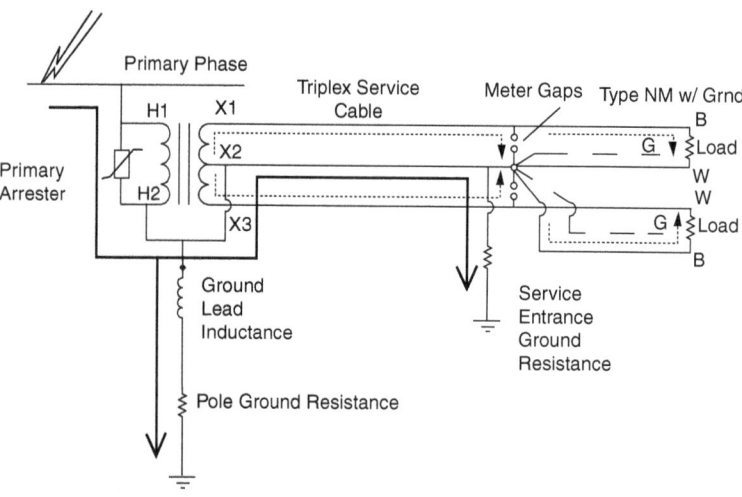

FIGURE 4.38   Primary arrester discharge current divides between pole and load ground.

dependent on the size of the pole ground resistance relative to the load ground. Inductive elements may play a significant role in the current division for the front of the surge, but the ground resistances basically dictate the division of the bulk of the stroke current.

The current that flows through the secondary cables causes a voltage drop in the neutral conductor that is only partially compensated by mutual inductive effects with the phase conductors. Thus, there is a net voltage across the cable, forcing current through the transformer secondary windings and into the load as shown by the dashed lines in Fig. 4.38. If there is a complete path, substantial surge current will flow. As it flows through the transformer secondary, a surge voltage is induced in the primary, sometimes causing a layer-to-layer insulation failure near the grounded end. If there is not a complete path, the voltage will build up across the load and may flash over somewhere on the secondary. It is common for the meter gaps to flash over, but not always before there is damage on the secondary because the meter gaps are usually 6 to 8 kV, or higher.

The amount of voltage induced in the cable is dependent on the rate of rise of the current, which is dependent on other circuit parameters as well as the lightning stroke.

The chief power quality problems this causes are

1. The impulse entering the load can cause failure or misoperation of load equipment.
2. The utility transformer will fail causing an extended power outage.
3. The failing transformer may subject the load to sustained steady-state overvoltages because part of the primary winding is shorted, decreasing the transformer turns ratio. Failure usually occurs in seconds but has been known to take hours.

The key to this problem is the amount of surge current traveling through the secondary service cable. Keep in mind that the same effect occurs regardless of the direction of the current. All that is required is for the current to get into the ground circuits and for a substantial portion to flow through the cable on its way to another ground. Thus, lightning strikes to either the utility system or the end-user facilities have the same effects. Transformer protection is more of an issue in residential services, but the secondary transients will appear in industrial systems as well.

Protecting the Transformer

There are two common ways for the utility to protect the transformer against low-side surges:

1. Use transformers with interlaced secondary windings.
2. Apply surge arresters at the X terminals in addition to surge arresters at the H-terminals.

Of course, the former is a design characteristic of the transformer and cannot be changed once the transformer has been made. If the transformer is a noninterlaced design, the only option is to add arresters to the low (X) voltage terminals.

Most distribution transformers are protected with utility arresters on the H side, and most are probably not protected on the X side. In lightning-prone areas, the failure rate of distribution transformer can be reduced further by X-side arresters. If the annual failure rate is 0.5 percent or higher, X-side arresters will likely be economically justifiable. A couple of field trials showed that a completely protected transformer would reduce failure rates to 0.1 percent or less.

Note that arresters at the load service entrance will not protect the transformer. In fact, they will virtually guarantee that there will be a surge current path and thereby cause additional stress on the transformer.

While interlaced transformers have a lower failure rate in lightning-prone areas than noninterlaced transformers, recent evidence suggests that low-voltage arresters have better success in preventing failures.[8] Figure 4.39 shows an example of a well-protected

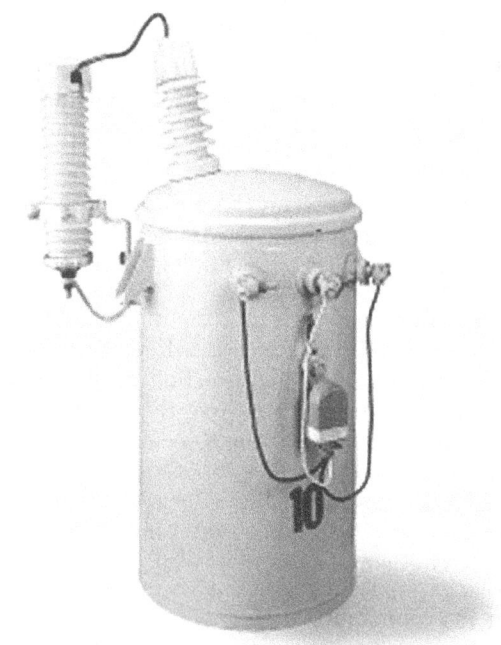

FIGURE 4.39 Example of a distribution transformer protected against lightning with tank-mounted primary and secondary arresters. (*Courtesy of Cooper Power Systems.*)

utility pole-top distribution transformer.[9] The primary arrester is mounted directly on the tank with very short lead lengths. With the evidence mounting that lightning surges have steeper wavefronts than previously believed, this is an ever increasing requirement for good protection practice.[10] It requires a special fuse in the cutout to prevent fuse damage on lightning current discharge. The transformer protection is completed by using a robust secondary arrester. This shows a heavy-duty, secondary arrester adapted for external mounting on transformers. Internally mounted arresters are also available. An arrester rating of 40-kA discharge current is recommended. The voltage discharge is not extremely critical in this application but is typically 3 to 5 kV. Transformer secondaries are generally assumed to have a BIL of 20 to 30 kV. Gap-type arresters also work in this application but cause voltage sags, which the MOV-type arresters avoid.

Impact on Load Circuits

Figure 4.40 shows a waveform of the open-circuit voltage measured at an electrical outlet location in a laboratory mock-up of a residential service.[12] For a relatively small stroke to the primary line (2.6 kA), the voltages at the outlet reached nearly 15 kV. In fact, higher-current strokes caused random flashovers of the test circuit, which made

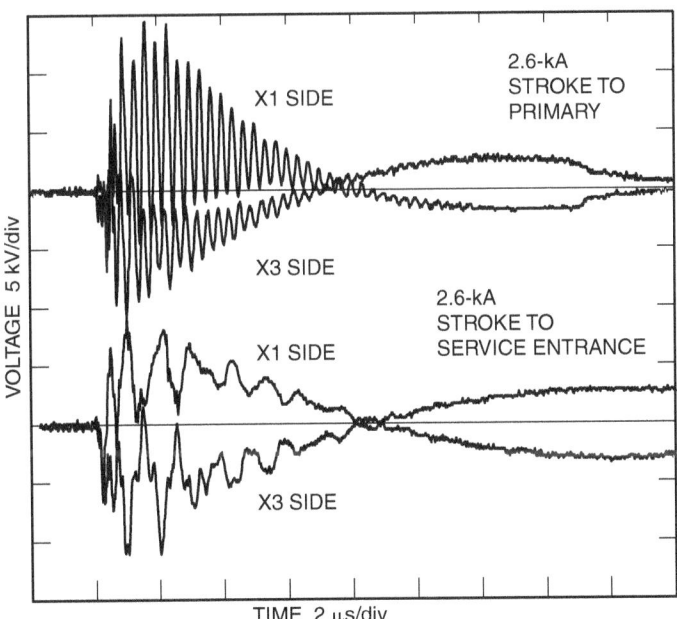

FIGURE 4.40   Voltage appearing at outlet due to low-side surge phenomena.

measurements difficult. This reported experience is indicative of the capacity of these surges to cause overvoltage problems.

The waveform is a very high frequency ringing wave riding on the main part of the low-side surge. The ringing is very sensitive to the cable lengths. A small amount of resistive load such as a light bulb would contribute greatly to the damping. The ringing wave differs depending on where the surge was applied, while the base low-side surge wave remains about the same; it is more dependent on the waveform of the current through the service cable.

One interesting aspect of this wave is that the ringing is so fast that it gets by the spark gaps in the meter base even though the voltage is 2 times the nominal sparkover value. In the tests, the outlets and lamp sockets could also withstand this kind of wave for about 1 μs before they flashed over. Thus, it is possible to have some high overvoltages propagating throughout the system. The waveform in this figure represents the available open-circuit voltage. In actual practice, a flashover would have occurred somewhere in the circuit after a brief time.

MOV arresters are not entirely effective against a ringing wave of this high frequency because of lead-length inductance. However, they are very effective for the lower-frequency portion of this transient, which contains the greater energy. Arresters should be applied both in the service entrance and at the outlets serving sensitive loads. Without the service entrance arresters to take most of the energy, arresters at the outlets are subject to failure. This is particularly true of single MOVs connected line to neutral. With the service entrance arresters, failure of outlet protectors and individual appliance protectors should be very rare unless lightning strikes the building structure closer to that location than the service entrance.

Service entrance arresters cannot be relied upon to protect the entire facility. They serve a useful purpose in shunting the bulk of the surge energy but cannot suppress the voltage sufficiently for remote loads. Likewise, the transformer arrester cannot be considered to take the place of the service entrance arrester although it may be only 50 ft (15 meters) away. This arrester is actually in *series* with the load for the low-side current surge. The basic guideline for arrester protection should always be followed: Place an arrester directly across the insulation structure that is to be protected. This becomes crucial for difficult-to-protect loads such as submersible pumps in deep water wells. The best protection is afforded by an arrester built directly into the motor rather than on the surface in the controller.

Some cases may not have as much to do with the surge voltage appearing at the outlet as with the differential voltage between two ground references. Such is the case for many TV receiver failures. Correct *bonding* of protective grounds is required as well as arrester protection.

The protective level of service entrance arresters for lightning impulses is typically about 2 kV. The lightning impulse current-carrying capability should be similar to the transformer secondary arrester, or approximately 40 kA. One must keep in mind that for low-frequency overvoltages, the arrester with the lowest discharge voltage is apt to take the brunt of the duty. MOV-type arresters will clamp the overvoltages without causing additional power quality problems such as interruptions and sags.

4.5.4 Cable Protection

One increasingly significant source of extended power outages on underground distribution (UD) systems is cable failures. The earliest utility distribution cables installed in the United States are now reaching the end of their useful life. As a cable ages, the insulation becomes progressively weaker and a moderate transient overvoltage causes breakdown and failure.

Many utilities are exploring ways of extending the cable life by arrester protection. Cable replacement is so costly that it is often worthwhile to retrofit the system with arresters even if the gain in life is only a few years. Depending on voltage class, the cable may have been installed with only one arrester at the riser pole or both a riser-pole arrester and an open-point arrester (see Fig. 4.41).

To provide additional protection, utilities may choose from a number of options:

1. Add an open-point arrester, if one does not exist.

2. Add a third arrester on the next-to-last transformer.

3. Add arresters at every transformer.

4. Add special low-discharge voltage arresters.

5. Inject an insulation-restoring fluid into the cable.

6. Employ a scout arrester scheme on the primary (see Sec. 4.5.5).

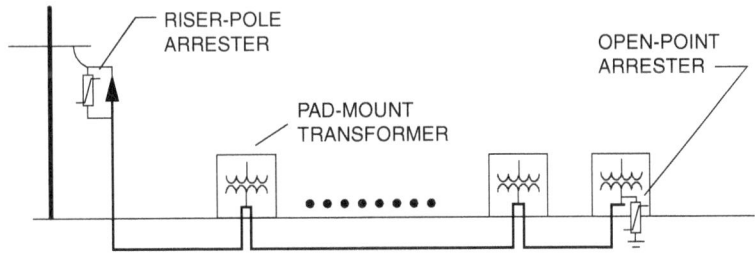

FIGURE 4.41 Typical UD cable arrester application.

The cable life is an exponential function of the number of impulses of a certain magnitude that it receives, according to Hopkinson.[11] The damage to the cable is related by

$$D = NV^c$$

where D = constant, representing damage to the cable
$\quad N$ = number of impulses
$\quad V$ = magnitude of impulses
$\quad c$ = empirical constant ranging from 10 to 15

Therefore, anything that will decrease the magnitude of the impulses only slightly has the potential to extend cable life a great deal.

Open-Point Arrester

Voltage waves double in magnitude when they strike an open point. Thus, the peak voltage appearing on the cable is about twice the discharge voltage of the riser-pole arrester. There is sufficient margin with new cables to get by without open-point arresters at some voltage classes. While open-point arresters are common at 35 kV, they are not used universally at lower voltage classes.

When the number of cable failures associated with storms begins to increase noticeably, the first option should be to add an arrester at the open point if there is not already one present.

Next-To-Last Transformer

Open-point arresters do not completely eliminate cable failures during lightning storms. With an open-point arrester, the greatest overvoltage stress is generally found at the next-to-last transformer. Figure 4.42 illustrates the phenomenon. Before the open-point arrester begins to conduct, it reflects the incoming wave just like an

VOLTAGE 40.00 kV/div

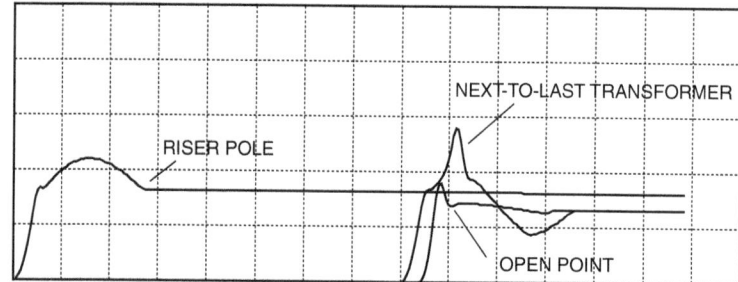

TIME 0.500 μs/div

FIGURE 4.42   Impulse voltages along a cable with an open-point arrester showing that the peak can occur at the next-to-last transformer. Simulation with UDSurge computer program. (*Courtesy of Cooper Power Systems.*)

open circuit. Therefore, there is a wave of approximately half the discharge voltage reflected back to the riser pole. This can be even higher if the wavefront is very steep and the arrester lead inductance aids the reflection briefly.

This results in a very short pulse riding on top of the voltage wave that dissipates fairly rapidly, as it flows toward the riser pole. However, at transformers within a few hundred feet of the open point there will be noticeable additional stress. Thus, we often see cable and transformer failures at this location.

The problem is readily solved by an additional arrester at the next-to-last transformer. In fact, this second arrester practically obliterates the impulse, providing effective protection for the rest of the cable system as well. Thus, some consider the most optimal UD cable protection configuration to be three arresters: a riser-pole arrester, an open-point arrester, and an arrester at the transformer next closest to the open point. This choice protects as well as having arresters at all transformers and is less costly, particularly in retrofitting.

Under-Oil Arresters
Transformer manufacturers can supply pad-mounted transformers for UD cable systems with the primary arresters inside the transformer compartment, under oil. If applied consistently, this achieves very good protection of the UD cable system by having arresters distributed along the cable. Of course, this protection comes at an incremental cost that must be evaluated to determine if it is economical for a utility to consider.

Elbow Arresters
The introduction of elbow arresters for transformer connections in UD cable systems has opened up protection options not previously economical. Previously, arrester installations on UD cable systems were adaptations of overhead arrester technology and were costly to implement. That is one reason why open-point arresters have not been used universally. The other alternative was under-oil arresters and it is also very costly to change out a pad-mount transformer just to get an open-point arrester. Now, the arrester is an integral part of the UD system hardware and installation at nearly any point on the system is practical. This is a particularly good option for many retrofit programs.

Lower-Discharge Arresters
The gapped MOV arrester technology described earlier in this chapter was developed specifically to improve the surge protection for UD cables and prolong their life. The arresters are able to achieve a substantially lower discharge voltage under lightning surge conditions while still providing the capability to withstand normal system conditions. By combining the gaps from the old SiC technology with fewer MOV blocks, a 20 to 30 percent gain could be made in the lightning protective margin. The gaps share the voltage with the MOV blocks during steady-state operation and prevent thermal runaway. Following

the logic of the Hopkinson formula, presented at the beginning of this section, converting to this kind of arrester in the UD cable system can be expected to yield a substantial increase in cable life.

Fluid Injection

This is a relatively new technology in which a restorative fluid is injected into a run of cable. The fluid fills the voids that have been created in the insulation by aging and gives the cable many more years of life. A vacuum is pulled on the receiving end and pressure is applied at the injection end. If there are no splices to block the flow, the fluid slowly penetrates the cable.

4.5.5   Scout Arrester Scheme

The idea of using a scout arrester scheme to protect utility UD cable runs goes back many years.[13] However, the idea has only been applied sporadically because of the additional initial expense. The concept is relatively simple: Place arresters on either side of the riser-pole arrester to reduce the lightning energy that can enter the cable. Figure 4.43 illustrates the basic scheme. The incoming lightning surge current from a strike downline first encounters a scout arrester. A large portion of the current is discharged into the ground at that location. A smaller portion proceeds on to the riser-pole arrester, which now produces a smaller discharge voltage. It is this voltage that is impressed upon the cable.

To further enhance the protection, the first span on either side of the riser pole can be shielded to prevent direct strokes to the line.

More recently, there has been a revival of interest in the scheme.[14] There is empirical evidence that the scout scheme helps prevent

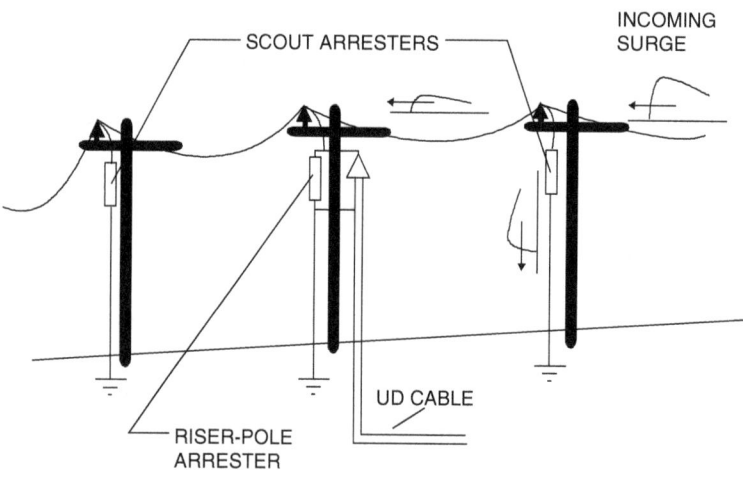

Figure 4.43   Scout arrester scheme.

open-point failures of both cables and transformers, and the expense of changing out a transformer far exceeds the additional cost of the scout arresters. Simulations suggest that while the nominal arrester discharge voltage may be reduced only a few percent, the greatest benefit of the scout scheme may be that it greatly reduces the rate of rise of surge voltages entering the cable. These steep-fronted surges reflect off the open point and frequently cause failures at the first or second pad-mount transformer from the end. Because of lead lengths, arresters are not always effective against such steep impulses. The scout scheme practically eliminates these from the cable.

Many distribution feeders in densely populated areas will have scout schemes by default. There are sufficient numbers of transformers that there are already arresters on either side of the riser pole.

4.6 Managing Ferroresonance

Ferroresonance in a distribution system occurs mainly when a lightly loaded, three-phase transformer becomes isolated on a cable with one or two open phases. This can happen both accidentally and intentionally. Strategies for dealing with ferroresonance include

- Preventing the open-phase condition
- Damping the resonance with load
- Limiting the overvoltages
- Limiting cable lengths
- Alternative cable-switching procedures

Most ferroresonance is a result of blown fuses in one or two of the phases in response to faults, or some type of single-pole switching in the primary circuit. A logical effective measure to guard against ferroresonance would be to use three-phase switching devices. For example, a three-phase recloser or sectionalizer could be used at the riser pole instead of fused cutouts. The main drawback is cost. Utilities could not afford to do this at every riser pole, but this could be done in special cases where there are particularly sensitive end users and frequent fuse blowings.

Another strategy on troublesome cable drops is to simply replace the fused cutouts with solid blades. This forces the upline recloser or breaker to operate to clear faults on the cable. Of course, this subjects many other utility customers to sustained interruptions when they would have normally seen only a brief voltage sag. However, it is an inexpensive way to handle the problem until a more permanent solution is implemented.

Manual, single-phase cable switching by pulling cutouts or cable elbows is also a major source of ferroresonance. This is a particular problem during new construction when there is a lot of activity and the transformers are not yet loaded. Some utilities have reported that line crews carry a "light board" or some other type of resistive load bank in their trucks for use in cable-switching activity when the transformers have no other load attached. One must be particularly careful when switching delta-connected transformers; such transformers should be protected because voltages may get extremely high. The common grounded wye-wye pad-mounted transformer may not be damaged internally if the exposure time is brief, although it may make considerable noise. When switching manually, the goal should be to open or close all three phases as promptly as possible.

Ferroresonance can generally be damped out by a relatively small amount of resistive load, although there are exceptions. For the typical case with one phase open, a resistive load of 1 to 4 percent of the transformer capacity can greatly reduce the effects of ferroresonance. The amount of load required is dependent on the length of cable and the design of the transformers. Also, the two-phase open case is sometimes more difficult to dampen with load. Figure 4.44 shows the effect of loading on ferroresonance overvoltages for a transformer connected to approximately 1.0 mi (1.61 km) of cable with one phase open. This was a particularly difficult case that damaged end-user equipment. Note the different characteristics of the phases. The transformer was of a five-legged

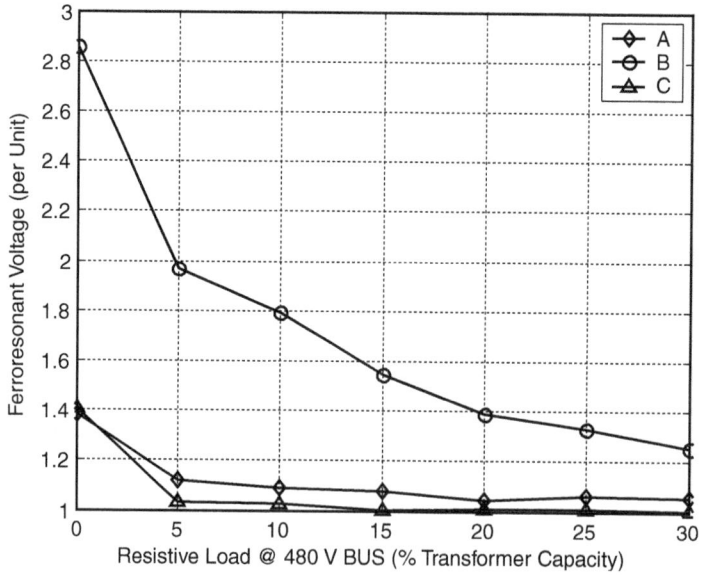

FIGURE 4.44 Example illustrating the impact of loading on ferroresonance.

core design, and the middle phase presents a condition that is more difficult to control with loading. Five percent resistive load reduces the overvoltage from approximately 2.8 to 2 pu. The transformer would have to be loaded approximately 20 to 25 percent of resistive equivalent load to limit ferroresonance overvoltages to 125 percent, the commonly accepted threshold. Since such a large load is required, a three-phase recloser was used to switch the cable.

On many utility systems, arresters are not applied on every pad-mounted distribution transformer due to costs. However, surge arresters can be an effective tool for suppressing the effects of ferroresonance. This is particularly true for transformers with ungrounded primary connections where the voltages can easily reach 3 to 4 pu if unchecked. Primary arresters will generally limit the voltages to 1.7 to 2.0 pu. There is some risk that arresters will fail if subjected to ferroresonance voltages for a long time. In fact, secondary arresters with protective levels lower than the primary-side arresters are frequent casualties of ferroresonance. Utility arresters are more robust, and there often is relatively little energy involved. However, if line crews encounter a transformer with arresters in ferroresonance, they should always deenergize the unit and allow the arresters to cool. An overheated arrester could fail violently if suddenly reconnected to a source with significant short-circuit capacity.

Ferroresonance occurs when the cable capacitance reaches a critical value sufficient to resonate with the transformer inductance (see Fig. 4.13). Therefore, one strategy to minimize the risk of frequent ferroresonance problems is to limit the length of cable runs. This is difficult to do for transformers with delta primary connections because with the high magnetizing reactance of modern transformers, ferroresonance can occur for cable runs of less than 100 ft. The grounded wye-wye connection will generally tolerate a few hundred feet of cable without exceeding 125 percent voltage during single-phasing situations. The allowable length of cable is also dependent on the voltage level with the general trend being that the higher the system voltage, the shorter the cable. However, modern trends in transformer designs with lower losses and exciting currents are making it more difficult to completely avoid ferroresonance at all primary distribution voltage levels.

The location of switching when energizing or deenergizing a transformer can play a critical role in reducing the likelihood of ferroresonance. Consider the two cable-transformer switching sequences in Fig. 4.45. Figure 4.45a depicts switching at the transformer terminals after the underground cable is energized, i.e., switch L is closed first, followed by switch R. Ferroresonance is less likely to occur since the equivalent capacitance seen from an open phase after each phase of switch R closes is the transformer's internal

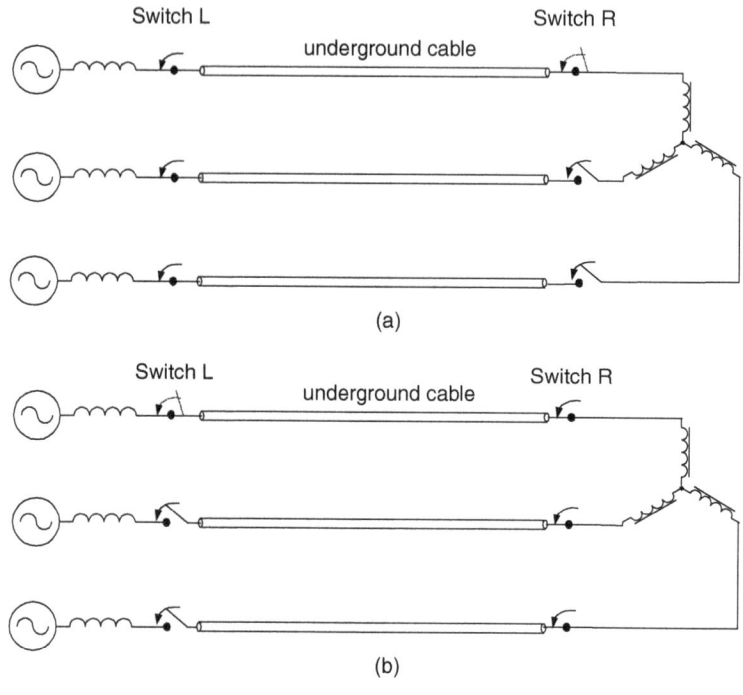

FIGURE **4.45** Switching at the transformer terminals (a) reduces the risk of isolating the transformer on sufficient capacitance to cause ferroresonance as opposed to (b) switching at some other location upline.

capacitance and does not involve the cable capacitance. Figure 4.45*b* depicts energization of the transformer remotely from another point in the cable system. The equivalent capacitance seen from switch L is the cable capacitance, and the likelihood of ferroresonance is much greater. Thus, one of the common rules to prevent ferro-resonance during cable switching is to switch the transformer by pulling the elbows at the primary terminals. There is little internal capacitance, and the losses of the transformers are usually sufficient to prevent resonance with this small capacitance. This is still a good general rule, although the reader should be aware that some modern transformers violate this rule. Low-loss transformers, particularly those built with an amorphous metal core, are prone to ferroresonance with their internal capacitances.

4.7   Switching Transient Problems with Loads

This section describes some transient problems related to loads and load switching.

4.7.1 Nuisance Tripping of Adjustable-Speed Drives (ASDs)

Most adjustable-speed drives typically use a voltage source inverter (VSI) design with a capacitor in the dc link. The controls are sensitive to dc overvoltages and may trip the drive at a level as low as 117 percent. Since transient voltages due to utility capacitor switching typically exceed 130 percent, the probability of nuisance tripping of the drive is high. One set of typical waveforms for this phenomenon is shown in Fig. 4.46.

The most effective way to eliminate nuisance tripping of small drives is to isolate them from the power system with ac line chokes. The additional series inductance of the choke will reduce the transient voltage magnitude that appears at the input to the

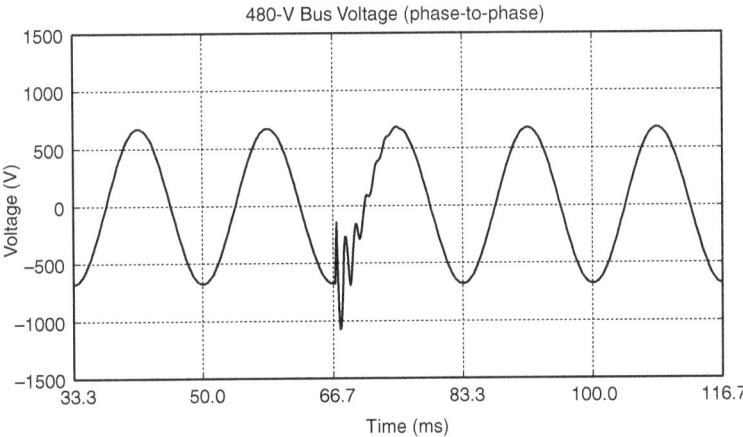

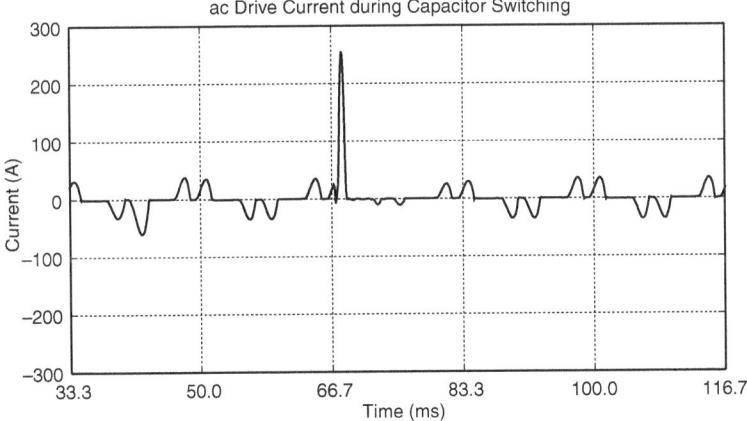

FIGURE 4.46 Effect of capacitor switching on adjustable-speed-drive ac current and dc voltage.

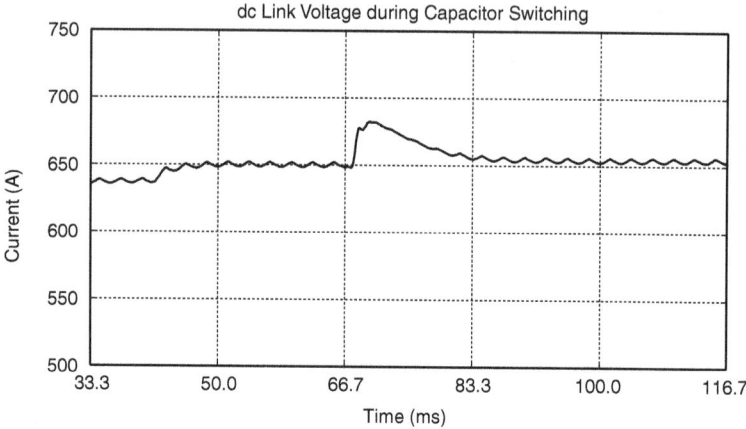

FIGURE 4.46 Effect of capacitor switching on adjustable-speed-drive ac current and dc voltage. (*Continued*)

adjustable-speed drive. Determining the precise inductor size required for a particular application (based on utility capacitor size, transformer size, etc.) requires a fairly detailed transient simulation. A series choke size of 3 percent based on the drive kVA rating is usually sufficient.

4.7.2 Transients from Load Switching

Deenergizing inductive circuits with air-gap switches, such as relays and contactors, can generate bursts of high-frequency impulses. Figure 4.47 shows an example. ANSI/IEEE C62.41–1991, *Recommended Practice for Surge Voltages in Low-Voltage AC Power Circuits,* cites a representative 15-ms burst composed of impulses having 5-ns rise times and 50-ns durations. There is very little energy in these types of transient due to the short duration, but they can interfere with the operation of electronic loads.

Such electrical fast transient (EFT) activity, producing spikes up to 1 kV, is frequently due to cycling motors, such as air conditioners and elevators. Transients as high as 3 kV can be caused by operation of arc welders and motor starters.

The duration of each impulse is short compared to the travel time of building wiring, thus the propagation of these impulses through the wiring can be analyzed with traveling wave theory. The impulses attenuate very quickly as they propagate through a building. Therefore, in most cases, the only protection needed is electrical separation. Physical separation is also required because the high rate of rise allows these transients to couple into nearby sensitive equipment.

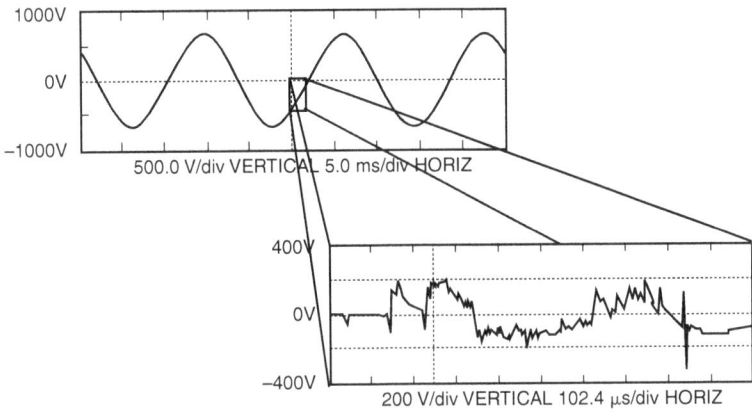

FIGURE 4.47 Fast transients caused by deenergizing an inductive load.

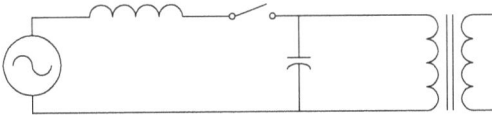

FIGURE 4.48 Energizing a capacitor and transformer simultaneously can lead to dynamic overvoltages.

EFT suppression may be required with extremely sensitive equipment in close proximity to a disturbing load, such as a computer room. High-frequency filters and isolation transformers can be used to protect against conduction of EFTs on power cables. Shielding is required to prevent coupling into equipment and data lines.

4.7.3 Transformer Energizing

Energizing a transformer produces inrush currents that are rich in harmonic components for a period lasting up to 1 s. If the system has a parallel resonance near one of the harmonic frequencies, a dynamic overvoltage condition results that can cause failure of arresters and problems with sensitive equipment. This problem can occur when large transformers are energized simultaneously with large power factor correction capacitor banks in industrial facilities. The equivalent circuit is shown in Fig. 4.48. A dynamic overvoltage waveform caused by a third-harmonic resonance in the circuit is shown in Fig. 4.49. After the expected initial transient, the voltage again swells to nearly 150 percent for many cycles until the losses and load damp out the oscillations. This can place severe stress on some arresters and has been known to significantly shorten the life of capacitors.

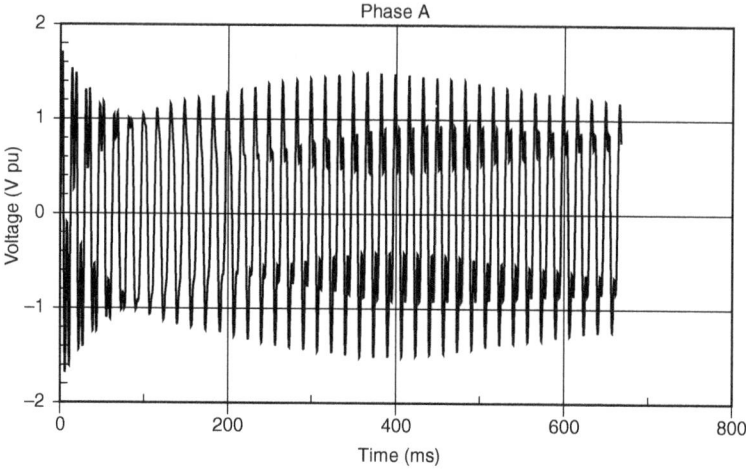

FIGURE **4.49**   Dynamic overvoltages during transformer energizing.

This form of dynamic overvoltage problem can often be eliminated simply by not energizing the capacitor and transformer together. One plant solved the problem by energizing the transformer first and not energizing the capacitor until load was about to be connected to the transformer.

4.8   Computer Tools for Transients Analysis

The most widely used computer programs for transients analysis of power systems are the Electromagnetic Transients Program, commonly known as EMTP, and its derivatives such as the Alternate Transients Program (ATP). EMTP was originally developed by Hermann W. Dommel at the Bonneville Power Administration (BPA) in the late 1960s[15] and has been continuously upgraded since. One of the reasons this program is popular is its low cost due to some versions being in the public domain. Some of the simulations presented in this book have been performed with a commercial analysis tool known as PSCAD/EMTDC, a program developed by the Manitoba HVDC Research Center. These programs feature a very sophisticated graphical user interface. The authors also use the EMPT-RV software. Some power system analysts use computer programs developed more for the analysis of electronic circuits, such as the well-known SPICE program[16] and its derivatives.

Although the programs just discussed continue to be used extensively, there are now many other capable programs available.

We will not attempt to list each one because there are so many and, also, at the present rate of software development, any such list would soon be outdated. The reader is referred to the Internet since all vendors of this type of software maintain websites.

Nearly all the tools for power systems solve the problem in the time domain, re-creating the waveform point by point. A few programs solve in the frequency domain and use the Fourier transform to convert to the time domain. Unfortunately, this essentially restricts the addressable problems to linear circuits. Time-domain solution is required to model nonlinear elements such as surge arresters and transformer magnetizing characteristics. The penalty for this extra capability is longer solution times, which with modern computers becomes less of a problem each day.

It takes considerably more modeling expertise to perform electromagnetic transients studies than to perform more common power system analyses such as of the power flow or of a short circuit. Therefore, this task is usually relegated to a few specialists within the utility organization or to consultants.

While transients programs for electronic circuit analysis may formulate the problem in any number of ways, power systems analysts almost uniformly favor some type of nodal admittance formulation. For one thing, the system admittance matrix is sparse allowing the use of very fast and efficient sparsity techniques for solving large problems. Also, the nodal admittance formulation reflects how most power engineers view the power system, with series and shunt elements connected to buses where the voltage is measured with respect to a single reference.

To obtain conductances for elements described by differential equations, transients programs discretize the equations with an appropriate numerical integration formula. The simple trapezoidal rule method appears to be the most commonly used, but there are also a variety of Runge-Kutta and other formulations used. Nonlinearities are handled by iterative solution methods. Some programs include the nonlinearities in the general formulation, while others, such as those that follow the EMTP methodology, separate the linear and nonlinear portions of the circuit to achieve faster solutions. This impairs the ability of the program to solve some classes of nonlinear problems but is not usually a significant constraint for most power system problems.

4.9   References

1. *Electrical Transmission and Distribution Reference Book,* 4th ed., Westinghouse Electric Corporation, East Pittsburgh, Pa., 1964.
2. *Electrical Distribution-System Protection,* 3d ed., Cooper Power Systems, Franksville, Wis., 1990.
3. K. Berger, R. B. Anderson, H. Kroninger, "Parameters of Lightning Flashes, " *Electra,* No. 41, July 1975, pp. 23–27.

4. R. Morrison and W. H. Lewis, *Grounding and Shielding in Facilities,* John Wiley & Sons, New York, 1990.
5. G. L. Goedde, L. J. Kojovic, M. B. Marz, J. J. Woodworth, "Series-Graded Gapped Arrester Provides Reliable Overvoltage Protection in Distribution Systems," *Conference Record,* 2001 IEEE Power Engineering Society Winter Meeting, Vol. 3, 2001, pp. 1122–1127.
6. R. A. Stansberry, "Protecting Distribution Circuits: Overhead Shield Wire versus Lightning Surge Arresters," *Transmission & Distribution,* April 1991, pp. 56ff.
7. IEEE Transformers Committee, "Secondary (Low-Side) Surges in Distribution Transformers," *Proceedings of the 1991 IEEE PES Transmission and Distribution Conference,* Dallas, September 1991, pp. 998–1008.
8. C. W. Plummer, et al., "Reduction in Distribution Transformer Failure Rates and Nuisance Outages Using Improved Lightning Protection Concepts," *Proceedings of the 1994 IEEE PES Transmission and Distribution Conference,* Chicago, April 1994, pp. 411–416.
9. G. L. Goedde, L. A. Kojovic, J. J. Woodworth, "Surge Arrester Characteristics That Provide Reliable Overvoltage Protection in Distribution and Low-Voltage Systems," *Conference Record,* 2000 IEEE Power Engineering Society Summer Meeting, Vol. 4, 2000, pp. 2375–2380.
10. P. Barker, R. Mancao, D. Kvaltine, D. Parrish, "Characteristics of Lightning Surges Measured at Metal Oxide Distribution Arresters," *IEEE Transactions on Power Delivery,* October 1993, pp. 301–310.
11. R. H. Hopkinson, "Better Surge Protection Extends URD Cable Life," *IEEE Transactions on Power Apparatus and Systems,* Vol. PAS-103, 1984, pp. 2827–2834.
12. G. L. Goedde, R. C Dugan, L. D. Rowe, "Full Scale Lightning Surge Tests of Distribution Transformers and Secondary Systems," *Proceedings of the 1991 IEEE PES Transmission and Distribution Conference,* Dallas, September 1991, pp. 691–697.
13. S. S. Kershaw, Jr., "Surge Protection for High Voltage Underground Distribution Circuits," *Conference Record, IEEE Conference on Underground Distribution,* Detroit, September 1971, pp. 370–384.
14. M. B. Marz, T. E. Royster, C. M. Wahlgren, "A Utility's Approach to the Application of Scout Arresters for Overvoltage Protection of Underground Distribution Circuits," *1994 IEEE Transmission and Distribution Conference Record,* Chicago, April 1994, pp. 417–425.
15. H. W. Dommel, "Digital Computer Solution of Electromagnetic Transients in Single and Multiphase Networks," *IEEE Transactions on Power Apparatus and Systems,* Vol. PAS-88, April 1969, pp. 388–399.
16. L. W. Nagel, "SPICE2: A Computer Program to Simulate Semiconductor Circuits," Ph. D. thesis, University of California, Berkeley, Electronics Research Laboratory, No. ERL-M520, May 1975.
17. Southern States, 15.5 kV–38 kV CapSwitcher®Vertical Interrupter Style Capacitor Switchers, Product Specification Guide, PSG-809-031209.
18. M. Beanland, T. Speas, J. Rostron, "Pre-insertion Resistors in High Voltage Capacitor Switching," Western Protective Relay Conference, Oct. 19–21, 2004, Spokane, WA.

CHAPTER 5

Fundamentals of Harmonics

A good assumption for most utilities in the United States is that the sine-wave voltage generated in central power stations is very good. In most areas, the voltage found on transmission systems typically has much less than 1.0 percent distortion. However, the distortion increases closer to the load. At some loads, the current waveform barely resembles a sine wave. Electronic power converters can chop the current into seemingly arbitrary waveforms.

While there are a few cases where the distortion is random, most distortion is periodic, or an integer multiple of the power system fundamental frequency. That is, the current waveform is nearly the same cycle after cycle, changing very slowly, if at all. This has given rise to the widespread use of the term *harmonics* to describe distortion of the waveform. This term must be carefully qualified to make sense. This chapter and Chap. 6 remove some of the mystery of harmonics in power systems.

When electronic power converters first became commonplace in the late 1970s, many utility engineers became quite concerned about the ability of the power system to accommodate the harmonic distortion. Many dire predictions were made about the fate of power systems if these devices were permitted to exist. While some of these concerns were probably overstated, the field of power quality analysis owes a great debt of gratitude to these people because their concern over this "new" problem of harmonics sparked the research that has eventually led to much of the knowledge about all aspects of power quality.

To some, harmonic distortion is still the most significant power quality problem. It is not hard to understand how an engineer faced with a difficult harmonics problem can come to hold that opinion. Harmonics problems counter many of the conventional rules of power system design and operation that consider only the fundamental frequency. Therefore, the engineer is faced with unfamiliar phenomena that require unfamiliar tools to analyze and unfamiliar equipment to solve. Although harmonic problems can be difficult, they are not actually very numerous on utility systems. Only a few percent of utility distribution feeders in the United States have a sufficiently severe harmonics problem to require attention.

In contrast, voltage sags and interruptions are nearly universal to every feeder and represent the most numerous and significant power quality deviations. The end-user sector suffers more from harmonic problems than does the utility sector. Industrial users with adjustable-speed drives (ASDs), arc furnaces, induction furnaces, and the like are much more susceptible to problems stemming from harmonic distortion.

Harmonic distortion is not a new phenomenon on power systems. Concern over distortion has ebbed and flowed a number of times during the history of ac electric power systems. Scanning the technical literature of the 1930s and 1940s, one will notice many articles on the subject. At that time the primary sources were the transformers and the primary problem was inductive interference with open-wire telephone systems. The forerunners of modern arc lighting were being introduced and were causing quite a stir because of their harmonic content—not unlike the stir caused by electronic power converters in more recent times.

Fortunately, if the system is properly sized to handle the power demands of the load, there is a low probability that harmonics will cause a problem with the power system, although they may cause problems with telecommunications. The power system problems arise most frequently when the capacitance in the system results in resonance at a critical harmonic frequency that dramatically increases the distortion above normal amounts. While these problems occur on utility systems, the most severe cases are usually found in industrial power systems because of the higher degree of resonance achieved.

5.1   Harmonic Distortion

Harmonic distortion is caused by nonlinear devices in the power system. A nonlinear device is one in which the current is not proportional to the applied voltage. Figure 5.1 illustrates this concept by the case of a sinusoidal voltage applied to a simple nonlinear resistor in which the voltage and current vary according to the curve shown. While the applied voltage is perfectly sinusoidal, the resulting current is distorted. Increasing the voltage by a few percent may cause the current to double and take on a different waveshape. This is the source of most harmonic distortion in a power system.

Figure 5.2 illustrates that any periodic, distorted waveform can be expressed as a sum of sinusoids. When a waveform is identical from one cycle to the next, it can be represented as a sum of pure sine

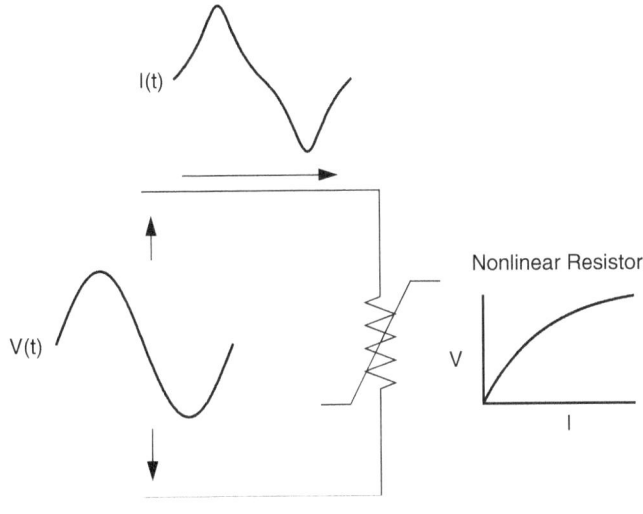

FIGURE **5.1**   Current distortion caused by nonlinear resistance.

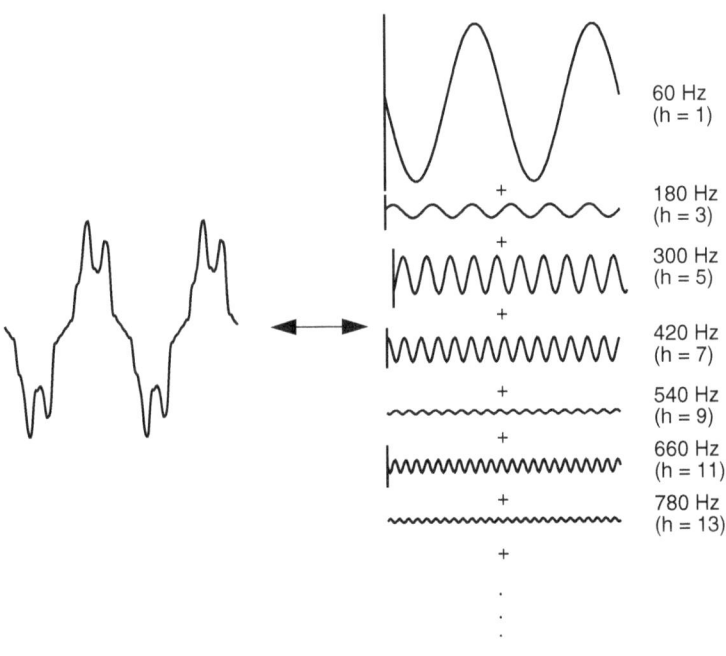

FIGURE **5.2**   Fourier series representation of a distorted waveform.

waves in which the frequency of each sinusoid is an integer multiple of the fundamental frequency of the distorted wave. This multiple is called a *harmonic* of the fundamental, hence the name of this subject matter. The sum of sinusoids is referred to as a *Fourier series,* named after the great mathematician who discovered the concept.

Because of the above-mentioned property, the Fourier series concept is universally applied in analyzing harmonic problems. The system can now be analyzed separately at each harmonic. In addition, finding the system response of a sinusoid of each harmonic individually is much more straightforward compared to that with the entire distorted waveforms. The outputs at each frequency are then combined to form a new Fourier series, from which the output waveform may be computed, if desired. Often, only the magnitudes of the harmonics are of interest.

When both the positive and negative half cycles of a waveform have identical shapes, the Fourier series contains only *odd* harmonics. This offers a further simplification for most power system studies because most common harmonic-producing devices look the same to both polarities. In fact, the presence of even harmonics is often a clue that there is something wrong—either with the load equipment or with the transducer used to make the measurement. There are notable exceptions to this such as half-wave rectifiers and arc furnaces when the arc is random.

Usually, the higher-order harmonics (above the range of the 25th to 50th, depending on the system) are negligible for power system analysis. While they may cause interference with low-power electronic devices, they are usually not damaging to the power system. It is also difficult to collect sufficiently accurate data to model power systems at these frequencies. A common exception to this occurs when there are system resonances in the range of frequencies. These resonances can be excited by notching or switching transients in electronic power converters. This causes voltage waveforms with multiple zero-crossings which disrupt timing circuits. These resonances generally occur on systems with underground cable but no power factor correction capacitors.

If the power system is depicted as series and shunt elements, as is the conventional practice, the vast majority of the nonlinearities in the system are found in *shunt* elements (i.e., loads). The series impedance of the power delivery system (i.e., the short-circuit impedance between the source and the load) is remarkably linear. In transformers, also, the source of harmonics is the shunt branch (magnetizing impedance) of the common "T" model; the leakage impedance is linear. Thus, the main sources of harmonic distortion will ultimately be end-user loads. This is not to say that all end users who experience harmonic distortion will themselves have significant sources of harmonics, but that the harmonic distortion generally originates with some end-user's load or combination of loads.

5.2   Voltage versus Current Distortion

The word *harmonics* is often used by itself without further qualification. For example, it is common to hear that an ASD or an induction furnace can't operate properly because of harmonics. What does that mean? Generally, it could mean one of the following three things:

1. The harmonic voltages are too great (the voltage too distorted) for the control to properly determine firing angles.

2. The harmonic currents are too great for the capacity of some device in the power supply system such as a transformer, and the machine must be operated at a lower than rated power.

3. The harmonic voltages are too great because the harmonic currents produced by the device are too great for the given system condition.

As suggested by this list, there are separate causes and effects for voltages and currents as well as some relationship between them. Thus, the term harmonics by itself is inadequate to definitively describe a problem.

Nonlinear loads appear to be sources of harmonic current in shunt with and injecting harmonic currents *into* the power system. For nearly all analyses, it is sufficient to treat these harmonic-producing loads simply as current sources. There are exceptions to this as will be described later.

As Fig. 5.3 shows, voltage distortion is the result of distorted currents passing through the linear, series impedance of the power delivery system, although, assuming that the source bus is ultimately a pure sinusoid, there is a nonlinear load that draws a distorted current. The harmonic currents passing through the impedance of the system cause a voltage drop for each harmonic. This results in voltage harmonics appearing at the load bus. The amount of voltage distortion depends on the impedance and the current. Assuming the

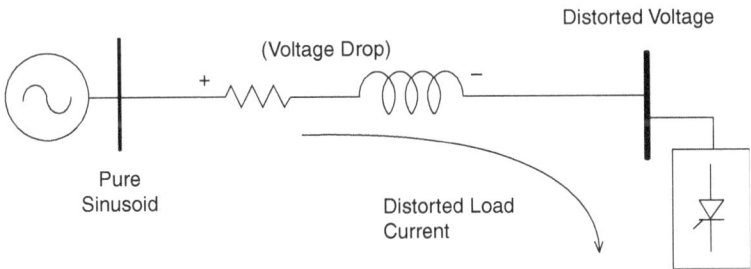

Figure 5.3 Harmonic currents flowing through the system impedance result in harmonic voltages at the load.

load bus distortion stays within reasonable limits (e.g., less than 5 percent), the amount of harmonic current produced by the load is generally constant.

While the load current harmonics ultimately cause the voltage distortion, it should be noted that load has no control over the voltage distortion. The same load put in two different locations on the power system will result in two different voltage distortion values. Recognition of this fact is the basis for the division of responsibilities for harmonic control that are found in standards such as IEEE Standard 519–1992, *Recommended Practices and Requirements for Harmonic Control in Electrical Power Systems*:

1. The control over the amount of harmonic current injected into the system takes place at the end-use application.

2. Assuming the harmonic current injection is within reasonable limits, the control over the voltage distortion is exercised by the entity having control over the system impedance, which is often the utility.

One must be careful when describing harmonic phenomena to understand that there are distinct differences between the causes and effects of harmonic voltages and currents. The use of the term harmonics should be qualified accordingly. By popular convention in the power industry, the majority of times when the term is used by itself to refer to the load apparatus, the speaker is referring to the harmonic currents. When referring to the utility system, the voltages are generally the subject. To be safe, make a habit of asking for clarification.

5.3   Harmonics versus Transients

Harmonic distortion is blamed for many power quality disturbances that are actually transients. A measurement of the event may show a distorted waveform with obvious high-frequency components. Although transient disturbances contain high-frequency components, transients and harmonics are distinctly different phenomena and are analyzed differently. Transient waveforms exhibit the high frequencies only briefly after there has been an abrupt change in the power system. The frequencies are not necessarily harmonics; they are the natural frequencies of the system at the time of the switching operation. These frequencies have no relation to the system fundamental frequency.

Harmonics, by definition, occur in the steady state and are integer multiples of the fundamental frequency. The waveform distortion that produces the harmonics is present continually, or at least for several seconds. Transients are usually dissipated within a few cycles. Transients are associated with changes in the system such as switching of a capacitor bank. Harmonics are associated with the continuing operation of a load.

One case in which the distinction is blurred is transformer energization. This is a transient event but can produce considerable waveform distortion for many seconds and has been known to excite system resonances.

5.4   Power System Quantities under Nonsinusoidal Conditions

Traditional power system quantities such as rms, power (reactive, active, apparent), power factor, and phase sequences are defined for the fundamental frequency context in a pure sinusoidal condition. In the presence of harmonic distortion the power system no longer operates in a sinusoidal condition, and unfortunately many of the simplifications power engineers use for the fundamental frequency analysis do not apply.

5.4.1   Active, Reactive, and Apparent Power

There are three standard quantities associated with power:

- *Apparent power S [voltampere (VA)].* The product of the rms voltage and current.
- *Active power P [watt (W)].* The average rate of delivery of energy.
- *Reactive power Q [voltampere – reactive (VAR)].* The portion of the apparent power that is out of phase, or in quadrature, with the active power.

The apparent power S applies to both sinusoidal and nonsinusoidal conditions. The apparent power can be written as follows:

$$S = V_{rms} \times I_{rms} \qquad (5.1)$$

where V_{rms} and I_{rms} are the rms values of the voltage and current. In a sinusoidal condition both the voltage and current waveforms contain only the fundamental frequency component; thus the rms values can be expressed simply as

$$V_{rms} = \frac{1}{\sqrt{2}} V_1 \quad \text{and} \quad I_{rms} = \frac{1}{\sqrt{2}} I_1 \qquad (5.2)$$

where V_1 and I_1 are the amplitude of voltage and current waveforms, respectively. The subscript "1" denotes quantities in the fundamental frequency. In a nonsinusoidal condition a harmonically distorted waveform is made up of sinusoids of harmonic frequencies with different amplitudes as shown in Fig. 5.2. The rms values of the waveforms are computed as the square root of the sum of rms squares of all individual components, i.e.,

$$V_{rms} = \sqrt{\sum_{h=1}^{h_{max}} \left(\frac{1}{\sqrt{2}} V_h \right)^2} = \frac{1}{\sqrt{2}} \sqrt{V_1^2 + V_2^2 + V_3^2 + \cdots + V_{h_{max}}^2} \qquad (5.3)$$

$$1_{rms} = \sqrt{\sum_{h=1}^{h_{max}} \left(\frac{1}{\sqrt{2}} I_h \right)^2} = \frac{1}{\sqrt{2}} \sqrt{I_1^2 + I_2^2 + I_3^2 + \cdots + I_{h_{max}}^2} \qquad (5.4)$$

where V_h and I_h are the amplitude of a waveform at the harmonic component h. In the sinusoidal condition, harmonic components of V_h and I_h are all zero, and only V_1 and I_1 remain. Equations (5.3) and (5.4) simplify to Eq. (5.2).

The active power P is also commonly referred to as the average power, real power, or true power. It represents useful power expended by loads to perform real work, i.e., to convert electric energy to other forms of energy. Real work performed by an incandescent light bulb is to convert electric energy into light and heat. In electric power, real work is performed for the portion of the current that is in phase with the voltage. No real work will result from the portion where the current is not in phase with the voltage. The active power is the rate at which energy is expended, dissipated, or consumed by the load and is measured in units of watts. P can be computed by averaging the product of the instantaneous voltage and current, i.e.,

$$P = \frac{1}{T} \int_0^T v(t) i(t) dt \qquad (5.5)$$

Equation (5.5) is valid for both sinusoidal and nonsinusoidal conditions. For the sinusoidal condition, P resolves to the familiar form,

$$P = \frac{V_1 I_1}{2} \cos \theta_1 = V_{1rms} I_{1rms} \cos \theta_1 = S \cos \theta_1 \qquad (5.6)$$

where θ_1 is the phase angle between voltage and current at the fundamental frequency. Equation (5.6) indicates that the average active power is a function only of the fundamental frequency quantities. In the nonsinusoidal case, the computation of the active power must include contributions from all harmonic components; thus it is the sum of active power at each harmonic. Furthermore, because the voltage distortion is generally very low on power systems (less than 5 percent), Eq. (5.6) is a good approximation regardless of how distorted the current is. This approximation cannot be applied when computing the apparent and reactive power. These two quantities are greatly influenced by the distortion. The apparent power S is a measure of the potential impact of the load on the thermal capability of the system. It is proportional to the rms of the distorted current, and its computation is straightforward, although slightly more complicated than the sinusoidal case. Also, many current probes can now directly report the true rms value of a distorted waveform.

The reactive power is a type of power that does no real work and is generally associated with reactive elements (inductors and capacitors). For example, the inductance of a load such as a motor causes the load current to lag behind the voltage. Power appearing across the inductance sloshes back and forth between the inductance itself and the power system source, producing no network. For this reason it is called imaginary or reactive power since no power is dissipated or expended. It is expressed in units of vars. In the sinusoidal case, the reactive power is simply defined as

$$Q = S\sin\theta_1 = \frac{V_1 I_1}{2}\sin\theta_1 = V_{1rms} I_{1rms}\sin\theta_1 \qquad (5.7)$$

which is the portion of power in quadrature with the active power shown in Eq. (5.6). Figure 5.4 summarizes the relationship between P, Q, and S in sinusoidal condition.

There is some disagreement among harmonics analysts on how to define Q in the presence of harmonic distortion. If it were not for the fact that many utilities measure Q and compute demand billing from the power factor computed by Q, it might be a moot point. It is more important to determine P and S; P defines how much active power is being consumed, while S defines the capacity of the power system required to deliver P. Q is not actually very useful by itself. However, Q_1, the traditional reactive power component at fundamental frequency, may be used to size shunt capacitors.

The reactive power when distortion is present has another interesting peculiarity. In fact, it may not be appropriate to call it reactive *power*. The concept of var flow in the power system is deeply

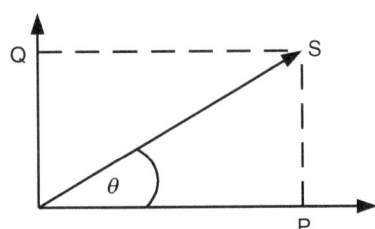

FIGURE 5.4   Relationship between P, Q, and S in sinusoidal condition.

ingrained in the minds of most power engineers. What many do not realize is that this concept is valid only in the sinusoidal steady state. When distortion is present, the component of S that remains after P is taken out is not conserved—that is, it does not sum to zero at a node. Power quantities are presumed to flow around the system in a conservative manner.

This does not imply that P is not conserved or that current is not conserved because the conservation of energy and Kirchoff's current laws are still applicable for any waveform. The reactive components actually sum in quadrature (square root of the sum of the squares). This has prompted some analysts to propose that Q be used to denote the reactive components that are conserved and introduce a new quantity for the components that are not. Many call this quantity D, for *distortion power* or, simply, *distortion voltamperes*. It has units of voltamperes, but it may not be strictly appropriate to refer to this quantity as *power*, because it does not flow through the system as power is assumed to do. In this concept, Q consists of the sum of the traditional reactive power values at each frequency. D represents all cross products of voltage and current at different frequencies, which yield no average power. P, Q, D, and S are related as follows, using the definitions for S and P previously given in Eqs. (5.1) and (5.5) as a starting point:

$$S = \sqrt{P^2 + Q^2 + D^2}$$
$$Q = \sum_k V_k I_k \sin \theta_k \tag{5.8}$$

Therefore, D can be determined after S, P, and Q by

$$D = \sqrt{S^2 - P^2 - Q^2} \tag{5.9}$$

Some prefer to use a three-dimensional vector chart to demonstrate the relationships of the components as shown in Fig. 5.5. P and Q contribute the traditional sinusoidal components to S, while D represents the additional contribution to the apparent power by the harmonics.

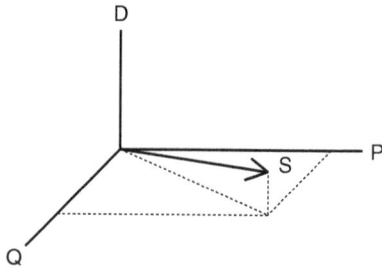

FIGURE 5.5 Relationship of components of the apparent power.

5.4.2   Power Factor: Displacement and True

Power factor (PF) is a ratio of useful power to perform real work (active power) to the power supplied by a utility (apparent power), i.e.,

$$PF = \frac{P}{S} \tag{5.10}$$

In other words, the power factor ratio measures the percentage of power expended for its intended use. Power factor ranges from zero to unity. A load with a power factor of 0.9 lagging denotes that the load can effectively expend 90 percent of the apparent power supplied (voltamperes) and convert it to perform useful work (watts). The term *lagging* denotes that the fundamental current lags behind the fundamental voltage by 25.84°.

In the sinusoidal case there is only one phase angle between the voltage and the current (since only the fundamental frequency is present; the power factor can be computed as the cosine of the phase angle and is commonly referred as the *displacement power factor*:

$$PF = \frac{P}{S} = \cos\theta \tag{5.11}$$

In the nonsinusoidal case the power factor cannot be defined as the cosine of the phase angle as in Eq. (5.11). The power factor that takes into account the contribution from all active power, including both fundamental and harmonic frequencies, is known as the *true power factor*. The true power factor is simply the ratio of total active power for all frequencies to the apparent power delivered by the utility as shown in Eq. (5.10).

Power quality monitoring instruments now commonly report both displacement and true power factors. Many devices such as switch-mode power supplies and PWM ASDs have a near-unity displacement power factor, but the true power factor may be 0.5 to 0.6. An ac-side capacitor will do little to improve the true power factor in this case because Q_1 is zero. In fact, if it results in resonance, the distortion may increase, causing the power factor to degrade. The true power factor indicates how large the power delivery system must be built to supply a given load. In this example, using only the displacement power factor would give a false sense of security that all is well.

The bottom line is that distortion results in additional current components flowing in the system that do not yield any net energy except that they cause losses in the power system elements they pass through. This requires the system to be built to a slightly larger capacity to deliver the power to the load than if no distortion were present.

5.4.3 Harmonic Phase Sequences

Power engineers have traditionally used symmetrical components to help describe three-phase system behavior. The three-phase system is transformed into three single-phase systems that are much simpler to analyze. The method of symmetrical components can be employed for analysis of the system's response to harmonic currents provided care is taken not to violate the fundamental assumptions of the method.

The method allows any unbalanced set of phase currents (or voltages) to be transformed into three balanced sets. The *positive-sequence* set contains three sinusoids displaced 120° from each other, with the normal A-B-C phase rotation (e.g., 0°, –120°, 120°). The sinusoids of the *negative-sequence* set are also displaced 120°, but have opposite phase rotation (A-C-B, e.g., 0°, 120°, –120°). The sinusoids of the *zero sequence* are in phase with each other (e.g., 0°, 0°, 0°).

In a perfect balanced three-phase system, the harmonic phase sequence can be determined by multiplying the harmonic number h with the normal positive-sequence phase rotation. For example, for the second harmonic, $h = 2$, we get 2 × (0, –120°, –120°) or (0°, 120°, –120°), which is the negative sequence. For the third harmonic, $h = 3$, we get 3 × (0°, –120°, –120°) or (0°, 0°, 0°), which is the zero sequence. Phase sequences for all other harmonic orders can be determined in the same fashion. Since a distorted waveform in power systems contains only odd-harmonic components (see Sec. 5.1), only odd-harmonic phase-sequence rotations are summarized here:

- Harmonics of order h = 1, 7, 13, ... are generally positive sequence.
- Harmonics of order h = 5, 11, 17, ... are generally negative sequence.
- Triplens ($h = 3, 9, 15, ...$) are generally zero sequence.

Impacts of sequence harmonics on various power system components are detailed in Sec. 5.10.

5.4.4 Triplen Harmonics

As previously mentioned, triplen harmonics are the odd multiples of the third harmonic ($h = 3, 9, 15, 21, ...$). They deserve special consideration because the system response is often considerably different for triplens than for the rest of the harmonics. Triplens become an important issue for grounded-wye systems with current flowing on the neutral. Two typical problems are overloading the neutral and telephone interference. One also hears occasionally of devices that misoperate because the line-to-neutral voltage is badly distorted by the triplen harmonic voltage drop in the neutral conductor.

For the system with perfectly balanced single-phase loads illustrated in Fig. 5.6, an assumption is made that fundamental and third-harmonic components are present. Summing the currents at node N, the fundamental current components in the neutral are found to be zero, but the third-harmonic components are 3 times those of the phase currents because they naturally coincide in phase and time.

Transformer winding connections have a significant impact on the flow of triplen harmonic currents from single-phase nonlinear loads. Two cases are shown in Fig. 5.7. In the wye-delta transformer (top), the triplen harmonic currents are shown entering the wye side. Since they are in phase, they add in the neutral. The delta winding provides ampere-turn balance so that they can flow, but they remain trapped in the delta and do not show up in the line currents on the delta side. When the currents are balanced, the triplen harmonic currents behave exactly as zero-sequence currents, which is precisely what they are. This type of transformer connection is the most common employed in utility distribution substations with the delta winding connected to the transmission feed.

Using grounded-wye windings on both sides of the transformer (bottom) allows balanced triplens to flow from the low-voltage system to the high-voltage system unimpeded. They will be present

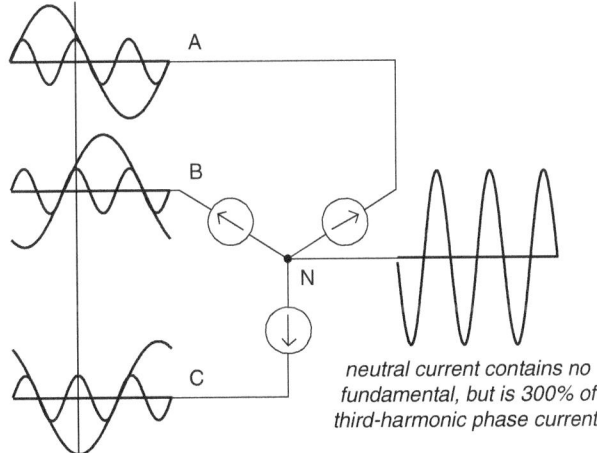

balanced fundamental currents sum to 0, but balanced third-harmonic currents coincide

neutral current contains no fundamental, but is 300% of third-harmonic phase current

Figure 5.6 High neutral currents in circuits serving single-phase nonlinear loads.

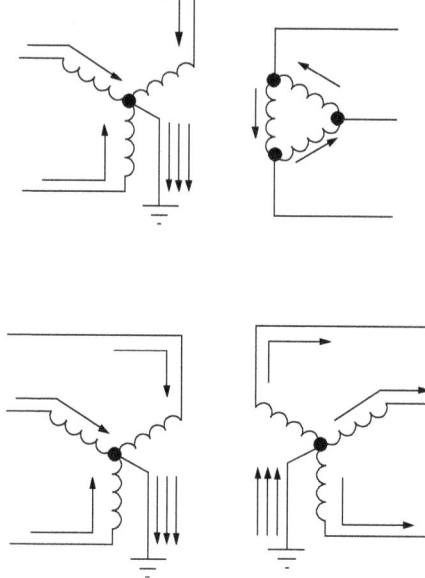

FIGURE 5.7 Flow of third-harmonic current in three-phase transformers.

in equal proportion on both sides. Many loads in the United States are served in this fashion.

Some important implications of this related to power quality analysis are

1. Transformers, particularly the neutral connections, are susceptible to overheating when serving single-phase loads on the wye side that have high third-harmonic content.

2. Measuring the current on the delta side of a transformer will not show the triplens and, therefore, not give a true idea of the heating the transformer is being subjected to.

3. The flow of triplen harmonic currents can be interrupted by the appropriate isolation transformer connection.

Removing the neutral connection in one or both wye windings blocks the flow of triplen harmonic current. There is no place for ampere-turn balance. Likewise, a delta winding blocks the flow from the line. One should note that three-legged core transformers behave as if they have a "phantom" delta tertiary winding. Therefore, a wye-wye connection with only one neutral point grounded will still be able to conduct the triplen harmonics from that side.

These rules about triplen harmonic current flow in transformers apply only to *balanced* loading conditions. When the phases are not balanced, currents of normal triplen harmonic frequencies may

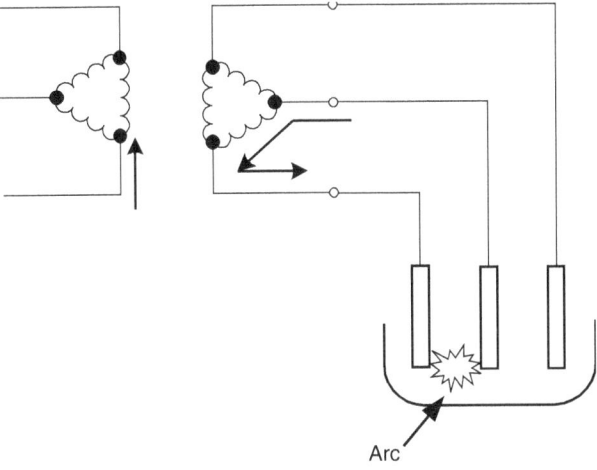

FIGURE 5.8   Arc furnace operation in an unbalanced mode allows triplen harmonics to reach the power system despite a delta-connected transformer.

very well show up where they are not expected. The normal mode for triplen harmonics is to be zero sequence. During imbalances, triplen harmonics may have positive- or negative-sequence components, too.

One notable case of this is a three-phase arc furnace. The furnace is nearly always fed by a delta-delta connected transformer to block the flow of the zero-sequence currents as shown in Fig. 5.8. Thinking that third harmonics are synonymous with zero sequence, many engineers are surprised to find substantial third-harmonic current present in large magnitudes in the line current. However, during scrap meltdown, the furnace will frequently operate in an unbalanced mode with only two electrodes carrying current. Large third-harmonic currents can then freely circulate in these two phases just as in a single-phase circuit. However, they are not zero-sequence currents. The third-harmonic currents have equal amounts of positive- and negative-sequence currents.

But to the extent that the system is *mostly* balanced, triplens *mostly* behave in the manner described.

5.5   Harmonic Indices

The two most commonly used indices for measuring the harmonic content of a waveform are the total harmonic distortion (THD) and the total demand distortion. Both are measures of the effective value of a waveform and may be applied to either voltage or current.

5.5.1 Total Harmonic Distortion

The THD is a measure of the *effective value* of the harmonic components of a distorted waveform. That is, it is the potential heating value of the harmonics relative to the fundamental. This index can be calculated for either voltage or current:

$$\text{THD} = \frac{\sqrt{\sum_{h>1}^{h_{\max}} M_h^2}}{M_1} \tag{5.12}$$

where M_h is the rms value of harmonic component h of the quantity M.

The rms value of a distorted waveform is the square root of the sum of the squares as shown in Eqs. (5.3) and (5.4). The THD is related to the rms value of the waveform as follows:

$$\text{RMS} = \sqrt{\sum_{h=1}^{h_{\max}} M_h^2} = M_1\sqrt{1+\text{THD}^2} \tag{5.13}$$

The THD is a very useful quantity for many applications, but its limitations must be realized. It can provide a good idea of how much extra heat will be realized when a distorted voltage is applied across a resistive load. Likewise, it can give an indication of the additional losses caused by the current flowing through a conductor. However, it is not a good indicator of the voltage stress within a capacitor because that is related to the peak value of the voltage waveform, not its heating value.

The THD index is most often used to describe voltage harmonic distortion. Harmonic voltages are almost always referenced to the fundamental value of the waveform at the time of the sample. Because fundamental voltage varies by only a few percent, the voltage THD is nearly always a meaningful number.

Variations in the THD over a period of time often follow a distinct pattern representing nonlinear load activities in the system. Figure 5.9 shows the voltage THD variation over a 1-week period where a daily cyclical pattern is obvious. The voltage THD shown in Fig. 5.9 was taken at a 13.2-kV distribution substation supplying a residential load. High-voltage THD occurs at night and during the early morning hours since the nonlinear loads are relatively high compared to the amount of linear load during these hours. A 1-week observation period is often required to come up with a meaningful THD pattern since it is usually the shortest period to obtain representative and reproducible measurement results.

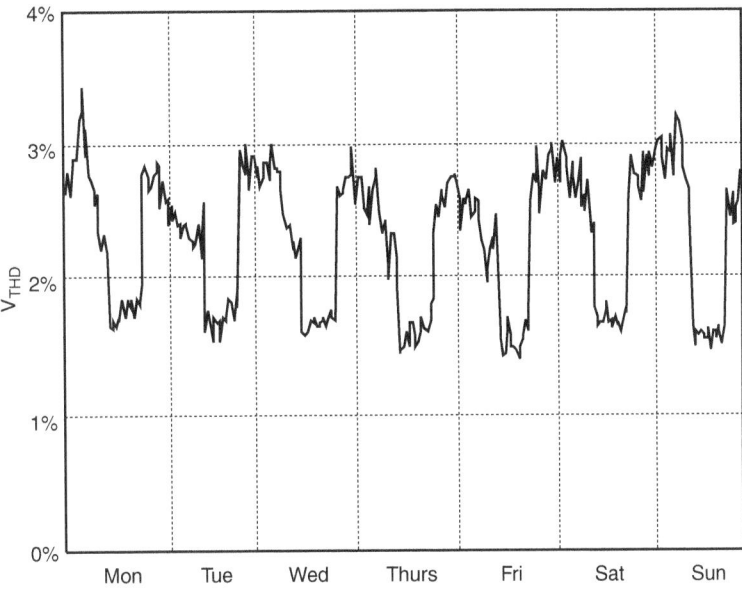

FIGURE 5.9   Variation of the voltage THD over a 1-week period.

5.5.2   Total Demand Distortion

Current distortion levels can be characterized by a THD value, as has been described, but this can often be misleading. A small current may have a high THD but not be a significant threat to the system. For example, many ASDs will exhibit high THD values for the input current when they are operating at very light loads. This is not necessarily a significant concern because the magnitude of harmonic current is low, even though its relative current distortion is high.

Some analysts have attempted to avoid this difficulty by referring THD to the fundamental of the peak demand load current rather than the fundamental of the present sample. This is called total demand distortion and serves as the basis for the guidelines in IEEE Standard 519–1992, *Recommended Practices and Requirements for Harmonic Control in Electrical Power Systems*. It is defined as follows:

$$\text{TDD} = \frac{\sqrt{\sum_{h=2}^{h_{\max}} I_h^2}}{I_L} \qquad (5.14)$$

I_L is the peak, or maximum, demand load current at the fundamental frequency component measured at the point of common coupling

(PCC). There are two ways to measure I_L. With a load already in the system, it can be calculated as the average of the maximum demand current for the preceding 12 months. The calculation can simply be done by averaging the 12-month peak demand readings. For a new facility, I_L has to be estimated based on the predicted load profiles.

5.6   Harmonic Sources from Commercial Loads

Commercial facilities such as office complexes, department stores, hospitals, and Internet data centers are dominated with high-efficiency fluorescent lighting with electronic ballasts, ASDs for the heating, ventilation, and air conditioning (HVAC) loads, elevator drives, and sensitive electronic equipment supplied by single-phase switch-mode power supplies. Commercial loads are characterized by a large number of small harmonic-producing loads. Depending on the diversity of the different load types, these small harmonic currents may add in phase or cancel each other. The voltage distortion levels depend on both the circuit impedances and the overall harmonic current distortion. Since power factor correction capacitors are not typically used in commercial facilities, the circuit impedance is dominated by the service entrance transformers and conductor impedances. Therefore, the voltage distortion can be estimated simply by multiplying the current by the impedance adjusted for frequency. Characteristics of typical nonlinear commercial loads are detailed in the following sections.

5.6.1   Single-Phase Power Supplies

Electronic power converter loads with their capacity for producing harmonic currents now constitute the most important class of nonlinear loads in the power system. Advances in semiconductor device technology have fueled a revolution in power electronics over the past decade, and there is every indication that this trend will continue. Equipment includes adjustable-speed motor drives, electronic power supplies, dc motor drives, battery chargers, electronic ballasts, and many other rectifier and inverter applications.

A major concern in commercial buildings is that power supplies for single-phase electronic equipment will produce too much harmonic current for the wiring. DC power for modern electronic and microprocessor-based office equipment is commonly derived from single-phase full-wave diode bridge rectifiers. The percentage of load that contains electronic power supplies is increasing at a dramatic pace, with the increased utilization of personal computers in every commercial sector.

There are two common types of single-phase power supplies. Older technologies use ac-side voltage control methods, such as transformers, to reduce voltages to the level required for the dc bus. The inductance of the transformer provides a beneficial side effect by smoothing the input current waveform, reducing harmonic content. Newer-technology switch-mode power supplies (see Fig. 5.10) use dc-to-dc conversion techniques to achieve a smooth dc output with small, lightweight components. The input diode bridge is directly connected to the ac line, eliminating the transformer. This results in a coarsely regulated dc voltage on the capacitor. This direct current is then converted back to alternating current at a very high frequency by the switcher and subsequently rectified again. Personal computers, printers, copiers, and most other single-phase electronic equipment now almost universally employ switch-mode power supplies. The key advantages are the light weight, compact size, efficient operation, and lack of need for a transformer. Switch-mode power supplies can usually tolerate large variations in input voltage.

Because there is no large ac-side inductance, the input current to the power supply comes in very short pulses as the capacitor C_1 regains its charge on each half cycle. Figure 5.11 illustrates the current waveform and spectrum for an entire circuit supplying a variety of electronic equipment with switch-mode power supplies.

A distinctive characteristic of switch-mode power supplies is a very high third-harmonic content in the current. Since third-harmonic current components are additive in the neutral of a three-phase system, the increasing application of switch-mode power supplies causes concern for overloading of neutral conductors, especially in older buildings where an undersized neutral may have been installed. There is also a concern for transformer overheating due to a combination of harmonic content of the current, stray flux, and high neutral currents.

5.6.2 Fluorescent Lighting

Lighting typically accounts for 40 to 60 percent of a commercial building load. According to the 1995 Commercial Buildings Energy Consumption study conducted by the U.S. Energy Information Administration, fluorescent lighting was used on 77 percent of commercial floor spaces, while only 14 percent of the spaces used incandescent lighting.[1] Fluorescent lights are a popular choice for energy savings.

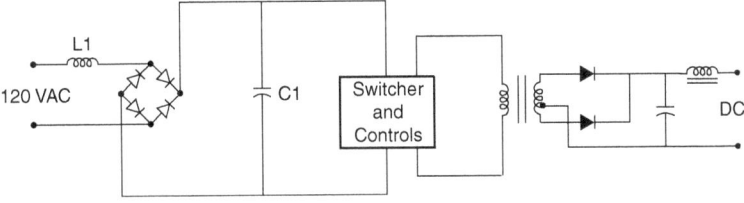

FIGURE **5.10** Switch-mode power supply.

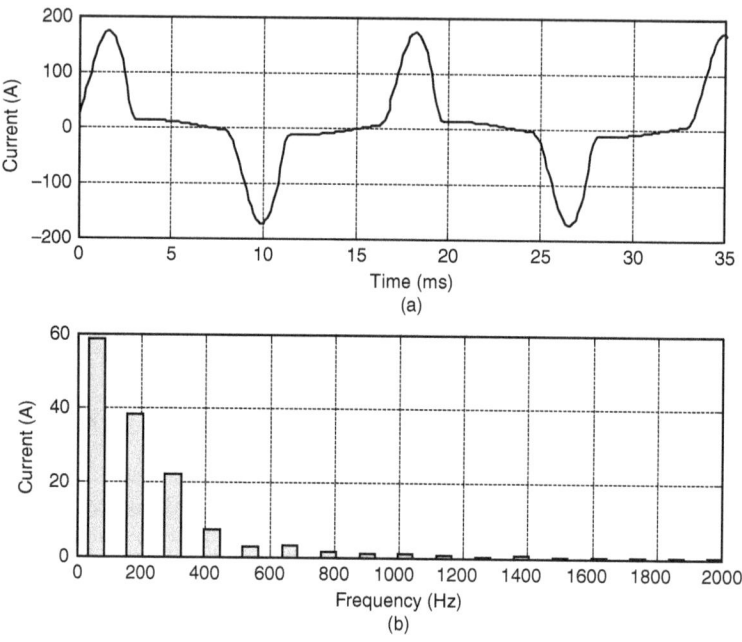

FIGURE **5.11**   SMPS current and harmonic spectrum.

Fluorescent lights are discharge lamps; thus they require a ballast to provide a high initial voltage to initiate the discharge for the electric current to flow between two electrodes in the fluorescent tube. Once the discharge is established, the voltage decreases as the arc current increases. It is essentially a short circuit between the two electrodes, and the ballast has to quickly reduce the current to a level to maintain the specified lumen output. Thus, a ballast is also a current-limiting device in lighting applications.

There are two types of ballasts, magnetic and electronic. A standard magnetic ballast is simply made up of an iron-core transformer with a capacitor encased in an insulating material. A single magnetic ballast can drive one or two fluorescent lamps, and it operates at the line fundamental frequency, i.e., 50 or 60 Hz. The iron-core magnetic ballast contributes additional heat losses, which makes it inefficient compared to an electronic ballast.

An electronic ballast employs a switch-mode–type power supply to convert the incoming fundamental frequency voltage to a much higher frequency voltage typically in the range of 25 to 40 kHz. This high frequency has two advantages. First, a small inductor is sufficient to limit the arc current. Second, the high frequency eliminates or greatly reduces the 100- or 120-Hz flicker associated with an iron-core magnetic ballast. A single electronic ballast typically can drive up to four fluorescent lamps.

Standard magnetic ballasts are usually rather benign sources of additional harmonics themselves since the main harmonic distortion comes from the behavior of the arc. Figure 5.12 shows a measured fluorescent lamp current and harmonic spectrum. The current THD is a moderate 15 percent. As a comparison, electronic ballasts, which employ switch-mode power supplies, can produce double or triple the standard magnetic ballast harmonic output. Figure 5.13 shows a fluorescent lamp with an electronic ballast that has a current THD of 144.

Other electronic ballasts have been specifically designed to minimize harmonics and may actually produce less harmonic distortion than the normal magnetic ballast-lamp combination. Electronic ballasts typically produce current THDs in the range of between 10 and 32 percent. A current THD greater than 32 percent is considered excessive according to ANSI C82.11–1993, *High-Frequency*

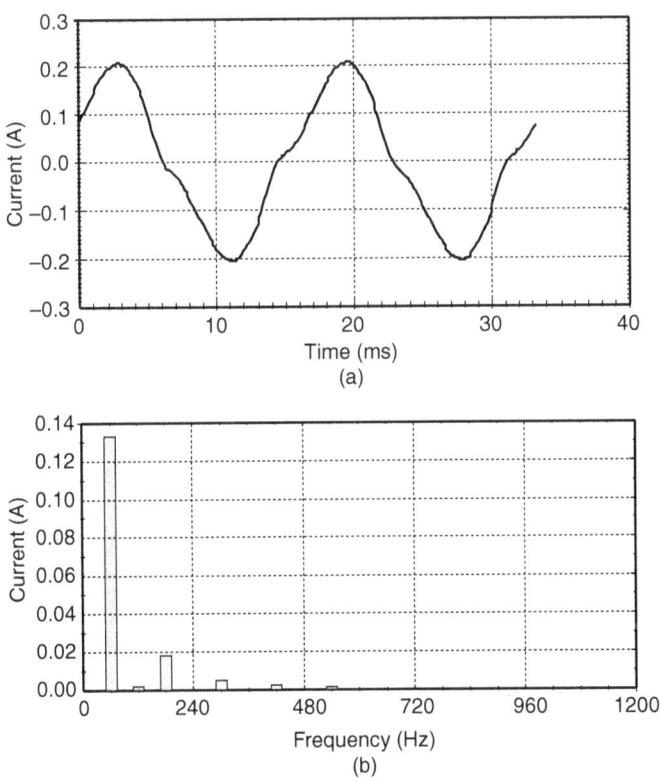

FIGURE 5.12 Fluorescent lamp with (*a*) magnetic ballast current waveform and (*b*) its harmonic spectrum.

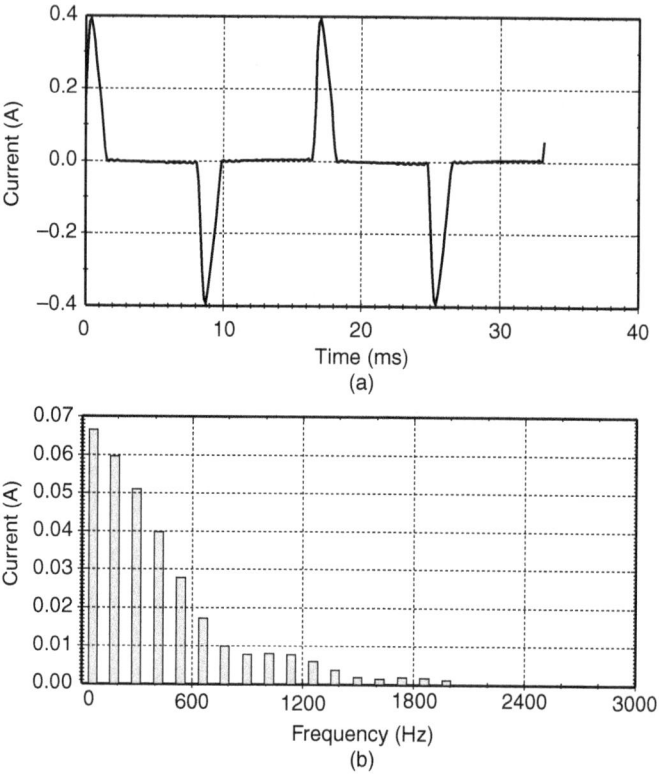

Figure 5.13   Fluorescent lamp with (a) electronic ballast current waveform and (b) its harmonic spectrum.

Fluorescent Lamp Ballasts. Most electronic ballasts are equipped with passive filtering to reduce the input current harmonic distortion to less than 20 percent.

Since fluorescent lamps are a significant source of harmonics in commercial buildings, they are usually distributed among the phases in a nearly balanced manner. With a delta-connected supply transformer, this reduces the amount of triplen harmonic currents flowing onto the power supply system. However, it should be noted that the common wye-wye supply transformers will not impede the flow of triplen harmonics regardless of how well balanced the phases are.

5.6.3   Adjustable-Speed Drives for HVAC and Elevators

Common applications of ASDs in commercial loads can be found in elevator motors and in pumps and fans in HVAC systems. An ASD consists of an electronic power converter that converts ac voltage and frequency into variable voltage and frequency. The variable voltage and frequency allows the ASD to control motor speed to match the application requirement such as slowing a pump or fan. ASDs also find many applications in industrial loads.

5.7 Harmonic Sources from Industrial Loads

Modern industrial facilities are characterized by the widespread application of nonlinear loads. These loads can make up a significant portion of the total facility loads and inject harmonic currents into the power system, causing harmonic distortion in the voltage. This harmonic problem is compounded by the fact that these nonlinear loads have a relatively low power factor. Industrial facilities often utilize capacitor banks to improve the power factor to avoid penalty charges. The application of power factor correction capacitors can potentially magnify harmonic currents from the nonlinear loads, giving rise to resonance conditions within the facility. The highest voltage distortion level usually occurs at the facility's low-voltage bus where the capacitors are applied. Resonance conditions cause motor and transformer overheating, and misoperation of sensitive electronic equipment.

Nonlinear industrial loads can generally be grouped into three categories: three-phase power converters, arcing devices, and saturable devices. Sections 5.7.1 to 5.7.3 detail the industrial load characteristics.

5.7.1 Three-Phase Power Converters

Three-phase electronic power converters differ from single-phase converters mainly because they do not generate third-harmonic currents. This is a great advantage because the third-harmonic current is the largest component of harmonics. However, they can still be significant sources of harmonics at their characteristic frequencies, as shown in Fig. 5.14. This is a typical current source type of ASD. The harmonic spectrum given in Fig. 5.14 would also be typical of a dc motor drive input current. Voltage source inverter (VSI) drives (such as PWM-type drives) can have much higher distortion levels as shown in Fig. 5.15.

The input to the PWM drive is generally designed like a three-phase version of the switch-mode power supply in computers. The rectifier feeds directly from the ac bus to a large capacitor on the dc bus. With little intentional inductance, the capacitor is charged in very short pulses, creating the distinctive "rabbit ear" ac-side current waveform with very high distortion. Whereas the switch-mode power supplies are generally for very small loads, PWM drives are now being applied for loads up to 500 horsepower (hp). This is a justifiable cause for concern from power engineers.

DC Drives

Rectification is the only step required for dc drives. Therefore, they have the advantage of relatively simple control systems. Compared

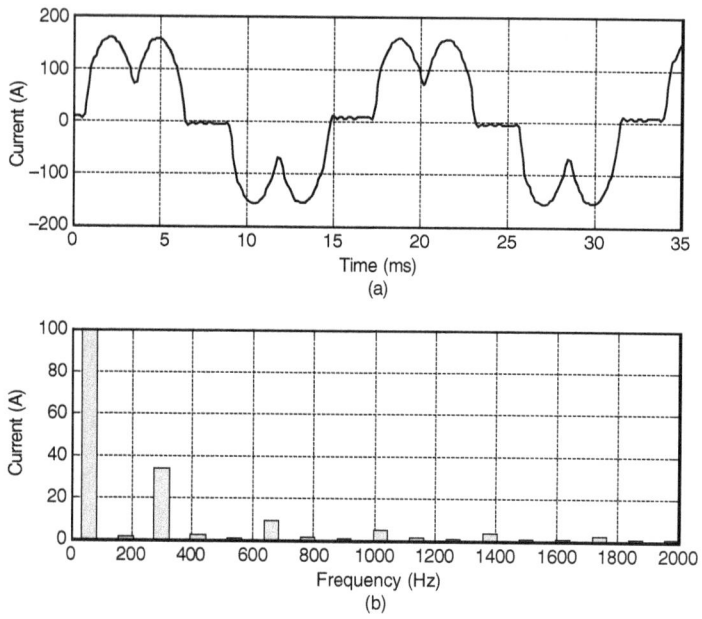

FIGURE 5.14 Current and harmonic spectrum for current source inverter (CSI)-type ASD.

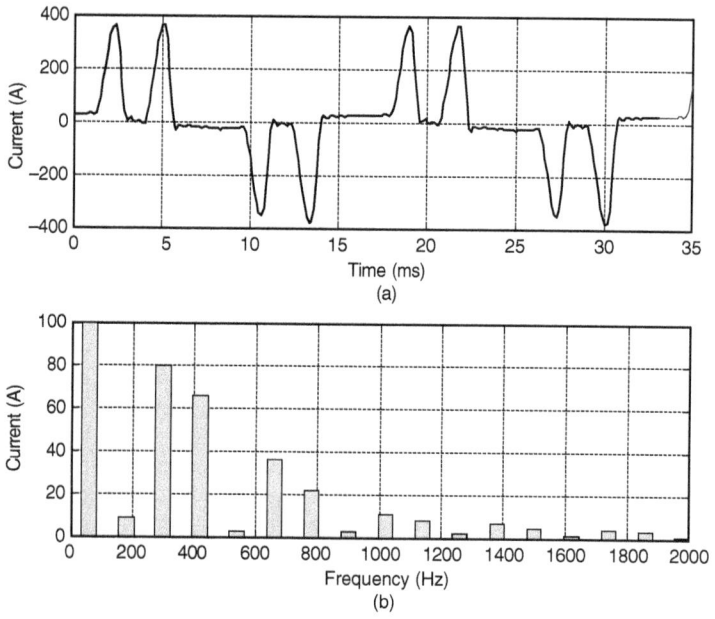

FIGURE 5.15 Current and harmonic spectrum for PWM-type ASD.

with ac drive systems, the dc drive offers a wider speed range and higher starting torque. However, purchase and maintenance costs for dc motors are high, while the cost of power electronic devices has been dropping year after year. Thus, economic considerations limit use of the dc drive to applications that require the speed and torque characteristics of the dc motor.

Most dc drives use the six-pulse rectifier shown in Fig. 5.16. Large drives may employ a 12-pulse rectifier. This reduces thyristor current duties and reduces some of the larger ac current harmonics. The two largest harmonic currents for the six-pulse drive are the fifth and seventh. They are also the most troublesome in terms of system response. A 12-pulse rectifier in this application can be expected to eliminate about 90 percent of the fifth and seventh harmonics, depending on system imbalances. The disadvantages of the 12-pulse drive are that there is more cost in electronics and another transformer is generally required.

AC Drives

In ac drives, the rectifier output is inverted to produce a variable-frequency ac voltage for the motor. Inverters are classified as voltage source inverters (VSIs) or current source inverters (CSIs). A VSI requires a constant dc (i.e., low-ripple) voltage input to the inverter stage. This is achieved with a capacitor or LC filter in the dc link. The CSI requires a constant current input; hence, a series inductor is placed in the dc link.

AC drives generally use standard squirrel cage induction motors. These motors are rugged, relatively low in cost, and require little maintenance. Synchronous motors are used where precise speed control is critical.

A popular ac drive configuration uses a VSI employing PWM techniques to synthesize an ac waveform as a train of variable-width dc pulses (see Fig. 5.17). The inverter uses either SCRs, gate turnoff (GTO) thyristors, or power transistors for this purpose. Currently, the

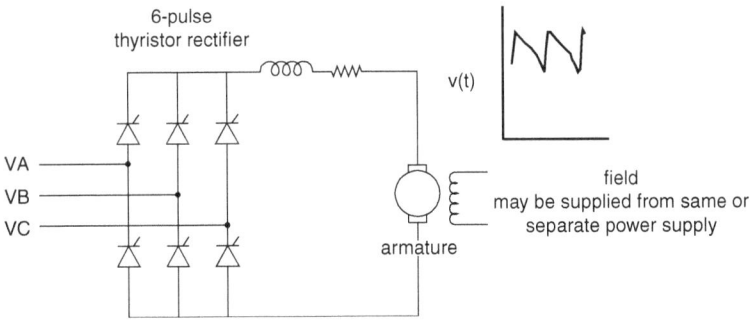

FIGURE 5.16 Six-pulse dc ASD.

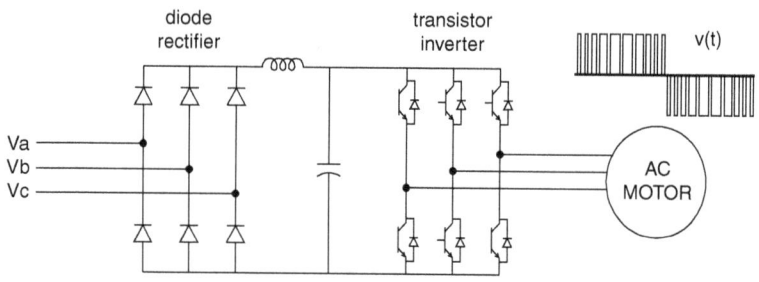

FIGURE 5.17   PWM ASD.

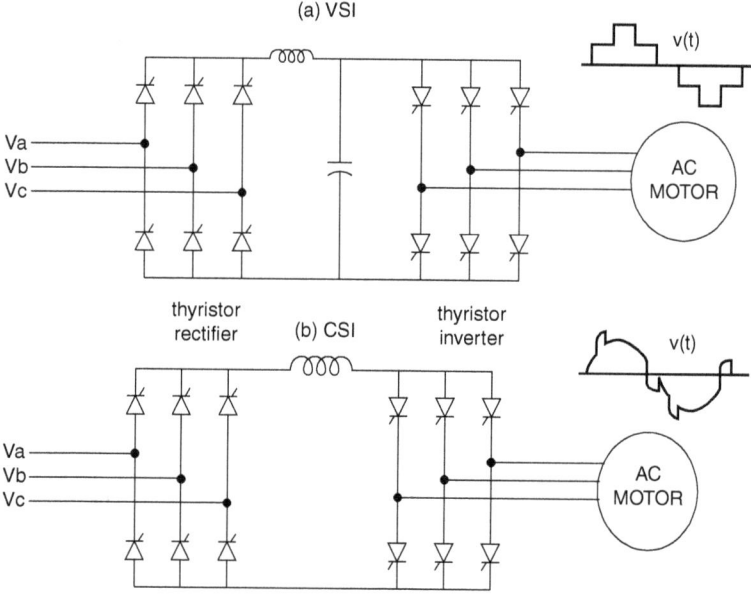

FIGURE 5.18   Large ac ASDs.

VSI PWM drive offers the best energy efficiency for applications over a wide speed range for drives up through at least 500 hp. Another advantage of PWM drives is that, unlike other types of drives, it is not necessary to vary rectifier output voltage to control motor speed. This allows the rectifier thyristors to be replaced with diodes, and the thyristor control circuitry to be eliminated.

Very high power drives employ SCRs and inverters. These may be 6-pulse, as shown in Fig. 5.18, or like large dc drives, 12-pulse. VSI drives (Fig. 5.18*a*) are limited to applications that do not require rapid changes in speed. CSI drives (Fig. 5.18*b*) have good acceleration/

deceleration characteristics but require a motor with a leading power factor (synchronous or induction with capacitors) or added control circuitry to commutate the inverter thyristors. In either case, the CSI drive must be designed for use with a specific motor. Thyristors in CSIs must be protected against inductive voltage spikes, which increases the cost of this type of drive.

Impact of Operating Condition

The harmonic current distortion in ASDs is not constant. The waveform changes significantly for different speed and torque values.

Figure 5.19 shows two operating conditions for a PWM ASD. While the waveform at 42-percent speed is much more distorted proportionately, the drive injects considerably higher magnitude harmonic currents at rated speed. The bar chart shows the amount of current injected. This will be the limiting design factor, not the highest THD. Engineers should be careful to understand the basis of data and measurements concerning these drives before making design decisions.

5.7.2   Arcing Devices

This category includes arc furnaces, arc welders, and discharge-type lighting (fluorescent, sodium vapor, mercury vapor) with magnetic (rather than electronic) ballasts. As shown in Fig. 5.20, the arc is

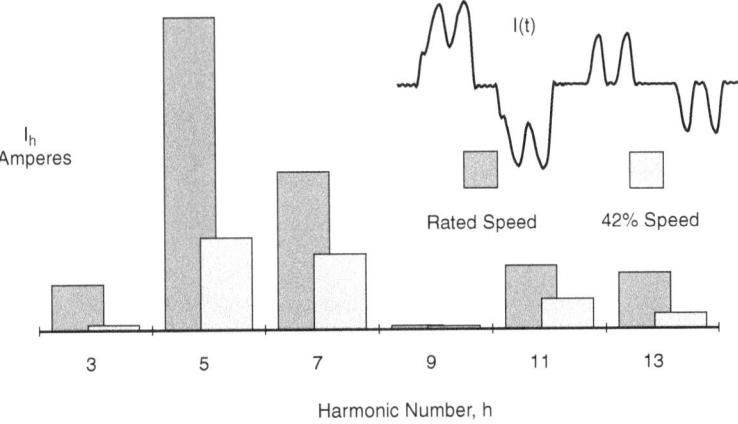

FIGURE **5.19**   Effect of PWM ASD speed on ac current harmonics.

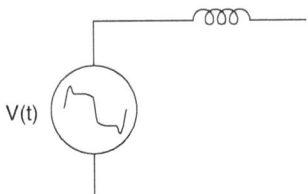

FIGURE **5.20**   Equivalent circuit for an arcing device.

basically a voltage clamp in series with a reactance that limits current to a reasonable value.

The voltage-current characteristics of electric arcs are nonlinear. Following arc ignition, the voltage decreases as the arc current increases, limited only by the impedance of the power system. This gives the arc the appearance of having a negative resistance for a portion of its operating cycle such as in fluorescent lighting applications.

In electric arc furnace applications, the limiting impedance is primarily the furnace cable and leads with some contribution from the power system and furnace transformer. Currents in excess of 60,000 A are common.

The electric arc itself is actually best represented as a source of voltage harmonics. If a probe were to be placed directly across the arc, one would observe a somewhat trapezoidal waveform. Its magnitude is largely a function of the length of the arc. However, the impedance of ballasts or furnace leads acts as a buffer so that the supply voltage is only moderately distorted. The arcing load thus appears to be a relatively stable harmonic current source, which is adequate for most analyses. The exception occurs when the system is near resonance and a Thevenin equivalent model using the arc voltage waveform gives more realistic answers.

The harmonic content of an arc furnace load and other arcing devices is similar to that of the magnetic ballast shown in Fig. 5.12. Three-phase arcing devices can be arranged to cancel the triplen harmonics through the transformer connection. However, this cancellation may not work in three-phase arc furnaces because of the frequent unbalanced operation during the melting phase. During the refining stage when the arc is more constant, the cancellation is better.

5.7.3   Saturable Devices

Equipment in this category includes transformers and other electromagnetic devices with a steel core, including motors. Harmonics are generated due to the nonlinear magnetizing characteristics of the steel (see Fig. 5.21).

Power transformers are designed to normally operate just below the "knee" point of the magnetizing saturation characteristic. The operating flux density of a transformer is selected based on a complicated optimization of steel cost, no-load losses, noise, and numerous other factors. Many electric utilities will penalize transformer vendors by various amounts for no-load and load losses, and the vendor will try to meet the specification with a transformer that has the lowest evaluated cost. A high-cost penalty on the no-load losses or noise will generally result in more steel in the core and a higher saturation curve that yields lower harmonic currents.

Although transformer exciting current is rich in harmonics at normal operating voltage (see Fig. 5.22), it is typically less than 1 percent of rated full-load current. Transformers are not as much of a concern as electronic power converters and arcing devices which can produce harmonic currents of 20 percent of their rating, or higher. However, their effect will be noticeable, particularly on utility distribution systems, which have hundreds of transformers. It is common to notice a significant increase in triplen harmonic currents during the early morning hours when the load is low and the voltage rises. Transformer exciting current is more visible then because there is insufficient load to obscure it and the increased voltage causes more current to be produced. Harmonic voltage distortion from transformer overexcitation is generally only apparent under these light load conditions.

Some transformers are purposefully operated in the saturated region. One example is a triplen transformer used to generate 180 Hz for induction furnaces.

Motors also exhibit some distortion in the current when overexcited, although it is generally of little consequence. There are, however, some fractional horsepower, single-phase motors that have a nearly triangular waveform with significant third-harmonic currents.

The waveform shown in Fig. 5.22 is for single-phase or wye-grounded three-phase transformers. The current obviously contains a large amount of third harmonic. Delta connections and ungrounded-wye connections prevent the flow of zero-sequence harmonic, which triplens tend to be. Thus, the line current will be void of these harmonics unless there is an imbalance in the system.

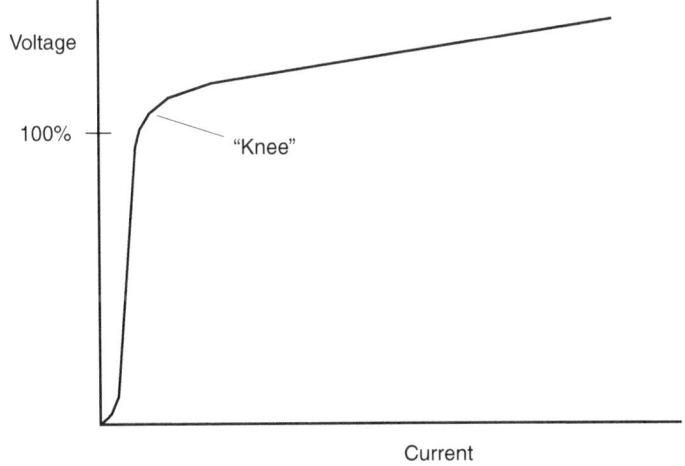

FIGURE 5.21 Transformer magnetizing characteristic.

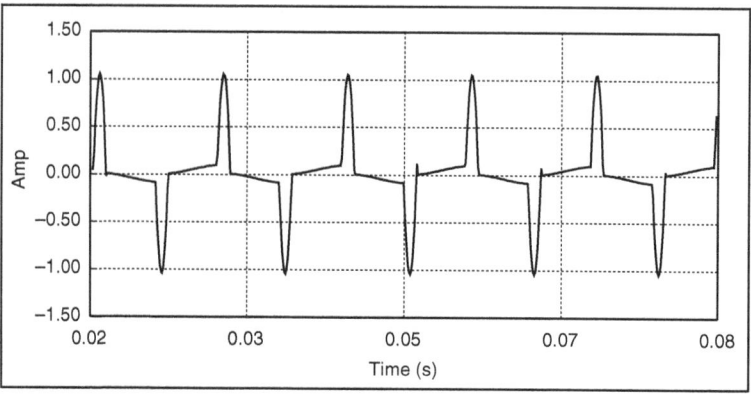

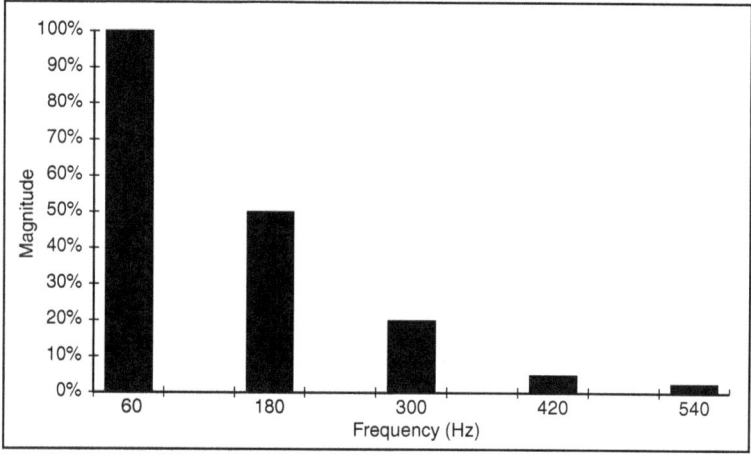

FIGURE 5.22   Transformer magnetizing current and harmonic spectrum.

5.8   Locating Harmonic Sources

On radial utility distribution feeders and industrial plant power systems, the main tendency is for the harmonic currents to flow from the harmonic-producing load to the power system source. This is illustrated in Fig. 5.23. The impedance of the power system is normally the lowest impedance seen by the harmonic currents. Thus, bulk of the current flows into the source.

This general tendency of harmonic current flows can be used to locate sources of harmonics. Using a power quality monitor capable of reporting the harmonic content of the current, simply measure the harmonic currents in each branch starting at the beginning of the circuit and trace the harmonics to the source.

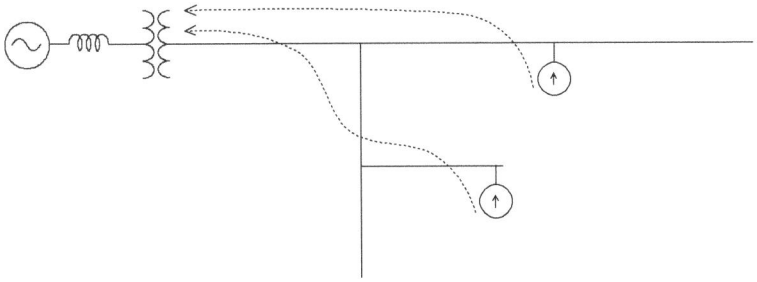

FIGURE **5.23**   General flow of harmonic currents in a radial power system.

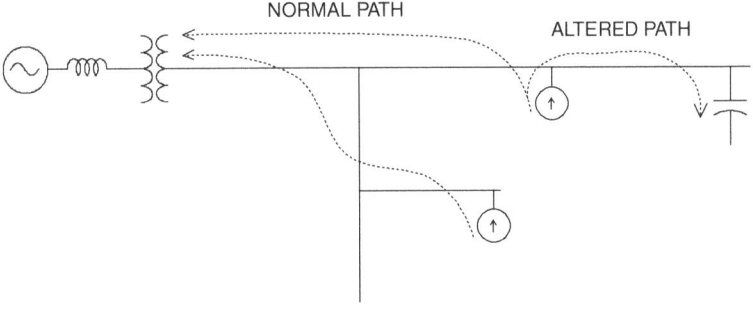

FIGURE **5.24**   Power factor capacitors can alter the direction of flow of one of the harmonic components of the current.

Power factor correction capacitors can alter this flow pattern for at least one of the harmonics. For example, adding a capacitor to the previous circuit as shown in Fig. 5.24 may draw a large amount of harmonic current into that portion of the circuit. In such a situation, following the path of the harmonic current will lead to a capacitor bank instead of the actual harmonic source. Thus, it is generally necessary to temporarily disconnect all capacitors to reliably locate the sources of harmonics.

It is usually straightforward to differentiate harmonic currents due to actual sources from harmonic currents that are strictly due to resonance involving a capacitor bank. A resonance current typically has only one dominant harmonic riding on top of the fundamental sine wave. Note that none of the harmonic sources presented earlier in this chapter produce a single harmonic frequency in addition to the fundamental. They all produce more than one single harmonic frequency. Waveforms of these harmonic sources have somewhat arbitrary waveshapes depending on the distorting phenomena, but they contain several harmonics in significant quantities. A single, large, significant harmonic nearly always signifies resonance.

This fact can be exploited to determine if harmonic resonance problems are likely to exist in a system with capacitors. Simply measure the current in the capacitors. If it contains a very large amount of one harmonic other than the fundamental, it is likely that the capacitor is participating in a resonant circuit within the power system. Always check the capacitor currents first in any installations where harmonic problems are suspected.

Another method to locate harmonic sources is by correlating the time variations of the voltage distortion with specific customer and load characteristics. Patterns from the harmonic distortion measurements can be compared to particular types of loads, such as arc furnaces, mill drives, and mass transits which appear intermittently. Correlating the time from the measurements and the actual operation time can identify the harmonic source.

5.9   System Response Characteristics

In power systems, the response of the system is equally as important as the sources of harmonics. In fact, power systems are quite tolerant of the currents injected by harmonic-producing loads unless there is some adverse interaction with the impedance of the system. Identifying the sources is only half the job of harmonic analysis. The response of the power system at each harmonic frequency determines the true impact of the nonlinear load on harmonic voltage distortion.

There are three primary variables affecting the system response characteristics, i.e., the system impedance, the presence of a capacitor bank, and the amount of resistive loads in the system. Sections 5.9.1 through 5.9.4 detail these variables.

5.9.1   System Impedance

At the fundamental frequency, power systems are primarily inductive, and the equivalent impedance is sometimes called simply the short-circuit reactance. Capacitive effects are frequently neglected on utility distribution systems and industrial power systems. One of the most frequently used quantities in the analysis of harmonics on power systems is the short-circuit impedance to the point on a network at which a capacitor is located. If not directly available, it can be computed from short-circuit study results that give either the short-circuit megavoltampere (MVA) or the short-circuit current as follows:

$$Z_{SC} = R_{SC} + jX_{SC} = \frac{kV^2}{MVA_{SC}} = \frac{kV \times 1000}{\sqrt{3}I_{SC}} \tag{5.15}$$

where Z_{SC} = short-circuit impedance
 R_{SC} = short-circuit resistance

X_{SC} = short-circuit reactance
k_V = phase-to-phase voltage, kV
MVA_{SC} = three-phase short-circuit MVA
I_{SC} = short-circuit current, A

Z_{SC} is a phasor quantity, consisting of both resistance and reactance. However, if the short-circuit data contain no phase information, one is usually constrained to assuming that the impedance is purely reactive. This is a reasonably good assumption for industrial power systems for buses close to the mains and for most utility systems. When this is not the case, an effort should be made to determine a more realistic resistance value because that will affect the results once capacitors are considered.

The inductive reactance portion of the impedance changes linearly with frequency. One common error made by novices in harmonic analysis is to forget to adjust the reactance for frequency. The reactance at the hth harmonic is determined from the fundamental impedance reactance X_1 by:

$$X_h = hX_1 \qquad (5.16)$$

In most power systems, one can generally assume that the resistance does not change significantly when studying the effects of harmonics less than the ninth. For lines and cables, the resistance varies approximately by the square root of the frequency once skin effect becomes significant in the conductor at a higher frequency. The exception to this rule is with some transformers. Because of stray eddy current losses, the apparent resistance of larger transformers may vary almost proportionately with the frequency. This can have a very beneficial effect on damping of resonance as will be shown later. In smaller transformers, less than 100 kVA, the resistance of the winding is often so large relative to the other impedances that it swamps out the stray eddy current effects and there is little change in the total apparent resistance until the frequency reaches about 500 Hz. Of course, these smaller transformers may have an X/R ratio of 1.0 to 2.0 at fundamental frequency, while large substation transformers might typically have a ratio of 20 to 30. Therefore, if the bus that is being studied is dominated by transformer impedance rather than line impedance, the system impedance model should be considered more carefully. Neglecting the resistance will generally give a conservatively high prediction of the harmonic distortion.

At utilization voltages, such as industrial power systems, the equivalent system reactance is often dominated by the service transformer impedance. A good approximation for X_{SC} may be based on the impedance of the service entrance transformer only:

$$X_{SC} \approx X_{tx} \qquad (5.17)$$

While not precise, this is generally at least 90 percent of the total impedance and is commonly more. This is usually sufficient to evaluate whether or not there will be a significant harmonic resonance problem. Transformer impedance in ohms can be determined from the percent impedance Z_{tx} found on the nameplate by

$$X_{tx} = \left(\frac{kV^2}{MVA_{3\phi}} \right) \times Z_{tx}(\%) \tag{5.18}$$

where $MVA_{3\phi}$ is the kVA rating of the transformer. This assumes that the impedance is predominantly reactive. For example, for a 1500-kVA, 6-percent transformer, the equivalent impedance on the 480-V side is

$$X_{tx} = \left(\frac{kV^2}{MVA_{3\phi}} \right) \times Z_{tx}(\%) = \left(\frac{0.480^2}{1.5} \right) \times 0.06 = 0.0092 \ \Omega$$

A plot of impedance versus frequency for an inductive system (no capacitors installed) would look like Fig. 5.25. Real power systems are not quite as well behaved. This simple model neglects capacitance, which cannot be done for harmonic analysis.

5.9.2 Capacitor Impedance

Shunt capacitors, either at the customer location for power factor correction or on the distribution system for voltage control, dramatically alter the system impedance variation with frequency. Capacitors do not create harmonics, but severe harmonic distortion can sometimes be attributed to their presence. While the reactance of inductive components increases proportionately to frequency, capacitive reactance X_C decreases proportionately:

$$X_C = \frac{1}{2\pi f C} \tag{5.19}$$

C is the capacitance in farads. This quantity is seldom readily available for power capacitors, which are rated in terms of kvar or Mvar at a given voltage. The equivalent line-to-neutral capacitive reactance at fundamental frequency for a capacitor bank can be determined by

$$X_C = \frac{kV^2}{Mvar} \tag{5.20}$$

For three-phase banks, use phase-to-phase voltage and the three-phase reactive power rating. For single-phase units, use the capacitor

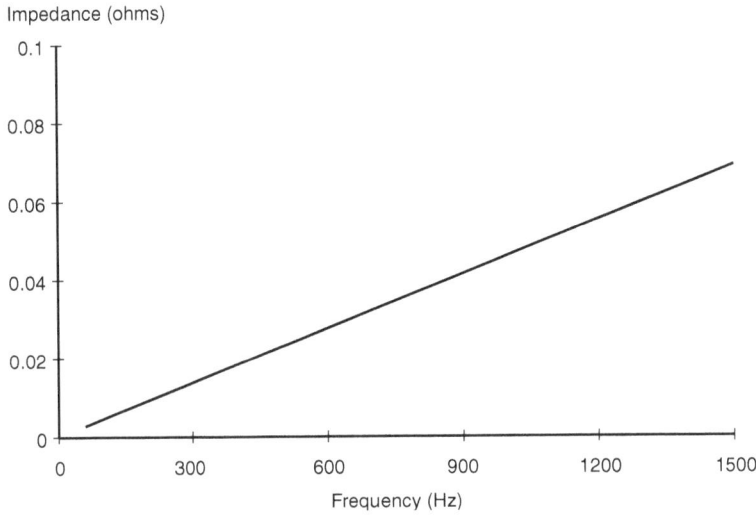

Impedance (ohms)

FIGURE 5.25   Impedance versus frequency for inductive system.

voltage rating and the reactive power rating. For example, for a three-phase, 1200-kvar, 13.8-kV capacitor bank, the positive-sequence reactance in ohms would be

$$X_C = \frac{kV^2}{Mvar} = \frac{13.8^2}{1.2} = 158.7\,\Omega$$

5.9.3   Parallel Resonance

All circuits containing both capacitances and inductances have one or more natural frequencies. When one of these frequencies corresponds to a frequency being produced by nonlinear loads, we can have development of resonance in which the voltage and current at that frequency continue to persist at very high values. This is the root of most problems with harmonic distortion on power systems.

In practice, it is not uncommon to observe an increase in harmonic distortion following capacitor energizing. It is often indicative of a parallel resonance frequency being near a harmonic frequency produced by nonlinear loads. Figure 5.26 shows voltage and current waveforms immediately before and after energizing a 12.47-kV three-phase distribution capacitor bank. By inspection, voltage and current waveforms before energizing show symptoms of harmonic distortions. Immediately after capacitor energizing, voltage and current waveforms become significantly more distorted. Profiling the fifth harmonic over time as shown in Figs. 5.27 and 5.28 indicates that the fifth harmonic voltage and current were already present

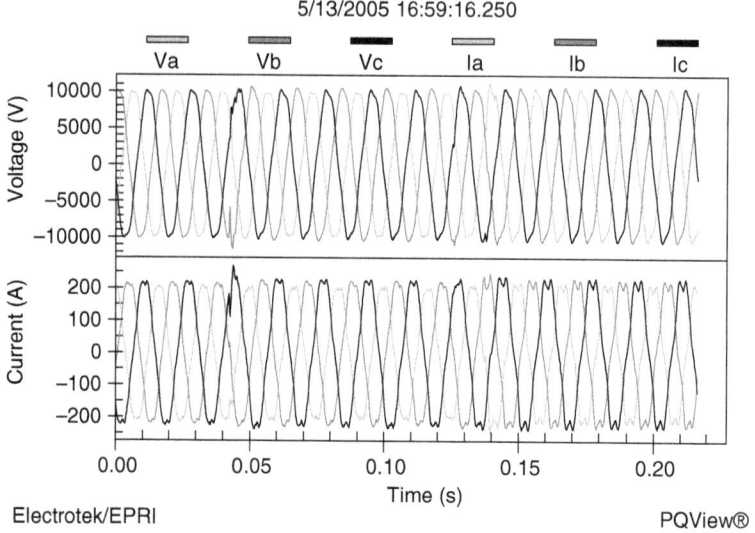

FIGURE 5.26   Increased harmonic distortion following capacitor energizing: (top) voltage waveforms *and* (bottom) current waveforms.

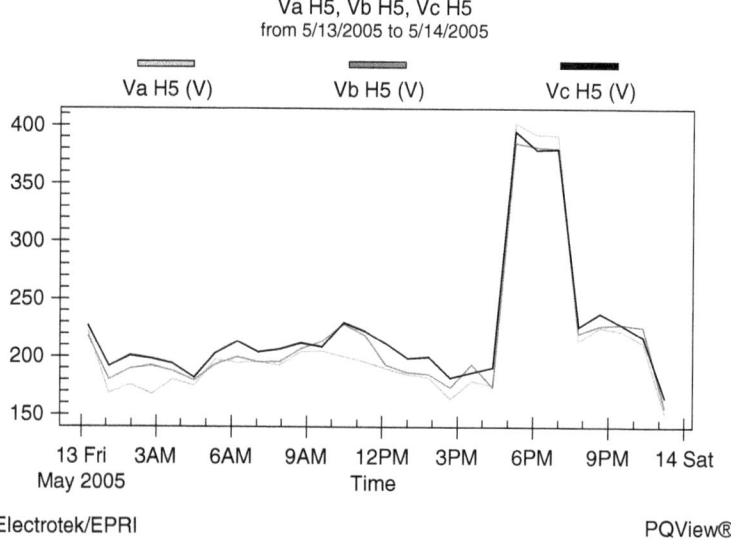

FIGURE 5.27   Fifth harmonic voltage daily profile showing distortion prior to capacitor energizing, with capacitor energized, and after the capacitor is de-energized.

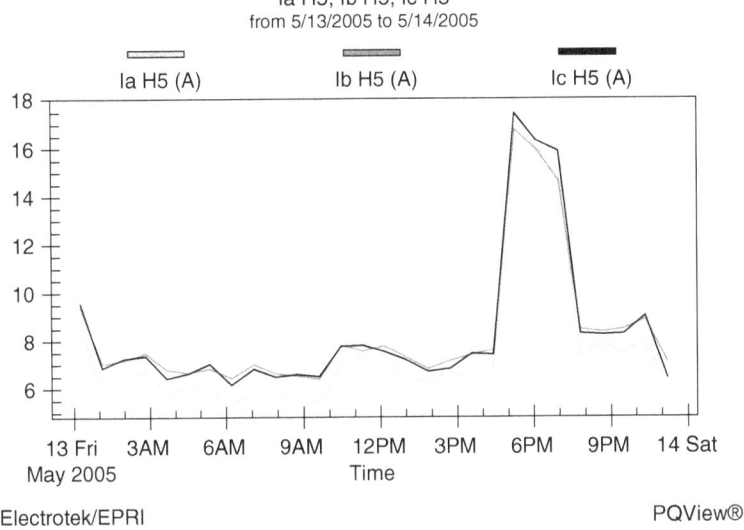

Electrotek/EPRI

PQView®

Figure 5.28 Fifth harmonic current daily profile showing distortion prior to capacitor energizing, with capacitor energized, and after the capacitor is de-energized.

prior to capacitor energizing at approximately 200 V (2.8 percent of rated voltage) and 5 A (5 percent of load current), respectively. Immediately after energizing, they increase to about 400 V (5 percent) and 17 A (12 percent) which are about double the pre-energizing distortion level. This high harmonic distortion persists while the capacitor is in the circuit. When the capacitor is switched off, the fifth harmonic distortion returns to its original level. This section analyzes the nature of parallel resonance in distribution systems.

Figure 5.29 shows a distribution system with potential parallel resonance problems. From the perspective of harmonic sources the shunt capacitor appears in parallel with the equivalent system inductance (source and transformer inductances) at harmonic frequencies as depicted in Fig. 5.30*b*. Furthermore, since the power system is assumed to have an equivalent voltage source of fundamental frequency only, the power system voltage source appears short circuited in the figure.

Parallel resonance occurs when the reactance of X_C and the distribution system cancel each other out. The frequency at which this phenomenon occurs is called the parallel resonant frequency. It can be expressed as follows:

$$f_p = \frac{1}{2\pi} \sqrt{\frac{1}{L_{eq}C} - \frac{R^2}{4L_{eq}^2}} \approx \frac{1}{2\pi} \sqrt{\frac{1}{L_{eq}C}} \qquad (5.21)$$

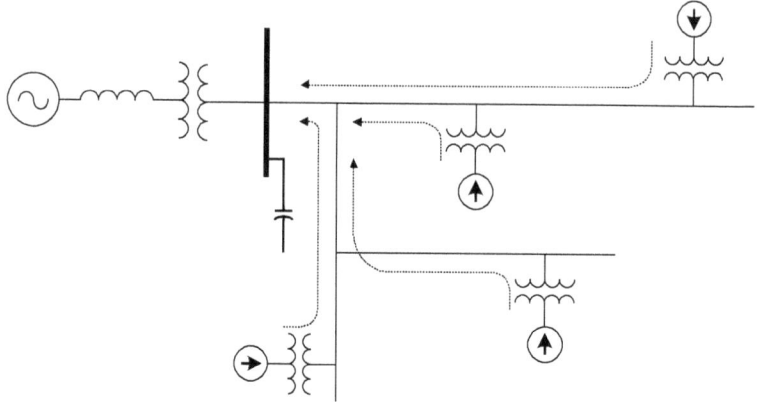

FIGURE 5.29   System with potential parallel resonance problems.

where R = resistance of combined equivalent source and transformer
  (not shown in Fig. 5.30)
   L_{eq} = inductance of combined equivalent source and transformer
   C = capacitance of capacitor bank

At the resonant frequency, the apparent impedance of the parallel combination of the equivalent inductance and capacitance as seen from the harmonic current source becomes very large, i.e.,

$$Z_p = \frac{X_C\left(X_{Leq}+R\right)}{X_C+X_{Leq}+R} = \frac{X_C\left(X_{Leq}+R\right)}{R}$$

$$\approx \frac{X_{Leq}^2}{R} = \frac{X_{C^2}}{R} = QX_{Leq} = QX_C$$

(5.22)

where $Q = X_L/R = X_C/R$ and $R \ll X_{Leq}$. Keep in mind that the reactances in this equation are computed at the resonant frequency.
  Q often is known as the quality factor of a resonant circuit that determines the sharpness of the frequency response. Q varies considerably by location on the power system. It might be less than 5 on a distribution feeder and more than 30 on the secondary bus of a large step-down transformer. From Eq. (5.22), it is clear that during parallel resonance, a small harmonic current can cause a large voltage drop across the apparent impedance, i.e., $V_p = QX_{Leq}I_h$. The voltage near the capacitor bank will be magnified and heavily distorted.

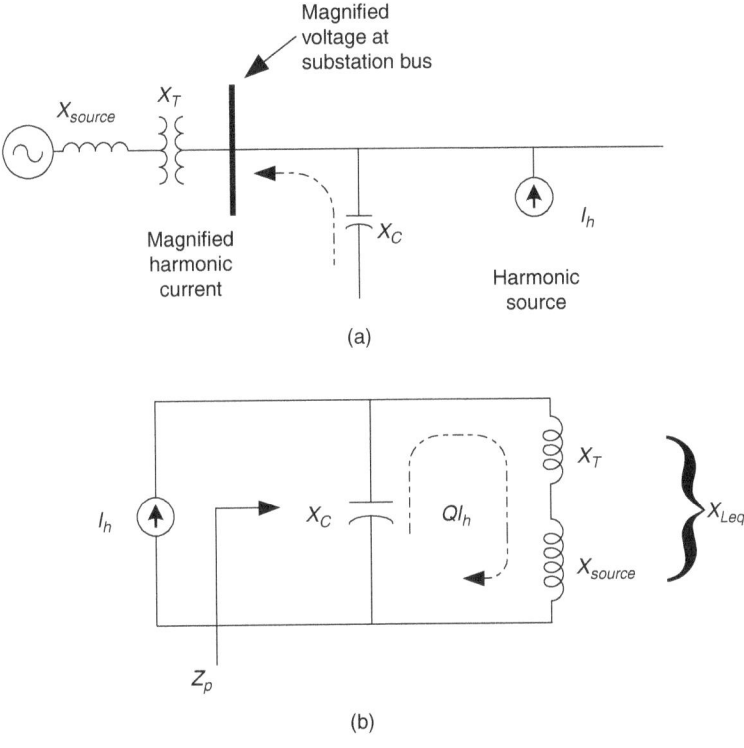

FIGURE 5.30 At harmonic frequencies, the shunt capacitor bank appears in parallel with the system inductance. (a) Simplified distribution circuit; (b) parallel resonant circuit as seen from the harmonic source.

Let us now examine current behavior during the parallel resonance. Let the current flowing in the capacitor bank or into the power system be $I_{\text{resonance}}$; thus,

$$I_{\text{resonance}} = \frac{V_p}{X_C} = \frac{QX_CI_h}{X_C} = QI_h$$

or

$$I_{\text{resonance}} = \frac{V_p}{X_{\text{Leq}}} = \frac{QX_{\text{Leq}}I_h}{X_{\text{Leq}}} = QL_h \tag{5.23}$$

From Eq. (5.23), it is clear that currents flowing in the capacitor bank and in the power system (i.e., through the transformer) will also be

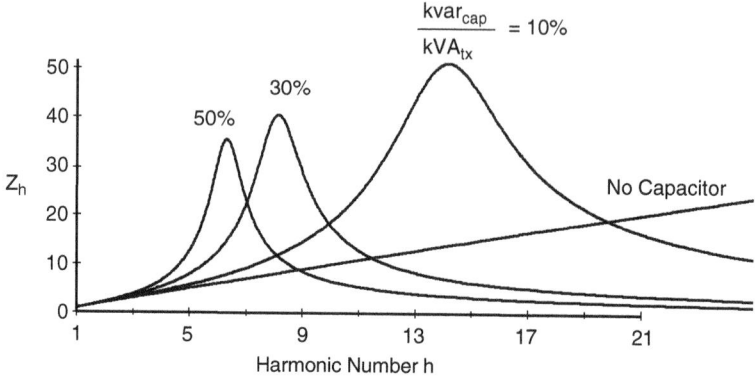

Figure 5.31 System frequency response as capacitor size is varied in relation to transformer.

magnified Q times. This phenomenon will likely cause capacitor failure, fuse blowing, or transformer overheating.

The extent of voltage and current magnification is determined by the size of the shunt capacitor bank. Figure 5.31 shows the effect of varying capacitor size in relation to the transformer on the impedance seen from the harmonic source and compared with the case in which there is no capacitor. The following illustrates how the parallel resonant frequency is computed.

Power systems analysts typically do not have L and C readily available and prefer to use other forms of this relationship. They commonly compute the resonant harmonic h_r based on fundamental frequency impedances and ratings using one of the following:

$$h_r = \sqrt{\frac{X_C}{X_{SC}}} = \sqrt{\frac{MVA_{SC}}{Mvar_{cap}}} \approx \sqrt{\frac{kVA_{tx} \times 100}{kvar_{cap} \times Z_{tx}(\%)}} \qquad (5.24)$$

where h_r = resonant harmonic
X_C = capacitor reactance
X_{SC} = system short-circuit reactance
MVA_{SC} = system short-circuit MVA
MVA_{cap} = Mvar rating of capacitor bank
kVA_{tx} = kVA rating of step-down transformer
Z_{tx} = step-down transformer impedance
kvarcap = kvar rating of capacitor bank

For example, for an industrial load bus where the transformer impedance is dominant, the resonant harmonic for a 1500-kVA, 6-percent transformer and a 500-kvar capacitor bank is approximately

$$h_r \approx \sqrt{\frac{\text{kVA}_{tx} \times 100}{\text{kvar}_{cap} \times Z_{tx}(\%)}} = \sqrt{\frac{1500 \times 100}{500 \times 6}} = 7.07$$

5.9.4 Series Resonance

There are certain instances when a shunt capacitor and the inductance of a transformer or distribution line may appear as a series LC circuit to a source of harmonic currents. If the resonant frequency corresponds to a characteristic harmonic frequency of the nonlinear load, the LC circuit will attract a large portion of the harmonic current that is generated in the distribution system. A customer having no nonlinear load, but utilizing power factor correction capacitors, may in this way experience high harmonic voltage distortion due to neighboring harmonic sources. This situation is depicted in Fig. 5.32.

During resonance, the power factor correction capacitor forms a series circuit with the transformer and harmonic sources. The simplified circuit is shown in Fig. 5.33. The harmonic source shown in this figure represents the total harmonics produced by other loads. The inductance in series with the capacitor is that of the service entrance transformer. The series combination of the transformer inductance and the capacitor bank is very small (theoretically zero) and only limited by its resistance. Thus the harmonic current corresponding to the resonant frequency will flow freely in this circuit. The voltage at the power factor correction capacitor is magnified and highly distorted. This is apparent from the following equation:

$$V_s \left(\text{at power factor capacitor bank} \right) = \frac{X_c}{X_T + X_C + R} V_h \approx \frac{X_C}{R} V_h \quad (5.25)$$

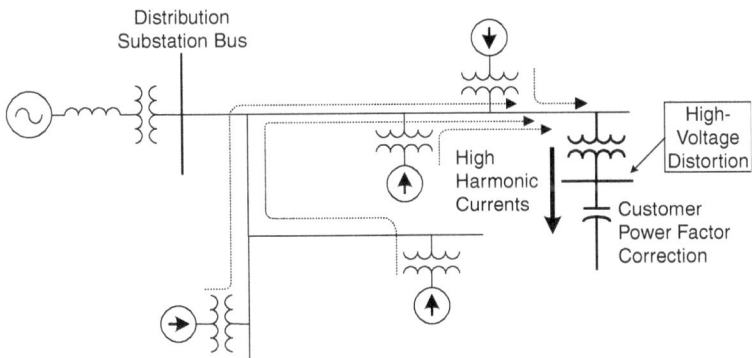

FIGURE 5.32 System with potential series resonance problems.

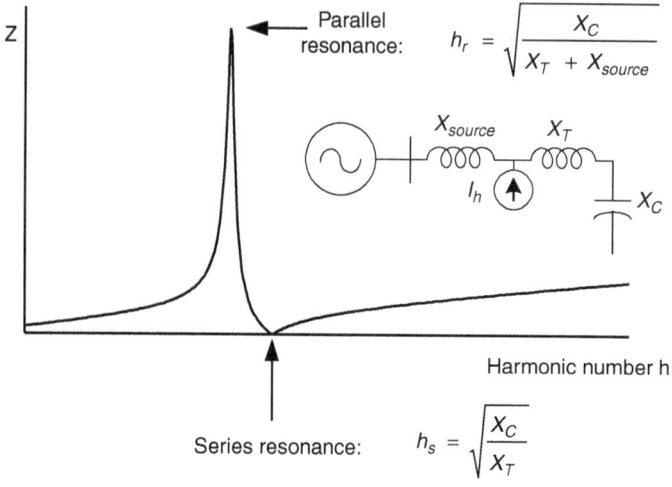

FIGURE **5.33**   Frequency response of a circuit with series resonance.

where V_h and V_s are the harmonic voltage corresponding to the harmonic current I_h and the voltage at the power factor capacitor bank, respectively. The resistance R of the series resonant circuit is not shown in Fig. 5.33, and it is small compared to the reactance.

The negligible impedance of the series resonant circuit can be exploited to absorb desired harmonic currents. This is indeed the principle in designing a notch filter.

In many systems with potential series resonance problems, parallel resonance also arises due to the circuit topology. One of these is shown in Fig. 5.33 where the parallel resonance is formed by the parallel combination between X_{source} and a series between X_T and X_C. The resulting parallel resonant frequency is always smaller than its series resonant frequency due to the source inductance contribution. The parallel resonant frequency can be represented by the following equation:

$$h_r = \sqrt{\frac{X_C}{X_T + X_{\text{source}}}} \tag{5.26}$$

5.9.5   Effects of Resistance and Resistive Load

Determining that the resonant harmonic aligns with a common harmonic source is not always cause for alarm. The damping provided by resistance in the system is often sufficient to prevent catastrophic voltages and currents. Figure 5.34 shows the parallel resonant circuit impedance characteristic for various amounts of resistive load in parallel with the capacitance. As little as 10 percent resistive loading can have a significant beneficial impact on peak impedance. Likewise, if there is a significant length of lines or cables between the capacitor bus and the nearest upline transformer, the resonance will be suppressed. Lines and cables can add a significant amount of the resistance to the equivalent circuit.

Loads and line resistances are the reasons why catastrophic harmonic problems from capacitors on utility distribution feeders are seldom seen. That is not to say that there will not be any harmonic problems due to resonance, but the problems will generally not cause physical damage to the electrical system components. The most troublesome resonant conditions occur when capacitors are installed on substation buses, either utility substations or in industrial facilities. In these cases, where the transformer dominates the system impedance and has a high X/R ratio, the relative resistance is low and the corresponding parallel resonant impedance peak is very sharp and high. This is a common cause of capacitor, transformer, or load equipment failure.

While utility distribution engineers may be able to place feeder banks with little concern about resonance, studies should always be performed for industrial capacitor applications and for utility substation applications. Utility engineers familiar with the problems indicate that about 20 percent of industrial installations for which no studies are performed have major operating disruptions or equipment failure due to resonance. In fact, selecting capacitor sizes from manufacturers' tables to correct the power factor based on average monthly billing data tends to result in a combination that tunes the system near the fifth harmonic. This is one of the worst harmonics to which to be tuned because it is frequently the largest component in the system.

It is a misconception that resistive loads damp harmonics because in the absence of resonance, loads of any kind will have little impact

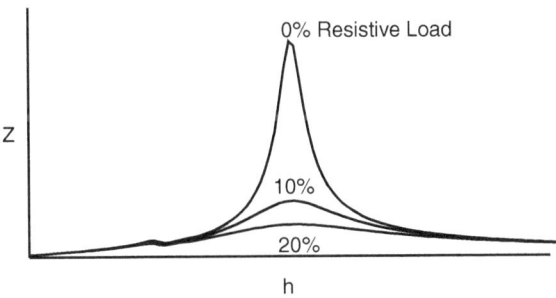

FIGURE 5.34   Effect of resistive loads on parallel resonance.

on the harmonic currents and resulting voltage distortion. Most of the current will flow back into the power source. However, it is very appropriate to say that resistive loads will damp *resonance,* which will lead to a significant reduction in the harmonic distortion.

Motor loads are primarily inductive and provide little damping. In fact, they may increase distortion by shifting the system resonant frequency closer to a significant harmonic. Small, fractional-horsepower motors may contribute significantly to damping because their apparent X/R ratio is lower than that of large three-phase motors.

5.10   Effects of Harmonic Distortion

Harmonic currents produced by nonlinear loads are injected back into the supply systems. These currents can interact adversely with a wide range of power system equipment, most notably capacitors, transformers, and motors, causing additional losses, overheating, and overloading. These harmonic currents can also cause interference with telecommunication lines and errors in power metering. Sections 5.10.1 through 5.10.5 discuss impacts of harmonic distortion on various power system components.

5.10.1   Impact on Capacitors

Problems involving harmonics often show up at capacitor banks first. As discussed in Secs. 5.9.3 and 5.9.4, a capacitor bank experiences high voltage distortion during resonance. The current flowing in the capacitor bank is also significantly large and rich in a monotonic harmonic. Figure 5.35 shows a current waveform of a capacitor bank in resonance with the system at the 11th harmonic. The harmonic current shows up distinctly, resulting in a waveform that is essentially the 11th harmonic riding on top of the fundamental frequency. This current waveform typically indicates that the system is in resonance and a capacitor bank is involved. In such a resonance condition, the rms current is typically higher than the capacitor rms current rating.

IEEE Standard for Shunt Power Capacitors (IEEE Standard 18–1992) specifies the following continuous capacitor ratings:

- 135 percent of nameplate kvar
- 110 percent of rated rms voltage (including harmonics but excluding transients)
- 180 percent of rated rms current (including fundamental and harmonic current)
- 120 percent of peak voltage (including harmonics)

Table 5.1 summarizes an example capacitor evaluation using a computer spreadsheet that is designed to help evaluate the various capacitor duties against the standards.

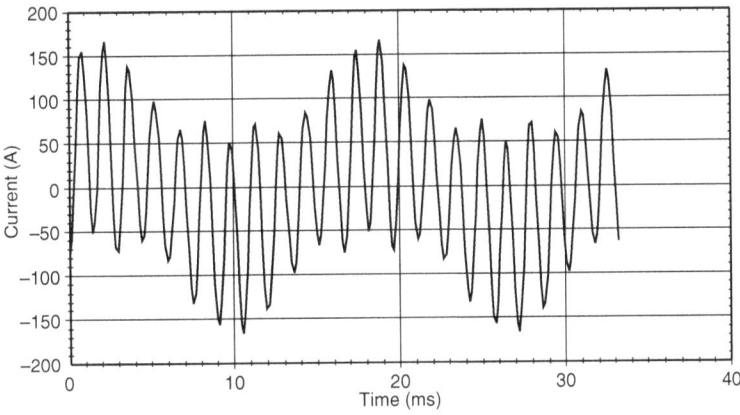

FIGURE 5.35 Typical capacitor current from a system in 11th-harmonic resonance.

The fundamental full-load current for the 1200-kvar capacitor bank is determined from

$$I_C = \frac{\text{kvar}_{3\phi}}{\sqrt{3} \times kV_{LL}} = \frac{1200}{\sqrt{3} \times 13.8} = 50.2 \text{ A}$$

The capacitor is subjected principally to two harmonics: the fifth and the seventh. The voltage distortion consists of 4 percent fifth and 3 percent seventh. This results in 20 percent fifth harmonic current and 21 percent seventh harmonic current. The resultant values all come out well below standard limits in this case, as shown in the box at the bottom of Table 5.1.

5.10.2 Impact on Transformers

Transformers are designed to deliver the required power to the connected loads with minimum losses at fundamental frequency. Harmonic distortion of the current, in particular, as well as of the voltage will contribute significantly to additional heating. To design a transformer to accommodate higher frequencies, designers make different design choices such as using continuously transposed cable instead of solid conductor and putting in more cooling ducts. As a general rule, a transformer in which the current distortion exceeds 5 percent is a candidate for derating for harmonics.

There are three effects that result in increased transformer heating when the load current includes harmonic components:

1. *RMS current.* If the transformer is sized only for the kVA requirements of the load, harmonic currents may result in

the transformer rms current being higher than its capacity. The increased total rms current results in increased conductor losses.

2. *Eddy current losses.* These are induced currents in a transformer caused by the magnetic fluxes. These induced currents flow in the windings, in the core, and in other conducting bodies subjected to the magnetic field of the transformer and cause

---

Recommended Practice for Establishing Capacitor Capabilities When Supplied by Nonsinusoidal Voltages   **IEEE Std 18-1980**

Capacitor Bank Data:

Bank Rating:	1200	kVAr
Voltage Rating:	13800	V (L-L)
Operating Voltage:	13800	V (L-L)
Supplied Compensation:	1200	kVAr

Fundamental Current Rating:	50.2	Amps
Fundamental Frequency:	60	Hz
Capacitive Reactance:	158.700	Ω

Harmonic Distribution of Bus Voltage:

Harmonic Number	Frequency (Hertz)	Volt Mag V_h (% of Fund.)	Volt Mag V_h (Volts)	Line Current I_h (% of Fund.)
1	60	100.00	7967.4	100.00
3	180	0.00	0.0	0.00
5	300	4.00	318.7	20.00
7	420	3.00	239.0	21.00
11	660	0.00	0.0	0.00
13	780	0.00	0.0	0.00
17	1020	0.00	0.0	0.00
19	1140	0.00	0.0	0.00
21	1260	0.00	0.0	0.00
23	1380	0.00	0.0	0.00
25	1500	0.00	0.0	0.00

Voltage Distortion (THD):	5.00	%
RMS Capacitor Voltage:	7977.39	Volts
Capacitor Current Distortion:	29.00	%
RMS Capacitor Current:	52.27	Amps

Capacitor Bank Limits:

	Calculated	Limit	Exceeds Limit
Peak Voltage:	107.0%	120%	No
RMS Voltage:	100.1%	110%	No
RMS Current:	104.1%	180%	No
kVAr:	104.3%	135%	No

TABLE 5.1   Example Capacitor Evaluation

additional heating. This component of the transformer losses increases with the square of the frequency of the current causing the eddy currents. Therefore, this becomes a very important component of transformer losses for harmonic heating.

3. *Core losses*. The increase in core losses in the presence of harmonics will be dependent on the effect of the harmonics on the applied voltage and the design of the transformer core. Increasing the voltage distortion may increase the eddy currents in the core laminations. The net impact that this will have depends on the thickness of the core laminations and the quality of the core steel. The increase in these losses due to harmonics is generally not as critical as the previous two.

Guidelines for transformer derating are detailed in ANSI/IEEE Standard C57.110–1998, *Recommended Practice for Establishing Transformer Capability When Supplying Nonsinusoidal Load Currents*.

The analysis represented in this table can be summarized as follows. The load loss P_{LL} can be considered to have two components: I^2R loss and eddy current loss P_{EC}:

$$P_{LL} = I^2R + P_{EC} \tag{5.27}$$

The I^2R loss is directly proportional to the rms value of the current. However, the eddy current is proportional to the square of the current and frequency, which is defined by

$$P_{EC} = K_{EC} \times I^2 \times h^2 \tag{5.28}$$

where K_{EC} is the proportionality constant.

Using per-unit currents, the per-unit full-load loss under harmonic current conditions is given by

$$P_{LL} = \sum I_h^2 + \left(\sum I_h^2 \times h^2 \right) P_{EC-R} \tag{5.29}$$

where P_{EC-R} is the eddy current loss factor under rated conditions.

The K factor[3] commonly found in power quality literature concerning transformer derating can be defined solely in terms of the harmonic currents as follows:

$$K = \frac{\sum \left(I_h^2 \times h^2 \right)}{\sum I_h^2} \tag{5.30}$$

Then, in terms of the K factor, the rms of the distorted current is derived to be

$$\sqrt{\sum I_h^2} = \sqrt{\frac{1+P_{EC-R}}{1+K \times P_{EC-R}}} \quad (\text{pu}) \tag{5.31}$$

where P_{EC-R} = eddy current loss factor
h = harmonic number
I_h = harmonic current

Thus, the transformer derating can be estimated by knowing the per-unit eddy current loss factor. This factor can be determined by

1. Obtaining the factor from the transformer designer
2. Using transformer test data and the procedure in ANSI/IEEE Standard C57.110
3. Typical values based on transformer type and size (see Table 5.2)

Exceptions

There are often cases with transformers that do not appear to have a harmonics problem from the criteria given in Table 5.2, yet are running hot or failing due to what appears to be overload. One common case found with grounded-wye transformers is that the line currents contain about 8 percent third harmonic, which is relatively low, and the transformer is overheating at less than rated load. Why would this transformer pass the heat run test in the factory, and, perhaps, an overload test also, and fail to perform as expected in practice? Discounting mechanical cooling problems, chances are good that there is some conducting element in the magnetic field that is being affected by the harmonic fluxes. Three of several possibilities are as follows:

- Zero-sequence fluxes will "escape" the core on three-legged core designs (the most popular design for utility distribution substation transformers). This is illustrated in Fig. 5.36. The

Type	MVA	Voltage	P_{EC-R}, %
Dry	≤1	—	3–8
	≥1.5	5 kV HV	12–20
	≤1.5	15 kV HV	9–15
Oil-filled	≤2.5	480 V LV	1
	2.5–5	480 V LV	1–5
	>5	480 V LV	9–15

TABLE 5.2 Typical Values of P_{EC-R}

3rd, 9th, 15th, etc., harmonics are predominantly zero sequence. Therefore, if the winding connections are proper to allow zero-sequence current flow, these harmonic fluxes can cause additional heating in the tanks, core clamps, etc., that would not necessarily be found under balanced three-phase or single-phase tests. The 8-percent line current previously mentioned translates to a neutral third-harmonic current of 24 percent of the phase current. This could add considerably to the leakage flux in the tank and in the oil and air space. Two indicators are charred or bubbled paint on the tank and evidence of heating on the end of a bayonet fuse tube (without blowing the fuse) or bushing end.

• DC offsets in the current can also cause flux to "escape" the confines of the core. The core will become slightly saturated on, for example, the positive half cycle while remaining normal for the negative half cycle. There are a number of electronic power converters that produce current waveforms that are nonsymmetrical either by accident or by design. This can result in a small dc offset on the load side of the transformer (it cannot be measured from the source side). Only a small amount of dc offset is required to cause problems with most power transformers.

• There may be a clamping structure, bushing end, or some other conducting element too close to the magnetic field. It may be sufficiently small in size that there is no notable effect

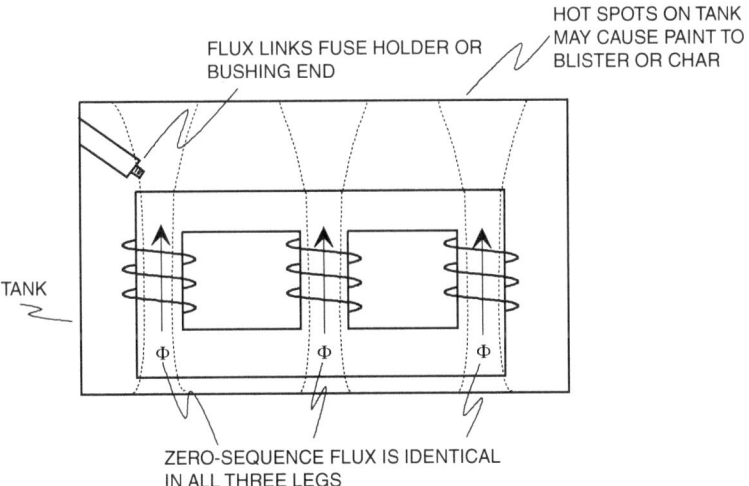

FIGURE 5.36 Zero-sequence flux in three-legged core transformers enters the tank and the air and oil space.

in stray losses at fundamental frequency but may produce a hot spot when subjected to harmonic fluxes.

5.10.3   Impact on Motors

Motors can be significantly impacted by the harmonic voltage distortion. Harmonic voltage distortion at the motor terminals is translated into harmonic fluxes within the motor. Harmonic fluxes do not contribute significantly to motor torque, but rotate at a frequency different than the rotor synchronous frequency, basically inducing high-frequency currents in the rotor. The effect on motors is similar to that of negative-sequence currents at fundamental frequency: The additional fluxes do little more than induce additional losses. Decreased efficiency along with heating, vibration, and high-pitched noises are indicators of harmonic voltage distortion.

At harmonic frequencies, motors can usually be represented by the blocked rotor reactance connected across the line. The lower-order harmonic voltage components, for which the magnitudes are larger and the apparent motor impedance lower, are usually the most important for motors.

There is usually no need to derate motors if the voltage distortion remains within IEEE Standard 519–1992 limits of 5 percent THD and 3 percent for any individual harmonic. Excessive heating problems begin when the voltage distortion reaches 8 to 10 percent and higher. Such distortion should be corrected for long motor life.

Motors appear to be in parallel with the power system impedance with respect to the harmonic current flow and generally shift the system resonance higher by causing the net inductance to decrease. Whether this is detrimental to the system depends on the location of the system resonance prior to energizing the motor. Motors also may contribute to the damping of some of the harmonic components depending on the X/R ratio of the blocked rotor circuit. In systems with many smaller-sized motors, which have a low X/R ratio, this could help attenuate harmonic resonance. However, one cannot depend on this for large motors.

5.10.4   Impact on Telecommunications

Harmonic currents flowing on the utility distribution system or within an end-user facility can create interference in communication circuits sharing a common path. Voltages induced in parallel conductors by the common harmonic currents often fall within the bandwidth of normal voice communications. Harmonics between 540 (ninth harmonic) and 1200 Hz are particularly disruptive. The induced voltage per ampere of current increases with frequency. Triplen harmonics (3rd, 9th, 15th) are especially troublesome in four-wire systems because they are in phase in all conductors of a

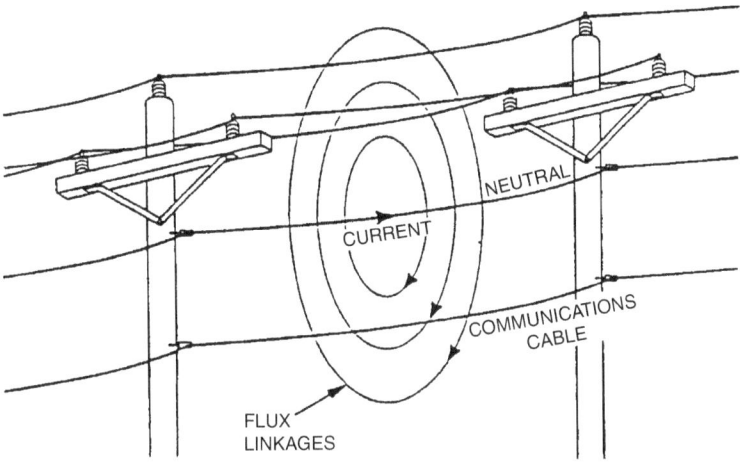

Figure 5.37   Inductive coupling of power system residual current to telephone circuit.

three-phase circuit and, therefore, add directly in the neutral circuit, which has the greatest exposure with the communications circuit.

Harmonic currents on the power system are coupled into communication circuits by either induction or direct conduction. Figure 5.37 illustrates coupling from the neutral of an overhead distribution line by induction. This was a severe problem in the days of open wire telephone circuits. Now, with the prevalent use of shielded, twisted-pair conductors for telephone circuits, this mode of coupling is less significant. The direct inductive coupling is equal in both conductors, resulting in zero net voltage in the loop formed by the conductors.

Inductive coupling can still be a problem if high currents are induced in the shield surrounding the telephone conductors. Current flowing in the shield causes an IR drop (Fig. 5.38), which results in a potential difference in the ground references at the ends of the telephone cable.

Shield currents can also be caused by direct conduction. As illustrated in Fig. 5.39, the shield is in parallel with the power system ground path. If local ground conditions are such that a relatively large amount of current flows in the shield, high shield IR drop will again cause a potential difference in the ground references at the ends of the telephone cable.

5.10.5   Impact on Energy and Demand Metering

Electric utility companies usually measure energy consumption in two quantities: the total cumulative energy consumed and the

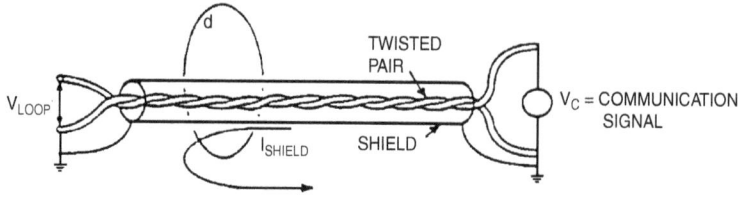

FIGURE 5.38 IR drop in cable shield resulting in potential differences in ground references at ends of cable.

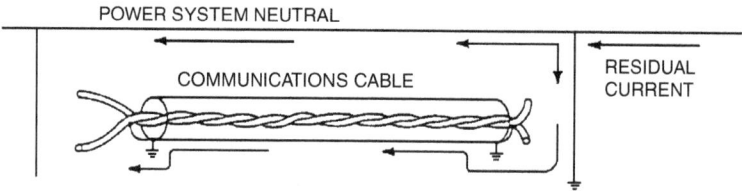

FIGURE 5.39 Conductive coupling through a common ground path.

maximum power used for a given period. Thus, there are two charges in any given billing period especially for larger industrial customers: energy charges and demand charges. Residential customers are typically charged for the energy consumption only. The energy charge represents the costs of producing and supplying the total energy consumed over a billing period and is measured in kilowatt-hours. The second part of the bill, the demand charge, represents utility costs to maintain adequate electrical capacity at all times to meet each customer's peak demand for energy use. The demand charge reflects the utility's fixed cost in providing peak power requirements. The demand charge is usually determined by the highest 15- or 30-min peak demand of use in a billing period and is generally, or in some cases VA depending on specific tariffs, measured in kilowatts.

Both energy and demand charges are measured using the so-called watthour and demand meters. A demand meter is usually integrated to a watthour meter with a timing device to register the peak power use and returns the demand pointer to zero at the end of each timing interval (typically 15 or 30 min).

Harmonic currents from nonlinear loads can impact the accuracy of watthour and demand meters adversely. Traditional watthour meters are based on the induction motor principle. The rotor element or the rotating disk inside the meter revolves at a speed proportional to the power flow. This disk in turn drives a series of gears that move dials on a register.

Conventional magnetic disk watthour meters tend to have a negative error at harmonic frequencies. That is, they register low for power at harmonic frequencies if they are properly calibrated for fundamental frequency. This error increases with increasing frequency. In general, nonlinear loads tend to inject harmonic power back onto the supply system and linear loads absorb harmonic power due to the distortion in the voltage. This is depicted in Fig. 5.40 by showing the directions on the currents.

Thus for the nonlinear load in Fig. 5.40, the meter would read

$$P_{measured} = P_1 - a_3P_3 - a_5P_5 - a_7P_7 - \cdots \qquad (5.32)$$

where a_3, a_5, and a_7 are multiplying factors (<1.0) that represent the inaccuracy of the meter at harmonic frequency. The measured power is a little greater than that actually used in the load because the meter does not subtract off quite all the harmonic powers. However, these powers simply go to feed the line and transformer losses, and some would argue that they should not be subtracted at all. That is, the customer injecting the harmonic currents should pay something additional for the increased losses in the power delivery system.

In the case of the linear load, the measured power is

$$P_{measured} = P_1 + a_3P_3 + a_5P_5 + a_7P_7 + \cdots \qquad (5.33)$$

The linear load absorbs the additional energy, but the meter does not register as much energy as is actually consumed. The question is, Does the customer really want the extra energy? If the load consists of motors, the answer is no, because the extra energy results in losses induced in the motors from harmonic distortion. If the load is resistive, the energy is likely to be efficiently consumed.

Fortunately, in most practical cases where the voltage distortion is within electricity supply recommended limits, the error is very small (much less than 1 percent). The latest electronic meters in use today are based on time-division and digital sampling. These electronic meters are much more accurate than the conventional watthour meter based on induction motor principle. Although these electronic watthour meters are able to measure harmonic components, they could be set to measure only the fundamental power. The user should be careful to ascertain that the meters are measuring the desired quantity.

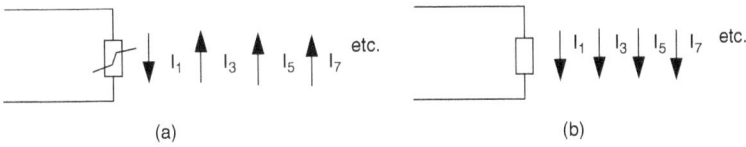

(a)   (b)

FIGURE 5.40   Nominal direction of harmonic currents in (a) nonlinear load and (b) linear load (voltage is distorted).

The greatest potential errors occur when metering demand. The metering error is the result of ignoring the portion of the apparent power that is due solely to the harmonic distortion. Some metering schemes accurately measure the active (P) and reactive power (Q), but basically ignore D. If Q is determined by a second watthour meter fed by a voltage that is phase-shifted from the energy meter, the D term is generally not accounted for—only Q at the fundamental is measured. Even some electronic meters do not account for the total apparent power properly, although many newer meters are certified to properly account for harmonics. Thus, the errors for this metering scheme are such that the measured kVA demand is less than actual. The error would be in favor of the customer.

The worst errors occur when the total current at the metering site is greatly distorted. The total kVA demand can be off by 10 to 15 percent. Fortunately, at the metering point for total plant load, the current distortion is not as greatly distorted as individual load currents. Therefore, the metering error is frequently fairly small. There are, however, some exceptions to this such as pumping stations where a PWM drive is the only load on the meter. While the energy meter should be sufficiently accurate given that the voltage has low distortion, the demand metering could have substantial error.

5.11 Interharmonics

According to the Fourier theory, a periodic waveform can be expressed as a sum of pure sine waves of different amplitudes where the frequency of each sinusoid is an integer multiple of the fundamental frequency of the periodic waveform. A frequency that is an integer multiple of the fundamental frequency is called a harmonic frequency, i.e., $f_h = hf_0$ where f_0 and h are the fundamental frequency and an integer number, respectively.

On the other hand, the sum of two or more pure sine waves with different amplitudes where the frequency of each sinusoid is not an integer multiple of the fundamental frequency does not necessarily result in a periodic waveform. This noninteger multiple of the fundamental frequency is commonly known as an *interharmonic frequency*, i.e., $f_{ih} = h_i f_0$ where h_i is a noninteger number larger than unity. Thus in practical terms, interharmonic frequencies are frequencies between two adjacent harmonic frequencies.

One primary source of interharmonics is the widespread use of electronic power converter loads capable of producing current distortion over a whole range of frequencies, i.e., characteristic and noncharacteristic frequencies.[4] Examples of these loads are ASDs in industrial applications and PWM inverters in UPS applications, active filters, and custom power conditioning equipment. As illustrated in Fig. 5.18, the front end of an ASD is typically a diode rectifier that converts an incoming ac voltage to a dc voltage. An inverter then converts the dc voltage to variable ac voltage with variable frequency. The inverter can produce interharmonics in the current especially when the inverter employs an asynchronous switching scheme. An asynchronous switching scheme is when the ratio of the switching frequency of the power electronic switches is an integer multiple of the fundamental frequency of the inverter voltage output.[5] If the harmonic current passes through the dc link and propagates into the supply system, interharmonic-related problems may arise.

Another significant source of interharmonic distortion commonly comes from rapidly changing load current such as in induction furnaces and cycloconverters. The rapid fluctuation of load current causes sideband frequencies to appear around the fundamental or harmonic frequencies. The generation of interharmonics is best illustrated using an induction furnace example.[6]

Induction furnaces have been widely used to heat ferrous and nonferrous stocks in the forging and extruding industry. Modern induction furnaces use electronic power converters to supply a variable frequency to the furnace induction coil as shown in Fig. 5.41. The frequency at the melting coil varies to match the type of material being melted and the amount of the material in the furnace. The furnace coil and capacitor form a resonant circuit, and the dc-to-ac inverter drives the circuit to keep it in resonance. The inductance of the coil varies depending on the type, temperature, and amount of material as the furnace completes one cycle to another such as from a melt to pour cycle. This situation results in a varying operating frequency for the furnace. The typical range of frequencies for induction furnaces is 150 to 1200 Hz.

We now present an example. An induction furnace has a 12-pulse current source design with reactors on the dc link to smooth the current into the inverter as shown in Fig. 5.41. Typical characteristic harmonics in the ac-side line currents are 11th, 13th, 23rd, 25th, ... , with some noncharacteristic harmonics such as the 5th and 7th also possibly present. However, there are also currents at noninteger frequencies due to the interaction with the inverter output frequency as the furnace goes from one cycle to another. The switching of the

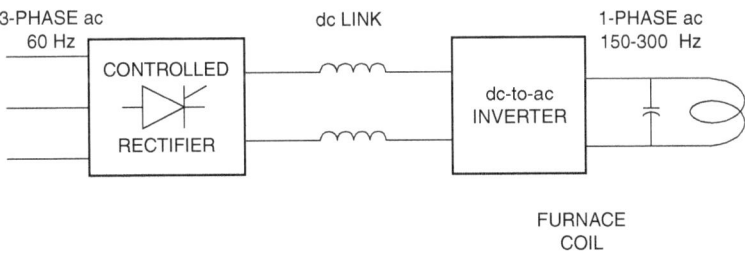

FIGURE 5.41 Block diagram of a modern induction furnace with a CSI.

inverter reflects the frequency of the furnace circuit to the ac-side power through small perturbations of the dc link current. This interaction results in interharmonic frequencies at the ac side and bears no relation to the power supply frequency. The interharmonics appear in pairs at the following frequencies:

$$2f_o \pm f_s, \, 4f_o \pm f_s, \, \dots \qquad (5.34)$$

where f_o and f_s are the furnace operating frequency and the fundamental of the ac main frequency, respectively. Thus, if the furnace operates at 160 Hz, the first interharmonic currents will appear at 260 and 380 Hz. The second pair of lesser magnitude will appear at 580 and 700 Hz. A typical spectrum of induction furnace current is shown in Fig. 5.42. In this particular example, the fifth harmonic was noncharacteristic but was found in significant amounts in nearly all practical power systems. The interharmonic frequencies move slowly, from several seconds to several minutes, through a wide frequency range as the furnace completes its melt and pour cycle. The wide range of the resulting interharmonics can potentially excite resonances in the power supply system.

Our example illustrates how interharmonics are produced in modern induction furnaces. Cycloconverters, ASDs, induction motors with wound rotor using subsynchronous converter cascades, and arcing devices also produce interharmonics in a similar fashion.

Since interharmonics can assume any values between harmonic frequencies, the interharmonic spectrum must have sufficient frequency resolution. Thus, a single-cycle waveform sample is no longer adequate to compute the interharmonic spectrum since it only

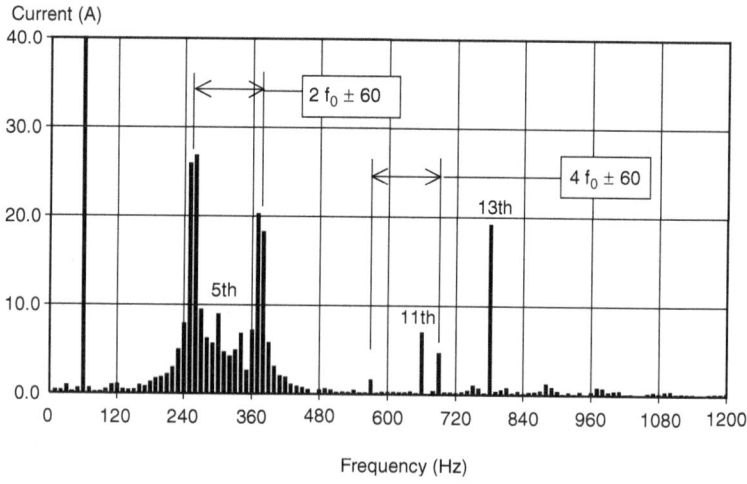

Figure 5.42 Typical spectrum of induction furnace current.

provides a frequency resolution of 50 or 60 Hz. Any frequency in between harmonic frequencies is lost. The one-cycle waveform though is commonly used to compute the harmonic spectrum since there is no frequency between harmonic frequencies.

A 12- or 10-cycle waveform is then recommended for a 60- or 50-Hz power system to achieve higher frequency resolution. The resulting frequency resolution is 5 Hz.[7]

Impacts of interharmonics are similar to those of harmonics such as filter overloading, overheating, power line carrier interference, ripple, voltage fluctuation, and flicker.[7,8] However, solving interharmonic problems can be more challenging, especially when interharmonic frequencies vary from time to time as do those in induction furnaces. Broadband filters are usually used to mitigate interharmonic problems. In the next chapter (Sec. 6.7), a case study of interharmonics causing an electric clock to go faster is presented.

5.12   References

1. Energy Information Agency, *A Look at Commercial Buildings in 1995: Characteristics, Energy Consumption and Energy Expenditures*, DOE/EIA-0625(95), October 1998.
2. D. E. Rice, "Adjustable-Speed Drive and Power Rectifier Harmonics. Their Effects on Power System Components," *IEEE Trans. on Industrial Applications*, IA-22(1), January/February 1986, pp. 161–177.
3. J. M. Frank, "Origin, Development and Design of K-Factor Transformers," in *Conference Record*, 1994 IEEE Industry Applications Society Annual Meeting, Denver, October 1994, pp. 2273–2274.
4. IEC 61000-4-7, *Electromagnetic Compatibility (EMC)*—Part 4-7, "Testing and Measurement Techniques—General Guide on Harmonics and Interharmonics Measurements and Instrumentation, for Power Supply Systems and Equipment Connected Thereto," SC77A, 2000, Draft.
5. N. Mohan, T. M. Undeland, W. P. Robbins, *Power Electronics: Converters, Applications, and Design*, 2d ed., John Wiley & Sons, New York, 1995.
6. R. C. Dugan and L. E Conrad, "Impact of Induction Furnace Interharmonics on Distribution Systems," *Proceedings of the 1999 IEEE Transmission and Distribution Conference*, April 1999, pp. 791–796.
7. WG1 TF3 CD for IEC 61000-1-4, *Electromagnetic Compatibility (EMC)*: "Rationale for Limiting Power-Frequency Conducted Harmonic and Interharmonic Current Emissions from Equipment in the Frequency Range Up to 9 kHz," SC77A, 2001, Draft.
8. IEEE Interharmonic Task Force, "Interharmonics in Power Systems," Cigre 36.05/CIRED 2 CC02 Voltage Quality Working Group, 1997.

5.13   Bibliography

Acha, Enrique, and Madrigal, Manuel, *Power Systems Harmonics: Computer Modelling and Analysis*, John Wiley & Sons, New York, 2001.
Arrillaga, J., Watson, Neville R., Wood, Alan R., Smith, B.C., *Power System Harmonic Analysis*, John Wiley & Sons, New York, 1997.
Dugan, R. C., McGranaghan, M. R., Rizy, D. T., Stovall, J. P., *Electric Power System Harmonics Design Guide*, ORNL/Sub/81–95011/3, Oak Ridge National Laboratory, U.S. DOE, September 1986.

CHAPTER **6**

Applied Harmonics

C hapter 5 showed how harmonics are produced and how they impact various power system components. This chapter shows ways to deal with them, i.e., how to

- Evaluate harmonic distortion
- Properly control harmonics
- Perform a harmonic study
- Design a filter bank

This chapter will also present representative case studies.

6.1 Harmonic Distortion Evaluations

As discussed in Chap. 5, harmonic currents produced by nonlinear loads can interact adversely with the utility supply system. The interaction often gives rise to voltage and current harmonic distortion observed in many places in the system. Therefore, to limit both voltage and current harmonic distortion, IEEE Standard 519-1992[2] proposes to limit harmonic current injection from end users so that harmonic voltage levels on the overall power system will be acceptable if the power system does not inordinately accentuate the harmonic currents. This approach requires participation from both end users and utilities.[1-3]

1. *End users.* For individual end users, IEEE Standard 519-1992 limits the level of harmonic current injection at the point of common coupling (PCC). This is the quantity end users have control over. Recommended limits are provided for both individual harmonic components and the total demand distortion. The concept of PCC is illustrated in Fig. 6.1. These limits are expressed in terms of a percentage of the end user's maximum demand current level, rather than as a percentage of the fundamental. This is intended to provide a common basis for evaluation over time.

2. *The utility.* Since the harmonic voltage distortion on the utility system arises from the interaction between distorted load currents and the utility system impedance, the utility is mainly responsible

for limiting the voltage distortion at the PCC. The limits are given for the maximum individual harmonic components and for the total harmonic distortion (THD). These values are expressed as the percentage of the fundamental voltage. For systems below 69 kV, the THD should be less than 5 percent. Sometimes the utility system impedance at harmonic frequencies is determined by the resonance of power factor correction capacitor banks. This results in a very high impedance and high harmonic voltages. Therefore, compliance with IEEE Standard 519-1992 often means that the utility must ensure that system resonances do not coincide with harmonic frequencies present in the load currents.

Thus, in principle, end users and utilities share responsibility for limiting harmonic current injections and voltage distortion at the PCC.

Since there are two parties involved in limiting harmonic distortions, the evaluation of harmonic distortion is divided into two parts: measurements of the currents being injected by the load and calculations of the frequency response of the system impedance. Measurements should be taken continuously over a sufficient period of time so that time variations and statistical characteristics of the harmonic distortion can be accurately represented. Sporadic measurements should be avoided since they do not represent harmonic characteristics accurately given that harmonics are a continuous phenomenon. The minimum measurement period is usually 1 week since this provides a representative loading cycle for most industrial and commercial loads.

6.1.1   Concept of Point of Common Coupling

Evaluations of harmonic distortion are usually performed at a point between the end user or customer and the utility system where another customer can be served. This point is known as the point of common coupling.[1]

The PCC can be located at either the primary side or the secondary side of the service transformer depending on whether or not multiple customers are supplied from the transformer. In other words, if multiple customers are served from the primary of the transformer, the PCC is then located at the primary. On the other hand, if multiple customers are served from the secondary of the transformer, the PCC is located at the secondary. Figure 6.1 illustrates these two possibilities.

Note that when the primary of the transformer is the PCC, current measurements for verification can still be performed at the transformer secondary. The measurement results should be referred to the transformer high side by the turns ratio of the transformer, and the effect of transformer connection on the zero-sequence components must be taken into account. For instance, a delta-wye connected transformer will not allow zero-sequence current components to

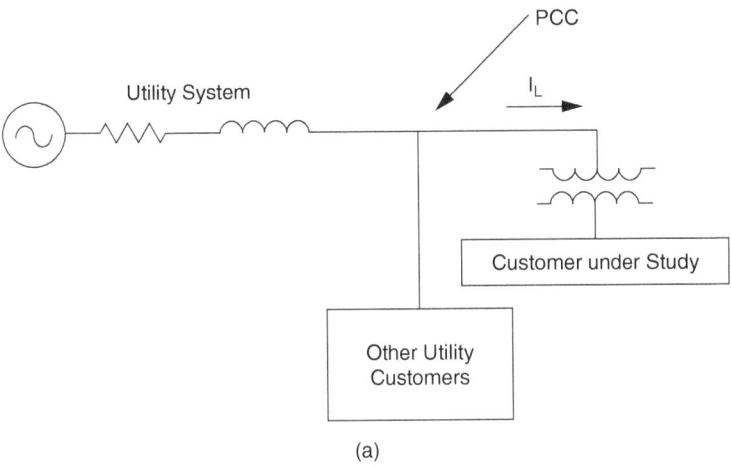

(a)

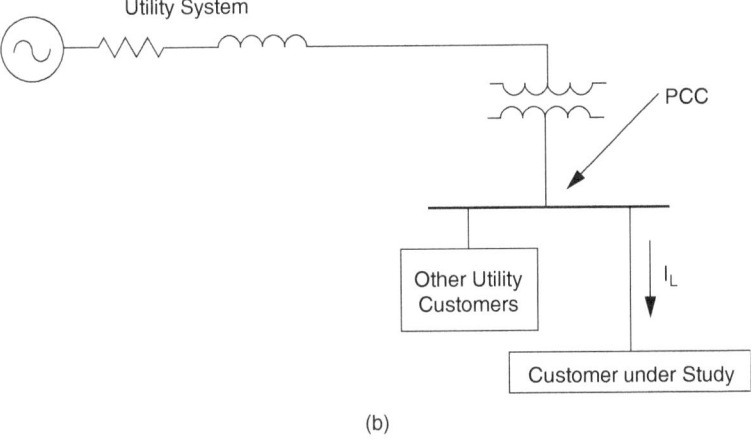

(b)

FIGURE 6.1   PCC selection depends on where multiple customers are served.
(a) PCC at the transformer primary where multiple customers are served.
(b) PCC at the transformer secondary where multiple customers are served.

flow from the secondary to the primary system. These secondary components will be trapped in the primary delta winding. Therefore, zero-sequence components (which are balanced triplen harmonic components) measured on the secondary side would not be included in the evaluation for a PCC on the primary side.

6.1.2   Harmonic Evaluations on the Utility System

Harmonic evaluations on the utility system involve procedures to determine the acceptability of the voltage distortion for all customers.

Bus voltage at PCC, V_n (kV)	Individual harmonic voltage distortion (%)	Total voltage distortion, $THDV_n$ (%)
$V_n \le 69$	3.0	5.0
$69, V_n \le 161$	1.5	2.5
$V_n > 161$	1.0	1.5

SOURCE: IEEE Standard 519-1992, table 11.1.

TABLE 6.1   Harmonic Voltage Distortion Limits in Percent of Nominal Fundamental Frequency Voltage

Should the voltage distortion exceed the recommended limits, corrective actions will be taken to reduce the distortion to a level within limits. IEEE Standard 519-1992 provides guidelines for acceptable levels of voltage distortion on the utility system. These are summarized in Table 6.1. Note that the recommended limits are specified for the maximum individual harmonic component and for the THD.

Note that the definition of the THD in Table 6.1 is slightly different than the conventional definition. The THD value in this table is expressed as a function of the *nominal* system rms voltage rather than of the fundamental frequency voltage magnitude at the time of the measurement. The definition used here allows the evaluation of the voltage distortion with respect to fixed limits rather than limits that fluctuate with the system voltage. A similar concept is applied for the current limits.

There are two important components for limiting voltage distortion levels on the overall utility system:

1. Harmonic currents injected from individual end users on the system must be limited. These currents propagate toward the supply source through the system impedance, creating voltage distortion. Thus by limiting the amount of injected harmonic currents, the voltage distortion can be limited as well. This is indeed the basic method of controlling the overall distortion levels proposed by IEEE Standard 519-1992.

2. The overall voltage distortion levels can be excessively high even if the harmonic current injections are within limits. This condition occurs primarily when one of the harmonic current frequencies is close to a system resonance frequency. This can result in unacceptable voltage distortion levels at some system locations. The highest voltage distortion will generally occur at a capacitor bank that participates in the resonance. This location can be remote from the point of injection.

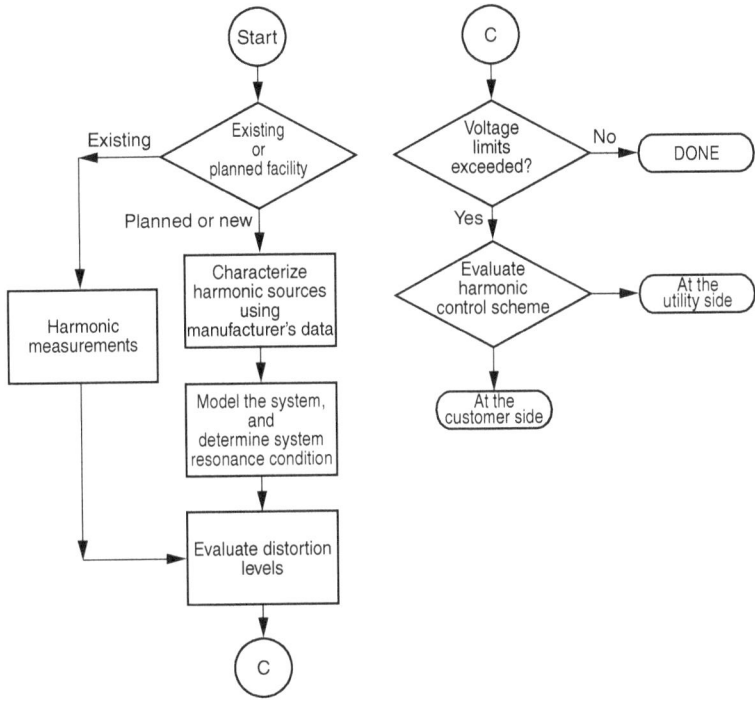

Figure 6.2 Voltage limit evaluation procedure.

Voltage Limit Evaluation Procedure

The overall procedure for utility system harmonic evaluation is described here. This procedure is applicable to both existing and planned installations. Figure 6.2 shows a flowchart of the evaluation procedure.

1. *Characterization of harmonic sources.* Characteristics of harmonic sources on the system are best determined with measurements for existing installations. These measurements should be performed at facilities suspected of having offending nonlinear loads. The duration of measurements is usually at least 1 week so that all the cyclical load variations can be captured. For new or planned installations, harmonic characteristics provided by manufacturers may suffice.

2. *System modeling.* The system response to the harmonic currents injected at end-user locations or by nonlinear devices on the power system is determined by developing a computer model of the system. Distribution and transmission system models are developed as described in Sec. 6.4.

3. *System frequency response.* Possible system resonances should be determined by a frequency scan of the entire power delivery system.

Frequency scans are performed for all capacitor bank configurations of interest since capacitor configuration is the main variable that will affect the resonant frequencies.

4. *Evaluate expected distortion levels.* Even with system resonance close to characteristic harmonics, the voltage distortion levels around the system may be acceptable. On distribution systems, most resonances are significantly damped by the resistances on the system, which reduces magnification of the harmonic currents. The estimated harmonic sources are used with the system configuration yielding the worst-case frequency-response characteristics to compute the highest expected harmonic distortion. This will indicate whether or not harmonic mitigation measures are necessary.

5. *Evaluate harmonic control scheme.* Harmonic control options consist of controlling the harmonic injection from nonlinear loads, changing the system frequency-response characteristics, or blocking the flow of harmonic currents by applying harmonic filters. Design of passive filters for some systems can be difficult because the system characteristics are constantly changing as loads vary and capacitor banks are switched. Section 6.2 discusses harmonic controls in detail.

6.1.3   Harmonic Evaluation for End-User Facilities

Harmonic problems are more common at end-user facilities than on the utility supply system. Most nonlinear loads are located within end-user facilities, and the highest voltage distortion levels occur close to harmonic sources. The most significant problems occur when there are nonlinear loads and power factor correction capacitors that result in resonant conditions.

IEEE Standard 519-1992 establishes harmonic current distortion limits at the PCC. The limits, summarized in Table 6.2, are dependent on the customer load in relation to the system short-circuit capacity at the PCC.

The variables and additional restrictions to the limits given in Table 6.2 are:

- I_h is the magnitude of individual harmonic components (rms amps).
- I_{SC} is the short-circuit current at the PCC.
- I_L is the fundamental component of the maximum demand load current at the PCC. It can be calculated as the average of the maximum monthly demand currents for the previous 12 months or it may have to be estimated.
- The individual harmonic component limits apply to the odd-harmonic components. Even-harmonic components are limited to 25 percent of the limits.

I_{SC}/I_L	$h < 11$	$11 \leq h < 17$	$17 \leq h < 23$	$23 \leq h < 35$	$35 \leq h$	TDD
<20	4.0	2.0	1.5	0.6	0.3	5.0
20–50	7.0	3.5	2.5	1.0	0.5	8.0
50–100	10.0	4.5	4.0	1.5	0.7	12.0
100–1000	12.0	5.5	5.0	2.0	1.0	15.0
>1000	15.0	7.0	6.0	2.5	1.4	20.0
<20*	2.0	1.0	0.75	0.3	0.15	2.5
20–50	3.5	1.75	1.25	0.5	0.25	4.0
50–100	5.0	2.25	2.0	0.75	0.35	6.0
100–1000	6.0	2.75	2.5	1.0	0.5	7.5
>1000	7.5	3.5	3.0	1.25	0.7	10.0
<50	2.0	1.0	0.75	0.3	0.15	2.5
≥50	3.0	1.50	1.15	0.45	0.22	3.75

*All power generation equipment applications are limited to these values of current distortion regardless of the actual short-circuit current ratio I_{SC}/I_L.
SOURCE: IEEE Standard 519-1992, tables 10.3, 10.4, 10.5.

TABLE 6.2 Harmonic Current Distortion Limits (I_h) in Percent of I_L

- Current distortion which results in a dc offset at the PCC is not allowed.
- The total demand distortion (TDD) is expressed in terms of the maximum demand load current, i.e.,

$$TDD = \frac{\sqrt{\sum_{2} I_h^2}}{I_L} \times 100\% \qquad (6.1)$$

- If the harmonic-producing loads consist of power converters with pulse number q higher than 6, the limits indicated in Table 6.2 are increased by a factor equal to $\sqrt{q/6}$.

In computing the short-circuit current at the PCC, the normal system conditions that result in minimum short-circuit capacity at the PCC should be used since this condition results in the most severe system impacts.

A procedure to determine the short-circuit ratio is as follows:

1. Determine the three-phase short-circuit duty I_{SC} at the PCC. This value may be obtained directly from the utility and expressed in amperes. If the short-circuit duty is given in

megavoltamperes, convert it to an amperage value using the following expression:

$$I_{SC} = \frac{1000 \times MVA}{\sqrt{3}\,kV} \text{ A} \qquad (6.2)$$

where MVA and kV represent the three-phase short-circuit capacity in megavoltamperes and the line-to-line voltage at the PCC in kV, respectively.

2. Find the load average kilowatt demand P_D over the most recent 12 months. This can be found from billing information.

3. Convert the average kilowatt demand to the average demand current in amperes using the following expression:

$$I_L = \frac{kW}{PF\,\sqrt{3}\,kV} \text{ A} \qquad (6.3)$$

where PF is the average billed power factor.

4. The short-circuit ratio is now determined by:

$$\text{Short-circuit ratio} = \frac{I_{SC}}{I_L} \qquad (6.4)$$

This is the short-circuit ratio used to determine the limits on harmonic currents in IEEE Standard 519-1992.

In some instances, the average of the maximum demand load current at the PCC for the previous 12 months is not available. In such circumstances, this value must be estimated based on the predicted load profiles. For seasonal loads, the average should be over the maximum loads only.

Current Limit Evaluation Procedure

This procedure involves evaluation of the harmonic generation characteristics from individual end-user loads with respect to IEEE Standard 519-1992 limits. However, special consideration is required when considering power factor correction equipment:

1. Define the PCC. For industrial and commercial end users, the PCC is usually at the primary side of a service transformer supplying the facility.

2. Calculate the short-circuit ratio at the PCC and find the corresponding limits on individual harmonics and on the TDD.

3. Characterize the harmonic sources. Individual nonlinear loads in the facility combine to form the overall level of harmonic current generation. The best way to characterize harmonic current in an existing facility is to perform measurements at the PCC over a period of time (at least 1 week). For planning studies, the harmonic current can be estimated knowing the characteristics of individual nonlinear loads and the percentage of the total load made up by these nonlinear loads. Typical characteristics of individual harmonic sources were presented in Secs. 5.6 and 5.7.

4. Evaluate harmonic current levels with respect to current limits using Table 6.2. If these values exceed limits, the facility does not meet the limit recommended by IEEE Standard 519-1992 and mitigation may be required.

6.2   Principles for Controlling Harmonics

Harmonic distortion is present to some degree on all power systems. Fundamentally, one needs to control harmonics only when they become a problem. There are three common causes of harmonic problems:

1. The source of harmonic currents is too great.

2. The path in which the currents flow is too long (electrically), resulting in either high-voltage distortion or telephone interference.

3. The response of the system magnifies one or more harmonics to a greater degree than can be tolerated.

When a problem occurs, the basic options for controlling harmonics are:

1. Reduce the harmonic currents produced by the load.

2. Add filters to either siphon the harmonic currents off the system, block the currents from entering the system, or supply the harmonic currents locally.

3. Modify the frequency response of the system by filters, inductors, or capacitors.

These options are described in Secs. 6.2.1 through 6.2.3.

6.2.1   Reducing Harmonic Currents in Loads

There is often little that can be done with existing load equipment to significantly reduce the amount of harmonic current it is producing unless it is being misoperated. While an overexcited transformer can be brought back into normal operation by lowering the applied voltage to the correct range, arcing devices and most electronic power converters are locked into their designed characteristics.

Pulse-width modulated (PWM) drives that charge the dc bus capacitor directly from the line without any intentional impedance are one exception to this. Adding a line reactor or transformer in series (as shown in Sec. 5.7.1) will significantly reduce harmonics, as well as provide transient protection benefits.

Transformer connections can be employed to reduce harmonic currents in three-phase systems. Phase-shifting half of the 6-pulse power converters in a plant load by 30° can approximate the benefits of 12-pulse loads by dramatically reducing the fifth and seventh harmonics. Delta-connected transformers can block the flow of zero-sequence harmonics (typically triplens) from the line. Zigzag and grounding transformers can shunt the triplens off the line.

Purchasing specifications can go a long way toward preventing harmonic problems by penalizing bids from vendors with high harmonic content. This is particularly important for such loads as high-efficiency lighting.

6.2.2   Filtering

The shunt filter works by short-circuiting harmonic currents as close to the source of distortion as practical. This keeps the currents out of the supply system. This is the most common type of filtering applied because of economics and because it also tends to correct the load power factor as well as remove the harmonic current.

Another approach is to apply a series filter that blocks the harmonic currents. This is a parallel-tuned circuit that offers a high impedance to the harmonic current. It is not often used because it is difficult to insulate and the load voltage is very distorted. One common application is in the neutral of a grounded-wye capacitor to block the flow of triplen harmonics while still retaining a good ground at fundamental frequency.

Active filters work by electronically supplying the harmonic component of the current into a nonlinear load. More information on filtering is given in Sec. 6.5.

6.2.3   Modifying the System Frequency Response

There are a number of methods to modify adverse system responses to harmonics:

1. Add a shunt filter. Not only does this shunt a troublesome harmonic current off the system, but it completely changes the system response, most often, but not always, for the better.

2. Add a reactor to detune the system. Harmful resonances generally occur between the system inductance and shunt

power factor correction capacitors. The reactor must be added between the capacitor and the supply system source. One method is to simply put a reactor in series with the capacitor to move the system resonance without actually tuning the capacitor to create a filter. Another is to add reactance in the line.

3. Change the capacitor size. This is often one of the least expensive options for both utilities and industrial customers.

4. Move a capacitor to a point on the system with a different short-circuit impedance or higher losses. This is also an option for utilities when a new bank causes telephone interference—moving the bank to another branch of the feeder may very well resolve the problem. This is frequently not an option for industrial users because the capacitor cannot be moved far enough to make a difference.

5. Remove the capacitor and simply accept the higher losses, lower voltage, and power factor penalty. If technically feasible, this is occasionally the best economic choice.

6.3   Where to Control Harmonics

The strategies for mitigating harmonic distortion problems differ somewhat by location. The following techniques are ways for controlling harmonic distortion on both the utility distribution feeder and end-user power system.

6.3.1   On Utility Distribution Feeders

The X/R ratio of a utility distribution feeder is generally low. Therefore, the magnification of harmonics by resonance with feeder banks is usually minor in comparison to what might be found inside an industrial facility. Utility distribution engineers are accustomed to placing feeder banks where they are needed without concern about harmonics. However, voltage distortion from the resonance of feeder banks may exceed limits in a few cases and require mitigation. When problems do occur, the usual strategy is to first attempt a solution by moving the offending bank or changing the capacitor size or neutral connection.

Some harmonic problems associated with feeder capacitor banks are due to increasing the triplen harmonics in the neutral circuit of the feeder. To change the flow of zero-sequence harmonic currents, changes are made to the neutral connection of wye-connected banks. To block the flow, the neutral is allowed to float. In other cases, it is more advantageous to aid the flow by putting a reactor in the neutral to convert the bank into a tuned resonant shunt for a zero-sequence harmonic.

FIGURE 6.3   Filter installation on an overhead distribution feeder. Oil-insulated, iron-core reactors are mounted on a separate pole from the capacitor bank and switches. (*Courtesy of Gilbert Electrical Systems.*)

Harmonic problems on distribution feeders often exist only at light load. The voltage rises, causing the distribution transformers to produce more harmonic currents and there is less load to damp out resonance. Switching the capacitors off at this time frequently solves the problem.

Should harmonic currents from widely dispersed sources require filtering on distribution feeders, the general idea is to distribute a few filters toward the ends of the feeder. While this is not done frequently, the number of feeder filter installations is growing. Figure 6.3 shows one example of a filter installed on an overhead distribution feeder. This shortens the average path for the harmonic currents, reducing the opportunity for telephone interference and reducing the harmonic voltage drop in the lines. The filters appear as nearly a short circuit to at least one harmonic component. This keeps the voltage distortion on the feeder to a minimum. With the ends of the feeder "nailed down" by filters with respect to the voltage distortion, it is more difficult for the voltage distortion to rise above limits elsewhere.

Harmonic flow studies should always be performed when large capacitor banks are installed in distribution substations. One cannot count on system losses to damp out resonance at this point on the system, and magnification by resonance can be severe.

6.3.2   In End-User Facilities

When harmonic problems arise in an end-user facility, the first step is to determine if the main cause is resonance with power factor capacitors in the facility. When it is, first attempt a simple solution by using a different capacitor size. With automatic power factor controllers, it may be possible to select a control scheme that avoids the configuration that causes problems. In other cases, there will be so many capacitors switched at random with loads that it will be impossible to avoid resonant conditions. Filtering will be necessary.

Installation of filters on end-user low-voltage systems is generally more practical and economical than on utility distribution systems. The criteria for filter installation are more easily met, and filtering equipment is more readily available on the market.

When the magnitude of harmonic currents injected by loads is excessive, industrial users should also investigate means of reducing harmonics by using different transformer connections and line chokes. In office buildings, zigzag transformers and triplen harmonic filters can reduce the impact of triplen harmonic currents on neutral circuits.

Studies should be performed on all capacitors installed on the main bus in industrial systems. At this location, there are insufficient line losses to dampen resonance. Thus, when resonance coincides with a harmonic frequency that is a strong component in the load current, the resulting voltage distortion is often severe.

Resonance problems are often less severe when capacitors are located out on the plant floor on motors and in motor control centers. This also has the benefit of reducing the losses in the system compared to simply placing the capacitor on the main bus. Of course, this solution can be more costly than use of a single capacitor on the main bus due to the numerous installations required. This assumes that the cables are sufficiently long to introduce enough resistance into the circuit to dampen the resonance. In plants with short cables, it may not be possible to achieve significant harmonic reduction benefit.

6.4   Harmonic Studies

Harmonic studies play an important role in characterizing and understanding the extent of harmonic problems. Harmonic studies are often performed when

1. Finding a solution to an existing harmonic problem
2. Installing large capacitor banks on utility distribution systems or industrial power systems
3. Installing large nonlinear devices or loads
4. Designing a harmonic filter
5. Converting a power factor capacitor bank to a harmonic filter

Harmonic studies provide a means to evaluate various possible solutions and their effectiveness under a wide range of conditions before implementing a final solution. In this section, methods for carrying out harmonic studies are presented.

6.4.1 Harmonic Study Procedure

The ideal procedure for performing a power systems harmonics study can be summarized as follows:

1. Determine the objectives of the study. This is important to keep the investigation on track. For example, the objective might be to identify what is causing an existing problem and solve it. Another might be to determine if a new plant expansion containing equipment such as adjustable-speed drives (ASDs) and capacitors is likely to have problems.

2. If the system is complex, make a premeasurement computer simulation based on the best information available. Measurements are expensive in terms of labor, equipment, and possible disruption to plant operations. It will generally be economical to have a good idea what to look for and where to look before beginning the measurements.

3. Make measurements of the existing harmonic conditions, characterizing sources of harmonic currents and system bus voltage distortion.

4. Calibrate the computer model using the measurements.

5. Study the new circuit condition or existing problem.

6. Develop solutions (filter, etc.) and investigate possible adverse system interactions. Also, check the sensitivity of the results to important variables.

7. After the installation of proposed solutions, perform monitoring to verify the correct operation of the system.

Admittedly, it is not always possible to perform each of these steps ideally. The most often omitted steps are one, or both, measurement steps due to the cost of engineering time, travel, and equipment charges. An experienced analyst may be able to solve a problem without measurements, but it is strongly recommended that the initial measurements be made if at all possible because there are many unpleasant surprises lurking in the shadows of harmonics analysis.

6.4.2 Developing a System Model

There are two fundamental issues that need to be considered in developing a system model for harmonic simulation studies. The first issue is the extent of the system model to be included in the simulation. Secondly, one must decide whether the model should be represented as a single-phase equivalent or a full three-phase model.

As an example of model extent, suppose a utility plans to install a large capacitor bank on a distribution feeder and would like to evaluate the frequency response associated with the bank. Representing the entire distribution system is usually not practical because it would be time-consuming to develop the model and it would strain computational resources to run simulations. One approach would be to start developing a model one or two buses back from the bus of interest and include everything in between. Another approach would be to start with a small simple circuit that accurately represents the phenomena and add more of the system details to determine the impact on the solution result. At the point when adding more system details does not change the analysis results, the physical system is sufficiently represented by the simulation model.

In modeling distribution systems for harmonic studies, it is usually sufficient to represent the upstream transmission system with a short-circuit equivalent at the high-voltage side of the substation transformer. The leakage impedance of the transformer dominates the short-circuit equivalent and effectively isolates the transmission and distribution for many studies. However, if there is a capacitor bank near the high-voltage side of the transformer, part of the transmission system must be modeled to include the capacitor bank. The combination of the transformer and the capacitor bank may behave as a filter for some frequency as seen from the low-voltage side of the transformer.

Distribution system components downstream from the substation transformer (or at the low-voltage side) such as feeder lines, capacitor banks, key service transformers, and end-user capacitor banks must be represented. Since the feeder capacitor banks dominate the system capacitance, it is usually acceptable to neglect capacitance from overhead feeder lines. However, if there is a significant amount of underground distribution cable, cable capacitance should be represented, especially if the study is concerned with higher-order harmonics.

The analyst must then decide if the model should be represented as a complete three-phase model or a single-phase equivalent. A single-phase equivalent model is generally simpler and less complicated to develop compared to a three-phase model. However, it is often inadequate to analyze unbalanced phenomena or systems with numerous single-phase loads.

Fortunately, there is a rule that permits the simplified positive-sequence modeling for many three-phase industrial loads. Determining the response of the system to positive-sequence harmonics is straightforward since both utility and industrial power engineers

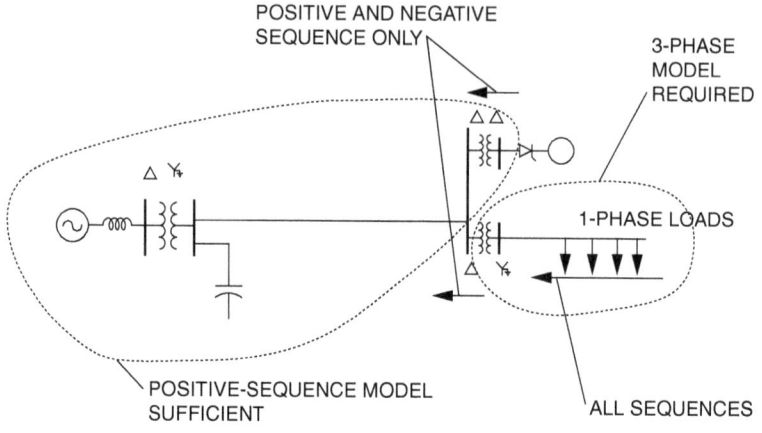

POSITIVE AND NEGATIVE SEQUENCE ONLY

3-PHASE MODEL REQUIRED

1-PHASE LOADS

POSITIVE-SEQUENCE MODEL SUFFICIENT

ALL SEQUENCES

Figure 6.4 Effects of transformer connection on the modeling requirements for analyzing harmonic flows in networks.

are accustomed to doing such modeling in their load flow and voltage drop analyses. The rule may be simply stated:

> When there is a delta winding in a transformer anywhere in series with the harmonic source and the power system, only the positive-sequence circuit need be represented to determine the system response. It is impossible for zero-sequence harmonics to be present; they are blocked.

Figure 6.4 illustrates this principle, showing what models apply to different parts of the system.

Both the positive- and negative-sequence networks are generally assumed to have the same response to harmonics. Sometimes measurements will show triplen harmonics in the line upstream from a delta winding. One normally assumes these harmonics are zero sequence. They may be, depending on what other sources are in system. However, they can also be due to unbalanced harmonic sources, one example of which would be an arc furnace. Only the triplens that are in phase are zero sequence and are blocked by the delta winding. Therefore, it is common to include triplen harmonics when performing analysis using a positive-sequence model.

The symmetrical component technique fails to yield an advantage when analyzing four-wire utility distribution feeders with numerous single-phase loads. Both the positive- and zero-sequence networks come into play. It is generally impractical to consider analyzing the system manually, and most computer programs capable of accurately modeling these systems simply set up the coupled three-phase equations and solve them directly. Fortunately, some computer tools now make it almost as easy to develop a three-phase model as to make a single-phase equivalent. It takes no more time to solve the complete three-phase model than to solve the sequence networks because they would have to be coupled also. Not only does the symmetrical component technique fail to yield an advantage in this case, but analysts often make errors and inadvertently violate the assumptions of the method. It is not generally recommended that harmonic analysis of unbalanced circuits be done using symmetrical components. It should be attempted only by those who are absolutely certain of their understanding of the method and its assumptions.

6.4.3   Modeling Harmonic Sources

Most harmonic flow analysis on power systems is performed using steady-state, linear circuit solution techniques. Harmonic sources, which are nonlinear elements, are generally considered to be injection sources into the linear network models. They can be represented as current injection sources or voltage sources.

For most harmonic flow studies, it is suitable to treat harmonic sources as simple sources of harmonic currents. This is illustrated in Fig. 6.5 where an electronic power converter is replaced with a current source in the equivalent circuit. The voltage distortion at the service bus is generally relatively low, less than 5 percent. Therefore, the current distortion for many nonlinear devices is relatively constant and independent of distortion in the supply system.

Values of injected current should be determined by measurement. In the absence of measurements and published data, it is common to assume that the harmonic content is inversely proportional to the

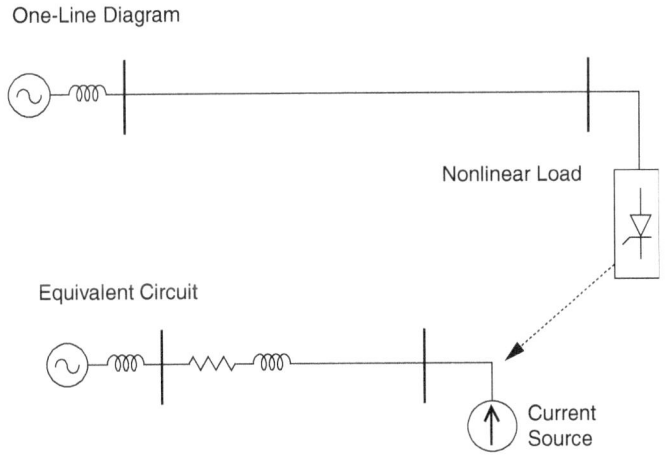

FIGURE 6.5   Representing a nonlinear load with a harmonic current source for analysis.

Harmonic	Six-pulse ASD	PWM drive	Arc lighting	SMPS
1	100	100	100	100
3			20*	70
5	18	90	7	40
7	12	80	3	15
9			2.4*	7
11	6	75	1.8	5
13	4	70	0.8	3

*For single-phase or unbalanced three-phase modeling;
otherwise assume triplen is zero.
ASD = adjustable-speed drive, PWM = pulse-width modulated,
SMPS = switch-mode power supply.

TABLE 6.3 Typical Percent Harmonic Distortion of Common Harmonic Sources: Odd Harmonics, 1 through 13

harmonic number. That is, the fifth-harmonic current is one-fifth, or 20 percent, of the fundamental, etc. This is derived from the Fourier series for a square wave, which is at the foundation of many nonlinear devices. However, it does not apply very well to the newer technology PWM drives and switch-mode power supplies, which have a much higher harmonic content. Table 6.3 shows typical values to assume for analysis of several types of devices.

When the system is near resonance, a simple current source model will give an excessively high prediction of voltage distortion. The model tries to inject a constant current into a high impedance, which is not a valid representation of reality. The harmonic current will not remain constant at a high-voltage distortion. Often, this is inconsequential because the most important thing is to know that the system cannot be successfully operated in resonance, which is readily observable from the simple model. Once the resonance is eliminated by, for example, adding a filter, the model will give a realistic answer.

For the cases where a more accurate answer is required during resonant conditions, a more sophisticated model must be used. For many power system devices, a Thevenin or Norton equivalent is adequate (see Fig. 6.6). The additional impedance moderates the response of the parallel resonant circuit.

A Thevenin equivalent is obtained in a straightforward manner for many nonlinear loads. For example, an arc furnace is well represented by a square-wave voltage of peak magnitude approximately 50 percent of the nominal ac system voltage. The series impedance is

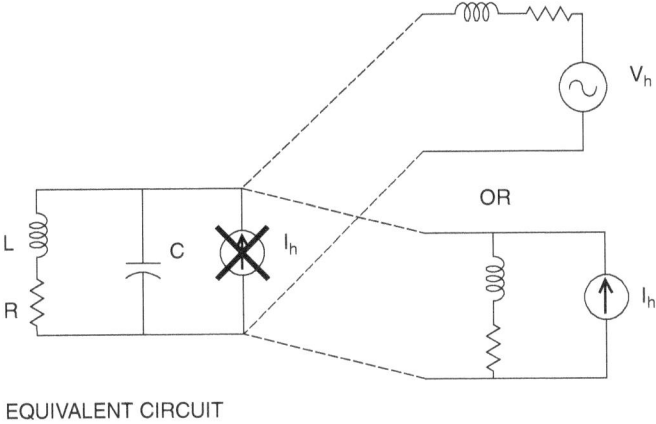

EQUIVALENT CIRCUIT

FIGURE 6.6 Replacing the simple current source model with a Thevenin or Norton equivalent for better source models of resonant conditions.

simply the short-circuit impedance of the furnace transformer and leads (the lead impedance is the larger of the two). Unfortunately, it is difficult to determine clear-cut equivalent impedances for many nonlinear devices. In these cases, a detailed simulation of the internals of the harmonic-producing load is necessary. This can be done with computer programs that iterate on the solution or through detailed time-domain analysis.

Fortunately, it is seldom essential to obtain such great accuracy during resonant conditions and analysts do not often have to take these measures. However, modeling arcing devices with a Thevenin model is recommended regardless of need.

6.4.4   Computer Tools for Harmonics Analysis

The preceding discussion has given the reader an idea of the types of functions that must be performed for harmonics analysis of power systems. It should be rather obvious that for anything but the simplest of circuits, a sophisticated computer program is required. The characteristics of such programs and the heritage of some popular analysis tools are described here.

First, it should be noted that one circuit appears frequently in simple industrial systems that does lend itself to manual calculations (Fig. 6.7). It is basically a one-bus circuit with one capacitor. Two things may be done relatively easily:

1. *Determine the resonant frequency.* If the resonant frequency is near a potentially damaging harmonic, either the capacitor must be changed or a filter designed.

2. *Determine an estimate of the voltage distortion due to the current* I_h. The voltages V_h are given by

$$V_h = \left(\frac{R + j\omega L}{1 - \omega^2 LC + j\omega RC}\right) I_h \qquad (6.5)$$

where $\omega = 2\pi(hf_1)$
  $h = 2, 3, 4, \ldots$
  $f_1 =$ fundamental frequency of power system

Given that the resonant frequency is not near a significant harmonic and that projected voltage distortion is low, the application will probably operate successfully.

Unfortunately, not all practical cases can be represented with such a simple circuit. In fact, adding just one more bus with a capacitor to the simple circuit in Fig. 6.7 makes the problem a real challenge to even the most skilled analysts. However, a computer can perform the chore in milliseconds.

To use the computer tools commonly available, the analyst must describe the circuit configuration, loads, and the sources to the program. Data that must be collected include

- Line and transformer impedances
- Transformer connections
- Capacitor values and locations (critical)
- Harmonic spectra for nonlinear loads
- Power source voltages

These values are entered into the program, which automatically adjusts impedances for frequency and computes the harmonic flow throughout the system.

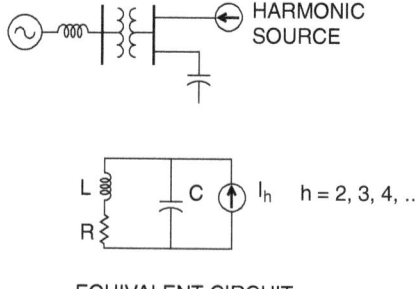

EQUIVALENT CIRCUIT

FIGURE 6.7 A simple circuit that may be analyzed manually.

Capabilities for Harmonics Analysis Programs

Acceptable computer software for harmonics analysis of power systems should have the following characteristics:

1. It should be capable of handling large networks of at least several hundred nodes.

2. It should be capable of handling multiphase models of arbitrary structure. Not all circuits, particularly those on utility distribution feeders, are amenable to accurate solution by balanced, positive-sequence models.

3. It should also be capable of modeling systems with positive-sequence models. When there can be no zero-sequence harmonics, there's no need to build a full three-phase model.

4. It should be able to perform a frequency scan at small intervals of frequency (e.g., 10 Hz) to develop the system frequency-response characteristics necessary to identify resonances.

5. It should be able to perform simultaneous solution of numerous harmonic sources to estimate the actual current and voltage distortion.

6. It should have built-in models of common harmonic sources.

7. It should allow both current source and voltage source models of harmonic sources.

8. It should be able to automatically adjust phase angles of the sources based on the fundamental frequency phase angles.

9. It should be able to model any transformer connection.

10. It should be able to display the results in a meaningful and user-friendly manner.

6.4.5 Harmonic Analysis by Computer—Historical Perspective

The most common type of computer analysis of power systems performed today is some form of power flow calculation. Most power engineers have some experience with this class of tool. Other common computer tools include short-circuit programs and, at least for transmission systems, dynamics (transient stability) programs. Harmonics and electromagnetic transient tools have traditionally been in the domain of specialists due to the modeling complexities.

While power flow tools are familiar, their formulation is generally unsuitable for harmonics analysis. Of the tools in common usage, the circuit model in short-circuit programs is closer to what is needed for harmonic flow analysis in networks. In fact, prior to the advent of special power systems harmonic analysis tools, many analysts would use short-circuit programs to compute harmonic distortion, manually adjusting the impedances for frequency. This is an interesting learning experience for the student, but not one that the practitioner will want to repeat often. Of course, one could also perform the analysis in the time domain using electromagnetic transients programs, but this generally is more time-consuming and is excessive for most problems.

Today, most power system harmonics analysis is performed in the sinusoidal steady state using computer programs specially developed for the purpose. It is encouraging to see many vendors of power system analysis software providing some harmonics analysis capabilities in their packages, although the main application in the package may be a power flow program. It is useful to see how this has evolved. Unlike power flow algorithms, few of the developers have written technical papers documenting their efforts. Therefore, it is difficult to trace the history of harmonics analysis in power systems through the literature. This book gives us the opportunity to acknowledge the contributions of several of the pioneers in this field, and this will help the reader understand the history behind some of the major computer tools available today.

Prior to the widespread use of computers for harmonics analysis, power systems harmonic studies were frequently performed on analog simulators such as a transient network analyzer (TNA). The few TNAs in the United States in the mid-1970s were located at large equipment manufacturers, primarily at General Electric Co., Westinghouse Electric, and McGraw-Edison Power Systems. Because of the inconvenience and high cost, harmonic studies were generally performed only on very special cases such as large arc furnace installations that might impact utility transmission systems. TNAs usually had at least two variable-frequency sources. Therefore, the general procedure was to use one source for the power frequency and the second source to represent the nonlinear load, one frequency at a time. One tricky part of this procedure was to sweep the frequency through system resonances fast enough to avoid burning up the power supply or damaging inductors and capacitors.

Our involvement in harmonics analysis began in 1975, when author Dugan, then with McGraw-Edison, constructed the first electronic arc model for a TNA to eliminate the need to use the second source.[4] In that same year, to overcome the limitations of harmonic analysis by analog simulator, Dugan teamed up with Dr. Sarosh N. Talukdar and William L. Sponsler at McGraw-Edison to develop one of the first commercial computer programs specifically designed to automate analysis of harmonic flows on large-scale power systems. Dubbed the Network Frequency Response Analysis

Program (NFRAP), it was developed for the Virginia Electric Power Company to study the impacts of adding 220-kV capacitor banks on the transmission system. These were some of the first capacitors applied at this voltage level.

The NFRAP program techniques, which are direct nodal admittance matrix solution techniques that treat nonlinearities as sources, evolved into what is probably the most prolific family of harmonic analysis programs. From 1977 to 1979, EPRI sponsored an investigation of harmonics on utility distribution feeders.[5,6] One of the products of this research was the Distribution Feeder Harmonics Analysis program. It was the first program designed specifically for analyzing harmonics on unbalanced distribution systems and had specific models of power systems elements to help the user develop models. It became the prototype for the modern harmonic analysis program. Key investigators on this project, RP 1024-1, were Robert E. Owen and author McGranaghan, and the key software designers were again Dugan and Sponsler.

The next generation of software tool based on the NFRAP program methodology was the McGraw-Edison Harmonic Analysis Program (MEHAP), under development from 1980 to 1984. It was written in Fortran for minicomputers and had the distinction of being interactive with graphical output. All previous efforts had been batch-mode programs with tabular output. For its time, it fit the definition of being user-friendly. The developers included Dugan, McGranaghan, and Jack A. King.

At about this same time, however, the personal computer (PC) revolution took place. Erich W. Gunther recoded the algorithms in the Pascal language and created the V-HARM program.[7] To our knowledge, this was the first commercial harmonic analysis program written expressly for the PC environment. Gunther subsequently has written the latest generation in this heritage of harmonic analysis tools in the C++ language for the Microsoft Windows environment. It is called the SuperHarm program and can be licensed from Electrotek Concepts, Inc.

The CYMHARMO program was developed first at Hydro Quebec's Research Institute (IREQ) in Montreal in 1983 and now can be licensed through CYME International Inc. The program was originally written in Fortran for the mainframe and was ported to the PC shortly afterward in 1984. It is now written in a mixture of Fortran and C languages. The principal authors of the software are Dr. Chinh Nguyen and Dr. Ali Moshref. The Canadian Electric Association (CEA) has supported the development of this program, which uses analysis techniques similar to the previously mentioned programs.

Beginning in 1981, EPRI also sponsored the development of the HARMFLO program, which takes a different approach to the network solution. Drs. G. L. Hedyt, D. Xia, and W. Mack Grady[8,9] developed the program at Purdue University and based it on the Newton-Raphson power flow techniques. The program was the first to adjust the harmonic current output of the load for the harmonic voltage distortion. The Fortran program was originally developed for mainframe batch computers and is now also available for the PC.

Of course, harmonics problems can be solved on electromagnetic transients analysis programs such as the EMTP, originally developed by the Bonneville Power Administration. Another one is the PSCAD/EMTDC program from the Manitoba HVDC Research Centre. The special-purpose steady-state programs are generally more efficient for the usual power system harmonics problems, but, occasionally, a very difficult problem will arise that requires simulation in the time domain.

The contributions of these pioneers have made it much easier for subsequent generations of power engineers to perform harmonics analysis on power systems, and harmonics analysis is now becoming commonplace.

6.5   Devices for Controlling Harmonic Distortion

There are a number of devices available to control harmonic distortion. They can be as simple as a capacitor bank or a line reactor, or as complex as an active filter.

As described in Sec. 6.2, a simple mitigation action such as adding, resizing, or relocating a shunt capacitor bank can effectively modify an unfavorable system frequency response, and thus bring the harmonic distortion to an acceptable level. Similarly, a reactor can perform the same function by detuning the system off harmful resonances. The effectiveness of such simple solutions in controlling harmonic distortion should be explored prior to considering a more complex device.

The following material first discusses the effectiveness of a simple in-line reactor, or choke, in mitigating harmonic distortion. Then, two general classes of harmonic filters, i.e., passive and active filters, are discussed. The former are based on passive elements, while the latter are based on power electronic devices.

6.5.1   In-Line Reactors or Chokes

A simple, but often successful, method to control harmonic distortion generated by ASDs involves a relatively small reactor, or choke, inserted at the line input side of the drive. This is particularly effective for PWM-type drives.

The inductance slows the rate at which the capacitor on the dc bus can be charged and forces the drive to draw current over a longer time period. The net effect is a lower-magnitude current with much less harmonic content while still delivering the same energy.

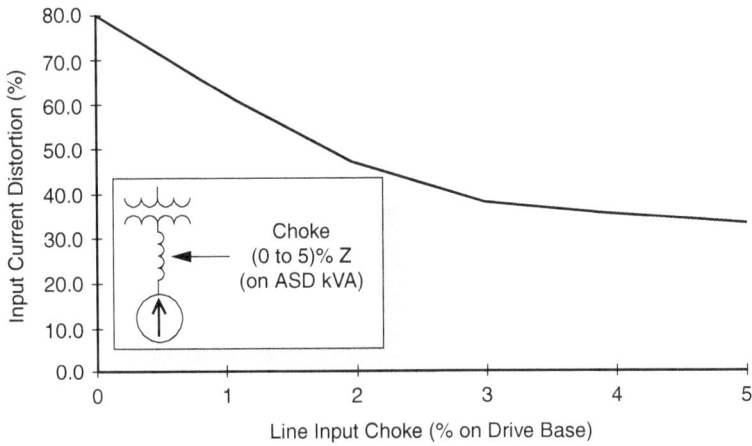

FIGURE 6.8 Harmonic reduction for a PWM-type ASD as a function of input choke size.

FIGURE 6.9 Three-phase line chokes for ASD applications. (*Courtesy of MTE Corp.*)

A typical 3-percent input choke can reduce the harmonic current distortion for a PWM-type drive from approximately 80 to 40 percent. This impressive harmonic reduction is illustrated in Fig. 6.8. Additional harmonic reduction is rather limited when the choke size is increased beyond 3 percent. The choke size is computed on the drive kVA base. Figure 6.9 shows typical line chokes used in 480-V ASD applications.

Figure 6.10 compares the effectiveness of a 3-percent choke in reducing harmonic current distortion to the condition without a

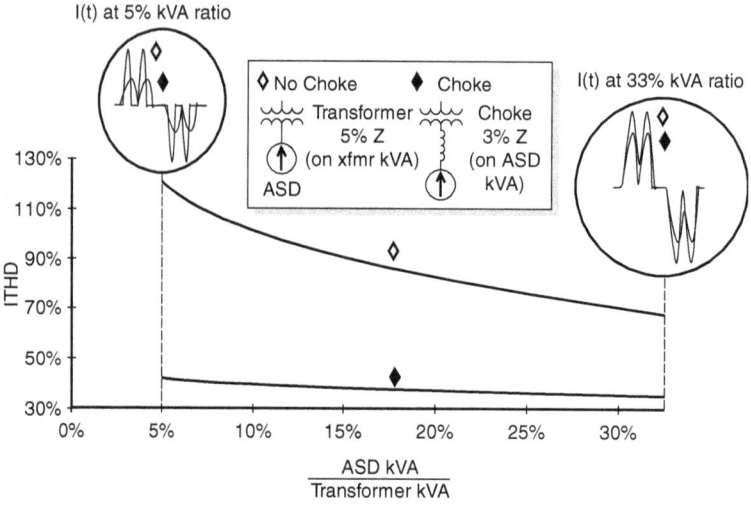

FIGURE 6.10   Effect of ac line chokes on ASD current harmonics.

choke for various ASD sizes (ASD sizes are normalized to the service transformer kVA). Representative waveforms for each end of the range are shown. The larger waveform is without the choke. As is clear from Fig. 6.10, a substantial improvement is achieved by inserting a choke in the ASD line. The current THD drops from the 80 to 120 percent range down to approximately 40 percent. Better reduction is obtained when the size of the ASD is significantly smaller than the service transformer. When the size of the ASD is 5 percent of the transformer, the current THD drops from 125 to 40 percent.

It is also important to note that there are other advantages of the choke in ASD applications. The effect of slowing the dc capacitor charging rate also makes the choke very effective in blocking some high-frequency transients. This helps avoid nuisance drive tripping during capacitor energization operations on the utility system.

Isolation transformers can provide the same benefit as a choke but may be more costly. However, isolation transformers with multiple drives offer the advantage of creating effective 12-pulse operation. Figure 6.11 illustrates this concept.

A 12-pulse configuration can be achieved by supplying one drive through a delta-wye connected transformer, and another drive through a delta-delta connected transformer. Figure 6.11 shows the current waveforms for two separate six-pulse ASDs. When the two waveforms are added together on the primary, the resulting waveform injected onto the utility system has much lower distortion, primarily because the fifth and seventh harmonics are cancelled out. These

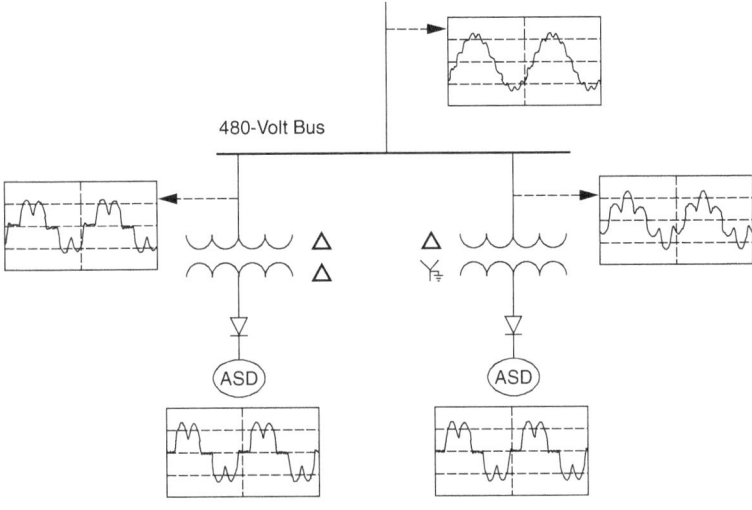

FIGURE 6.11   A 12-pulse configuration as a method to control harmonics from two ASDs.

two harmonics are responsible for most of the distortion for six-pulse drives.

6.5.2   Zigzag Transformers

Zigzag transformers are often applied in commercial facilities to control zero-sequence harmonic components. A zigzag transformer acts like a filter to the zero-sequence current by offering a low-impedance path to neutral. This reduces the amount of current that flows in the neutral back toward the supply by providing a shorter path for the current. To be effective, the transformer must be located near the load on the circuit that is being protected.

The two most important problems in commercial facilities are overloaded neutral conductors and transformer heating. Both of these problems can be solved with proper zigzag transformer placement. Some new commercial buildings include zigzag transformers on the 480/208-V supply transformer secondaries to prevent transformer overheating. A zigzag transformer located at the supply transformer secondary does not provide any benefit for neutral conductors supplying the loads.

Typical results with a zigzag transformer show that it can shunt about 50 percent of the third-harmonic current away from the main circuit neutral conductors. Thus, the zigzag transformer can almost always reduce neutral currents due to zero-sequence harmonics to acceptable levels. The largest zero-sequence harmonic will nearly always be the third harmonic in office buildings with many computers and related equipment.

Zigzag transformers are an excellent choice for existing facilities where neutral conductor problems and possible transformer heating are concerns, assuming that there is a convenient place to install the transformer between the neutral circuit of concern and the actual loads. In new facilities, it may be better to simply design the circuits with sufficient current-carrying capacity in the neutrals and with higher-capacity transformers.

6.5.3   Passive Filters

Passive filters are inductance, capacitance, and resistance elements configured and tuned to control harmonics. They are commonly used and are relatively inexpensive compared with other means for eliminating harmonic distortion. However, they have the disadvantage of potentially interacting adversely with the power system, and it is important to check all possible system interactions when they are designed. They are employed either to shunt the harmonic currents off the line or to block their flow between parts of the system by tuning the elements to create a resonance at a selected frequency. Figure 6.12 shows several types of common filter arrangements.

Shunt Passive Filters

The most common type of passive filter is the single-tuned "notch" filter. This is the most economical type and is frequently sufficient for the application. The notch filter is series-tuned to present a low impedance to a particular harmonic current and is connected in shunt with the power system. Thus, harmonic currents are diverted from their normal flow path on the line through the filter. Notch filters can provide power factor correction in addition to harmonic suppression. In fact, power factor correction capacitors may be used to make notch filters.

Figure 6.13 shows one example of such a filter designed for medium-voltage applications. The dry-type iron-core reactor is

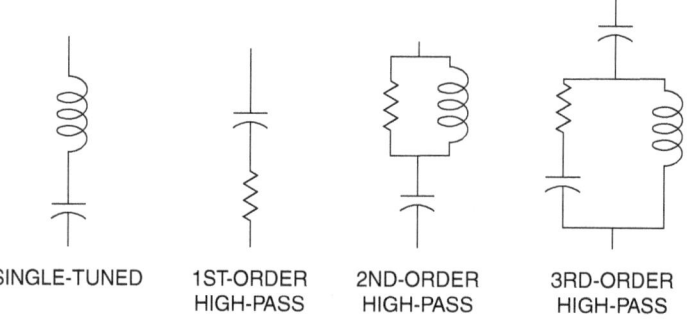

SINGLE-TUNED   1ST-ORDER   2ND-ORDER   3RD-ORDER
                HIGH-PASS   HIGH-PASS   HIGH-PASS

Figure 6.12   Common passive filter configurations.

FIGURE 6.13 One phase of a three-phase metal-enclosed filter. (*Courtesy of Northeast Power Systems, Inc.*)

positioned atop the capacitors, which are connected in a wye, or star, configuration with the other phases (not shown). Each capacitor can is fused with a current-limiting fuse to minimize damage in case of a can failure. In outdoor installations it is often more economical to use air-core reactors. Iron-core reactors may also be oil-insulated.

Figure 6.14 shows another design for industrial site applications. Here the reactors are placed on top of the cabinet housing the capacitors and switchgear.

An example of a common 480-V filter arrangement is illustrated in Fig. 6.15. The figure shows a delta-connected low-voltage capacitor bank converted into a filter by adding an inductance in series with

FIGURE 6.14   Filter for industrial power system applications. (*Courtesy of Gilbert Electrical Systems.*)

the phases. In this case, the notch harmonic h_{notch} is related to the fundamental frequency reactances by

$$h_{\text{notch}} = \sqrt{\frac{X_C}{3X_F}}$$

(6.6)

Note that X_C in this case is the reactance of one leg of the delta rather than the equivalent line-to-neutral capacitive reactance. If phase-to-phase voltage and three-phase kvar are used to compute X_C, as previously described, factor 3 would be omitted.

One important side effect of this type of filter is that it creates a sharp parallel resonance point at a frequency below the notch frequency (Fig. 6.15c). This resonant frequency must be safely away from any significant harmonic or other frequency component that may be produced by the load. Filters are commonly tuned slightly

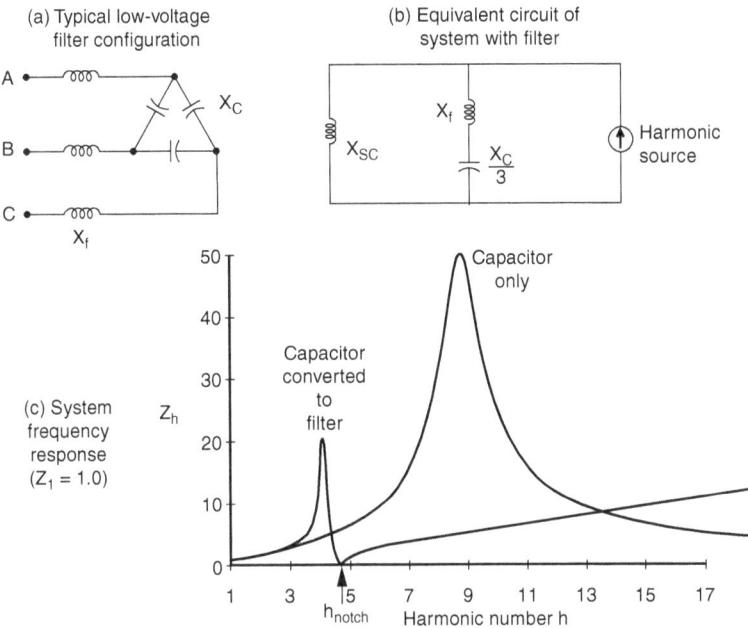

(a) Typical low-voltage filter configuration

(b) Equivalent circuit of system with filter

(c) System frequency response $(Z_1 = 1.0)$

Figure 6.15 Creating a fifth-harmonic notch filter and its effect on system response.

lower than the harmonic to be filtered to provide a margin of safety in case there is some change in system parameters that would raise the notch frequency. If they were tuned exactly to the harmonic, changes in either capacitance or inductance with temperature or failure might shift the parallel resonance higher into the harmonic being filtered. This could present a situation worse than one without a filter because the resonance is generally very sharp.

To avoid problems with this resonance, filters are added to the system starting with the lowest significant harmonic found in the system. For example, installing a seventh-harmonic filter usually requires that a fifth-harmonic filter also be installed. The new parallel resonance with a seventh-harmonic filter alone is often very near the fifth, which is generally disastrous.

The filter configuration of Fig. 6.15a does not admit zero-sequence currents because the capacitor is delta-connected, which makes it ineffective for filtering zero-sequence triplen harmonics. Because 480-V capacitors are usually delta-configured, other solutions must be employed when it becomes necessary to control zero-sequence third-harmonic currents in many industrial and commercial building facilities. In contrast, capacitors on utility distribution systems are more commonly wye-connected. This gives the option of controlling the zero-sequence triplen harmonics simply by changing the neutral connection.

Placing a reactor in the neutral of a capacitor is a common way to force the bank to filter only zero-sequence harmonics. This technique is often employed to eliminate telephone interference. A tapped reactor is installed in the neutral and the tap adjusted to minimize the telephone.

Passive filters should always be placed on a bus where the short-circuit reactance X_{SC} can be expected to remain constant. While the notch frequency will remain fixed, the parallel resonance will move with system impedance. For example, the parallel resonant frequency for running with standby generation by itself is likely to be much lower than when interconnected with the utility because the generator impedance is much higher than the utility impedance. This could magnify a harmonic that is normally insignificant. Thus, filters are often removed for operation with standby generation.

Also, filters must be designed with the capacity of the bus in mind. The temptation is to size the current-carrying capability based solely on the load that is producing the harmonic. However, a small amount of background voltage distortion on a very strong bus may impose excessive duty on the filter.

Series Passive Filters

Unlike a notch filter which is connected in shunt with the power system, a series passive filter is connected in series with the load. The inductance and capacitance are connected in parallel and are tuned to provide a high impedance at a selected harmonic frequency. The high impedance then blocks the flow of harmonic currents at the tuned frequency only. At fundamental frequency, the filter would be designed to yield a low impedance, thereby allowing the fundamental current to follow with only minor additional impedance and losses. Figure 6.16 shows a typical series filter arrangement.

Series filters are used to block a single harmonic current (such as the third harmonic) and are especially useful in a single-phase circuit where it is not possible to take advantage of zero-sequence characteristics. The use of the series filters is limited in blocking multiple harmonic currents. Each harmonic current requires a series

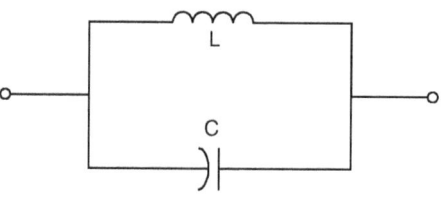

FIGURE 6.16   A series passive filter.

filter tuned to that harmonic. This arrangement can create significant losses at the fundamental frequency.

Furthermore, like other series components in power systems, a series filter must be designed to carry a full rated load current and must have an overcurrent protection scheme. Thus, series filters are much less commonly applied than shunt filters.

Low-Pass Broadband Filters

Multiple stages of both series and shunt filters are often required in practical applications. For example, in shunt filter applications, a filter for blocking a seventh-harmonic frequency would typically require two stages of shunt filters, the seventh-harmonic filter itself and the lower fifth-harmonic filter. Similarly, in series filter applications, each frequency requires a series filter of its own; thus, multiple stages of filters are needed to block multiple frequencies.

In numerous power system conditions, harmonics can appear not only in a single frequency but can spread over a wide range of frequencies. A six-pulse converter generates characteristic harmonics of 5th, 7th, 11th, 13th, etc. Electronic power converters can essentially generate time-varying interharmonics covering a wide range of frequencies. Designing a shunt or series filter to eliminate or reduce these widespread and time-varying harmonics would be very difficult using shunt filters. Therefore, an alternative harmonic filter must be devised.

A low-pass broadband filter is an ideal application to block multiple or widespread harmonic frequencies. Current with frequency components below the filter cutoff frequency can pass; however, current with frequency components above the cutoff frequency is filtered out. Since this type of low-pass filter is typically designed to achieve a low cutoff frequency, it is then called a low-pass broadband filter. A typical configuration of a low-pass broadband filter is shown in Fig. 6.17.

In distribution system applications, the effect of low-pass broadband filters can be obtained by installing a capacitor bank on the low-voltage side of a transformer as shown in Fig. 6.18a. The size of the capacitor bank would have to be so selected to provide the

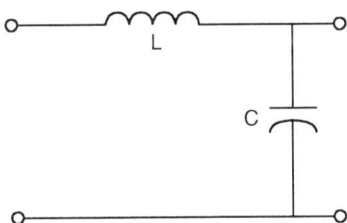

FIGURE 6.17   A low-pass broadband filter configuration.

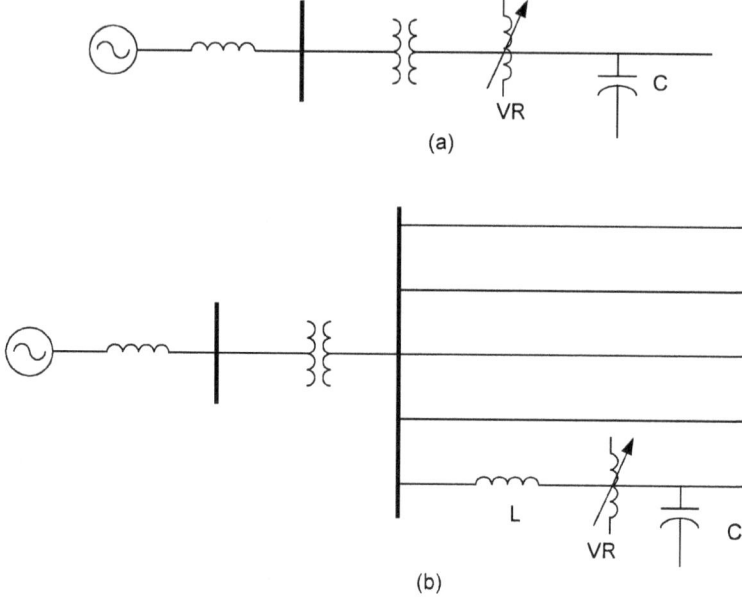

FIGURE 6.18 A low-pass broadband filter application in power service for industrial systems. Filter arrangement in a substation with (a) dedicated and (b) multiple feeders.

desired cutoff frequency when combined with the transformer leakage inductance and the system impedance. It is then capable of preventing harmonics above the cutoff frequency from penetrating the high-voltage side of the transformer. Since the cutoff frequency can be sometimes quite low, the size of the capacitor bank may be fairly large. This will result in a significant voltage rise. Should the voltage remain high, a voltage regulator or transformer load tap changer (LTC) must be used to lower the voltage to an acceptable level.

In a substation serving multiple feeders, a line reactor and voltage regulator can be installed at the beginning of the feeder to isolate the portion of the system subject to high voltage. This arrangement will allow voltage levels at other feeders to be maintained at normal values. The combination of the transformer leakage inductance, the line reactor, the voltage regulator, and the capacitor bank yields the desired cutoff frequency. Figure 6.18b depicts this arrangement.

In industrial system applications, commercial low-pass broadband filters have been used to prevent harmonics produced by nonlinear loads from entering the ac system. The typical design is illustrated in Fig. 6.19.[10] Figure 6.20 shows a typical system intended for this application. A line reactor installed in series with the main ac line is

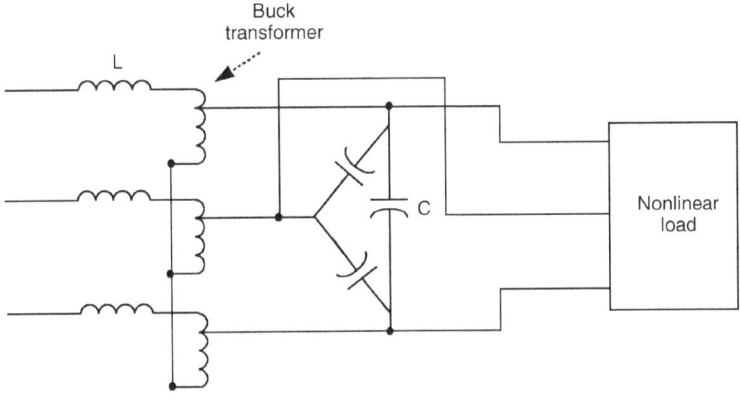

FIGURE 6.19 A low-pass broadband filter application in an industrial system.

FIGURE 6.20 A commercial 600-V, three-phase broadband filter. (*Courtesy of MTE Corp.*)

used to provide an electrical separation between the ac system and the nonlinear load. A capacitor bank is installed in shunt to form a low-pass broadband filter configuration. Since the presence of the capacitor bank increases the voltage at the input of the ASD, a buck transformer is needed to bring the voltage at the line reactor output down to a level where the voltage at the capacitor is acceptable. The optimum performance of a low-pass broadband filter in ASD applications is obtained when there is no series inductor between the filter capacitor banks and the ASD dc bus capacitor. Any impedance in between reduces the charging capability of the dc bus capacitor since it is charged from the filter capacitor.

The cutoff frequency for a low-pass broadband filter for ASD applications is typically designed at a low harmonic frequency, such as at 100 to 200 Hz on a 60-Hz system. With this low tuning frequency, the filter is unlikely to excite any undesired resonance with the rest of the system and can filter out much of the harmonic currents. In ASD applications the filter can generally reduce the overall current harmonic distortion from the 90 to 100 percent range down to the 9 to 12 percent range under rated load conditions. This performance is certainly much better than a simple ac line choke, which only reduces the overall current distortion down to the 30 to 40 percent range. However, the cost of an ac line choke is less than a low-pass broadband filter.

C Filters

C filters are an alternative to low-pass broadband filters in reducing multiple harmonic frequencies simultaneously in industrial and utility systems. They can attenuate a wide range of steady-state and time-varying harmonic and interharmonic frequencies generated by electronic converters, induction furnaces, cycloconverters, and the like.

The configuration of a C filter is nearly identical to that of the second-order high-pass filter shown earlier in Fig. 6.12. The main distinction between the two configurations is that the C filter possesses an auxiliary capacitor C_a in series with the inductor L_m. A typical configuration of a C filter is shown in Fig. 6.21. The auxiliary capacitor C_a is sized in such a way that its capacitive reactance cancels out L_m at the fundamental frequency, bypassing the damping resistance R. For this reason, the losses associated with R are practically eliminated, allowing a C filter to be tuned to a low frequency.

The impedance frequency response of a C filter is also essentially identical to that of a second-order high-pass filter. At high-order harmonic frequencies, the reactance of C_a is small, while that of L_m is large. Therefore, the impedance of the series L_m and C_a branch is dominated by the reactance of L_m. The high-frequency responses of the C filter and second-order high-pass filters are similar (see Fig. 6.21).

In designing a C filter,[11] it is necessary to specify $I_{SF}(h_T)$, the maximum harmonic current allowed to flow into the system at h_T, the tuned harmonic frequency. It is also assumed that the requirement for the reactive power compensation is known, thus establishing the nominal size of capacitor C_m. Figure 6.22 shows an equivalent circuit for deriving filter components R, C_a, and L_m. The short-circuit reactance is denoted as X_S. Filter components can be computed as follows:

$$R = \frac{R_F\left(h_T\right)^2 + \left(\dfrac{X_{C_m}}{h_T}\right)^2}{R_F\left(h_T\right)} \qquad X_{LM} = X_{C_a} = \frac{R_F\left(h_T\right)^2 + \left(\dfrac{X_{C_m}}{h_T}\right)^2}{\left(\dfrac{X_{C_m}}{h_T}\right)\left(\dfrac{h_T - 1}{h_T}\right)}$$

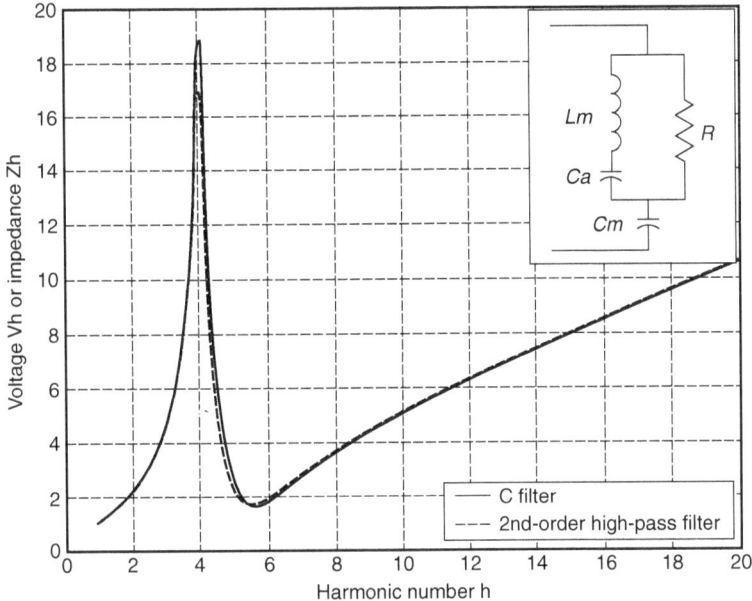

Figure 6.21 A typical C-filter configuration and its impedance frequency response (solid line).

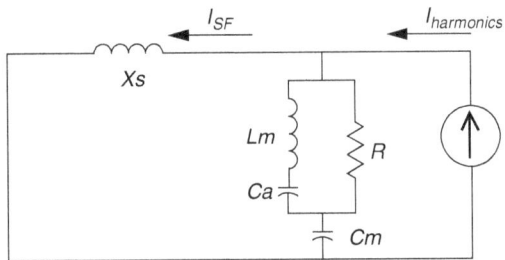

Figure 6.22 Equivalent circuit for deriving C-filter specifications.

where $R_F = \dfrac{h_T X_s}{\sqrt{\dfrac{1}{I_{SF}(h_T)^2} - 1}}$

X_S = short-circuit reactance at fundamental frequency
X_{Lm} = reactance of L_m at fundamental frequency
X_{Ca} = reactance of C_a at fundamental frequency

Example. Let us consider a C filter applied at a 13.8-kV bus. The filter is designed to deliver 5 Mvar at the fundamental frequency (X_{Cm} = 13.8²/5 = 38.1 Ω) and to attenuate 70 percent of injected harmonics at the tuned harmonic h_T of 5.5.

Thus, the maximum current allowed to flow in the system at the tuned frequency would be I_{SF} (h_T = 5.5) = 0.3 pu (30 percent). The short-circuit reactance of the system is assumed to be 1.0 V. From the above equations, it can be computed that $R = 29.5$ Ω and $X_{Ca} = X_{Lm} = 1.383$ Ω. Figure 6.23 shows the harmonic current allowed to flow in the system, i.e., I_{SF} as a function of frequency for C filters with I_{SF} (h_T = 5.5) of 10, 30, and 50 percent.

From Fig. 6.23, it is observed that at lower harmonic frequencies the I_{SF} (h_T) is smaller, and the damping of the filter would be less. For example, a C filter with I_{SF} (h_T = 5.5) = 0.1 has less damping compared to one with I_{SF} (h_T = 5.5) = 0.3 or 0.5 because the resistive component R has a larger value.

At higher harmonic frequencies, the amount of current flowing in the system is approximately identical in all cases shown in Fig. 6.23 since the C_m reactance dominates the overall filter impedance. For this reason, higher attenuation at high harmonic frequencies cannot

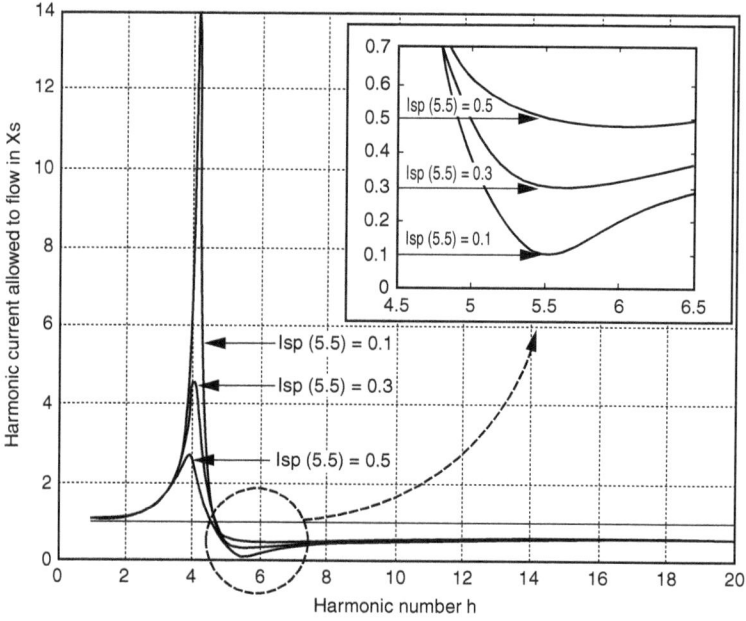

FIGURE 6.23 An example of a C filter where the maximum harmonic current allowed to flow in the system is 10, 30, and 50 percent at the tuned harmonic order of 5.5.

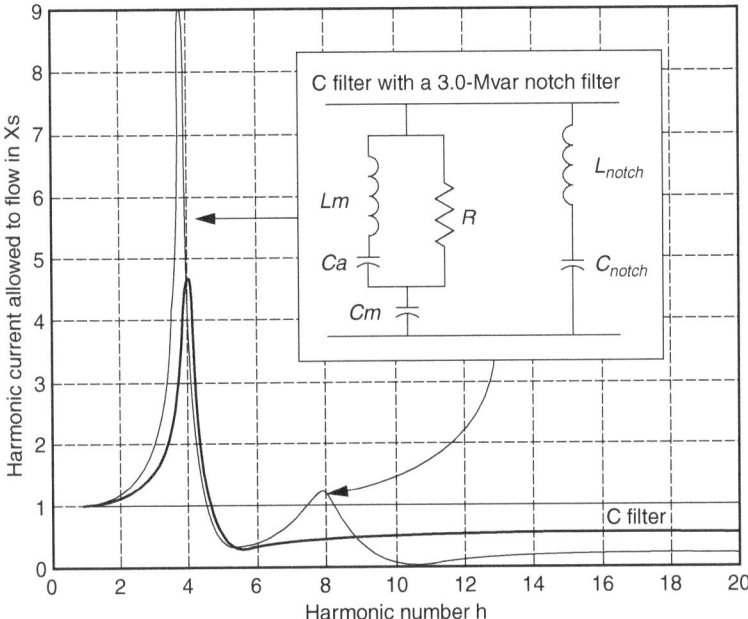

FIGURE 6.24 A C filter with and without a notch filter.

be achieved by having a C filter with lower I_{SF} (h_T). Instead, an additional filter would be required.

Figure 6.24 shows the configuration of a C filter coupled with a notch filter to achieve more attenuation at higher frequencies. The drawback with this arrangement is that there is a new parallel resonance associated with the notch filter. This parallel resonance must be so selected that it is not excited by any harmonic currents present in the system.

The notch filter is typically tuned higher than a C filter. If the notch filter is tuned below the C filter, the size of the auxiliary capacitor C_a would be significantly larger, making the C filter impractical.

6.5.4 Active Filters

Active filters are relatively new types of devices for eliminating harmonics. They are based on sophisticated power electronics and are much more expensive than passive filters. However, they have the distinct advantage that they do not resonate with the system. Active filters can work independently of the system impedance characteristics. Thus, they can be used in very difficult circumstances where passive filters cannot operate successfully because of parallel resonance problems. They can also address more than one harmonic

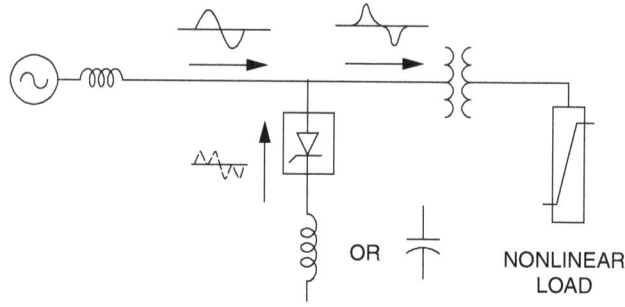

Figure 6.25   Application of an active filter at a load.

at a time and combat other power quality problems such as flicker. They are particularly useful for large, distorting loads fed from relatively weak points on the power system.

The basic idea is to replace the portion of the sine wave that is missing in the current in a nonlinear load. Figure 6.25 illustrates the concept. An electronic control monitors the line voltage and/or current, switching the power electronics very precisely to track the load current or voltage and force it to be sinusoidal. As shown, there are two fundamental approaches: one that uses an inductor to store current to be injected into the system at the appropriate instant and one that uses a capacitor. Therefore, while the load current is distorted to the extent demanded by the nonlinear load, the current seen by the system is much more sinusoidal.

Active filters can typically be programmed to correct for the power factor as well as harmonics.

6.6   Harmonic Filter Design: A Case Study

This section illustrates a procedure for designing harmonic filters for industrial applications. This procedure can also be used to convert an existing power factor correction capacitor into a harmonic filter. As described in Sec. 4.1.2, power factor correction capacitors are used widely in industrial facilities to lower losses and utility bills by improving power factor. On the other hand, power factor correction capacitors may produce harmonic resonance and magnify utility capacitor-switching transients. Therefore, it is often desirable to implement one or more capacitor banks in a facility as a harmonic filter.

Filter design procedures are detailed in the steps shown below. The best way to illustrate the design procedures is through an example.

A single-tuned notch filter will be designed for an industrial facility and applied at a 480-V bus. The load where the filter will be installed is approximately 1200 kVA with a relatively poor displacement power factor of 0.75 lagging. The total harmonic current produced by this load is approximately 30 percent of the fundamental current, with a maximum of 25 percent fifth harmonic. The facility is supplied by a 1500-kVA transformer with 6.0 percent of impedance. The fifth-harmonic background voltage distortion on the utility side of the transformer is 1.0 percent of the fundamental when there is no load. Figure 6.7 shown earlier depicts the industrial facility where the filter will be applied. The harmonic design procedures are provided in the following steps.

1. Select a tuned frequency for the filter.   The tuned frequency is selected based on the harmonic characteristics of the loads involved. Because of the nature of a single-tuned filter, the filtering should start at the lowest harmonic frequency generated by the load. In this case, that will be the fifth harmonic. The filter will be tuned slightly below the harmonic frequency of concern to allow for tolerances in the filter components and variations in system impedance. This prevents the filter from acting as a direct short circuit for the offending harmonic current, reducing duty on the filter components. It also minimizes the possibility of dangerous harmonic resonance should the system parameters change and cause the tuning frequency to shift.

In this example, the filter is designed to be tuned to the 4.7th. This is a common choice of notch frequency since the resulting parallel resonant frequency will be located around the fourth harmonic, a harmonic frequency that is not produced by most nonlinear loads. The notch filter is illustrated in Fig. 6.26.

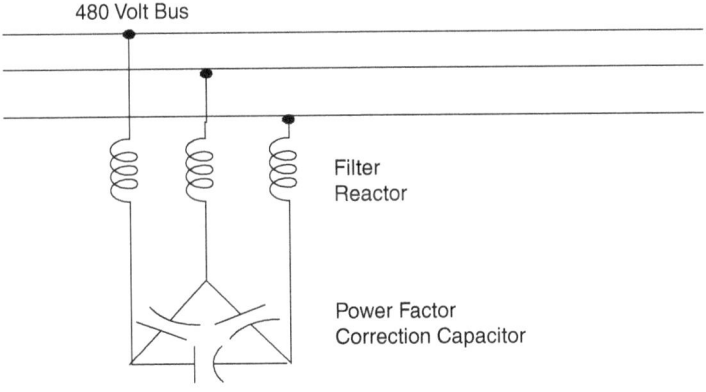

480 Volt Bus

Filter
Reactor

Power Factor
Correction Capacitor

Figure 6.26   Example of low-voltage figure configuration.

2. Compute capacitor bank size and the resonant frequency. As a general rule, the filter size is based on the load reactive power requirement for power factor correction. When an existing power factor correction capacitor is converted to a harmonic filter, the capacitor size is given. The reactor size is then selected to tune the capacitor to the desired frequency. However, depending on the tuned frequency, the voltage rating of the capacitor bank may have to be higher than the system voltage to allow for the voltage rise across the reactor. Therefore, one may have to change out the capacitor anyway.

This example assumes that no capacitor is installed and that the desired power factor is 96 percent. Thus, the net reactive power from the filter required to correct from 75 to 96 percent power factor can be computed as follows:

- Active power demand of a 1200-kVA load with power factor of 0.75 lagging

  $1200 \times 0.75 = 900$ kW

- Required compensation from the filter:

  $900 \tan[\text{acos}(0.75)] - 900 \tan[\text{acos}(0.96)] = 531.22$ kVar

For a nominal 480-V system, the net wye-equivalent filter reactance (capacitive), X_{Filt}, is determined by

$$X_{\text{Filt}} = \frac{kV^2(1000)}{kvar} = \frac{0.48^2(1000)}{531.22} = 0.4337 \; \Omega$$

X_{Filt} is the difference between the capacitive reactance and the inductive reactance at fundamental frequency:

$$X_{\text{Filt}} = X_{\text{Cap}} - X_L$$

For tuning at the 4.7th harmonic,

$$X_{\text{Cap}} = h^2 X_L = 4.7^2 X_L$$

Thus, the desired capacitive reactance can be determined by

$$X_{\text{Cap}} = \frac{X_{\text{Filt}} h^2}{h^2 - 1} = \frac{0.4337 \times 4.7^2}{4.7^2 - 1} = 0.4543 \; \Omega$$

At this point, it is not known whether the filter capacitor can be rated the same as the system, 480 V, or would have to be rated one step higher at 600 V. To achieve this reactance at a 480-V rating, the capacitor would have to be rated

$$kvar = \frac{kV^2(1000)}{X_{Cap}} = \frac{0.48^2 \times 1000}{0.4543} = 507 \text{ kvar}$$

Similarly, at 600 V, the capacitor would have to be rated 792 kvar. For now, the filter will be designed using a 480-V capacitor rated 540 kvar, which is a commonly available size near the desired value. For this capacitor rating,

$$X_{Cap} = 0.4267 \ \Omega$$

Note that the capacitor reactance is given in wye equivalent. The rated capacitor current is

$$I_{Cap,rated} = \frac{540 \times 10^3}{\sqrt{3} \times 480} = 649.5 \text{ A}$$

3. Compute filter reactor size. The filter reactor size can now be selected to tune the capacitor to the desired frequency. From Step 1, the desired frequency is at 4.7th harmonic or 282 Hz. The filter reactor size is computed from the equivalent wye capacitive reactance above as follows:

$$X_{L(fund)} = \frac{X_{Cap(wye)}}{h^2} = \frac{0.4267}{4.7^2} = 0.0193 \ \Omega$$

or

$$L = \frac{X_{L(fund)}}{2\pi \times 60} = 0.0512 \text{ mH}$$

Alternatively, the reactor size can be computed by solving for L in the following equation:

$$f_h = \frac{1}{2\pi \sqrt{LC_{(wye)}}}$$

where $f_h = 4.7 \times 60 = 282$ Hz.
    The next step is to evaluate the duty requirements for the capacitor and reactor.

4. Evaluate filter duty requirements. Evaluation of filter duty requirements typically involves capacitor bank duties. These duties include peak voltage, current, kvar produced, and rms voltage. IEEE Standard 18-1990 is used as the limiting standard to evaluate these duties. Computation of the duties are fairly lengthy; therefore, they are divided into three steps: i.e., computation for fundamental duties, harmonic duties, and rms current and peak voltage duties.

5. Computation of fundamental duty requirements. In this step, a fundamental frequency operating voltage across the capacitor bank is determined. The computation is as follows:

a. The apparent reactance of the combined capacitor and reactor at the fundamental frequency is:

$$X_{fund} = \left| X_L - X_{Cap(wye)} \right| = \left| 0.0193 - 0.4267 \right| = 0.4074 \ \Omega.$$

b. The fundamental frequency filter current is

$$I_{fund} = \frac{kV_{actual}/\sqrt{3}}{X_{fund}} = \frac{480/\sqrt{3}}{0.4074} = 680.3 \ A$$

c. The fundamental frequency operating voltage across the capacitor bank is

$$V_{LL,Cap(fund)} = \sqrt{3} \times I_{fund} \times X_{Cap(wye)} = 502.8 \ V$$

This is the nominal fundamental voltage across the capacitor. It should be adjusted for any contingency conditions (maximum system voltage), and it should be less than 110 percent of the capacitor rated voltage.

d. Due to the fact that the filter draws more fundamental current than the capacitor alone, the actual reactive power produced is larger than the capacitor rating:

$$kvar_{fund} = \sqrt{3} \times I_{fund} \times V_{LL,Cap(fund)} = 592.4 \ kvar$$

6. Computation of harmonic duty requirements. In this step, the maximum harmonic current expected in the filter is computed. This current has two components: the harmonic current produced by the nonlinear load (as computed in a) and harmonic current from the utility side (as computed in b).

a. Because the nonlinear load produces 25 percent fifth harmonic of the fundamental current, the harmonic current in amperes produced by the load would be

$$I_{h(\text{amps})} = I_h(\text{pu})\frac{\text{kVA}}{\sqrt{3} \times kV_{\text{actual}}} = 0.25\frac{1200}{\sqrt{3} \times 0.48} = 360.8 \text{ A}$$

b. Harmonic current contributed to the filter from the source side is estimated as follows. It will be assumed that the 1 percent fifth harmonic voltage distortion present on the utility system will be limited only by the impedances of the service transformer and the filter, and the utility impedance will be neglected.

- Fundamental frequency impedance of the service transformer:

$$X_{T(\text{fund})} = Z_T(\%)\frac{kV_{\text{actual}}^2}{\text{MVA}_{\text{Xfmr}}} = 0.06\frac{0.48^2}{1.5} = 0.0092 \ \Omega$$

- The fifth harmonic impedance of the service transformer (where the transformer is inductive):

$$X_{T(\text{harm})} = hX_{T(\text{fund})} = 5 \times 0.0092 = 0.0461 \ \Omega$$

- The harmonic impedance of the capacitor bank is

$$X_{\text{Cap(wye),harm}} = \frac{X_{\text{Cap(wye)}}}{h} = \frac{0.4267}{5} = 0.0853 \ \Omega$$

- The harmonic impedance of the reactor is

$$X_{L(\text{harm})} = hX_{L(\text{fund})} = 5 \times 0.0193 = 0.0966 \ \Omega$$

- Given that the voltage distortion on the utility system is 0.01 pu, the estimated amount of fifth harmonic current contributed to the filter from the source side would be

$$I_{h(\text{utility})} = \frac{V_{h(\text{utility})}(\text{pu}) \times kV_{\text{actual}}}{\sqrt{3} \times \left(X_{T(\text{harm})} - X_{\text{Cap(wye),harm}} + X_{L(\text{harm})}\right)}$$

$$= \frac{0.01 \times 480}{\sqrt{3} \times (0.0461 - 0.0853 + 0.0966)} = 48.34 \text{ A}$$

c. The maximum harmonic current is the sum of the harmonic current produced by the load and that contributed from the utility side:

$$I_{h(\text{total})} = 360.8 + 48.34 = 409.19 \text{ A}$$

d. The harmonic voltage across the capacitor can be computed as follows:

$$V_{Cap(L\text{-}L,rms\text{-}harm)} = \sqrt{3}I_{h(total)}\frac{X_{Cap(wye)}}{h}$$

$$= \sqrt{3} \times 409.19 \times \frac{0.4267}{5} = 60.48 \text{ V}$$

7. Evaluate total rms current and peak voltage requirements.   These two quantities are computed as follows:

a. Total rms current passing through the filter:

$$I_{rms,total} = \sqrt{I_{fund}^2 + I_{h(total)}^2} = \sqrt{680.3^2 + 409.19^2} = 793.9 \text{ A}$$

This is the total rms current rating that is required for the filter reactor.

b. Assuming the harmonic and fundamental components add together, the maximum peak voltage across the capacitor is

$$V_{L-L,Cap(maxPeak)} = \sqrt{2}\left(V_{L-L,Cap(fund)} + V_{Cap(L\text{-}L,rms\text{-}harm)}\right)$$

$$= \sqrt{2}(502.8 + 60.48) = 563.3\sqrt{2} \text{ V}.$$

c. The rms voltage across the capacitor is

$$V_{L-L,Cap(rms\ total)} = \sqrt{V_{L-L,Cap(fund)}^2 + V_{Cap(L-L,rms\text{-}harm)}^2}$$

$$= \sqrt{502.8^2 + 60.48^2} = 506.4 \text{ V}$$

d. The total kvar seen by the capacitor:

$$kvar_{Cap(wye),total} = \sqrt{3}I_{rms,total} \times kV_{L-L,Cap(rms\ total)}$$

$$= \sqrt{3} \times 793.9 \times 0.506 = 696 \text{ kvar}$$

8. Evaluate capacitor rating limits.   The duties (peak voltage, rms voltage and current, and kvar produced) for the proposed filter capacitor are compared to the IEEE standard limits in Table 6.4. This would be a very marginal application because the capacitor duties are essentially at the maximum limits. There is no

Duty	Definition	Limit in %	Actual values	Actual values in %
Peak voltage	$\dfrac{V_{L-L,Cap(maxPeak)}}{kV_{rated}}$	120	$\dfrac{563.3\sqrt{2}\ \text{V}}{480\sqrt{2}\ \text{V}}$	118
rms voltage	$\dfrac{V_{L-L,Cap(rms\ total)}}{kV_{rated}}$	110	$\dfrac{506\ \text{V}}{480\ \text{V}}$	106
rms current	$\dfrac{I_{rms,total}}{I_{Cap(rated)}}$	180	$\dfrac{793\ \text{A}}{649.5\ \text{A}}$	123
kvar	$\dfrac{\text{kvar}_{Cap(wye),total}}{\text{kvar}_{rated}}$	135	$\dfrac{696\ \text{kvar}}{540\ \text{kvar}}$	129

TABLE 6.4   Comparison Table for Evaluating Filter Duty Limit

tolerance for any deviation in assumptions or increases in service voltage. A 480-V capacitor will likely have a short life in this application.

When this happens, a capacitor rated for higher voltage must be used. At 600 V, the equivalent capacitor rating would be

$$540 \times \frac{600^2}{480^2} = 843 \ \text{kvar}$$

A nominal rating of 840 kvar with the reactor values as computed in the above-mentioned steps would provide essentially the same filter within normal manufacturing tolerances. The 600-V capacitor would be well within its rating in this application.

9. Evaluate filter frequency response.   The filter frequency response is now evaluated to make sure that the filter does not create a new resonance at a frequency that could cause additional problems. The harmonics at which the parallel resonance below the notch frequency will occur is computed as follows:

$$h_0' = \sqrt{\frac{X_{Cap(wye)}}{X_{T(fund)} + X_{L(fund)}}} = \sqrt{\frac{0.4267}{0.0092 + 0.0193}} = 3.86$$

This assumes that the service transformer reactance dominates the source impedance, including the utility system impedance, which will lower the frequency.

This filter results in a resonance very near the fourth harmonic, which is an interesting case. Normally, there are very few significant sources of an even harmonic during steady-state operation and this filter would work acceptably. However, there are significant fourth harmonic currents during events such as transformer energization. If the filter is in service, when a large transformer is energized, and there is very little load to dampen the resonance, there can be overvoltages that persist well past the usual inrush transient period. In this case, the designer should first include the utility system impedance in the calculation. To gain additional margin from the fourth, the basic filter size would have to be increased.

10. Evaluate the effect of filter parameter variations within specified tolerance. Filter designers generally assume that capacitors are designed with a tolerance of +15 percent of the nominal capacitance value. Reactors are assumed to have tolerance of ±5 percent of the nominal inductance. These tolerances can significantly affect the filter performance should the frequency response over this range create a harmful resonance. Therefore, the final step is to check the filter design for various extremes. This is automatically done using some filter design software.

The preceding steps illustrate a typical single-tuned filter design. Multiple single-tuned filters might be necessary when a single-tuned filter does not control harmonics to acceptable levels. For example, 5th, 7th, and 11th harmonic filters may be needed for some large 6-pulse loads. The general procedure is the same except that the reactive power requirement is first divided between the filter stages. Evaluating the effect of component tolerance is particularly important because there are multiple filters involved.

The tuning characteristic of the filter is described by its quality factor, Q, a measure of the sharpness of tuning, and, for series filter resistance, is defined as

$$Q = \frac{nX_L}{R}$$

where R = series resistance of filter elements
$\quad\quad n$ = tuning harmonic
$\quad\quad X_L$ = reactance of filter reactor at fundamental frequency

Typically, the value of R consists of only the resistance of the inductor. This usually results in a very large value of Q and a very sharp filtering action. This is normally satisfactory for the typical single filter application and results in a filter that is very economical to operate (small-energy consumption). However, sometimes it is desirable to introduce some intentional losses to help dampen the response of the system. A resistor is commonly added in *parallel* with the reactor to create a high-pass filter. In this case, Q is defined as the inverse of the above series case so that large numbers reflect sharp tuning. High-pass filters are generally used only at the 11th and 13th harmonics, and higher. It is usually not economical to operate such a filter at the 5th and 7th harmonics because of the amount of losses and the size of the resistor (for which a C-filter might be applicable).

The reactors used for larger filter applications are generally built with an air core, which provides linear characteristics with respect to frequency and current. Reactors for smaller filters and filters that must fit into a confined space or near steel structures are built with a steel core. As stated in Step #10, 5-percent tolerance in the reactance is usually acceptable for industrial applications. The 60-Hz X/R ratio is usually between 50 and 150. A series resistor may be used to lower this ratio, if desired, to produce a filter with more damping. The reactor should be rated to withstand a short circuit at the junction of the reactor and capacitor. A design Q for the high-pass configuration might typically be 1 or 2 to achieve a flat response above the tuned frequency.

Filters for many high-power, three-phase applications such as static var systems almost always include 5th and 7th harmonics because those are the largest harmonics produced by the 6-pulse bridge. Occasionally this will cause a system resonance near the third that may require a third harmonic filter. Normally, one will not think that the third harmonic would be a problem in a three-phase bridge, but imbalances in the operation of the bridge and in system parameters will create small amounts of uncharacteristic harmonics. Analysts commonly assume the uncharacteristic harmonics attenuate 90 to 95 percent of the theoretical maximum. If the system responds to those harmonics, filters may have to be applied despite the assumption that these harmonics would be canceled. In three-phase loads that can operate while single-phased (e.g., arc furnaces), no attenuation of the uncharacteristic harmonics can be assumed.

6.7   Case Studies

Two additional case studies are presented which describe (1) the evaluation of neutral conductor loading and transformer derating and (2) interharmonics caused by induction furnaces.

6.7.1   Evaluation of Neutral Loading and Transformer Derating

Loads in a data center facility are dominated by hundreds of single-phase computer servers and networking equipment. The phase currents in the low-voltage circuits have the harmonic characteristics shown in Fig. 6.27. Since these loads are rich in the third harmonic, there is a good likelihood the neutral conductor may be overloaded. The problem is to estimate the neutral conductor loading in amperes and in percent of the rms phase current. In addition, the amount that the transformer supplying this load must be derated is to be

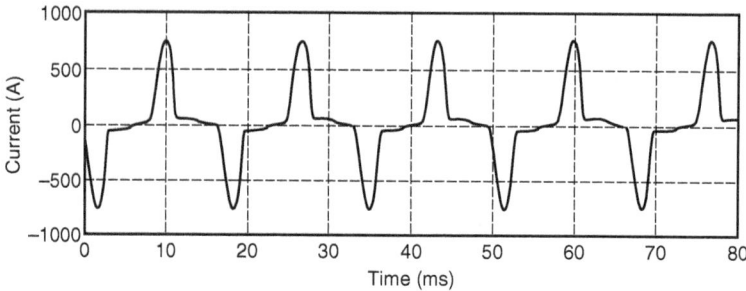

Harm	%	Phase
Fund	100.0	−37
3rd	65.7	−97
5th	37.7	−166
7th	12.7	113
9th	4.4	−46
11th	5.3	−158
13th	2.5	92
15th	1.9	−51

Harm	%	Phase
17th	1.8	−151
19th	1.1	84
21st	0.6	−41
23rd	0.8	−148
25th	0.4	64
27th	0.2	−25
29th	0.2	−122
31st	0.2	102

Figure 6.27 Phase current and its harmonic characteristics. Fundamental amps: 285.5 A. Phase angles are in degrees.

determined assuming the eddy-current loss factor under rated load $P_{EC\text{-}R}$ is 8 percent.

The system is assumed to be balanced. Therefore, the sum of all phase currents results in mostly third-harmonic current in the neutral conductor. The rms phase current is

$$I_{\text{rms}} = \sqrt{\sum_{h=1,3,5,N}^{N=31} I_h^2} = 1.26 I_1 = 359.1 \text{ A}$$

The third-harmonic current is 65.7 percent, giving a neutral current of

$$I_{\text{neutral}} = 3I_{3\text{rd}} = 3 \times 0.657 I_1 = 562.72 \text{ A}$$
$$= 1.56 I_{\text{rms}}$$

Based on this estimate, the neutral conductor will be loaded to approximately 156 percent of the phase conductor. This phenomenon has been responsible for neutral overloading. Common solutions are to use a

- Separate neutral conductor for each phase
- Double neutral conductor size

- Zigzag transformer close to the loads to shorten the return path for the third-harmonic currents and relieve the overloaded neutral
- Series filter tuned to the third harmonic in the neutral circuit at the transformer

The transformer derating can be estimated by first computing the K factor[12] by using Eq. (5.30) presented in Sec. 5.10.2. Table 6.5 shows this computation and yields $K = 6.34$. From IEEE Standard 57.110-1998, *Recommended Practice for Establishing Transformer Capability When Supplying Nonsinusoidal Load Currents,* the standard derating for this waveform is 0.85 pu for $P_{EC-R} = 8$ percent.

Harmonic	Current, %	Frequency, Hz	Current, pu	I^2	I^2h^2
1	100.00	60	1.000	1.000	1.000
3	65.70	180	0.657	0.432	3.885
5	37.70	300	0.377	0.142	3.553
7	12.70	420	0.127	0.016	0.790
9	4.40	540	0.044	0.002	0.157
11	5.30	660	0.053	0.003	0.340
13	2.50	780	0.025	0.001	0.106
15	1.90	900	0.019	0.000	0.081
17	1.80	1020	0.018	0.000	0.094
19	1.10	1140	0.011	0.000	0.044
21	0.60	1260	0.006	0.000	0.016
23	0.80	1380	0.008	0.000	0.034
25	0.40	1500	0.004	0.000	0.010
27	0.20	1620	0.002	0.000	0.003
29	0.20	1740	0.002	0.000	0.003
31	0.20	1860	0.002	0.000	0.004
			Total	1.596	10.119
				K factor 6.34	

Standard derating (ANSI/IEEE C57.110–1986)0.85 pu

Assumed eddy current loss factor P_{EC-R} 8%

TABLE 6.5 Computation for Transformer Derating

6.7.2   Interharmonics Caused by Induction Furnaces

The key symptom of this problem was that residential customers in a widespread area complained about their clocks running faster at about the same time each weekday. Other timekeeping instruments also behaved erratically.

The clocks that experienced the problem count time by detecting zero-crossings in the voltage waveform. The time between two adjacent zero-crossings is a half cycle of the power system fundamental frequency. Since the frequency error of the power system is negligible over long time periods, these clocks are very accurate.

Fast-clock phenomena occur when there are more zero-crossings than expected within a half cycle due to high-frequency distortion in the voltage waveform. The high-frequency signal appears as a sawtooth or sinusoid superimposed on the fundamental frequency signal. Figure 6.28 shows a typical voltage waveform measured on customer premises. It is clear that there will be instances where there are multiple zero-crossings within a half cycle.

Figure 6.28*b* shows that the high-frequency distortion occurs at the 29th (1740 Hz) and the 35th (2100 Hz) harmonics. Further investigation revealed that these frequencies were produced by induction furnaces located at a steel-grinding facility. The distortion affected residential customers several miles away. Both the grinding

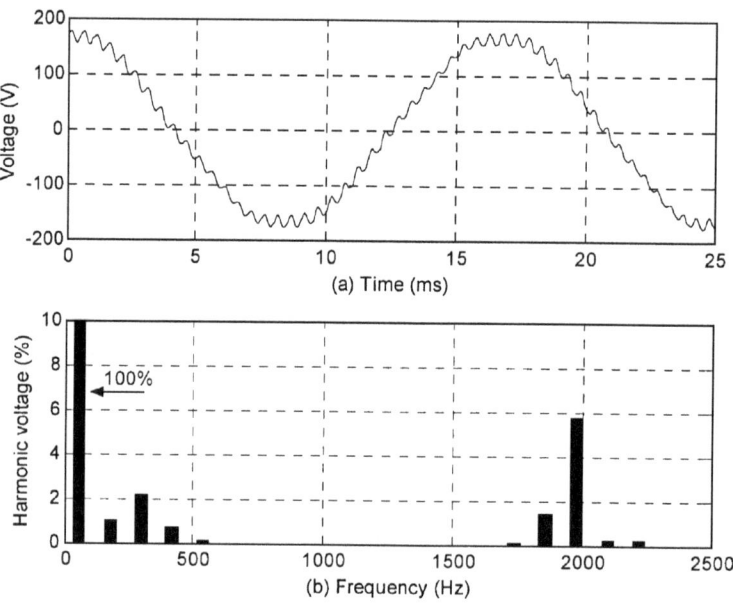

Figure 6.28   Voltage waveform causing fast-clock problems due to high-frequency distortion and its harmonic spectrum.

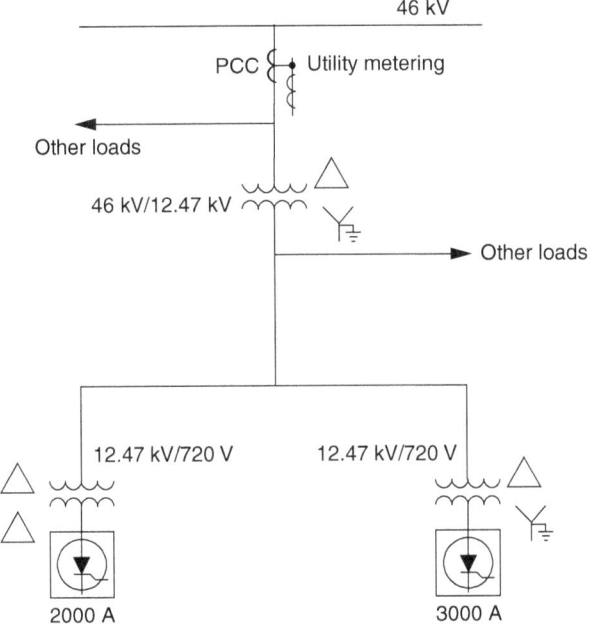

46 kV

PCC

Utility metering

Other loads

46 kV/12.47 kV

Other loads

12.47 kV/720 V

12.47 kV/720 V

2000 A

3000 A

FIGURE 6.29   Steel-grinding facility one-line diagram showing source, metering, and loads.

facility and residential customers were supplied from the same 46-kV distribution system, shown in the one-line diagram of the facility in Fig. 6.29.

The operating frequency of the two induction furnaces varies between 800 to 1000 Hz depending on the amount and type of material being melted. The harmonic characteristics of these furnaces were described in Sec. 5.11. Assuming the operating frequency at a particular operation stage is 950 Hz, the resulting line current computed using Eq. (5.34) would contain the following pairs of currents: (1840 Hz, 1960 Hz), (3740 Hz, 3860 Hz), etc. These currents are interharmonic currents since they are not integer multiples of the fundamental frequency. The first pair are the strongest interharmonic components and are more prominent in the voltage. Since the furnace operating frequency varies between 800 and 1000 Hz, the first pair of the resulting interharmonic current varies between 1540 (25.67th harmonic) and 2060 Hz (or 34.33th harmonic). This varying harmonic distortion makes the application of passive shunt filters impossible.

The PCC for this facility was at the high-voltage side of the 46/12.47-kV transformer. Figure 6.30 shows the voltage waveform at the PCC where the high-frequency distortion is clearly visible on top of the fundamental frequency waveform.

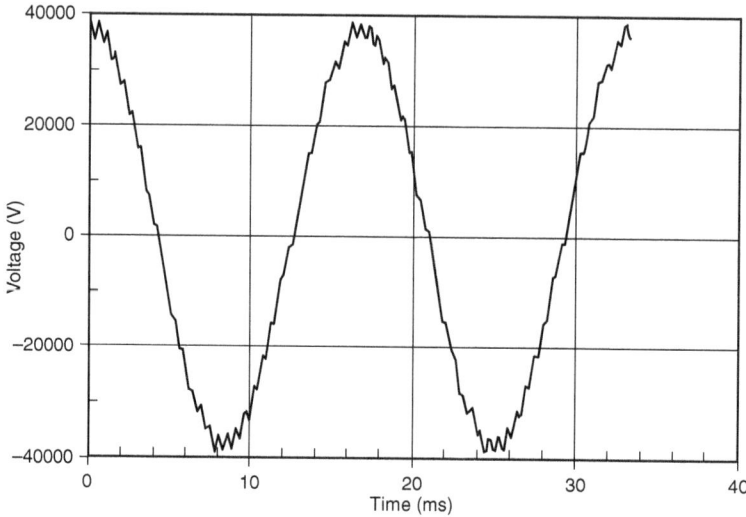

FIGURE 6.30   Voltage waveform at the PCC for steel-grinding facility (46 kV).

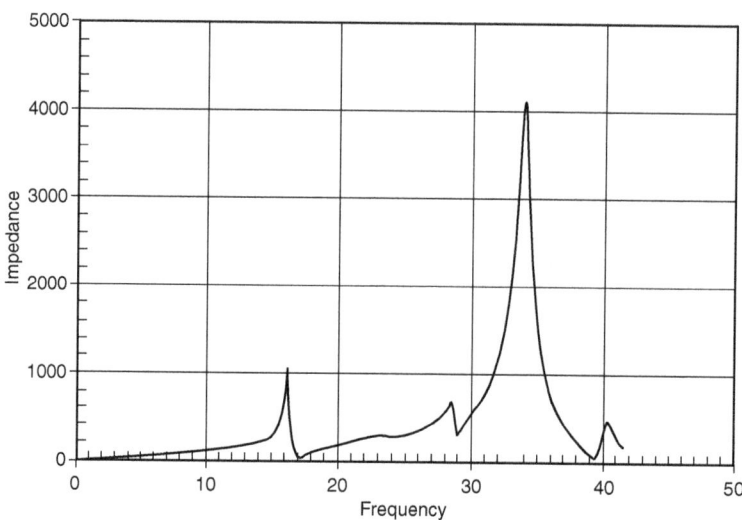

FIGURE 6.31   Impedance scan at the PCC (46 kV).

To understand why the distortion appeared throughout the 46-kV system, a frequency scan of the system looking from the PCC was performed. The resulting impedance characteristic is shown in Fig. 6.31. The scan indicated that the dominant resonance frequency was approximately at the 34th harmonic. When the frequency components produced by a nonlinear load line up with the system natural frequency, the distortion will be magnified. This is exactly what happened in this problem. The interharmonic frequencies produced by the induction furnaces varied between 25th and 34th harmonics, the upper end of this range coinciding with the system natural frequency. Thus, it was not surprising to find voltage distortion over a wide area.

Since the high-frequency distortion varied with time and the system frequency response accentuated the distortion, solutions employing single-tuned shunt filters (even with multiple stages) would not work. There were two possible filter solutions:

1. Modifying the frequency response at the 46-kV bus so that its natural frequency did not align with the induction furnace interharmonic frequencies

2. Placing a broadband filter at the facility main bus to prevent the distorted currents from entering the 46-kV system

The first approach requires a careful selection of a 46-kV capacitor bank. The new frequency response should not contain any resonance that aligns with a harmonic produced by the nonlinear loads. With simulations, it was estimated that a capacitor bank of approximately 3 Mvar would be required to move the existing system natural frequency from the 35th harmonic down to the 8th harmonic. The eighth harmonic was selected since there were no known nonlinear loads producing harmonic currents of this order. This solution was feasible; however, installing a 3-Mvar capacitor bank would be overcompensating much of the time. In addition, if the target tuning drops below the eighth harmonic due to line outages that would weaken the system, there is increased risk of causing problems with the fifth and seventh harmonics.

The second approach requires a mechanism to prevent high-frequency interharmonics from entering the 46-kV system. As described in the first approach, multiple stages of single-tuned shunt filter banks would not work well since the interharmonics are varying. Active filters would solve the problem; however, they are expensive. A more economical solution would be a low-pass broadband filter like that described in Sec. 6.5. Also, there is more control over the short-circuit impedance at the filter location. The solution is illustrated in Fig. 6.32.

It is easy to accomplish the attenuation of frequencies above the 30th harmonic with this approach. The problem is to find a capacitor size that will not result in a resonance that aligns with other harmonic frequencies produced by the furnaces, particularly, the 5th, 7th, 11th, and 13th. The eighth harmonic was again chosen as a target tuning frequency. The next best frequency might be the fourth harmonic; however, the resulting voltage rise due to a larger capacitor bank size

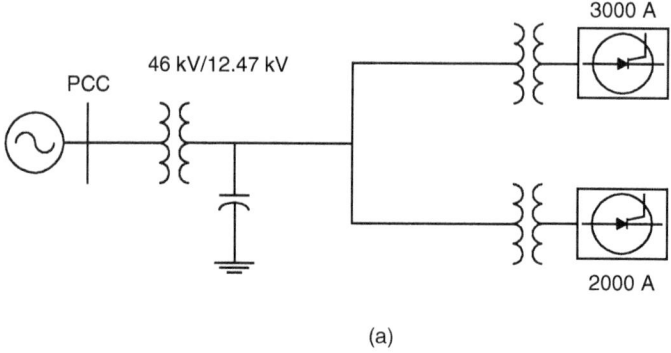

(a)

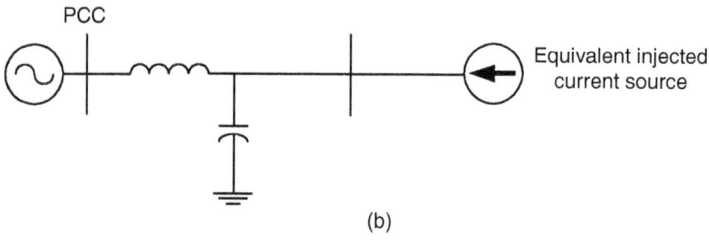

(b)

Figure 6.32   Solution at the 12.47-kV side (a) and its equivalent low-pass broadband filter effect (b).

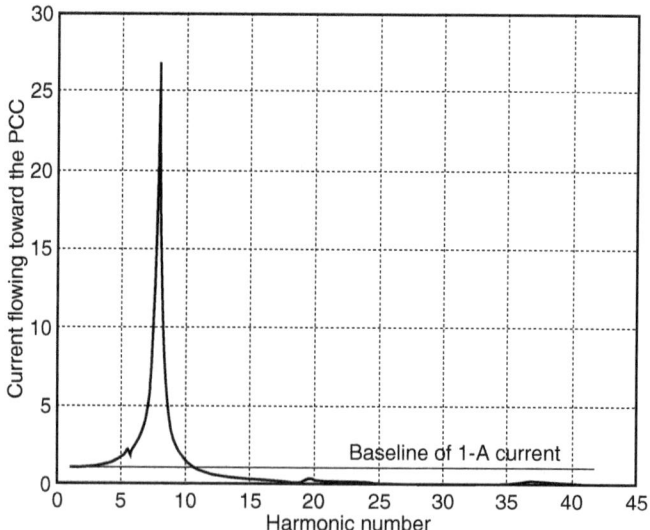

Figure 6.33   Current flowing toward the PCC when 1 A of current at various frequencies was injected from the 12.47-kV bus.

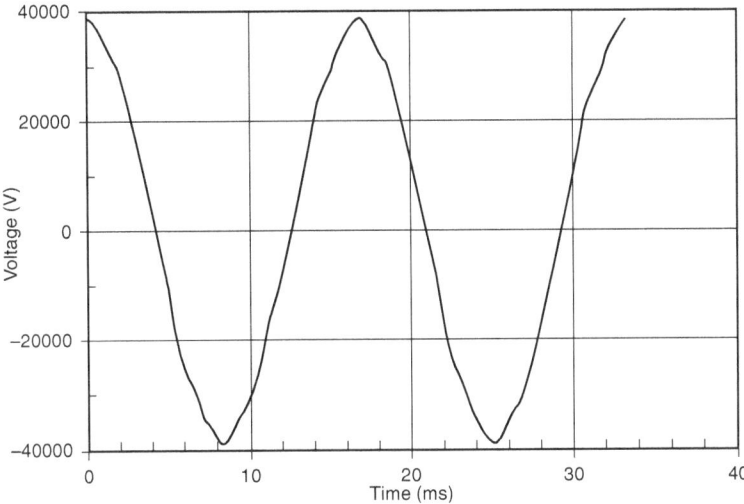

FIGURE **6.34** Voltage waveform at the PCC after installing a 1200-kvar bank rated at 13.2 kV at the 12.47-kV bus.

would require adding a voltage regulator to buck the voltage down. This would make the solution much more costly.

It was determined by simulation that a common 1200-kvar bank rated at 13.2 kV provides a good solution. Using a capacitor rated higher than nominal shifts the tuning slightly higher, giving less magnification of the seventh harmonic. Figure 6.33 shows the current flowing toward the PCC for 1 A of current. The high-frequency interharmonic currents above the 30th harmonic are greatly attenuated and are prevented from flowing through the transformer into the 46-kV system.

Figure 6.34 shows the resulting voltage waveform at the PCC. While some minor distortion remains (mostly fifth and seventh harmonics), this is acceptable. Thus, this problem can be solved simply by applying a relatively inexpensive, commonly available capacitor bank.

6.8 Standards on Harmonics

There are various organizations on the national and international levels working in concert with engineers, equipment manufacturers, and research organizations to come up with standards governing guidelines, recommended practices, and harmonic limits. The primary objective of the standards is to provide a common ground for all involved parties to work together to ensure compatibility between end-use equipment and the system equipment is applied. An example of compatibility (or lack of compatibility) between end-use equipment and the system equipment is the fast-clock problem in the case study given in Sec. 6.7.2. The end-use equipment is the clock with voltage zero-crossing detection technology, while the system yields a voltage distorted with harmonics between 30th and 35th. This illustrates a mismatch of compatibility that causes misoperation of the end-use equipment.

This section focuses on standards governing harmonic limits, including IEEE 519-1992, IEC 61000-2-2, IEC 61000-3-2, IEC 61000-3-4, IEC 61000-3-6, NRS 048-2,[13] and EN50160.[14]

6.8.1 IEEE Standard 519-1992

The limits on harmonic voltage and current based on IEEE Standard 519-1992 are described in Sec. 6.1. It should be emphasized that the philosophy behind this standard seeks to limit the harmonic injection from individual customers so that they do not create unacceptable voltage distortion under normal system characteristics and to limit the overall harmonic distortion in the voltage supplied by the utility. The voltage and current distortion limits should be used as system design values for the worst case of normal operating conditions lasting more than 1 h. For shorter periods, such as during start-ups, the limits may be exceeded by 50 percent.

This standard divides the responsibility for limiting harmonics between both end users and the utility. End users will be responsible for limiting the harmonic current injections, while the utility will be primarily responsible for limiting voltage distortion in the supply system.

The harmonic current and voltage limits are applied at the PCC. This is the point where other customers share the same bus or where new customers may be connected in the future. The standard seeks a fair approach to allocating a harmonic limit quota for each customer. The standard allocates current injection limits based on the size of the load with respect to the size of the power system, which is defined by its short-circuit capacity. The short-circuit ratio is defined as the ratio of the maximum short-circuit current at the PCC to the maximum demand load current (fundamental frequency component) at the PCC as well.

The basis for limiting harmonic injections from individual customers is to avoid unacceptable levels of voltage distortions. Thus the current limits are developed so that the total harmonic injections from an individual customer do not exceed the maximum voltage distortion shown in Table 6.6.

Table 6.6 shows harmonic current limits for various system voltages. Smaller loads (typically larger short-circuit ratio values) are

Short-circuit ratio at PCC	Maximum individual frequency voltage harmonic (%)	Related assumption
10	2.5–3.0	Dedicated system
20	2.0–2.5	1–2 large customers
50	1.0–1.5	A few relatively large customers
100	0.5–1.0	5–20 medium-size customers
1000	0.05–0.10	Many small customers

SOURCE: From IEEE Standard 519-1992, table 10.1.

TABLE 6.6   Basis for Harmonic Current Limits

allowed a higher percentage of harmonic currents than larger loads with smaller short-circuit ratio values. Larger loads have to meet more stringent limits since they occupy a larger portion of system load capacity. The current limits take into account the diversity of harmonic currents in which some harmonics tend to cancel out while others are additive.

The harmonic current limits at the PCC are developed to limit individual voltage distortion and voltage THD to the values shown in Table 6.1. Since voltage distortion is dependent on the system impedance, the key to controlling voltage distortion is to control the impedance. The two main conditions that result in high impedance are when the system is too weak to supply the load adequately or the system is in resonance. The latter is more common. Therefore, keeping the voltage distortion low usually means keeping the system out of resonance. Occasionally, new transformers and lines will have to be added to increase the system strength.

IEEE Standard 519-1992 represents a consensus of guidelines and recommended practices by the utilities and their customers in minimizing and controlling the impact of harmonics generated by nonlinear loads.

6.8.2 Overview of IEC Standards on Harmonics

The International Electrotechnical Commission (IEC), currently with headquarters in Geneva, Switzerland, has defined a category of electromagnetic compatibility (EMC) standards that deal with power quality issues. The term *electromagnetic compatibility* includes concerns for both radiated and conducted interference with end-use equipment. The IEC standards are broken down into six parts:

- *Part 1: General.* These standards deal with general considerations such as introduction, fundamental principles, rationale, definitions, and terminologies. They can also describe the application and interpretation of fundamental definitions and terms. Their designation number is IEC 61000-1-x.

- *Part 2: Environment.* These standards define characteristics of the environment where equipment will be applied, the classification of such environment, and its compatibility levels. Their designation number is IEC 61000-2-x.

- *Part 3: Limits.* These standards define the permissible levels of emissions that can be generated by equipment connected to the environment. They set numerical emission limits and also immunity limits. Their designation number is IEC 61000-3-x.

- *Part 4: Testing and measurement techniques.* These standards provide detailed guidelines for measurement equipment and test procedures to ensure compliance with the other parts of the standards. Their designation number is IEC 61000-4-x.

- *Part 5: Installation and mitigation guidelines.* These standards provide guidelines in application of equipment such as earthing and cabling of electrical and electronic systems for ensuring electromagnetic compatibility among electrical and electronic apparatus or systems. They also describe protection concepts for civil facilities against the high-altitude electromagnetic pulse (HEMP) due to high-altitude nuclear explosions. They are designated with IEC 61000-5-x.

- *Part 6: Miscellaneous.* These standards are generic standards defining immunity and emission levels required for equipment in general categories or for specific types of equipment. Their designation number is IEC 61000-6-x.

IEC standards relating to harmonics generally fall in parts 2 and 3. Unlike the IEEE standards on harmonics where there is only a single publication covering all issues related to harmonics, IEC standards on harmonics are separated into several publications. There are standards dealing with environments and limits which are further broken down based on the voltage and current levels. These key standards are as follows:

- IEC 61000-2-2 (1993): *Electromagnetic Compatibility (EMC).* Part 2: Environment. Section 2: Compatibility Levels for Low-Frequency Conducted Disturbances and Signaling in Public Low-Voltage Power Supply Systems.

- IEC 61000-3-2 (2000): *Electromagnetic Compatibility (EMC).* Part 3: Limits. Section 2: Limits for Harmonic Current Emissions

(Equipment Input Current Up to and Including 16 A per Phase).

- IEC 61000-3-4 (1998): *Electromagnetic Compatibility (EMC)*. Part 3: Limits. Section 4: Limitation of Emission of Harmonic Currents in Low-Voltage Power Supply Systems for Equipment with Rated Current Greater Than 16 A.

- IEC 61000-3-6 (1996): *Electromagnetic Compatibility (EMC)*. Part 3: Limits. Section 6: Assessment of Emission Limits for Distorting Loads in MV and HV Power Systems. Basic EMC publication.

Prior to 1997, these standards were designated by a 1000 series numbering scheme. For example, IEC 61000-2-2 was known as IEC 1000-2-2. These standards on harmonics are generally adopted by the European Community (CENELEC); thus, they are also designated with the EN 61000 series. For example, IEC 61000-3-2 is also known as EN 61000-3-2.

6.8.3 IEC 61000-2-2

IEC 61000-2-2 defines compatibility levels for low-frequency conducted disturbances and signaling in public low-voltage power supply systems such as 50- or 60-Hz single- and three-phase systems with nominal voltage up 240 and 415 V, respectively. Compatibility levels are defined empirically such that they reduce the number of complaints of misoperation to an acceptable level.[15] These levels are not rigid and can be exceeded in a few exceptional conditions. Compatibility levels for individual harmonic voltages in the low-voltage network are shown in Table 6.7. They are given in percentage of the fundamental voltage.

6.8.4   IEC 61000-3-2 and IEC 61000-3-4

Both IEC 61000-3-2 and 61000-3-4 define limits for harmonic current emission from equipment drawing input current of up to and including 16 A per phase and larger than 16 A per phase, respectively. These standards are aimed at limiting harmonic emissions from equipment connected to the low-voltage public network so that compliance with the limits ensures that the voltage in the public network satisfies the compatibility limits defined in IEC 61000-2-2.

The IEC 61000-3-2 is an outgrowth from IEC 555-2 (EN 60555-2). The standard classifies equipment into four categories:

- Class A: Balanced three-phase equipment and all other equipment not belonging to classes B, C, and D
- Class B: Portable tools
- Class C: Lighting equipment including dimming devices
- Class D: Equipment having an input current with a "special waveshape" and an active input power of less than 600 W

Not multiple of 3		Multiple of 3			
Odd order h	Harmonic voltage h	Odd order (%)	Harmonic voltage h	Even order (%)	Harmonic voltage h
5	6	3	5	2	2
7	5	9	1.5	4	1
11	3.5	15	0.3	6	0.5
13	3	21	0.2	8	0.5
17	2	>21	0.2	10	0.2
19	1.5			12	0.2
23	1.5			>12	0.2
25	1.5				
>25	0.2 + 1.3 × 25/h				

*The THD of the supply voltage including all harmonics up to the 40th order shall be less than 8 percent.

TABLE **6.7**   Compatibility Levels for Individual Harmonic Voltages in the Low-Voltage Public Network According to IEC 61000-2-2*

Figure 6.35 can be used for classifying equipment in IEC 61000-3-2. It should be noted that equipment in classes B and C and provisionally motor-driven equipment are not considered class D equipment regardless of their input current waveshapes. The half-cycle waveshape of class D equipment input current should be within the envelope of the inverted T-shape shown in Fig. 6.36 for at least 95 percent of the time. The center line at p/2 lines up with the peak value of the input current I_{pk}.

Maximum permissible harmonic currents for classes A, B, C, and D are given in actual amperage measured at the input current of the equipment. Note that harmonic current limits for class B equipment are 150 percent of those in class A. Harmonic current limits according to IEC 61000-3-2 are shown in Tables 6.8 through 6.10. Note that harmonic current limits for class D equipment are specified in absolute numbers and in values relative to active power. The limits only apply to equipment operating at input power up to 600 W.

IEC 61000-3-4 limits emissions from equipment drawing input current larger than 16 A and up to 75 A. Connections of this type of equipment do not require consent from the utility. Harmonic current limits based on this standard are shown in Table 6.11.

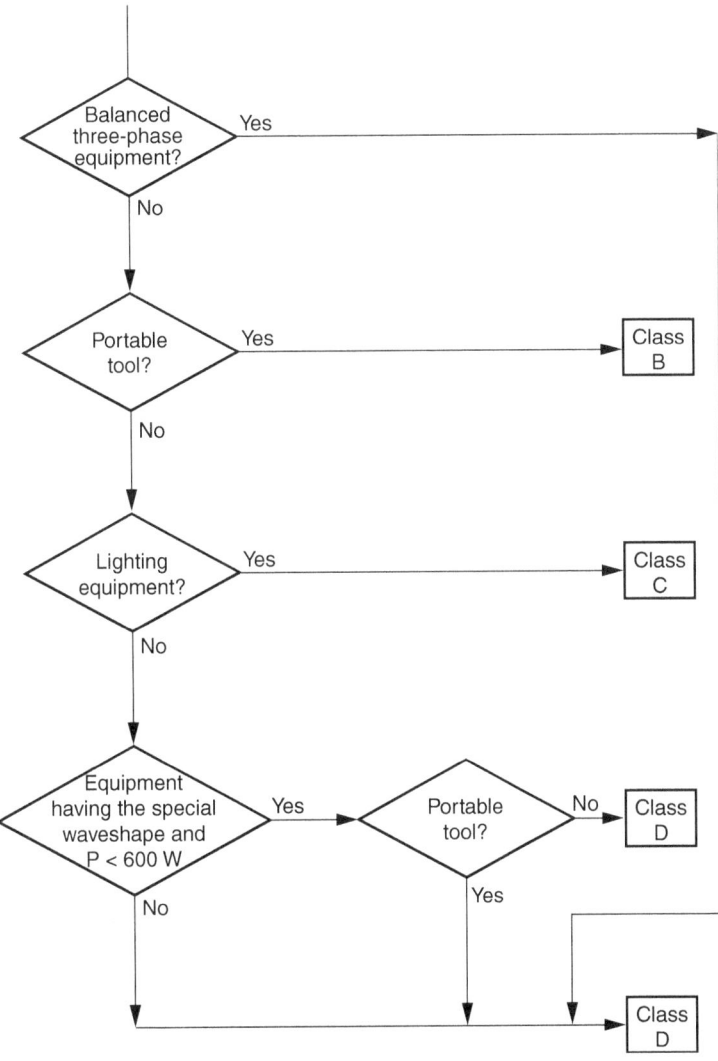

FIGURE **6.35** Flowchart for classifying equipment according to IEC 61000-3-2.

6.8.5 IEC 61000-3-6

IEC 61000-3-6 specifies limits of harmonic current emission for equipment connected to medium-voltage (MV) and high-voltage (HV) supply systems. In the context of the standard, MV and HV refer to voltages between 1 and 35 kV, and between 35 and 230 kV, respectively. A voltage higher than 230 kV is considered extra high voltage (EHV), while a voltage less than 1 kV is considered low voltage (LV).

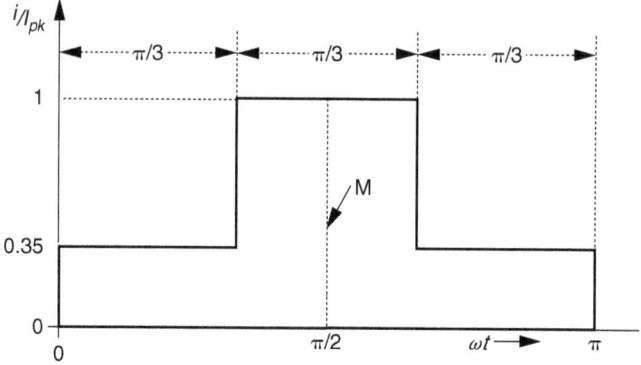

FIGURE 6.36 Envelope of the input current to define the special waveshape for class D equipment.

Odd order h	Max. permissible harmonic current order (A)	Even order h	Max. permissible harmonic order (A)
3	2.3	2	1.08
5	1.14	4	0.43
7	0.77	6	0.3
9	0.4	8–40	0.23 × 8/h
11	0.33		
13	0.21		
15–39	0.15 × 15/h		

TABLE 6.8 Harmonic Current Limits for Class A Equipment

Harmonic order h	Max. permissible harmonic current* (%)
2	2
3	30 × circuit power factor
5	10
7	7
9	5
11–39	3

*Percent of the fundamental input current.

TABLE 6.9 Harmonic Current Limits for Class C Equipment

Harmonic order h	Max. permissible harmonic current Per watt (mA/W)	(A)
2	3.4	2.3
5	1.9	1.14
7	1.0	0.77
9	0.50	0.40
13	0.35	0.33
11–39	3.86/h	See Table 6.8

TABLE 6.10   Harmonic Current Limits for Class D Equipment

Harmonic order h	Max. permissible harmonic current* (%)	Harmonic order h	Max. permissible harmonic current* (%)
3	21.6	19	1.1
5	10.7	21	0.6
7	7.2	23	0.9
9	3.8	25	0.8
11	3.1	27	0.6
13	2	29	0.7
15	0.7	31	0.7
17	1.2	33	0.6

* Percent of the fundamental input current.

TABLE 6.11   Harmonic Current Limits According to IEC 61000-3-4

The standard argues that emission limits for individual equipment connected to the MV and HV systems should be evaluated on the voltage distortion basis. This is to ensure that harmonic current injections from harmonic-producing equipment do not result in excess voltage distortion levels. The standard provides compatibility levels and planning levels for harmonic voltages in the LV and MV systems. The compatibility level refers to a level where the compatibility between the equipment and its environment is achieved. The compatibility level is usually established empirically so that a piece of equipment is compatible with its environment *most of the time.* Compatibility levels are generally based on the 95-percent probability level, i.e., 95 percent of the time, the compatibility can be achieved. Table 6.12 shows compatibility levels for harmonic voltages as a percentage of the fundamental voltage in both LV and MV system.

| Odd harmonics | | | | Even harmonics | |
| Not multiple of 3 | | Multiple of 3 | | | |
Order h	Harmonic voltage (%)	Order h	Harmonic voltage (%)	Order h	Harmonic voltage (%)
5	6	3	5	2	2
7	5	9	1.5	4	1
11	3.5	15	0.3	6	0.5
13	3	21	0.2	8	0.5
17	2	>21	0.2	10	0.5
19	1.5			12	0.2
23	1.5			>12	0.2
25	1.5				
>25	0.2 + 1.3 × 25/h				

*Limit for THD is 8 percent.

TABLE 6.12   Compatibility Levels for Harmonic Voltages (in Percent of Fundamental) for LV and MV Systems

Planning levels are design criteria or levels specified by the utility company. Planning levels are more stringent than compatibility levels. Thus, their levels are lower than the compatibility levels. Planning levels for harmonic voltage expressed in the percentage of the fundamental voltage for MV, HV, and EHV systems are given in Tables 6.13 and 6.14.

The IEC 61000-3-6 provides evaluation guidelines to determine admissibility of equipment connected to MV and HV systems. There are three stages for evaluating equipment admissibility:

- *Stage 1.* Simplified evaluation of disturbance emission
- *Stage 2.* Emission limits relative to actual network characteristics
- *Stage 3.* Acceptance of higher emission levels on an exceptional and precarious basis

In stage 1, equipment can be connected to MV or HV systems without conducting harmonic studies as long as its size is considered small in relation to the system short-circuit capacity. For small appliances, manufacturers are responsible for limiting their harmonic emissions.

If the equipment does meet stage 1 criteria, the harmonic characteristics of the equipment should be evaluated in detail along with the available system absorption capacity. Upon evaluation,

Odd harmonics				Even harmonics	
Not multiple of 3		Multiple of 3			
Order *h*	Harmonic voltage (%)	Order *h*	Harmonic voltage (%)	Order *h*	Harmonic voltage (%)
5	5	3	4	2	1.6
7	4	9	1.2	4	1
11	3	15	0.3	6	0.5
13	2.5	21	0.2	8	0.4
17	1.6	>21	0.2	10	0.4
19	1.2			12	0.2
23	1.2			>12	0.2
25	1.2				
>25	$0.2 + 0.5 \times 25/h$				

*Limit for THD is 6.5 percent.

TABLE 6.13 Planning Levels for Harmonic Voltages (in Percent of Fundamental) for MV Systems

Odd harmonics				Even harmonics	
Not multiple of 3		Multiple of 3			
Order *h*	Harmonic voltage (%)	Order *h*	Harmonic voltage (%)	Order *h*	Harmonic voltage (%)
5	2	3	2	2	1.6
7	2	9	1	4	1
11	1.5	15	0.3	6	0.5
13	1.5	21	0.2	8	0.4
17	1	>21	0.2	10	0.4
19	1			12	0.2
23	0.7			>12	0.2
25	0.7				
>25	$0.2 + 0.5 \times 25/h$				

*Limit for THD is 6.5 percent.

TABLE 6.14 Planning Levels for Harmonic Voltages (in Percent of Fundamental) for HV and EHV Systems

individual equipment will be allocated with appropriate system absorption capacity according to its size. Thus, if the system absorption capacity has been fully allocated to all equipment, and this equipment injects its harmonic currents up to its limits, the system voltage distortion should be within its planning levels.

If equipment does not meet stage 2 criteria, it may be allowed to be connected to the system if the end user and utility agree to make special arrangement to facilitate such a connection.

6.8.6 NRS 048-02

The *Quality of Supply Standard*, NRS 048, is the South African standard for dealing with the quality of electricity supply and has been implemented since July 1, 1997. This standard requires electricity suppliers to measure and report their quality of supply to the National Electricity Regulator.

The NRS 048 is divided into five parts. It is, perhaps, the most thorough standard dealing with all aspects of quality of supply. It covers the minimum standards of quality of supply (QOS), measurement and reporting of QOS, application and implementation guidelines for QOS, and instrumentation for voltage quality monitoring and recording.

Part 2 of NRS 048 sets minimum standards for the quality of the electrical product supplied by South African utilities to end users. The minimum standards include limits for voltage harmonics and interharmonics, voltage flicker, voltage unbalance, voltage dips, voltage regulation, and frequency.

NRS 048-02 adopts IEC 61000-2-2 harmonic voltage limits shown in Table 6.7 as its compatibility standards for LV and MV systems. For South African systems, the nominal voltage for LV systems is less than 1 kV, while the nominal voltage for MV systems ranges between 1 and 44 kV.

NRS 048 has not established limits for harmonic voltages for HV systems yet. However, it adopts IEC 61000-3-6 planning levels for harmonic voltages for HV and EHV systems (shown in Table 6.14) as its recommended planning limits for HV systems (the nominal voltage is between 200 and 400 kV).

6.8.7 EN 50160

EN 50160 is a European standard for dealing with supply quality requirements for European utilities. The standard defines specific levels of voltage characteristics that must be met by utilities and methods for evaluating compliance. EN 50160 was approved by the European Committee for Electrotechnical Standardization (CENELEC) in 1994.

EN 50160 specifies voltage characteristics at the customer's supply terminals or in public LV and MV electricity distribution systems under normal operating conditions. In other words, EN 50160

| Odd harmonics | | | | Even harmonics | |
| Not multiple of 3 | | Multiple of 3 | | | |
Order *h*	Harmonic voltage (%)	Order *h*	Harmonic voltage (%)	Order *h*	Harmonic voltage (%)
5	6	3	5	2	2
7	5	9	1.5	4	1
11	3.5	15	0.3	6–24	0.5
13	3	21	0.2		
17	2				
19	1.5				
23	1.5				
25	1.5				

TABLE **6.15**   Harmonic Voltage Limits at the Supply Terminals

confines itself to voltage characteristics at the PCC and does not specify requirements for power quality within the supply system or within customer facilities.

Harmonic voltage limits for EN 50160 are given in percentage of the fundamental voltage. The limits apply to systems supplied at both LV and MV levels, i.e., from a nominal 230 V up to 35 kV. Medium voltage is between 1 and 35 kV. The harmonic voltage limits are shown in Table 6.15. The total harmonic distortion of the supply voltage including all harmonics up to order 40 should not exceed 8 percent. Values for higher-order harmonics are not specified since they are too small to use as a practical measure to establish a meaningful reference value.

Note that limits in EN 50160 are nearly identical to the IEC 61000-3-6 compatibility levels for harmonic voltages for its corresponding LV and MV systems, except for the absence of higher-order harmonic limits in EN 50160.

6.9 References

1. M. F. McGranaghan, "Overview of the Guide for Applying Harmonic Limits on Power Systems—IEEE P519A," *Eighth International Conference on Harmonics and Quality of Power,* ICHQP 1998, Athens, Greece, pp. 462–469.
2. IEEE 519-1992, *Recommended Practices and Requirements for Harmonic Control in Electric Power Systems.*
3. IEEE P519A-2000, *Guide for Applying Harmonic Limits on Power Systems.*
4. R. C. Dugan, "Simulation of Arc Furnace Power Systems," *IEEE Transactions on Industry Applications,* November/December 1980, pp. 813–818.
5. M. F. McGranaghan, J. H. Shaw, R. E. Owen, "Measuring Voltage and Current Harmonics on Distribution Systems," *IEEE Transactions on Power Apparatus and Systems,* Vol. 101, No. 7, July 1981.

6. M. F. McGranaghan, R. C. Dugan, W. L. Sponsler, "Digital Simulation of Distribution System Frequency Response Characteristics," *IEEE Transactions on Power Apparatus and Systems*, Vol. 101, No. 3, March 1981.
7. M. F. McGranaghan, and E. W. Gunther, "Design of a PC-Based Harmonic Simulation Program," *Second International Conference on Harmonics in Power Systems*, Winnipeg, Manitoba, October 1986.
8. D. Xia and G. T. Heydt, "Harmonic Power Flow Studies Part I—Formulation and Solution," *IEEE Transactions on Power Apparatus and Systems*, June 1982, pp. 1257–1265.
9. W. M. Grady, "Harmonic Power Flow Studies," Ph.D. thesis, Purdue University, May 1983.
10. M. M. Swamy, "Harmonic Reduction Using Broad Band Harmonic Filters," MTE Corporation Technical Articles, Menomonee Falls, Wisconsin.
11. R. Dwyer, H. V. Nguyen, S. G. Ashmore, "C Filters for Wide-bandwidth Harmonic Attenuation with Low Losses," *Conference Record*, IEEE Power Engineering Society Meeting, Winter 2000, Singapore.
12. J. M. Frank, "Origin, Development and Design of K-Factor Transformers," *Conference Record*, 1994 IEEE Industry Applications Society Annual Meeting, Denver, October 1994, pp. 2273–2274.
13. NRS 048, *Electricity Supply—The Quality of Supply Standard.*
14. EN 50160, *Voltage Characteristics of Electricity Supplied by Public Distribution Systems.*
15. IEC 61000-1-4, *Electromagnetic Compatibility* (EMC): "Rationale for Limiting Power-Frequency Conducted Harmonic and Interharmonic Current Emissions from Equipment, in the Frequency Range up to 9 kHz."

6.10   Bibliography

Dwyer, R. V., Gunther, E. W., Adapa, R. "A Comparison of Solution Techniques for the Calculation of Harmonic Distortion Due to Adjustable Speed DC Drives," *Fourth International Conference on Harmonic Systems*, Budapest, Hungary, October 1990.

Grebe, T. E., McGranaghan, M. F., Samotyj, M. "Solving Harmonic Problems in Industrial Plants and Harmonic Mitigation Techniques for Adjustable-Speed Drives," *Electrotech 92*, Montreal, Canada, 1992.

McGranaghan, M. F., Grebe, T. E., Samotyj, M. "Solving Harmonic Problems in Industrial Plants—Case Studies," *First International Conference on Power Quality*, PQA '91, Paris, France, 1991.

McGranaghan, M. F., and Mueller, D. R. "Designing Harmonic Filters for Adjustable-Speed Drives to Comply with New IEEE-519 Harmonic Limits," *IEEE/IAS Annual Conference, Petroleum and Chemical Industry Technical Conference*, 1993.

Schwabe, R. J., Melhorn, C. J., Samotyj, M. "Effect of High Efficiency Lighting on Power Quality in Public Buildings," *Third International Conference on Power Quality*, PQA '93, San Diego, Calif.

Zavadil, R., McGranaghan, M. F., Hensley, G., Johnson, K. "Analysis of Harmonic Distortion Levels in Commercial Buildings," *First International Conference on Power Quality*, PQA '91, Paris, France, 1991.

CHAPTER 7

Long-Duration Voltage Variations

U tilities generally try to maintain the service voltage supplied to an end user within ±5 percent of nominal. Under emergency conditions, for short periods, ANSI Standard C84.1 permits the utilization voltage to be in the range of +6 to −13 percent of the nominal voltage. Some sensitive loads have more stringent voltage limits for proper operation and, of course, equipment generally operates more efficiently at near nominal voltage. This chapter addresses the fundamental problems behind voltage regulation and the general types of devices available to correct the problem.

7.1 Principles of Regulating the Voltage

The root cause of most voltage regulation problems is that there is too much impedance in the power system to properly supply the load (Fig. 7.1). Another way of describing this is to say that the power system is too weak for the load. Therefore, the voltage drops too low under heavy load. Conversely, when the source voltage is boosted to overcome the impedance, there can be an overvoltage condition when the load drops too low. The corrective measures usually involve either compensating for the impedance Z or compensating for the voltage drop $IR + jIX$ caused by the impedance.

Some common options for improving power system voltage regulation, in the approximate order of priority that a utility might apply, are

1. Add shunt capacitors to reduce the current I and shift it to be more in phase with the voltage.

2. Add voltage regulators, which boost the apparent V_1.

3. Reconductor lines to a larger size to reduce the impedance Z.

4. Change substation or service transformers to larger sizes to reduce impedance Z.

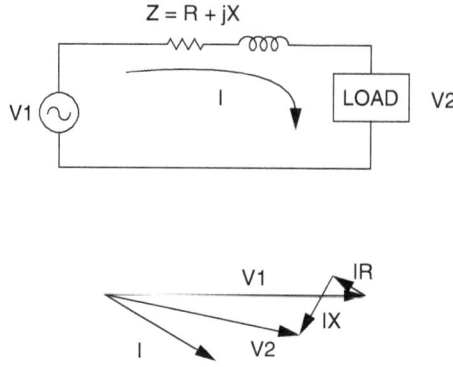

FIGURE 7.1 Voltage drop across the system impedance is the root cause of voltage regulation problems.

5. Add some kind of dynamic reactive power (var) compensation, which serves the same purpose as capacitors for rapidly changing loads.

6. Add series capacitors to cancel the inductive impedance drop IX.

7.2   Devices for Voltage Regulation

There are a variety of voltage regulation devices in use on utility and industrial power systems. We have divided these into three major classes:

1. Tap-changing transformers

2. Isolation devices with separate voltage regulators

3. Impedance compensation devices, such as capacitors

There are both mechanical and electronic tap-changing transformers. Tap-changing transformers are often autotransformer designs, although two- and three-winding transformers may also be equipped with tap changers. The mechanical devices are for the slower-changing loads, while the electronic ones can respond very quickly to voltage changes.

Isolation devices include UPS systems, ferroresonant (constant-voltage) transformers, and motor-generator sets. These are devices that essentially isolate the load from the power source by performing some sort of energy conversion. Therefore, the load side of the device can be separately regulated and can maintain constant voltage regardless of what is occurring at the power supply. The downside of using such devices is that they are costly, introduce more losses, and can cause harmonic distortion problems on the power supply system.

Shunt capacitors help maintain the voltage by reducing the current in the lines. Also, by overcompensating inductive circuits, a voltage rise can be achieved. To maintain a more constant voltage, the capacitors can be switched in conjunction with the load, sometimes in small incremental steps to follow the load more closely. If the objective is simply to maintain the voltage at a higher value to avoid an undervoltage condition, the capacitors are often fixed (not switched).

Series capacitors are relatively rare in utility distribution systems, but are useful for some impulse loads like rock crushers and tire testers.[1] Many potential users will shy away from them because of the extra care in engineering required for the series capacitor installation to function properly. However, they are very effective in certain system conditions, primarily with rapidly changing large loads that are causing excessive flicker (voltage fluctuations).

The series capacitors compensate for most of the inductance in the system leading up to the load. If the system is highly inductive, this will represent a significant reduction in the impedance. If the system is not highly inductive, but has a high proportion of resistance, series capacitors will not be very effective. This is typically the case in many industrial plant power systems that have long lengths of cable between the transformer and the load. To achieve a significant reduction in the impedance, the size of the cables and transformers must be increased.

Another approach to flicker-causing loads is to apply devices that are commonly called *static var compensators*. These can react within a few cycles to maintain a nearly constant voltage by rapidly controlling the reactive power production. Such devices are commonly used on arc furnaces, stone crushers, and other randomly varying loads where the system is weak and the resulting voltage fluctuations are affecting nearby customers.

7.2.1   Utility Step-Voltage Regulators

The typical utility tap-changing regulator can regulate from −10 to +10 percent of the incoming line voltage in 32 steps of 5/8 percent. There are some variations, but the majority are of this type. Distribution substation transformers commonly have three-phase load tap changers (LTCs) while line regulators installed out on the feeders are typically single-phase in North America. When installed on a three-phase feeder, line regulators are generally installed in banks of three. However, there are also many installations of open-delta regulator banks on lightly loaded three-phase feeders branches. This requires only two regulators and is less costly than a full three-phase bank.

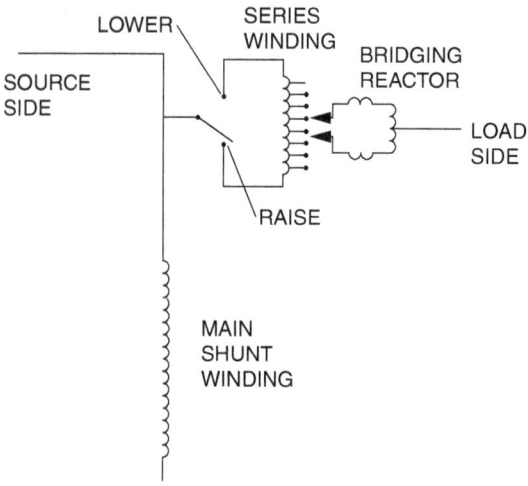

FIGURE **7.2** Schematic diagram of one type of utility voltage regulator commonly applied on distribution lines.

Figure 7.2 shows a schematic of a utility step-voltage regulator. Although the concept of a tap-changing autotransformer is simple, a utility voltage regulator is a fairly complicated apparatus in order to achieve a durable and highly reliable tap-changing mechanism.

Utility line voltage regulators and substation LTCs are relatively slow. The time delay when the voltage goes out of band is at least 15 s and is commonly 30 or 45 s. Thus, it is of little benefit where voltages may vary in matters of cycles or seconds. Their main application is boosting voltage on long feeders where the load is changing slowly over several minutes or hours. The voltage band typically ranges from 1.5 to 3.0 V on a 120-V base. The control can be set to maintain voltage at some point downline from the feeder by using the *line drop compensator*. This results in a more level average voltage response and helps prevent overvoltages on customers near the regulator.

7.2.2   Ferroresonant Transformers

On the end-user side, ferroresonant transformers are not only useful in protecting equipment from voltage sags (see Chap. 3), but they can also be used to attain very good voltage regulation (±1 percent output). Figure 7.3 shows the steady-state input/output characteristics of a 120-VA ferroresonant transformer with a 15-VA load. As the input voltage is reduced down to 30 V, the output voltage stays constant. If the input voltage is reduced further, the output voltage begins to collapse. In addition, as the input voltage is reduced, the current

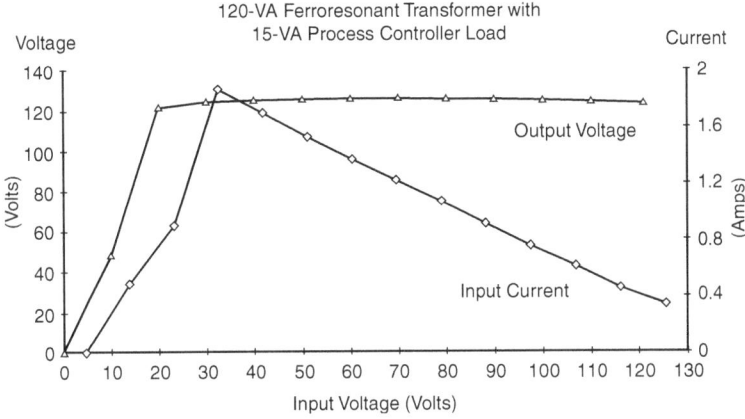

FIGURE **7.3**   Ferroresonant transformer steady-state characteristics.

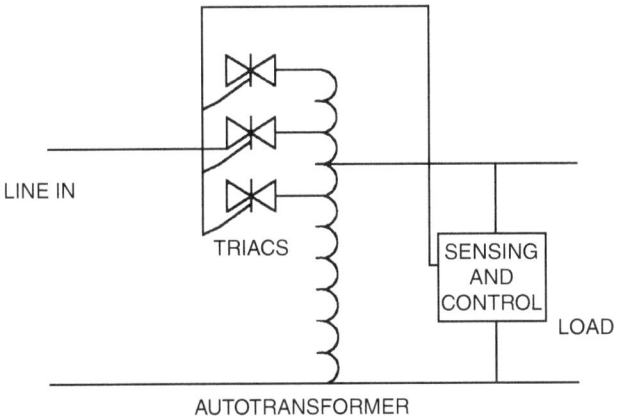

FIGURE **7.4**   Electronic tap-switching regulator.

drawn by the ferroresonant transformer increases substantially from 0.4 to 2 A. Thus, ferroresonant transformers tend to be lossy and inefficient.

7.2.3   Electronic Tap-Switching Regulators

Electronic tap-switching regulators (Fig. 7.4) can also be used to regulate voltage. They are more efficient than ferroresonant transformers and use SCRs or triacs to quickly change taps, and hence voltage. Tap-switching regulators have a very fast response time of a half cycle and are popular for medium-power applications.

7.2.4 Magnetic Synthesizers

Magnetic synthesizers, although intended for short-duration voltage sags (see Chap. 3), can also be used for steady-state voltage regulation. One manufacturer, for example, states that for input voltages of ±40 percent, the output voltage will remain within ±5 percent at full load.

7.2.5 On-Line UPS Systems

On-line UPS systems intended for protection against sags and brief interruptions can also be used for voltage regulation provided the source voltage stays sufficiently high to keep the batteries charged. This is a common solution for small, critical computer or electronic control loads in an industrial environment that has large, fluctuating loads causing the voltage to vary.

7.2.6 Motor-Generator Sets

Motor-generator sets (Fig. 7.5) are also used for voltage regulation. They completely decouple the load from the electric power system, shielding the load from electrical transients. Voltage regulation is provided by the generator control. The major drawback of motor-generator sets is their response time to large load changes. Motor-generator sets can take several seconds to bring the voltage back up to the required level, making this device too slow for voltage regulation of certain loads, especially rapidly varying loads. Motor-generator sets can also be used to provide "ride through" from input voltage variations, especially voltage sags, by storing energy in a flywheel.

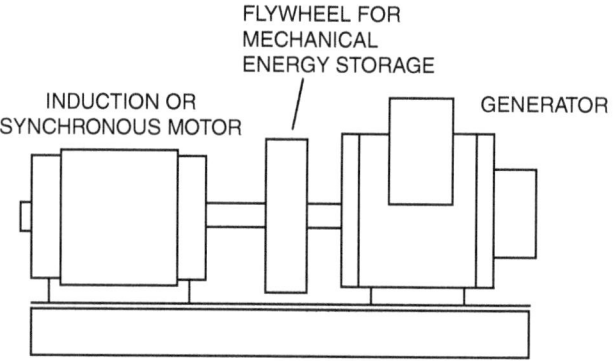

FIGURE 7.5 Motor-generator set.

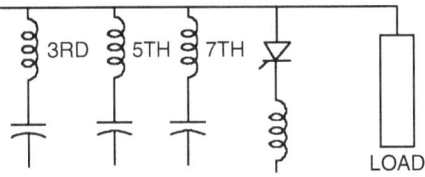

THYRISTOR-CONTROLLED REACTOR

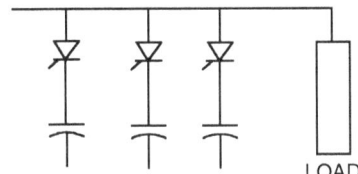

THYRISTOR-SWITCHED CAPACITOR

Figure 7.6   Common static var compensator configurations.

7.2.7   Static Var Compensators

Static var compensators can be applied to either utility systems or industrial systems. They help regulate the voltage by responding very quickly to supply or consume reactive power. This acts with the system impedance to either raise or lower the voltage on a cycle-by-cycle basis.

There are two main types of static var compensators in common usage, as shown in Fig. 7.6. The thyristor-controlled reactor (TCR) scheme is probably the most common. It employs a fixed capacitor bank to provide leading reactive power and a thyristor-controlled inductance that is gated on in various amounts to cancel all or part of the capacitance. The capacitors are frequently configured as filters to clean up the harmonic distortion caused by the thyristors.

The thyristor-switched capacitor operates by switching multiple steps of capacitors quickly to match the load requirements as closely as possible. This is a more coarse regulation than a TCR but is often adequate. The capacitors are generally gated fully on so there are no harmonics in the currents. The switching point is controlled so that there are no switching transients.

7.3   Utility Voltage Regulator Application

Figure 7.7 shows a photograph of a typical three-phase 32-step voltage regulator bank used by U.S. utilities. While this photo shows a substation installation, single-phase regulators are frequently pole-mounted, either one to a single pole or three on a platform between

FIGURE 7.7 Typical utility three-phase 32-step voltage regulator bank. (*Courtesy of Cooper Power Systems.*)

two poles. They may be connected in wye-grounded, leading delta, lagging delta, or open delta. The controls are integral to the device and each phase is generally controlled separately. Features on some regulator controls allow for ganged operation of all three phases such that all regulators are on the same tap.

It should be noted that ganged operation does not necessarily mean the voltage phase balance will be better. Utility distribution lines are commonly unbalanced in their construction and no attempt is made to transpose them. Also, numerous single-phase loads create significant imbalance in the phase currents. Therefore, three independently controlled regulators may very well yield better balance between the phase voltages than ganged operation.

Volumes could be written on the application of regulators, but we will restrict our discussion here to a few topics particularly relevant to power quality: use of the line drop compensator for leveling voltage profiles and load rejection with respect to the application of regulators in series.

7.3.1 Line Drop Compensator

Regulators are very effective in alleviating low-voltage conditions on distribution feeders when the load has outgrown the capability of the feeder at peak load conditions. Because it is time-consuming to determine the correct settings for line drop compensation, the R and

X settings are often set to zero and the voltage regulation set point is set near the maximum allowable (125 or 126 V on a 120-V base). This results in the feeder voltage being near the maximum most of the time because the load is at peak for only a small percentage of the hours each year. This is adequate in most respects except that

1. Transformers operate higher on their saturation curve, producing more harmonic currents (and losses), contributing more to the harmonic distortion on the feeder, which can be particularly troublesome at low loads.

2. Customers may experience more frequent replacement of incandescent lamps.

3. The higher voltage creates increased power demand, which may be undesirable at times and may translate into reduced energy efficiency (excessively low voltages will also yield inefficiencies).

The purpose of the line drop compensator is to level out the voltage profile so that it provides the necessary voltage boost at peak load yet keeps the voltage closer to nominal at lower loads. This is illustrated in Fig. 7.8. To simplify the discussion, we've assumed there is no LTC in the substation and the only regulator of concern is a feeder regulator at the substation. In Fig. 7.8a, no compensation is used and the voltage setting is 5 percent high, or 126 V on 120-V systems. Since there is some bandwidth on the control, the voltage may actually go higher than this. In Fig. 7.8b, the voltage setting is 120 V (100 percent) with the line drop compensator set some distance out on the feeder as shown. At peak load the voltage at the regulator rises to 105 percent, which is necessary to keep the end of the feeder at the proper voltage. However, at low load, the feeder voltage profile is closer to 100 percent voltage.

There are numerous practices for determining line drop compensator settings. Manufacturers provide computer programs for computing the settings given the current transformer (CT) ratings and potential transformer (PT) ratios. These vary with regulator sizes and must be specifically known before the proper setting can be computed. Of course, this also requires the user to model the feeders on a computer program, for which the data may not be readily available. Manufacturers' guide books also have simple formulas and rule-of-thumb procedures for determining settings.

The line drop compensator settings are called R and X for the resistive and reactive components of the compensator. However, the units are volts on a 120-V base instead of ohms. To convert from actual line impedance in ohms to the R and X settings, the basic formula is

$$\left(R + jX\right)_{setting} = \left(R + jX\right)_{ohms}\left(\frac{CT\,rating}{PT\,ratio}\right)$$

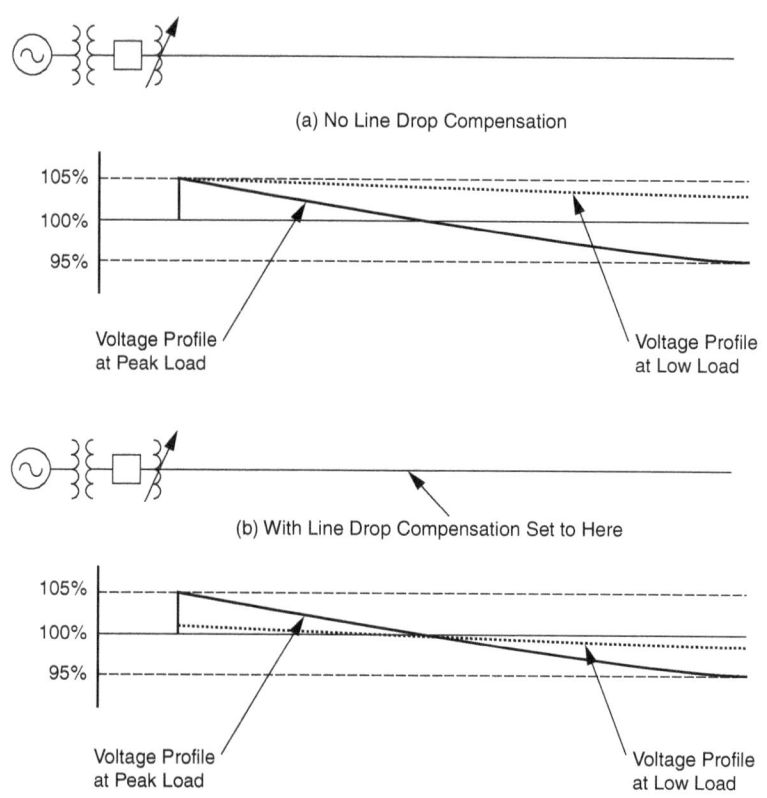

(a) No Line Drop Compensation

Voltage Profile at Peak Load

Voltage Profile at Low Load

(b) With Line Drop Compensation Set to Here

Voltage Profile at Peak Load

Voltage Profile at Low Load

FIGURE 7.8 The effect of line drop compensation on the voltage profile.

where the CT is specified by the line current *rating* and the PT *ratio* is the nominal line-to-neutral voltage divided by 120 V.

These R and X values are used directly for wye-connected regulators. For delta-connected regulators, these values must be modified to account for the 30° phase shift in the voltage with respect to the line current. For a leading delta connection, multiply by $1 \angle -30°$; for a lagging delta, multiply by $1 \angle +30°$.

Some utilities have developed average standard settings that they have found to be effective. Many determine the R and X settings experimentally by sending a line technician to the low-voltage point on the feeder while another adjusts the R and X settings. Ideally, this should be done at the peak load so that a voltage setting and line drop setting may be found that are successful in meeting this condition. It will, in all likelihood, meet the lower load conditions satisfactorily, although switched capacitor banks downline from the regulator may fool the control when they switch to a different state. Therefore, the voltage profile should be monitored at one or two key locations for a few days to make certain the setting is adequate.

Obviously, this process takes time, and it is often not convenient to send a crew to check a regulator setting when the peak load occurs. Often, at this time, the crews will be busy with more urgent matters such as changing out overloaded transformers to get customers back in service. There is a definite benefit to the power quality if the regulator is set properly, so some effort should be made. Fortunately, manufacturers are now supplying controls with telecommunications capability so that the settings can be adjusted more conveniently from a control center.

Many manufacturers also offer sophisticated controls with a choice of load-following algorithms. In the case of power quality complaints with the voltage going out of band or too many tap changes, consult the user's manual and experiment with other algorithms to achieve a smoother regulation.

7.3.2 Regulators in Series

In sparsely populated areas it is not uncommon to find two or more regulator banks in series on extremely long lines feeding remote loads. Two notable applications are service to irrigation and mining loads where lines extend for miles with only an occasional load. These applications require special considerations to avoid power quality problems.

One important consideration for coordinating the regulators in series is properly setting the initial time delay. The regulator nearest the substation is set with the shortest time delay, typically 15 or 30 s. Regulators further downline are set with a time delay of 15 s longer. This minimizes tap changing on the downline regulators, keeping the voltage variations to a minimum and extending contact life.

Perhaps, the greatest power quality problem in this situation is load rejection. The sudden loss of load, which can happen after a fault, will result in greatly excessive voltages because the regulator boosting will be cumulative (see Fig. 7.9). Overvoltages of 20 percent or more can occur. Transformer saturation and the remaining load will help hold the voltage down, but it will still exceed normal limits by a considerable margin.

To minimize damage to loads, regulators employ a "rapid runback" control scheme that bypasses the normal time delay and runs the regulators back down as quickly as possible. This is typically 2 to 4 s per tap change.

7.4 Capacitors for Voltage Regulation

Capacitors may be used for voltage regulation on the power system in either the shunt or series configuration (Fig. 7.10). We will discuss each class of application separately.

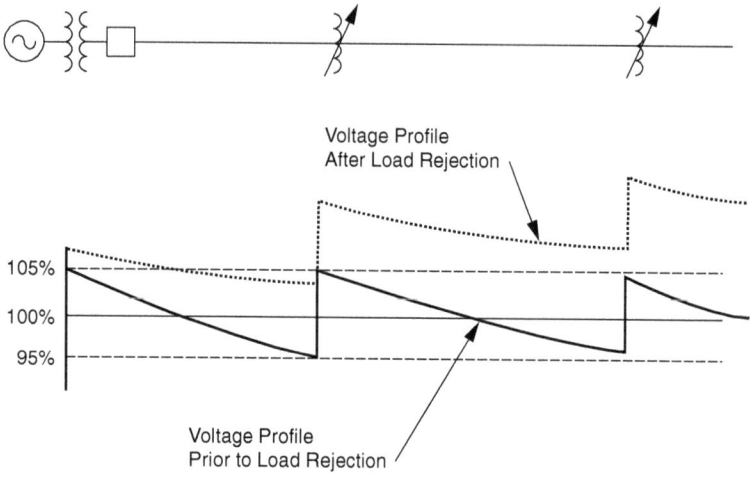

FIGURE 7.9   Illustration of overvoltage resulting from load rejection on regulators in series.

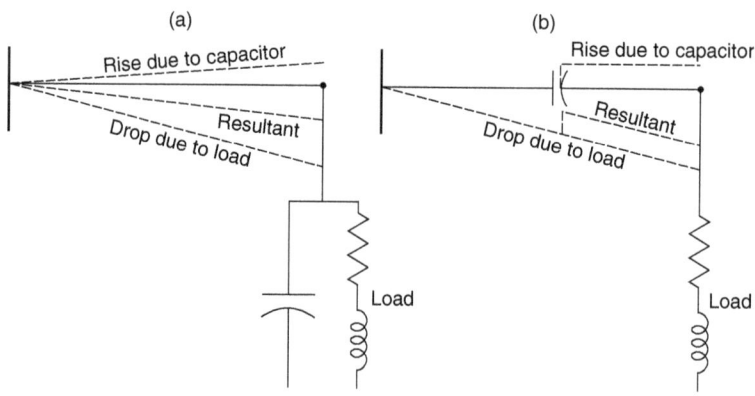

FIGURE 7.10   Feeder voltage rise due to shunt (a) and series (b) capacitors.

7.4.1   Shunt Capacitors

As shown in Fig. 7.10a, the presence of a shunt capacitor at the end of a feeder results in a gradual change in voltage along the feeder. Ideally, the percent voltage rise at the capacitor

$$\%\Delta V = \frac{100\left(V_{\text{with cap}} - V_{\text{no cap}}\right)}{V_{\text{with cap}}}$$

would be zero at no load and rise to maximum at full load. However, with shunt capacitors, percent voltage rise is essentially independent of load. Therefore, automatic switching is often employed in order to deliver the desired regulation at high loads, but prevent excessive voltage at low loads. Switching may result in transient overvoltages inside customer facilities, as described in Chap. 4.

Application of shunt capacitors may also result in a variety of harmonic problems (see Chaps. 5 and 6).

7.4.2   Series Capacitors

Unlike the shunt capacitor, a capacitor connected in series with the feeder results in a voltage rise at the end of the feeder that varies directly with load current. Voltage rise is zero at no load and maximum at full load. Thus, series capacitors do not need to be switched in response to changes in load. Moreover, a series capacitor will require far smaller kV and kvar ratings than a shunt capacitor delivering equivalent regulation.

But series capacitors have several disadvantages. Firstly, they cannot provide reactive compensation for feeder loads and do not significantly reduce system losses. Series capacitors can only release additional system capacity if it is limited by excessive feeder voltage drop. Shunt capacitors, on the other hand, are effective when system capacity is limited by high feeder current as well.

Secondly, series capacitors cannot tolerate fault current. This would result in a catastrophic overvoltage and must be prevented by bypassing the capacitor through an automatic switch. An arrester must also be connected across the capacitor to divert current until the switch closes.

There are several other concerns that must be evaluated in a series capacitor application. These include resonance and/or hunting with synchronous and induction motors, and ferroresonance with transformers. Because of these concerns, the application of series capacitors on distribution systems is very limited. One area where they have proved to be advantageous is where feeder reactance must be minimized, e.g., to reduce flicker.

7.5   End-User Capacitor Application

The reasons that an end user might decide to apply power factor correction capacitors are to

- Reduce electric utility bill
- Reduce I^2R losses and, therefore, heating in lines and transformers
- Increase the voltage at the load, increasing production and/ or the efficiency of the operation

- Reduce current in the lines and transformers, allowing additional load to be served without building new circuits

The primary motivation is generally economics to eliminate utility power factor penalties, but there are technical benefits related to power quality as well.

There can be power quality problems as a result of adding capacitors. The most common are harmonics problems. While power factor correction capacitors are not harmonic sources, they can interact with the system to accentuate the harmonics that are already there (see Chaps. 5 and 6). There are also switching transient side effects such as magnification of utility capacitor-switching transients (see Chap. 4).

7.5.1   Location for Power Factor Correction Capacitors

The benefits realized by installing power factor correction capacitors include the reduction of reactive power flow on the system. Therefore, for best results, power factor correction should be located as close to the load as possible. However, this may not be the most economical solution or even the best engineering solution, due to the interaction of harmonics and capacitors.

Often, capacitors will be installed with large induction motors (C3 in Fig. 7.11). This allows the capacitor and motor to be switched as a unit. Large plants with extensive distribution systems often install capacitors at the primary voltage bus (C1) when utility billing encourages power factor correction. Many times, however, power

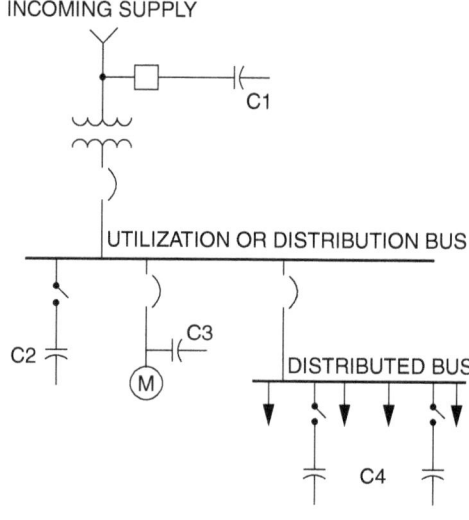

FIGURE 7.11   Location of power factor correction.

factor correction and harmonic distortion reduction must be accomplished with the same capacitors. Location of larger harmonic filters on the main distribution bus (C2) provides the required compensation and a low-impedance path for harmonic currents to flow, keeping the harmonic currents off the utility system.

One disadvantage of placing capacitors only at the utilization or main distribution bus is that there is no reduction of current and line losses within the plant. Loss and current reduction are achieved when the capacitors (C4) are distributed throughout the system. Some industrial end users install capacitors at the motor control centers, which is often more economical than putting the capacitors on each motor. The capacitors' controls can be tied in with the motor controls so that the capacitors are switched when needed.

Another disadvantage is related to harmonic performance. The main distribution bus often has a high X/R ratio, and if the capacitor installation results in harmonic resonance, very high harmonic voltages and currents can result. This is the most common cause of damage from harmonics in industrial plants. If there is sufficient resistance in the cables to the loads, a distributed capacitor application may have some natural suppression of harmonic resonance. On the other hand, the resonance may be more convenient to alleviate when the offending capacitors are in one location. Thus, the optimal design for capacitor compensation is often a compromise between conflicting objectives.

7.5.2   Voltage Rise

The voltage rise from placing capacitors on an inductive circuit is a two-edged sword from the power quality standpoint. If the voltage is low, then the capacitors provide an increase to bring the voltage back into tolerable limits. However, if the capacitors are left energized when the load is turned off, the voltage can rise too high, resulting in a sustained overvoltage.

The voltage rise by the end user from the installation of capacitors is approximated from

$$\%\Delta V = \frac{\mathrm{kvar_{cap}} \times Z_{tx}\,(\%)}{\mathrm{kVA_{tx}}}$$

where %ΔV = percent voltage rise
      kvar$_{cap}$ = capacitor bank rating
      kVA$_{tx}$ = step-down transformer rating
        Z_{tx} = step-down transformer impedance, %

This formula assumes that the transformer is the bulk of the total impedance of the power system up to the point at which the capacitor is applied.

As mentioned, one power quality problem that arises is that the voltage rises too high when the capacitors remain energized at low load levels. One common symptom of this is loud humming in the supply transformer and, in some cases, overheating due to overexcitation of the core. Another symptom is the loss of excessive numbers of incandescent light bulbs coincident with the installation of a capacitor bank. Thus, this formula should be applied to investigate whether it is feasible to leave the capacitors energized. If not, some control strategy must be devised to switch the capacitors off at light loads.

7.5.3   Reduction in Power System Losses

The reduction in power system losses is estimated from

$$\% \, loss_{reduction} = 100\left[1 - \left(\frac{pf_{original}}{pf_{corrected}}\right)^2\right]$$

$$\% \, power \, loss \propto 100\left(\frac{pf_{original}}{pf_{corrected}}\right)^2$$

where % $loss_{reduction}$ = percent reduction in losses
$pf_{original}$ = original power factor (pu)
$pf_{corrected}$ = corrected power factor (pu)

This formula basically applies to a single capacitor on a radial feed. However, it is also approximately correct if the capacitors are well distributed throughout the plant so that each major branch circuit sees approximately the same percentage loss improvement.

Keep in mind that this formula gives the percent reduction possible over the present losses *upline* from the capacitors. There is no reduction in losses in the lines and transformers between the capacitor and the load.

7.5.4   Reduction in Line Current

The percent line current reduction can be approximated from

$$\% \Delta I = 100\left[1 - \left(\frac{\cos\theta_{before}}{\cos\theta_{after}}\right)\right]$$

where %ΔI = percent current reduction
$\cos\theta_{before}$ = power factor angle before correction
$\cos\theta_{after}$ = power factor angle after correction

Again, this applies only to currents upline from the capacitor.

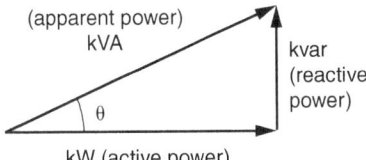

FIGURE 7.12 Displacement power factor triangle.

7.5.5   Displacement Power Factor versus True Power Factor

The traditional concepts of selecting power factor correction are based on the assumption that loads on the system have linear voltage-current characteristics and that harmonic distortion can be ignored. With these assumptions, the power factor is equal to the *displacement power factor* (DPF). The DPF is calculated using the traditional power factor triangle method (Fig. 7.12) and is often written as

$$DPF = \frac{kW}{kVA} = \cos\theta$$

where kW and kVA are the fundamental frequency quantities only.

Harmonic distortion in the voltage and current caused by nonlinear loads on the system changes the way the power factor must be calculated. The *true power factor* (TPF) is defined as the ratio of real power to the total voltamperes in the circuit:

$$TPF = \frac{kW}{kVA} = \frac{P}{V_{rms}I_{rms}}$$

As before, the power factor is defined as the ratio of kW to kVA, but in this case, the kVA includes harmonic distortion voltamperes. The total kVA (apparent power) is determined by multiplying the true rms voltage by the true rms current. It can be significantly higher than the fundamental frequency kVA. The active power P is generally increased only marginally by the distortion.

The TPF is the true measure of the efficiency with which the real power is being used. In the trivial case of no distortion, it defaults to the DPF. Capacitors basically compensate only for the fundamental frequency reactive power (vars) and cannot completely correct the true power factor to unity when there are harmonics present. In fact, capacitors can make the TPF worse by creating resonance conditions which magnify the harmonic distortion. On typical power systems, the I_{rms} term in the given TPF equation is generally the one most affected by harmonic distortion, although the V_{rms} term may also be increased. Assuming the voltage THD is zero, the maximum to which you can correct the TPF can be approximated by

$$\text{TPF} \approx \sqrt{\frac{1}{1 + \text{THD}^2_{\text{current}}}}$$

where THD is in pu.

The DPF is still very important to most industrial customers because utility billing for power factor penalties is generally based on it. Most revenue metering schemes currently account only for the DPF. However, this could change because modern electronic meters certainly have the capability to compute the TPF, which will be considerably lower for some types of industrial loads.

7.5.6   Selecting the Amount of Capacitance

As reference for those wishing to apply capacitors to correct the power factor, the kvar of capacitance required to correct a load to a desired power factor is given by

$$\text{kvar} = \text{kW}\left(\tan\theta_{\text{orig}} - \tan\theta_{\text{new}}\right)$$

$$\text{kW} = \sqrt{\frac{1}{\text{PF}^2_{\text{orig}}} - 1} - \sqrt{\frac{1}{\text{PF}^2_{\text{new}}} - 1}$$

where kvar = required compensation in kvar
kW = real power in kW
θ_{orig} = original power factor phase angle
θ_{new} = desired power factor phase angle
PF_{orig} = original power factor
PF_{new} = desired power factor

Table 7.1 summarizes the equation in tabular form.

After selecting estimated capacitor sizes, two power quality checks should be done:

1. Determine the no-load voltage rise to make sure that the voltage will not rise above 110 percent when the load is minimum. If it does, you will have to switch some of the capacitors off or apply fewer capacitors.

2. Determine the impact of the capacitors on harmonics (see Chap. 5).

If harmonics prove to be a problem, typical options are

1. Change the amount of capacitors, if possible. Avoid certain switching combinations. This is generally the least-cost solution.

	Corrected power factor										
Original PF	0.80	0.82	0.84	0.86	0.88	0.90	0.92	0.94	0.96	0.98	1.00
0.50	0.982	1.034	1.086	1.139	1.192	1.248	1.306	1.369	1.440	1.529	1.732
0.52	0.893	0.945	0.997	1.049	1.103	1.158	1.217	1.280	1.351	1.440	1.643
0.54	0.809	0.861	0.913	0.965	1.019	1.074	1.133	1.196	1.267	1.356	1.559
0.56	0.729	0.781	0.834	0.886	0.940	0.995	1.053	1.116	1.188	1.276	1.479
0.58	0.655	0.707	0.759	0.811	0.865	0.920	0.979	1.042	1.113	1.201	1.405
0.60	0.583	0.635	0.687	0.740	0.794	0.849	0.907	0.970	1.042	1.130	1.333
0.62	0.515	0.567	0.620	0.672	0.726	0.781	0.839	0.903	0.974	1.062	1.265
0.64	0.451	0.503	0.555	0.607	0.661	0.716	0.775	0.838	0.909	0.998	1.201
0.66	0.388	0.440	0.492	0.545	0.599	0.654	0.712	0.775	0.847	0.935	1.138
0.68	0.328	0.380	0.432	0.485	0.539	0.594	0.652	0.715	0.787	0.875	1.078
0.70	0.270	0.322	0.374	0.427	0.480	0.536	0.594	0.657	0.729	0.817	1.020
0.72	0.214	0.266	0.318	0.370	0.424	0.480	0.538	0.601	0.672	0.761	0.964
0.74	0.159	0.211	0.263	0.316	0.369	0.425	0.483	0.546	0.617	0.706	0.909
0.76	0.105	0.157	0.209	0.262	0.315	0.371	0.429	0.492	0.563	0.652	0.855

TABLE 7.1 kW Multiplier to Determine kvar Requirement

Original PF	Corrected power factor										
	0.80	0.82	0.84	0.86	0.88	0.90	0.92	0.94	0.96	0.98	1.00
0.78	0.052	0.104	0.156	0.209	0.263	0.318	0.376	0.439	0.511	0.599	0.802
0.80	0.000	0.052	0.104	0.157	0.210	0.266	0.324	0.387	0.458	0.547	0.750
0.82		0.000	0.052	0.105	0.158	0.214	0.272	0.335	0.406	0.495	0.698
0.84			0.000	0.053	0.106	0.162	0.220	0.283	0.354	0.443	0.646
0.86				0.000	0.054	0.109	0.167	0.230	0.302	0.390	0.593
0.88					0.000	0.055	0.114	0.177	0.248	0.337	0.540
0.90						0.000	0.058	0.121	0.193	0.281	0.484
0.92							0.000	0.063	0.134	0.223	0.426
0.94								0.000	0.071	0.160	0.363
0.96									0.000	0.089	0.292
0.98										0.000	0.203
1.00											0.000

TABLE 7.1 kW Multiplier to Determine kvar Requirement (continued)

2. Convert some of the capacitors to one or more filters, usually placed at the main bus.

3. Employ an adaptive control to monitor the harmonic distortion and switch the capacitors to avoid resonance. This might be appropriate for large industrial loads where there are numerous switched capacitors coming on and off line randomly.

7.6   Regulating Utility Voltage with Distributed Resources

It is becoming more popular for utility distribution planners to consider distributed generation (DG) and storage devices to defer investments in substations and transmission lines until the load has grown to a sufficient size to warrant the larger investment. This concept is particularly useful when there are a relatively few number of hours each year when the load approaches the system capacity limits. The movement toward utility deregulation in recent years has created renewed interest in distributed resources, and many of the issues related to power quality are addressed in Chap. 9. Here, we will restrict our discussion to the potential of using distributed generators for distribution feeder voltage regulation.

Most of the utility-owned installations have been located in utility distribution substations. This offers load relief for the substation and transmission facilities, but contributes little else to the quality of power for the distribution feeder. Now, many distribution engineers are considering the benefits of moving the devices out onto the feeder to gain additional system capacity, loss reduction, improved reliability, and voltage regulation. These generators will often be owned by end users, but could be contracted to operate for utility system benefits as well. While this option may be too expensive to consider for voltage regulation alone, it is a useful side effect of dispersed sources justified on the basis of deferment of capital expansion.

While few utility distribution planners will rely on customer-owned generation for base capacity, it is more palatable to employ them to help cover contingencies. One example is illustrated in Fig. 7.13. Utilities usually have sectionalizing switches installed so that portions of a distribution feeder can be served from different feeders or substations during emergencies. If the fault occurs at the time of peak load, it may be impossible to pick up any more load from other feeders in the normal manner simply by closing a switch. However, a generator located near the switch tie point can potentially provide enough power to support the additional load at a satisfactory voltage. If the generator is of sufficient size, it could be employed to help regulate the voltage.

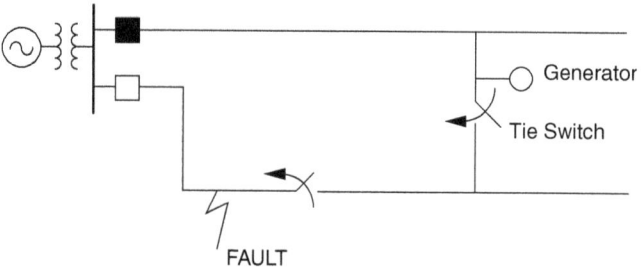

FIGURE 7.13   Using a generator to support restoration of service to the unfaulted portion of a feeder.

One advantage of using a generator to regulate the voltage is that its controls respond much faster and more smoothly than discrete tap-changing devices like regulators and LTCs. While this potential exists, most utility distribution protection engineers are reluctant to allow this type of operation without careful study and costly control equipment. The issue is that operating with automative voltage regulation makes the system more susceptible to sustaining an inadvertent island. Therefore, some means of direct transfer trip is generally required to ensure that the generator disconnects from the system when certain utility breakers operate.

A more normal connection of DG is to use power and power factor control. This minimizes the risk of islanding. Although the DG no longer attempts to regulate the voltage, it is still useful for voltage regulation purposes during constrained loading conditions by displacing some active and reactive power. Alternatively, customer-owned DG may be exploited simply by operating off-grid and supporting part or all of the customer's load off-line. This avoids interconnection issues and provides some assistance to voltage regulation by reducing the load.

The controls of distributed sources must be carefully coordinated with existing line regulators and substation LTCs. Reverse power flow can sometimes fool voltage regulators into moving the tap changer in the wrong direction. Also, it is possible for the generator to cause regulators to change taps constantly, causing early failure of the tap-changing mechanism. Fortunately, some regulator manufacturers have anticipated these problems and now provide sophisticated microcomputer-based regulator controls that are able to compensate.

To exploit dispersed sources for voltage regulation, one is limited in options to the types of devices with steady, controllable outputs such as reciprocating engines, combustion turbines, fuel cells, and battery storage. Randomly varying sources such as wind turbines and photovoltaics are unsatisfactory for this role and often must be placed on a relatively stiff part of the system or have special regulation to avoid voltage regulation difficulties. DG used for voltage regulation must also be large enough to accomplish the task.

Not all technologies are suitable for regulating voltage. They must be capable of producing a controlled amount of reactive power. Manufacturers of devices requiring inverters for interconnection sometimes program the inverter controls to operate only at unity power factor while grid-connected. Simple induction generators consume reactive power like an induction motor, which can cause low voltage.

7.7 Flicker[*]

Although voltage flicker is not technically a long-term voltage variation, it is included in this chapter because the root cause of problems is the same: The system is too weak to support the load. Also, some of the solutions are the same as for the slow-changing voltage regulation problems. The voltage variations resulting from flicker are often within the normal service voltage range, but the changes are sufficiently rapid to be irritating to certain end users.

Flicker is a relatively old subject that has gained considerable attention recently due to the increased awareness of issues concerning power quality. Power engineers first dealt with flicker in the 1880s when the decision of using ac over dc was of concern.[2] Low-frequency ac voltage resulted in a "flickering" of the lights. To avoid this problem, a higher 60-Hz frequency was chosen as the standard in North America.

The term flicker is sometimes considered synonymous with voltage fluctuations, voltage flicker, light flicker, or lamp flicker. The phenomenon being referred to can be defined as a fluctuation in system voltage that can result in observable changes (flickering) in light output. Because flicker is mostly a problem when the human eye observes it, it is considered to be a problem of perception.

In the early 1900s, many studies were done on humans to determine observable and objectionable levels of flicker. Many curves, such as the one shown in Fig. 7.14, were developed by various companies to determine the severity of flicker. The flicker curve shown in Fig. 7.14 was developed by C. P. Xenis and W. Perine in 1937 and was based upon data obtained from 21 groups of observers. In order to account for the nature of flicker, the observers were exposed to various waveshape voltage variations, levels of illumination, and types of lighting.[3]

Flicker can be separated into two types: cyclic and noncyclic. Cyclic flicker is a result of periodic voltage fluctuations on the system, while noncyclic is a result of occasional voltage fluctuations.

---

*This section was contributed by Jeff W. Smith.

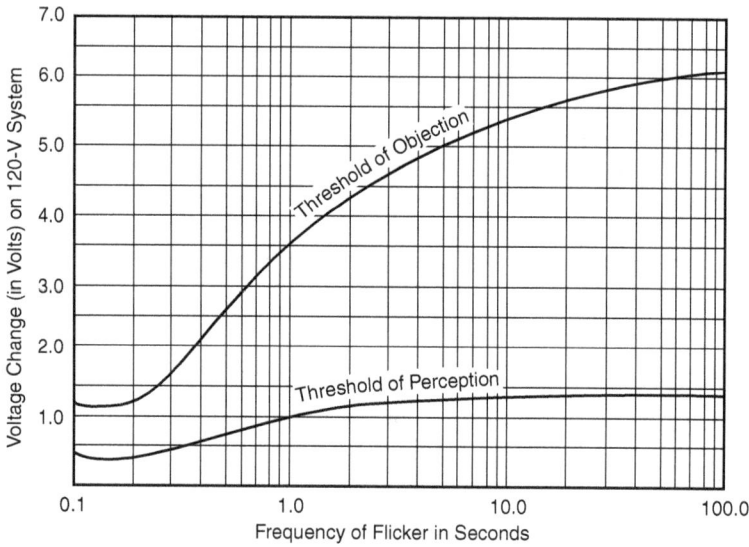

FIGURE **7.14** General flicker curve.

An example of sinusoidal-cyclic flicker is shown in Fig. 7.15. This type of flicker is simply amplitude modulation where the main signal (60 Hz for North America) is the carrier signal and flicker is the modulating signal. Flicker signals are usually specified as a percentage of the normal operating voltage. By using a percentage, the flicker signal is independent of peak, peak-to-peak, rms, line-to-neutral, etc. Typically, percent voltage modulation is expressed by

$$\text{Percent voltage modulation} = \frac{V_{max} - V_{min}}{V_0} \times 100\%$$

where V_{max} = maximum value of modulated signal
V_{min} = minimum value of modulated signal
V_0 = average value of normal operating voltage

The usual method for expressing flicker is similar to that of percent voltage modulation. It is usually expressed as a percent of the total change in voltage with respect to the average voltage ($\Delta V/V$) over a certain period of time.

The frequency content of flicker is extremely important in determining whether or not flicker levels are observable (or objectionable). Describing the frequency content of the flicker signal in terms of modulation would mean that the flicker frequency is essentially the frequency of the modulating signal. The typical

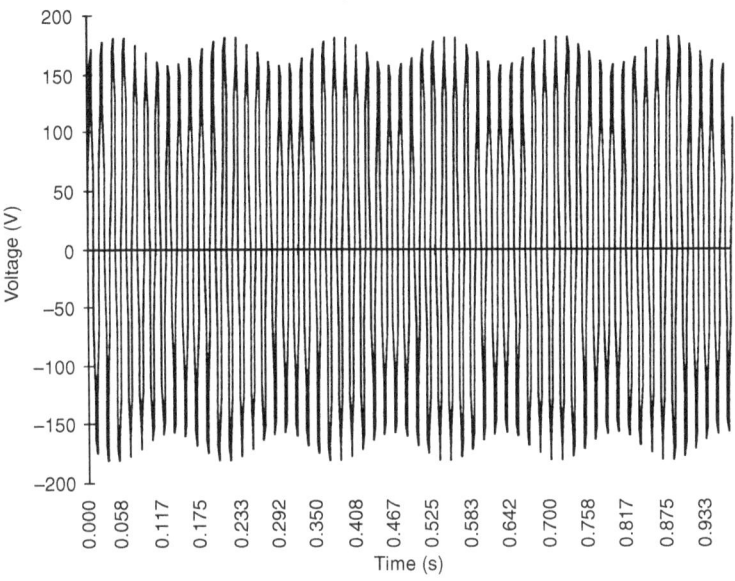

FIGURE 7.15   Example flicker waveform.

frequency range of observable flicker is from 0.5 to 30.0 Hz, with observable magnitudes starting at less than 1.0 percent.

As shown in Fig. 7.14, the human eye is more sensitive to luminance fluctuations in the 5- to 10-Hz range. As the frequency of flicker increases or decreases away from this range, the human eye generally becomes more tolerable of fluctuations.

One issue that was not considered in the development of the traditional flicker curve is that of multiple flicker signals. Generally, most flicker-producing loads contain multiple flicker signals (of varying magnitudes and frequencies), thus making it very difficult to accurately quantify flicker using flicker curves.

7.7.1   Sources of Flicker

Typically, flicker occurs on systems that are weak relative to the amount of power required by the load, resulting in a low short-circuit ratio. This, in combination with considerable variations in current over a short period of time, results in flicker. As the load increases, the current in the line increases, thus increasing the voltage drop across the line. This phenomenon results in a sudden reduction in bus voltage. Depending upon the change in magnitude of voltage and frequency of occurrence, this could result in observable amounts of flicker. If a lighting load were connected to the system in relatively close proximity to the fluctuating load, observers could see this as a dimming of the lights. A common situation, which could result in flicker, would be a large industrial plant located at the end of a weak distribution feeder.

Whether the resulting voltage fluctuations cause observable or objectionable flicker is dependent upon the following parameters:

- Size (VA) of potential flicker-producing source
- System impedance (stiffness of utility)
- Frequency of resulting voltage fluctuations

A common load that can often cause flicker is an electric arc furnace (EAF). EAFs are nonlinear, time-varying loads that often cause large voltage fluctuations and harmonic distortion. Most of the large current fluctuations occur at the beginning of the melting cycle. During this period, pieces of scrap steel can actually bridge the gap between the electrodes, resulting in a highly reactive short circuit on the secondary side of the furnace transformer. This meltdown period can generally result in flicker in the 1.0- to 10.0-Hz range. Once the melting cycle is over and the refining period is reached, stable arcs can usually be held on the electrodes resulting in a steady, three-phase load with high power factor.[4]

Large induction machines undergoing start-up or widely varying load torque changes are also known to produce voltage fluctuations on systems. As a motor is started up, most of the power drawn by the motor is reactive (see Fig. 7.16). This results in a large voltage drop across distribution lines. The most severe case would be when a motor is started across the line. This type of start-up can result in current drawn by the motor up to multiples of the full load current.

An example illustrating the impact motor starting and torque changes can have on system voltage is shown in Fig. 7.17. In this case,

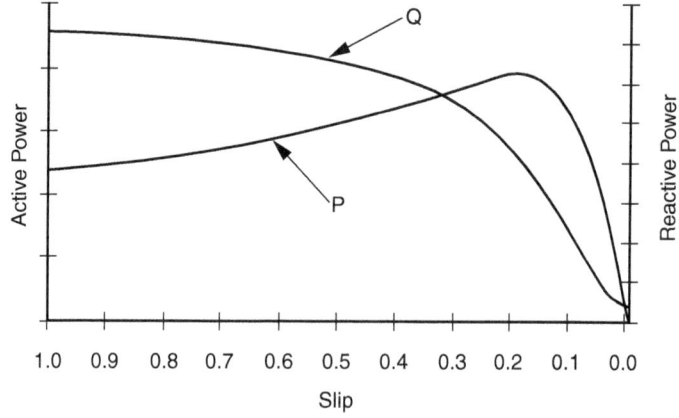

Figure 7.16   Active and reactive power during induction machine start-up.

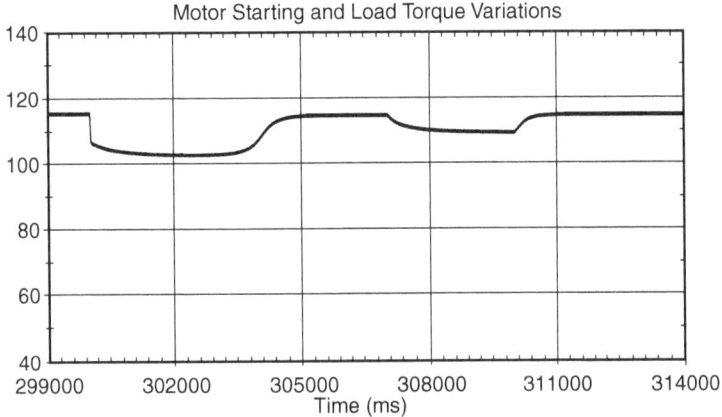

Figure **7.17**   Voltage fluctuations caused by induction machine operation.

a large industrial plant is located at the end of a weak distribution feeder. Within the plant are four relatively large induction machines that are frequently restarted and undergo relatively large load torque variations.[5]

Although starting large induction machines across the line is generally not a recommended practice, it does occur. To reduce flicker, large motors are brought up to speed using various soft-start techniques such as reduced-voltage starters or variable-speed drives.

In certain circumstances, superimposed interharmonics in the supply voltage can lead to oscillating luminous flux and cause flicker. Voltage interharmonics are components in the harmonic spectrum that are noninteger multiples of the fundamental frequency. This phenomenon can be observed with incandescent lamps as well as with fluorescent lamps. Sources of interharmonics include static frequency converters, cycloconverters, subsynchronous converter cascades, induction furnaces, and arc furnaces.[6]

7.7.2   Mitigation Techniques

Many options are available to alleviate flicker problems. Mitigation alternatives include static capacitors, power electronic-based switching devices, and increasing system capacity. The particular method chosen is based upon many factors such as the type of load causing the flicker, the capacity of the system supplying the load, and cost of mitigation technique.

Flicker is usually the result of a varying load that is large relative to the system short-circuit capacity. One obvious way to remove flicker from the system would be to increase the system capacity sufficiently to decrease the relative impact of the flicker-producing load. Upgrading the system could include any of the following: reconductoring, replacing existing transformers with higher kVA ratings, or increasing the operating voltage.

Motor modifications are also an available option to reduce the amount of flicker produced during motor starting and load variations. The motor can be rewound (changing the motor class) such that the speed-torque curves are modified. Unfortunately, in some cases this could result in a lower running efficiency. Flywheel energy systems can also reduce the amount of current drawn by motors by delivering the mechanical energy required to compensate for load torque variations.

Recently, series reactors have been found to reduce the amount of flicker experienced on a system caused by EAFs. Series reactors help stabilize the arc, thus reducing the current variations during the beginning of melting periods. By adding the series reactor, the sudden increase in current is reduced due to the increase in circuit reactance. Series reactors also have the benefit of reducing the supply-side harmonic levels.[7] The design of the reactor must be coordinated with power requirements.

Series capacitors can also be used to reduce the effect of flicker on an existing system. In general, series capacitors are placed in series with the transmission line supplying the load. The benefit of series capacitors is that the reaction time for the correction to load fluctuations is instantaneous in nature. The downside to series capacitors is that compensation is only available beyond the capacitor. Bus voltages between the supply and the capacitor are uncompensated. Also, series capacitors have operational difficulties that require careful engineering.

Fixed shunt-connected capacitor banks are used for long-term voltage support or power factor correction. A misconception is that shunt capacitors can be used to reduce flicker. The starting voltage sag is reduced, but the percent change in voltage ($\Delta V/V$) is not reduced, and in some cases can actually be increased.

A rather inexpensive method for reducing the flicker effects of motor starting would be to simply install a step-starter for the motor, which would reduce the amount of starting current during motor start-up. With the advances in solid-state technology, the size, weight, and cost of adjustable-speed drives have decreased, thus allowing the use of such devices to be more feasible in reducing the flicker effects caused by flicker-producing loads.

Static var compensators (SVCs) are very flexible and have many roles in power systems. SVCs can be used for power factor correction, flicker reduction, and steady-state voltage control, and also have the benefit of being able to filter out undesirable frequencies from the system. SVCs typically consist of a TCR in parallel with fixed capacitors (Fig. 7.18). The fixed capacitors are usually connected in ungrounded wye with a series inductor to implement a filter. The

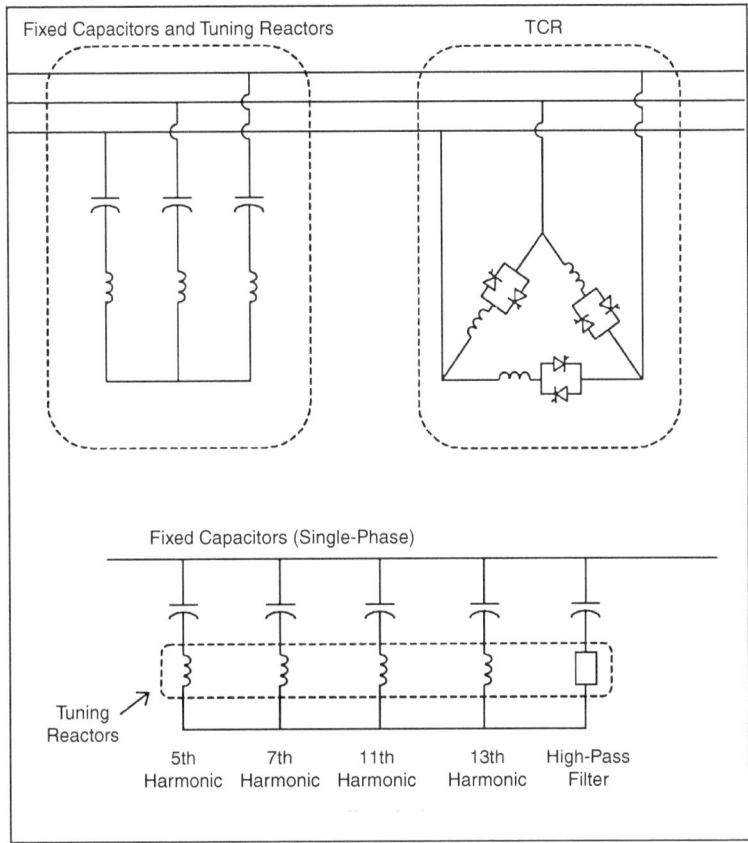

Fixed Capacitors and Tuning Reactors

TCR

Fixed Capacitors (Single-Phase)

Tuning Reactors

5th Harmonic   7th Harmonic   11th Harmonic   13th Harmonic   High-Pass Filter

FIGURE 7.18   Typical SVC configuration.

reactive power that the inductor delivers in the filter is small relative to the rating of the filter (approximately 1 to 2 percent). There are often multiple filter stages tuned to different harmonics. The controls in the TCR allow continuous variations in the amount of reactive power delivered to the system, thus increasing the reactive power during heavy loading periods and reducing the reactive power during light loading.

SVCs can be very effective in controlling voltage fluctuations at rapidly varying loads. Unfortunately, the price for such flexibility is high. Nevertheless, they are often the only cost-effective solution for many loads located in remote areas where the power system is weak. Much of the cost is in the power electronics on the TCR. Sometimes this can be reduced by using a number of capacitor steps. The TCR then need only be large enough to cover the reactive power gap between the capacitor stages.

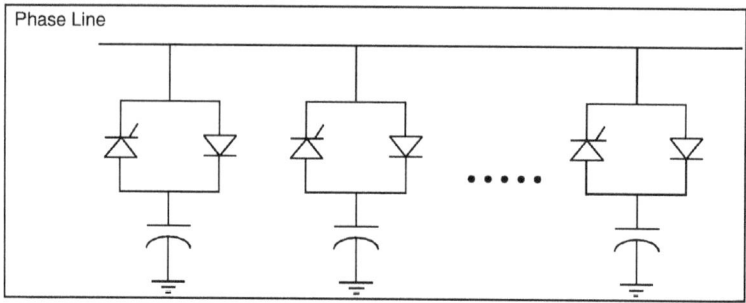

Phase Line

FIGURE 7.19   Typical TSC configuration.

Thyristor-switched capacitors (TSCs) can also be used to supply reactive power to the power system in a very short amount of time, thus being helpful in reducing the effects of quick load fluctuations. TSCs usually consist of two to five shunt capacitor banks connected in series with diodes and thyristors connected back to back. The capacitor sizes are usually equal to each other or are set at multiples of each other, allowing for smoother transitions and increased flexibility in reactive power control. Switching the capacitors in or out of the system in discrete steps controls the amount of reactive power delivered to the system by the TSC. This action is unlike that of the SVC, where the capacitors are static and the reactors are used to control the reactive power. An example diagram of a TSC is shown in Fig. 7.19.

The control of the TSC is usually based on line voltage magnitude, line current magnitude, or reactive power flow in the line. The control circuits can be used for all three phases or each phase separately. The individual phase control offers improved compensation when unbalanced loads are producing flicker.

7.7.3   Quantifying Flicker

Flicker has been a power quality problem even before the term *power quality* was established. However, it has taken many years to develop an adequate means of quantifying flicker levels. Chapter 11 provides an in-depth look at power quality monitoring, with a section that describes modern techniques for measuring and quantifying flicker.

7.8   References

1. L. Morgan, S. Ihara, "Distribution Feeder Modification to Service Both Sensitive Loads and Large Drives," *1991 IEEE PES Transmission and Distribution Conference Record*, Dallas, September 1991, pp. 686–690.

2. E. L. Owen, "Power Disturbance and Power Quality—Light Flicker Voltage Requirements," *Conference Record,* IEEE IAS Annual Meeting, Denver, October 1994, pp. 2303–2309.
3. C. P. Xenis and W. Perine, "Slide Rule Yields Lamp Flicker Data." *Electrical World,* Oct. 23, 1937, p. 53.
4. S. B. Griscom, "Lamp Flicker on Power Systems," Chap. 22, *Electrical Transmission and Distribution Reference Book,* 4th ed., Westinghouse Elec. Corp., East Pittsburgh, Pa., 1950.
5. S. M. Halpin, J. W. Smith, C. A. Litton, "Designing Industrial Systems with a Weak Utility Supply," *IEEE Industry Applications Magazine,* March/April 2001, pp. 63–70.
6. *Interharmonics in Power Systems,* IEEE Interharmonic Task Force, Cigre 36.05/CIRED 2 CC02, Voltage Quality Working Group.
7. S. R. Mendis, M. T. Bishop, T. R. Day, D. M. Boyd, "Evaluation of Supplementary Series Reactors to Optimize Electric Arc Furnace Operations," *Conference Record,* IEEE IAS Annual Meeting, Orlando, Fla., October 1995, pp. 2154–2161.

7.9   Bibliography

IEEE Standard 141–1993: *Recommended Practice for Power Distribution in Industrial Plants,* IEEE, 1993.

IEEE Standard 519–1992: *Recommended Practices and Requirements for Harmonic Control in Electrical Power Systems,* IEEE, 1993.

IEC 61000–4-15, *Electromagnetic Compatibility (EMC).* Part 4: Testing and Measuring Techniques. Section 15: Flickermeter—Functional and Design Specifications.

CHAPTER 8

Power Quality Benchmarking

Foreword

The original Electric Power Research Institute (EPRI) Distribution Power Quality (DPQ) project in the mid-1990s provided the first benchmarking of power quality levels on distribution systems across the United States. Subsequent efforts provided similar benchmarking of power quality on other power systems around the world. The work characterized steady-state power quality levels (voltage regulation, harmonics, unbalance) as well as voltage sags, momentary interruptions, and transient voltages. Benchmarking results from around the world were gathered by CIGRE C4.07 Working Group and published in a working group report.

Benchmarking results use the measurement procedures outlined in IEC 61000-4-30. This standard provides a convenient reference to make sure that different power systems are being characterized in a similar manner in terms of power quality characteristics.

Benchmarking of power quality continues to be very important just as benchmarking of reliability levels is critical or understanding the expected performance as a function of system characteristics and other parameters. The widespread deployment of continuous power quality monitoring has made power quality benchmarking more of a continuous process. EPRI continues to coordinate ongoing power quality benchmarking efforts for both transmission and distribution systems. Important objectives of these efforts include:

- *Provide a factual baseline for benchmarking of power delivery system performance required by regulatory bodies.*

- *Understand performance standards for transmission and distribution power quality and how these standards compare with national benchmarking baselines.*

- *Update existing benchmarks to include changes in load composition affecting the grid in ways that have not been adequately studied.*

- *Identify the ongoing effect of system design changes, such as increased number of variable loads (e.g., arc furnaces) and system*

> compensation (e.g., capacitors) that affect power quality levels and issues at both transmission and distribution levels.
>
> - Identify broad power quality trends for transmission and distribution due to power electronics and efficient technologies.
>
> EPRI also contributed to a new IEEE Guide (IEEE 1250—Guide for Identifying and Improving Power Quality in Power Systems, 2011) that provides a summary of expected power quality levels and recommendations for improving power quality when issues arise.
>
> This chapter provides a summary of benchmarking methods and results as a convenient reference on expected power quality levels.
>
> Bill Howe, Manager PQ Research Program, EPRI

8.1   Introduction

Because of sensitive customer loads, there is a need to define the quality of electricity provided in a common and succinct manner that can be evaluated by the electricity supplier as well as by consumers or equipment suppliers. This chapter describes recent developments in methods for benchmarking the performance of electricity supply.

One of the basic tenets of solving power quality problems is that disturbances in the electric power system are not restricted by legal boundaries. Power suppliers, power consumers, and equipment suppliers must work together to solve many problems. Before they can do that, they must understand the electrical environment in which end-use equipment operates. This is necessary to reduce the long-term economic impact of inevitable power quality variations and to identify system improvements that can mitigate power quality problems.[1-3]

A comprehensive set of power quality indices was defined for the Electric Power Research Institute (EPRI) Reliability Benchmarking Methodology (RBM) project[1] to serve as metrics for quantifying quality of service. The power quality indices are used to evaluate compatibility between the voltage as delivered by the electric utility and the sensitivity of the end user's equipment. The indices were patterned after the indices commonly used by utilities to describe reliability to reduce the learning curve. A few of the indices have become popular, and software has been developed to compute them from measured data and estimate them from simulations. We will examine the definitions of some of the indices and then look at how they might be included in contracts and planning.

8.2   Benchmarking Process

Electric utilities throughout the world are embracing the concept of benchmarking service quality. Utilities realize that they must understand the levels of service quality provided throughout their distribution systems and determine if the levels provided are appropriate. This is certainly becoming more prevalent as more utilities contract with specific customers to provide a specified quality of service over some period of time. The typical steps in the power quality benchmarking process are

1. *Select benchmarking metrics.* The EPRI RBM project defined several performance indices for evaluating the electric service quality.[4] A select group are described here in more detail.

2. *Collect power quality data.* This involves the placement of power quality monitors on the system and characterization of the performance of the system. A variety of instruments and monitoring systems have been recently developed to assist with this labor-intensive process (see Chap. 11).

3. *Select the benchmark.* This could be based on past performance, a standard adopted by similar utilities, or a standard established by a professional or standards organization such as the IEEE, IEC, ANSI, or NEMA.

4. *Determine target performance levels.* These are targets that are appropriate and economically feasible. Target levels may be limited to specific customers or customer groups and may exceed the benchmark values.

The benchmarking process begins with selection of the metrics to be used for benchmarking and evaluating service quality. The metrics could simply be estimated from historical data such as average number of faults per mile of line and assuming the fault resulted in a certain number of sags and interruptions. However, electricity providers and consumers are increasingly interested in metrics that describe the actual performance for a given time period. The indices developed as part of the EPRI RBM project are calculated from data measured on the system by specialized instrumentation.

Electric utilities throughout the world are deploying power quality monitoring infrastructures that provide the data required for accurate benchmarking of the service quality provided to consumers. These are permanent monitoring systems due to the time needed to obtain accurate data and the importance of power quality to the end users where these systems are being installed. For most utilities and consumers, the most important power quality variation is the voltage sag due to short-circuit faults. Although these events are not necessarily the most frequent, they have a tremendous economic impact on end users. The process of benchmarking voltage sag levels generally requires 2 to 3 years of sampling. These data can then be quantified to relate voltage sag performance with standardized indices that are understandable by both utilities and customers.

Finally, after the appropriate data have been acquired, the service provider must determine what levels of quality are appropriate and economically feasible. Increasingly, utilities are making these decisions in conjunction with individual customers or regulatory agencies. The economic law of diminishing returns applies to increasing the quality of electricity as it applies to most quality assurance programs. Electric utilities note that nearly any level of service quality can be achieved through alternate feeders, standby generators, UPS systems, energy storage, etc. However, at some point the costs cannot be economically justified and must be balanced with the needs of end users and the value of service to them.

Most utilities have been benchmarking *reliability* for several decades. In the context of this book, reliability deals with sustained interruptions. IEEE Standard 1366–1998 was established to define the benchmarking metrics for this area of power quality.[5] The metrics are defined in terms of system average or customer average indices regarding such things as the number of interruptions and the duration of interruption (SAIDI, SAIFI, etc.). However, the reliability indices do not capture the impact of loads tripping off-line for 70 percent voltage sags nor the loss of efficiency and premature equipment failure due to excessive harmonic distortion.

Interest in expanding the service quality benchmarking into areas other than traditional reliability increased markedly in the late 1980s. This was largely prompted by experiences with power electronic loads that produced significant harmonic currents and were much more sensitive to voltage sags than previous generations of electromechanical loads. In 1989, the EPRI initiated the EPRI Distribution Power Quality (DPQ) Project, RP 3098–1, to collect power quality data for distribution systems across the United States. Monitors were placed at nearly 300 locations on 100 distribution feeders, and data were collected for 27 months. The DPQ database contains over 30 gigabytes of power quality data and has served as the basis for standards efforts and many studies.[1,6] The results were made available to EPRI member utilities in 1996.

Upon completion of the DPQ project in 1995, it became apparent that there was no uniform way of benchmarking the performance of specific service quality measurements against these data. In 1996, the EPRI completed the RBM project, which provided the power quality indices to allow service quality to be defined in a consistent manner from one utility to another.[4] The indices were patterned after the traditional reliability indices with which utility engineers had already become comfortable. Indices were defined for

1. Short-duration rms voltage variations. These are voltage sags, swells, and interruptions of less than 1 min.

2. Harmonic distortion.

3. Transient overvoltages. This category is largely capacitor-switching transients, but could also include lightning-induced transients.

4. Steady-state voltage variations such as voltage regulation and phase balance.

This chapter describes methodologies for determining target levels of quality for various applications based on the statistical distribution of quality indices values calculated from actual measurement data. We will concentrate on the more popular indices for rms voltage variations and harmonics. Readers are referred to the documents cited in the references to this chapter for more details.

8.3   RMS Voltage Variation Indices

For many years, the only indices defined to quantify rms variation service quality were the sustained interruption indices (SAIFI, CAIDI, etc.). Sustained interruptions are in fact only one type of rms variation. IEEE Standard 1159–1995[7] defines a sustained interruption as a reduction in the rms voltage to less than 10 percent of nominal voltage for longer than 1 min (see Chap. 2).

Sustained interruptions are of great importance because all customers on the faulted section are affected by such disturbances. Indices for evaluating them have been in use informally by utilities for many years and were recently standardized by the IEEE in IEEE Standard 1366–1998.[5] Long before, some utilities had been required to report certain indices to regulatory agencies. The standard also defines indices quantifying momentary interruption performance, which quantifies another very important type of rms voltage variation. Momentary interruptions are due to clearing of temporary faults and the subsequent reclose operation (see Chap. 3). While they are not captured in the traditional reliability indices, they affect many end-user classes. The rms voltage variation indices take this one step farther and define metrics for voltage sags, which can also affect many end users adversely.

8.3.1   Characterizing RMS Variation Events

IEEE Standard 1159–1995[7] provides a common terminology that can be used to discuss and assess rms voltage variations, defining magnitude ranges for sags, swells, and interruptions. The standard suggests that the terms sag, swell, and interruption be preceded by a modifier describing the duration of the event (instantaneous, momentary, temporary, or sustained). These definitions are summarized in Chap. 2.

RMS variations are classified by the *magnitude* and *duration* of the disturbances. Therefore, before rms variation indices can be calculated, magnitude and duration characteristics must be extracted from the raw waveform data recorded for each event. *Character-ization* is a term used to describe the process of extracting from a measurement useful pieces of information which describe the event so that not every detail of the event has to be retained.

Characterization of rms variations can be very complicated. It is structured into three levels, each of which is identified as a type of event as follows:

1. Phase or component event

2. Measurement event

3. Aggregate event

Component Event Level

Each phase of each rms variation measurement may contain multiple components. Most rms variations have a simple rectangular shape and are accurately characterized by a single magnitude and duration. Approximately 10 percent of rms variations are nonrectangular[1] and have multiple components. Consider the rms variation shown in Fig. 8.1. It exhibits a voltage swell followed by two levels of voltage sag. This event was the result of clearing a temporary single-line-to-ground fault that evolved into a double-line-to-ground fault before the breaker tripped. The breaker then reclosed successfully in about 0.2 s. Note that only about 10 cycles of the initial voltage swell are shown in the waveform plot on the bottom. The entire event lasted nearly 1.5 s, although the

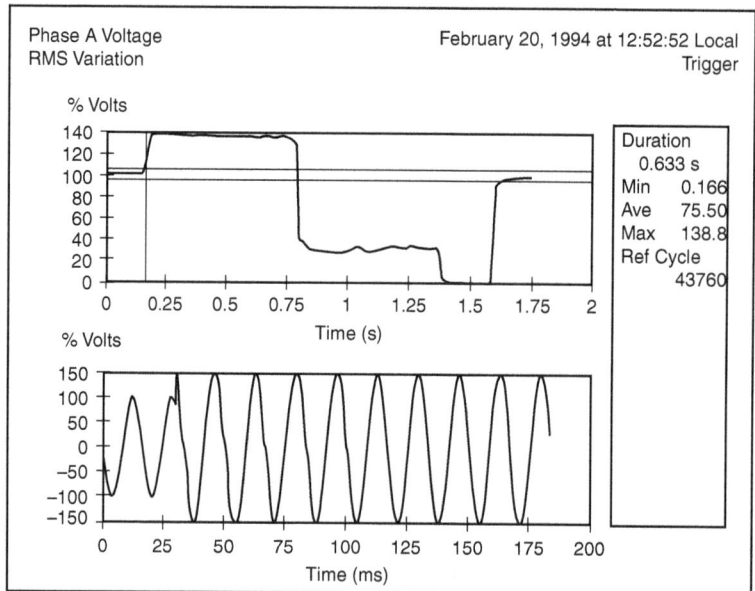

FIGURE 8.1 Multicomponent, nonrectangular rms variation.

instrument reports only the duration of the voltage swell. Other software is required to postprocess the waveform off-line to determine the other characteristics of this event. Variations like this are much more difficult to characterize because no single magnitude-duration pair completely represents the phase measurement.

Most of the methods for characterization agree that the magnitude reported must be the maximum deviation from nominal voltage. The difficulty lies in assigning a duration associated with the magnitude. The method defined here is called the *specified voltage method*. This method designates the duration as the period of time that the rms voltage exceeds a *specified threshold voltage level* used to characterize the disturbance.

Thus, events like the one in Fig. 8.1 would be assigned different duration values depending on the specified voltage threshold of interest. Figure 8.2 illustrates this concept for three voltage levels: 80, 50, and 10 percent. $T_{80\%}$ is the duration of the event for an assessment of sags having magnitudes #80 percent. Likewise, $T_{50\%}$ and $T_{10\%}$ are the durations associated with sags of the corresponding voltage levels. Notice that $T_{80\%}$ and $T_{50\%}$ are both 800 ms because both of the sag components of this nonrectangular event have magnitudes well below 50 percent. $T_{10\%}$, however, comprises only the duration of the second component, 200 ms.

Measurement Event Level

A power system occurrence such as a fault can affect one, two, or all three phases of the distribution system. The magnitude and duration of the resulting rms variation may differ substantially for different

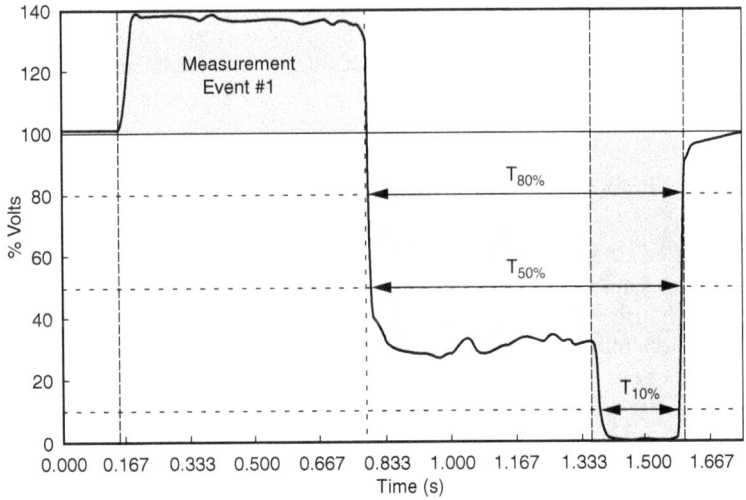

FIGURE 8.2 Illustration of specified voltage characterization of rms variation phase measurements.

phases. A determination must be made concerning how to report three-phase measurement events. For an assessment of single-phase performance, each of the three phases are reported separately. Thus, for some faults, three different rms variations are included in the indices. This will be inappropriate for loads that see this as a single event.

The method defined here for characterizing measurement events is a three-phase method. A single set of characteristics are determined for all affected phases. For each rms variation event, the magnitude and duration are designated as the magnitude and duration of the phase with the greatest voltage deviation from nominal voltage.

Aggregate Event Level

An aggregate event is the collection of all measurements associated with a single power system occurrence into a single set of event characteristics. For example, a single distribution system fault might result in several measurements as the overcurrent protection system operates to clear the faults and restore service. An aggregate event associated with this fault would summarize all the associated measurements into a single set of characteristics (magnitude, duration, etc.). While there may be many individual events, many end-user devices will trip or misoperate on the initial event. The succeeding rms variations have no further adverse effect on the end-user process. Thus, aggregation provides a truer assessment of service quality. RMS variation performance indices are usually based on aggregate events.

A good method of aggregating measurements is to consider all events that occur within a defined interval of the first event to be part of the same aggregate event. One minute is a typical time interval, which corresponds to the minimum length of a sustained interruption. The magnitude and duration of the aggregate event are determined from the measurement event most likely to result in customer equipment failure. This will generally be the event exhibiting the greatest voltage deviation.

8.3.2   RMS Variation Performance Indices

The rms variation indices are designed to assess the service quality for a specified circuit area. The indices may be scaled to systems of different sizes. They may be applied to measurements recorded across a utility's entire distribution system resulting in SAIFI-like system averages, or the indices may be applied to a single feeder or a single customer PCC.

There are many properties of rms variations that could be useful to quantify—properties such as the frequency of occurrence, the duration of disturbances, and the number of phases involved. Many rms variation indices were defined in the EPRI RBM project to address these various issues. Space does not permit a description of all of these, so we will concentrate on one index that has, perhaps, become the most popular. The papers and reports included in the references contain details on others.

System Average RMS (Variation) Frequency Index$_\text{Voltage}$ (SARFI$_x$)

SARFI$_x$ represents the average number of specified rms variation measurement events that occurred over the assessment period per customer served, where the specified disturbances are those with a magnitude less than x for sags or a magnitude greater than x for swells:

$$\text{SARFI}_x = \frac{\Sigma N_i}{N_T}$$

where x = rms voltage threshold; possible values are 140, 120, 110, 90, 80, 70, 50, and 10

N_i = number of customers experiencing short-duration voltage deviations with magnitudes above X percent for $X > 100$ or below X percent for $X < 100$ due to measurement event i

N_T = total number of customers served from section of system to be assessed

Notice that SARFI is defined with respect to the voltage threshold x. For example, if a utility has customers that are only susceptible to sags below 70 percent of nominal voltage, this disturbance group can be assessed using SARFI$_{70}$. The eight defined threshold values for the index are not arbitrary. They are chosen to coincide with the following:

140, 120, and 110. Overvoltage segments of the ITI curve.

90, 80, and 70. Undervoltage segments of ITI curve.

50. Typical break point for assessing motor contactors.

10. IEEE Standard 1159 definition of an interruption.

An increasing popular use of SARFI is to define the threshold as a curve. For example, SARFI$_{\text{ITIC}}$ would represent the frequency of rms variation events outside the ITI curve voltage tolerance envelope. Three such curve indices are commonly computed:

SARFI$_{\text{CBEMA}}$
SARFI$_{\text{ITIC}}$
SARFI$_{\text{SEMI}}$

This group of indices is similar to the System Average Interruption Frequency Index (SAIFI) value that many utilities have calculated for years. SARFI$_x$, however, assesses more than just interruptions. The frequency of occurrence of rms variations of varying magnitudes can be assessed using SARFI$_x$. Note that SARFI$_x$ is defined for short-duration variations as defined by IEEE Standard 1159.

There are three additional indices that are subsets of SARFIx. These indices assess variations of a specific IEEE Standard 1159 duration category:

1. System Instantaneous Average RMS (Variation) Frequency Index (SIARFI$_x$).

2. System Momentary Average RMS (Variation) Frequency Index (SMARFI$_x$).

3. System Temporary Average RMS (Variation) Frequency Index (STARFI$_x$).

8.3.3   SARFI for the EPRI DPQ Project

Table 8.1 shows the statistics for various forms of SARFI computed for the measurements taken by the EPRI DPQ project. These particular values are rms variation frequencies for substation sites in number of events per 365 days. One-minute temporal aggregation was used, and the data were treated using sampling weights. This can serve as a reference benchmark for distribution systems in the United States.

8.3.4   Example Index Computation Procedure

This example is based on actual data recorded on one of the feeders monitored during the EPRI DPQ project.[1] This illustrates some of the practical issues involved in computing the indices.

First, one must know how many customers experience a voltage exceeding the index threshold for each rms variation that occurs. Obviously, every customer will not be individually monitored. Consequently, one must approximate the voltage experienced by each customer during a disturbance. This is accomplished by segmenting the circuit into small areas across which all customers are assumed to experience the same voltage. Obviously, the smaller the segments, the better the approximation.

One method of determining voltages for many circuit segments based on a limited number of monitoring points is power quality state estimation. A special section (8.7) is included on this topic later. State estimation provides pseudomeasurements for those segments not containing a measuring instrument. Such state estimation requires a moderately detailed circuit model and known monitored data. Without the pseudomeasurements provided by state estimation, the number of physical monitoring locations becomes the number of constant-voltage segments upon which the indices that are calculated. This is referred to as *monitor-limited segmentation* (MLS) and results in only a few segments per circuit. Although the calculated index values are less accurate, MLS still yields indices that are informative.

Figure 8.3 illustrates the three MLS segments for the example calculation feeder corresponding to the three power quality monitors,

	SARFI$_{90}$	SARFI$_{80}$	SARFI$_{70}$	SARFI$_{50}$	SARFI$_{10}$	SARFI$_{CBEMA}$	SARFI$_{ITIC}$	SARFI$_{SEMI}$
Minimum	0.000	0.000	0.000	0.000	0.000	0.000	0.000	0.000
CP05†	11.887	5.594	0.000	0.000	0.000	5.316	2.791	2.362
CP50†	43.987	22.813	12.126	5.165	1.525	25.465	18.765	13.619
Mean	56.308	28.729	18.422	8.926	3.694	33.293	25.390	18.535
CP95†	135.185	66.260	51.000	27.037	13.519	71.413	51.500	38.238
Maximum	207.644	103.405	70.535	56.311	35.689	149.488	140.768	140.768

*Submitted for IEEE Standard P1564.[8]
†CP05, CP50, and CP95 are abbreviations that indicate that the value exceeds 5, 50, and 95 percent of the samples in the database. For example, 50 percent of the sites in the project had more than 18.765 events per year that were outside the ITI curve voltage tolerance envelope (SARFI$_{ITIC}$).

TABLE 8.1   SARFI Statistics from the EPRI DPQ Project*

367

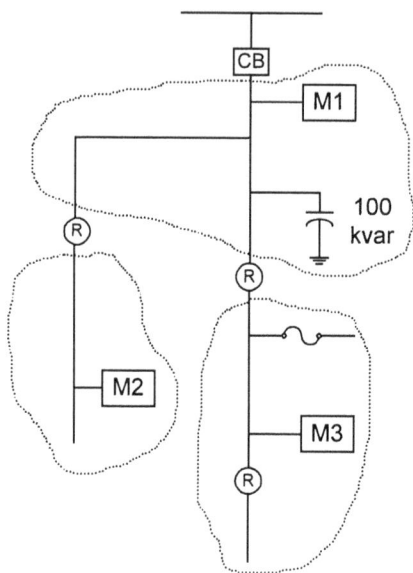

Figure 8.3 Circuit for example rms variation calculation.

x	SARFI$_x$	SIARFI$_x$	SMARFI$_x$	STARFI$_x$
140	0.0	0.0	0.0	0.0
120	0.0	0.0	0.0	0.0
110	0.5	0.5	0.0	0.0
90	27.5	22.7	4.3	0.5
80	13.6	8.8	4.3	0.5
70	7.3	2.5	4.3	0.5
50	4.8	0.5	3.8	0.5
10	4.3	Undefined	3.8	0.5

Table 8.2   Example RMS Variation Index Values Calculated for Circuit of Fig. 8.3 Based on 1 Year of Actual Monitored Data

M1, M2, and M3. The exact number of customers served from each MLS segment was not available, so values of 500, 100, and 400 were assumed for segments 1, 2, and 3, respectively, based on the load. With these assumptions, 1 year of monitoring data yielded the results summarized in Table 8.2.

The sag indices are typical of what would be expected. The number of customer disturbances decrease as the voltage threshold decreases. There were very few voltage swells on this feeder. The total number of sags per customer is estimated at 27.5 per year. Of these, only 7.3 are below 70 percent and 4.8 are below 50 percent. These two levels are typically where end users begin to experience problems, and utilities that use these indices typically set benchmark targets close to these values.

The $SARFI_{10}$ value of 4.3 cannot be compared to SAIFI because SAIFI reflects only sustained interruptions. The duration-based indices—SIARFI, SMARFI, and STARFI—are also quite interesting. The majority of the disturbances are classified as instantaneous by IEEE Standard 1159. Only 4.8 of the 27.5 sag disturbances are either momentary or temporary. However, these tend to be the more severe sags (magnitude of 50 percent and less).

8.3.5 Utility Applications

Utilities are using the discussed rms variation indices to improve their systems.[9] One productive use of the indices is to compute the separate indices for individual substations as well as the system index for several substations. The individual substation values are then compared to the system value. Those substations that exhibit significantly poor performance as compared to the system performance are targeted for maintenance efforts. Based on the sensitivity and needs of the customers served from the targeted substations, the economic viability of potential mitigating actions is assessed. The indices have also proven to be excellent tools for communicating performance of the power delivery system in a simplified manner to key industrial customers.

8.4 Harmonics Indices

Power electronic devices offer electrical efficiencies and flexibility but present a double-edged coordination problem with harmonics. Not only do they produce harmonics, but they also are typically more sensitive to the resulting distortion than more traditional electromechanical load devices. End users expecting an improved level of service may actually experience more problems. This section discusses power quality indices for assessing the quality of service with respect to harmonic voltage distortion. Before we get into the definition of the indices, some issues regarding sampling are discussed.

8.4.1 Sampling Techniques

Power quality engineers typically configure power quality monitors to periodically record a sample of voltage and current for each of the three phases and the neutral. The measurements typically consist of a single cycle, but longer samples may be needed to capture such phenomena as interharmonics. The power quality monitors take samples at intervals of 15 to 30 min and record thousands of measurements that are summarized by the indices. Besides harmonic distortion, the recorded waveforms yield information about other steady-state characteristics such as phase unbalance, power factor, form factor, and crest factor. We will focus here on harmonic content.

The fundamental quantity used to form the indices is the THD of the voltage. The definition of THD may be found in Chap. 5 and is repeated here in Eq. (8.1):

$$V_{THD} = \frac{\sqrt{\sum_{h=2}^{\infty} V_h^2}}{V_1} \tag{8.1}$$

Voltage distortion is not a constant value. On a typical system, the harmonic distortion follows daily, weekly, and seasonal patterns. An example of daily patterns of total harmonic voltage distortion for 1 week is shown in Fig. 8.4. This is typical for many residential feeders where the voltage distortion is highest late at night when the load is low.

A useful method of summarizing the THD samples of trends like that in Fig. 8.4 is to create a histogram like that shown in Fig. 8.5. Note the two distinct peaks in the distribution, which reflects the bimodal nature of the harmonic distortion trend.

Once the histogram is prepared, the cumulative frequency curve is computed. This is shown overlaying the histogram in Fig. 8.5 and has been pulled out separately in Fig. 8.6 to demonstrate the computation of the 95th percentile value, known as CP95. In this

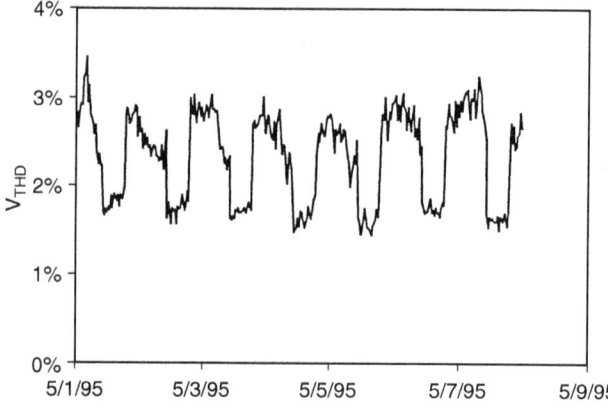

Figure 8.4 Trend of voltage total harmonic distortion demonstrating daily cycle for 1 week.

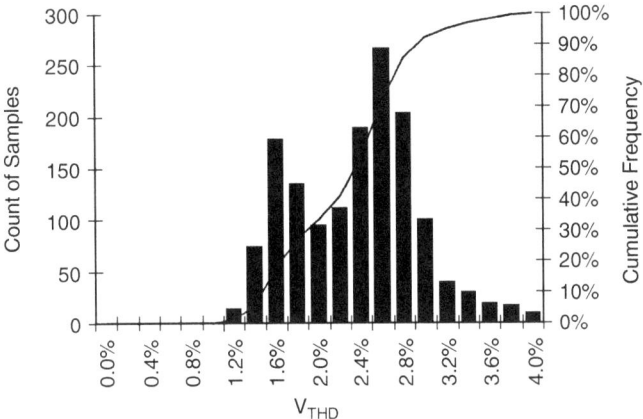

FIGURE 8.5 Histogram of voltage total harmonic distortion for 1 month demonstrating bimodal distribution.

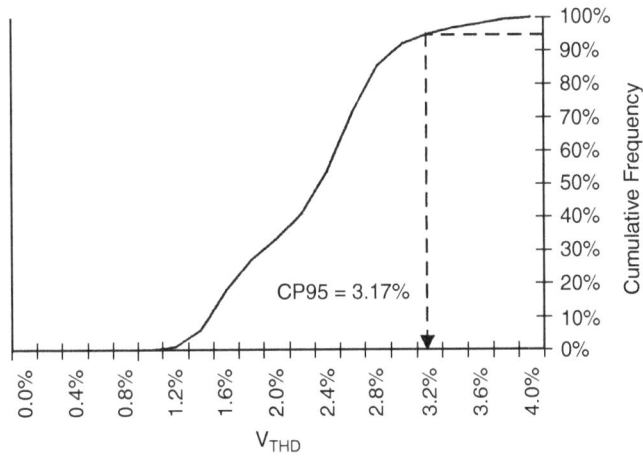

FIGURE 8.6 Demonstration of graphical method of calculating the CP95 of a distribution.

example, a voltage THD of 3.17 percent is larger than 95 percent of all other samples in the distribution. CP95 is frequently more valuable than the maximum value of a distribution because it is less sensitive to spurious measurements.

Usually an electric utility will collect measurements at more than one location and compute a different CP95 value for each monitoring

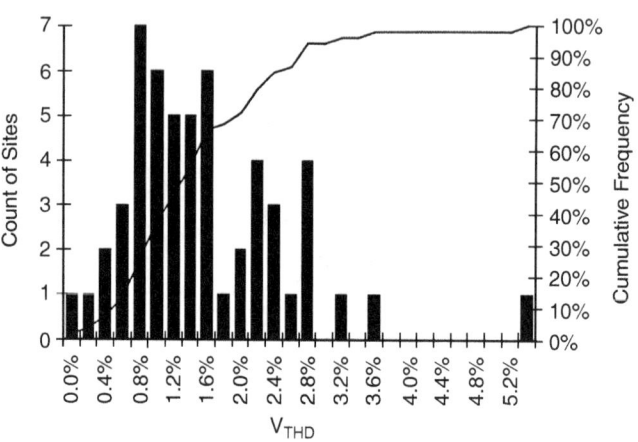

Figure 8.7 Histogram of CP95 values for voltage THD at 54 monitoring sites.

location. Figure 8.7 shows a histogram of CP95 values compiled from different sites, which serves to summarize the measurements both temporally and spatially. A CP95 value can also be determined from this histogram, which is a "statistic of a statistic" that can be used to provide a reference value for an entire utility system.

8.4.2 Characterization of Three-Phase Harmonic Voltage Measurements

Many distribution systems in the United States supply single-phase and other unbalanced loads. Therefore, the harmonic content of each primary phase voltage is generally different. This presents a problem in characterizing the harmonic distortion of a three-phase measurement which has varying distortion levels on each phase. There are two possible methods:

1. Consider each phase to be a separate measurement. The potential problem with this method is that a count of how often distortion levels exceed a specified level could be 3 times too large.

2. Average the distortion levels on the three phases. Each three-phase steady-state measurement contributes a single distortion level to the samples. A possible drawback is that a high distortion level on one phase is obscured if the other two phases exhibit low distortion levels.

The latter method has less potential for inaccuracy and is used for calculating the harmonic distortion indices presented here.

8.4.3   Definition of Harmonic Indices

Scalable indices have been developed to aid in the assessment of the quality of service related to harmonic voltage distortion for a specified circuit. As with other indices, they can be applied to various parts of the utility system. An index value for the whole system serves as a useful metric but is not intended as an exact representation of the quality of service provided to each individual customer. However, it can be used as a benchmark against which index values for selected areas within the distribution system can be compared.

System Total Harmonic Distortion CP95 (STHD95)

STHD95 represents the CP95 value of a weighted distribution of the individual circuit segment CP95 values for voltage THD (see earlier discussion concerning Fig. 8.7). STHD95 is defined by Eqs. (8.2) and (8.3):

$$\frac{\sum\limits_{-\infty}^{\text{STHD95}} f_t\left(\text{CP95}_s\right) \times L_s}{\sum\limits_{-\infty}^{\infty} f_t\left(\text{CP95}_s\right) \times L_s} = 0.95 \qquad (8.2)$$

$$\frac{\sum\limits_{-\infty}^{\text{CP95}_s} f_s\left(x_i\right)}{\sum\limits_{-\infty}^{\infty} fs\left(x_i\right)} = 0.95 \qquad (8.3)$$

where s = circuit segment number
   i = steady-state THD measurement number
   L_s = connected kVA served from circuit segment s
   $f_s\left(x_i\right)$ = probability distribution function comprised of sampled THD values for circuit segment s
   CP95_s = 95th percentile cumulative probability value; it is a statistical quantity representing the value of THD which is larger than exactly 95 percent of the samples comprising the THD distribution for segment s
   $f_t(\text{CP95}_s)$ = probability distribution function comprised of the individual circuit segment THD CP95 values

   Equipment degradation due to harmonics is often the result of sustained distortion over long periods of time. High THD levels that decrease after a relatively short duration may not affect end user or utility equipment as much. CP95 neglects the highest distortion samples. Thus, 5 percent of the samples can be very high, as might be the case for a circuit exhibiting short-duration harmonic increases, without significantly affecting the index value. IEEE Standard 519[10] specifies that THD limits are not to be exceeded for more than 1 h per day, which is approximately 4 percent of the time. Thus, the STHD95 value will approximately correspond to the allowable duration limits defined for excessive THD values in the IEEE standard.

System Average Total Harmonic Distortion (SATHD)

SATHD is based on the mean value of the distribution of voltage THD measurements recorded for each circuit segment rather than the CP95 value. SATHD represents the weighted average voltage THD experienced over the monitoring period normalized by the total connected kVA served from the assessed system. SATHD is defined by Eqs. (8.4) and (8.5):

$$\text{SATHD} = \frac{\sum_{s=1}^{k} L_s \times \text{MEANTHD}_s}{L_T} \tag{8.4}$$

$$\text{MEANTHD}_s = \frac{\sum_{i=1}^{N_{\text{MW}}} \text{THD}_i}{N_{\text{MW}}} \tag{8.5}$$

where s = circuit segment number
  k = total number of circuit segments in the system being assessed
  L_s = connected kVA served from circuit segment s
  L_T = total connected kVA served from the system being assessed
  i = steady-state measurement number
  THD_i = voltage total harmonic distortion calculated for measurement window i
  N_{MW} = total number of steady-state measurement windows collected for a given circuit segment over the duration of the monitoring period
MEANTHD_s = statistical mean of the THD values obtained from each of the steady-state measurement windows for circuit segment s

These harmonic distortion indices are weighted by connected kVA. This is one method to give more weight to data from monitoring sites deemed more important. This weighting may also be determined by the number of customers, the amount of actual load, sensitivity of customer loads, etc. Connected load is similar to weighting methods specified in IEC Standard 1000–3-6, *Assessment of Emission Limits for Distorting Loads in MV and HV Power Systems*.

System Average Excessive Total Harmonic Distortion Ratio Index$_{THD\ level}$ (SAETHDRI$_{THD}$)

SAETHDRI$_{THD}$ is a measure of the number of steady-state measurements that exhibit a THD value exceeding the specified threshold. It is more difficult to calculate frequency-of-occurrence indices for steady-state phenomena such as harmonics because they are not triggered measurements. Without continuously recording data, steady-state quantities can only be assessed using sampled data. An approximation of the amount of time the system THD exceeded a certain value can be calculated from the ratio of excessive THD samples to the total number of samples. This ratio is considered to be the frequency of occurrence of a specified threshold THD value.

For each circuit segment comprising the assessed system, the number of measurements exceeding the THD threshold is normalized by the total number of measurements. The system average is then computed by weighting each segment ratio by the load served from that segment. SAETHDRI$_{THD}$ is defined by

$$\text{SAETHDRI}_{THD} = \frac{\sum_{s=1}^{k} L_s \times \left(\frac{N_{THD_s}}{N_{MW_s}} \right)}{L_T} \tag{8.6}$$

where s = circuit segment number
k = total number of circuit segments in the system being assessed
L_s = connected kVA served from circuit segment s
L_T = total connected kVA served from the system being assessed
i = steady-state measurement number
THD = THD threshold specified for calculation of this index
N_{THD_s} = number of steady-state measurements that exhibit a THD value for segment s which exceeds the specified THD threshold value
N_{MW_s} = total number of steady-state measurements recorded for segment s over the assessment period

SAETHDRI$_{THD}$ provides a measure of the portion of the time that the designated system exceeds a specified THD value. For example, we can use SATHEDRI$_{5\%}$ to approximate the amount of time the defined system exceeds the IEEE Standard 519–1992 THD limit of 5 percent.

8.4.4 Harmonic Benchmark Data

Figure 8.8 shows how temporal average voltage THD is distributed for 277 sites in the EPRI DPQ Project. Nearly 18 percent of the monitoring sites had an average value of voltage THD of 1.2 percent for the period from 6/1/93 to 3/1/95. The data have been treated by

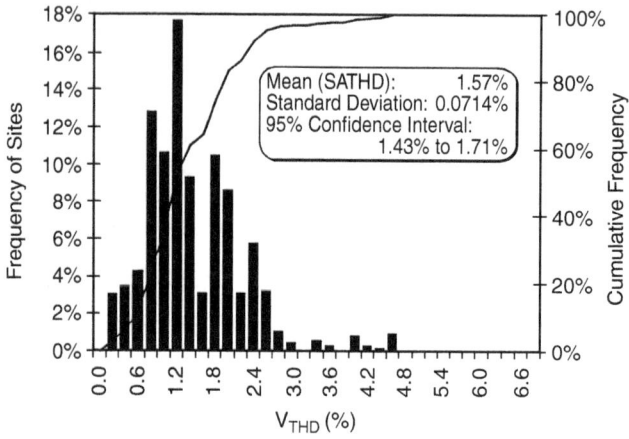

FIGURE 8.8   Histogram of average value for voltage THD at 277 monitoring sites from 6/1/93 to 3/1/95 (treated by sampling weights).

using sampling weights designed so that the charts represent an estimate of the entire distribution systems of the 24 EPRI DPQ utilities. The 95-percent confidence level means that we are 95 percent confident that the true mean of the sample population is between 1.43 and 1.71 percent. The mean of the distribution is the same as that of the SATHD index.

Figure 8.9 presents the distribution of the CP95 values of voltage THD at all the sites. The average CP95 value for the data sampled in the project was 2.18 percent. Computing the CP95 of the data presented in Fig. 8.9 gives the STHD95 index, 4.03 percent. The larger spread of samples between 0.2 and 6.4 percent THD causes the wider 95-percent confidence interval. The number of monitoring locations that exceeded the limits established by IEEE Standard 519–1992 can also be obtained from this chart. Of the sites monitored, 3.3 percent exceeded the 5 percent voltage THD limit for at least 5 percent of samples. (Only one monitored substation exceeded the IEEE Standard 519–1992 limit for more than 95 percent of samples.)

8.4.5   Seasonal Effects

Harmonic distortion varies seasonally as well as daily and weekly. Figure 8.10 shows the SATHD index for each of the 27 months of monitoring in the EPRI DPQ project. A simple average of site values was used without the sampling weights. A seasonal pattern is very evident. Voltage THD tends to be lower during the winter and summer months and peaks in the spring and fall months.

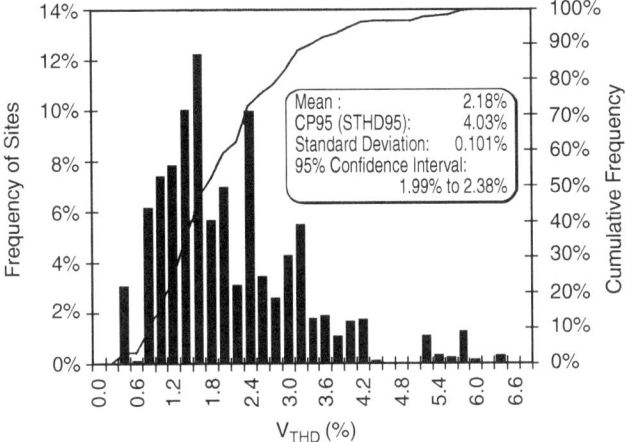

FIGURE 8.9   Histogram of CP95 value for voltage THD at 277 monitoring sites from 6/1/93 to 3/1/95 (treated by sampling weights).

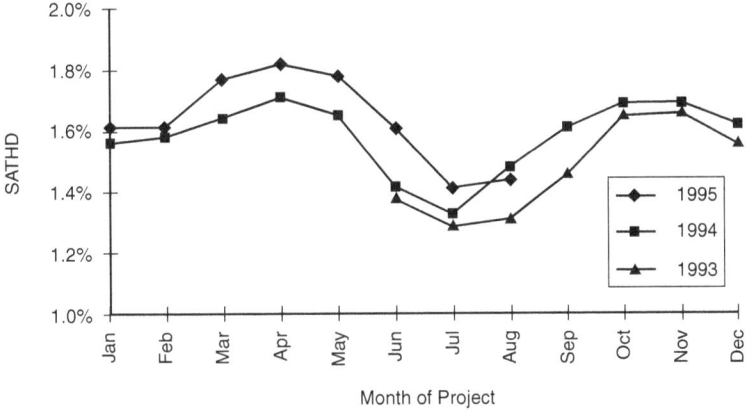

FIGURE 8.10   SATHD values by month, from 6/1/93 to 9/1/95, unweighted, all sites.

The periods of low THD correspond with peak loading periods of the year due to heating and air conditioning demand. Air conditioning load apparently provides for more damping effect than heating load, which might be counterintuitive. The capacitor configurations will also change for these seasons, which may also affect the distortion.

Note that the distortion trends are upward during the 2-year monitoring period. Each yearly trend is slightly higher than the previous. This would suggest that there is some validity to the increased concern about the proliferation of harmonic-producing power electronic equipment. Increases in background harmonic distortion attributable to increasing percentages of nonlinear loads have been observed by others as well.[11]

8.5 Power Quality Contracts

Once performance targets have been selected, utilities may enter into contractual agreements with end users with respect to power quality variations. While this is never an easy task, it was simpler when end users had to deal only with a single, vertically integrated utility company. The deregulation of the electric power utilities in many areas further complicates things. As Kennedy[12] points out regarding future trends, there now might be up to five entities involved:

1. The transmission provider (TRANSCO)
2. The local distributor (DISTCO), or the "wires" company
3. One or more independent power producers (IPPs) or market power producers (MPPs)
4. Retail energy marketers (RETAILCOs) or energy service companies (ESCOs)
5. The end user

To meet the performance requirements of the end user, there may have to be contracts between all these entities. While the bulk of the power quality variations may be the result of events on the local distributor's system, there will also be events on the transmission system that will affect large areas. Forced outages at power plants can cause spikes in the power market and, perhaps, voltage regulation issues if the power supply becomes constrained. The retail energy marketer, if any, may ultimately be responsible for financial obligations and could coordinate the agreements and present a single point of contact for the end user.

Detailed descriptions of service agreements are outside the scope of this book. Some important characteristics of contracts for rms voltage variations and harmonics are presented in Secs 8.5.1 and 8.5.2. More details may be found in the *Power Quality Standards and Specifications Workbook.*[13]

8.5.1 RMS Variations Agreements

Part of the purpose of an interconnection agreement would be to educate end users on the realities of power delivery by wire and the costs associated with mitigating voltage sags and interruptions. Another part would be the establishment of some formal means by which the utility records and evaluates the fault performance of its power delivery system.

Some of the key issues that should be addressed are

1. The number of interruptions expected each year.

2. The number of voltage sags below a certain level each year. The level can be defined in terms of a specific number such as 70 or 80 percent. Alternatively, it can be defined in terms of a curve such as the CBEMA or ITI curve.

3. The means by which end users can mitigate rms variations.

4. Responsibilities of utilities in analyzing the performance of the power delivery system, following up with fault events, etc.

5. Maintenance efforts to reduce the number of faults for events within the control of the utility.

8.5.2   Harmonics Agreements

Although harmonics problems are not as widespread as rms voltage variation problems, harmonics from ASDs and other electronic loads can have a severe impact on other end-user equipment. In some cases, the equipment will fail to operate properly, while in other cases, it may suffer premature failure. Therefore, agreements regarding harmonics can be very important.

The chief tool for the enforcement of harmonic emissions at the utility-customer interface is IEEE Standard 519–1992.[10] This is a two-edged sword: One part of the standard places limits on harmonic currents that can be injected by end-user loads onto the system, while another part effectively establishes minimum requirements for the utility. Agreements on harmonics should reflect this bilateral nature. Some of the key issues that should be addressed are

1. Definition of the PCC.

2. Limitation of the harmonic current distortion level at the PCC to that set by IEEE Standard 519–1992 or to another value allowed by a specified exception.

3. Periodic maintenance schedules for filters and other mitigating equipment. Some equipment will require constant monitoring by permanently installed devices.

4. Responsibilities of utilities, such as

   a. Keeping the system out of harmonic resonance

   b. Keeping records about new loads coming onto the system (this is getting tougher to do with deregulation)

   c. Performing engineering analyses when new loads come onto the system to prevent exacerbation of existing problems

d. Educating end users about mitigation options

e. Periodic monitoring or constant monitoring by permanently installed devices to verify proper operation of the system

5. Definition of responsibilities for mitigation costs when limits are exceeded. Is the last end user who created the excess load responsible or is the cost shared among a class of end users and the utility?

8.5.3   Example Contract

One of the most widely publicized examples of a power quality contract is the one between Detroit Edison and the "big three" automobile manufacturers. Detloff and Sabin[14] report that in 1995 Detroit Edison entered into long-term pricing and service quality agreements with Chrysler Corporation (now DaimlerChrysler), Ford Motor Company, and General Motors Corporation. The terms were specified in an agreement known as the Special Manufacturing Contract (SMC). The service agreement covered voltage interruptions and voltage sags and established service guarantees with compensation.

In response to competitive pressures, Detroit Edison entered into the 10-year agreement with these customers as a sole supplier of power. The service guarantees were created in response to the customers' concern that the utility might not have as much incentive to resolve power quality problems if the customers were locked into a 10-year sole-supplier agreement. Therefore, the parties devised a method of compensating the customers for power quality events if the annual schedule of power quality targets was exceeded. Compensation levels were negotiated that were related to the cost of customer production losses following an interruption. These levels were fixed for the duration of the SMC. Initially, only interruptions of power were included in the guarantees. In 1998, voltage sag guarantees were added that made Detroit Edison liable if voltage sag measures exceeded the negotiated performance targets.

Therefore, monitoring of the power quality and computation of the service indices are of very high importance. Detroit Edison installed a power quality monitoring system at over 50 of the three customers' locations throughout its territory. The power quality monitoring system allows Detroit Edison to determine the frequency and severity of voltage sags that occur at the customer locations. Some of the key details follow.

Interruption Targets

The interruption targets for the DaimlerChrysler and General Motors locations are either 0 or 1. This means that only one interruption is allowed at some of these locations and none at other locations in each calendar year. The service guarantee payment amounts (SGPAs) negotiated for these two companies range between $2000 and $297,000 and are based on the type of process that is being served. Several of these locations operate with their services in parallel so that they usually do not experience a zero-voltage event.

The Ford agreement was a little different. The locations were split into six groups with interruption targets ranging from one to nine per year. The utility would pay for interruptions in excess of these targets at the rate specified by the negotiated SGPA. Interestingly, the targets decrease by 5 percent each calendar year, rounded to the nearest whole number, requiring continually improving performance by the utility.

Voltage Sag Targets

The 1994 agreement specified that voltage sags would be included in the compensation scheme at a later date. The delay was considered reasonable because Detroit Edison had no means in place for taking statistically accurate power quality measurements. Installations of power quality monitors began in 1995 with a total of 138 monitors installed at the SMC locations. The utility and the customers finalized the voltage sag agreement ahead of schedule at the insistence of the customers in August 1998, having only about 2 years of measurements for the basis of the initial sag targets. It would have been preferable to have about 3 years to establish the targets. The sag agreement was made retroactive to January 1, 1998.

The monitors are connected to measure phase-to-ground voltages at the 13.2-kV bus and phase-to-phase voltages at the 4.8-kV buses. Each power quality monitor is automatically polled several times a day using three modems, each dedicated to one of the three customers. The measurements are then stored in the main monitoring database on a server. A workstation running the EPRI/Electrotek PQView program queries the database and imports the measurements into a database on a Web server. The rms voltage variations are characterized and the voltage sag indices computed.

The 1998 SMC amendment states five rules that establish a subset of sags that qualify for payment.

1. The rms voltage on any of the three phases must drop below 0.75 pu. There is no minimum duration for qualifying voltage sags; all durations are eligible. The threshold was established based on the ITI curve and discussions with the customers. Actual experience is not a factor in the sag qualification.

2. Voltage sags that are caused by the customer are excluded from the qualifying sag list.

3. Voltage sags that are measured on a nonloaded feeder are not qualifying. This is automatically determined in the PQView program from the maximum load current. Rules 2 and 3 are in place to ensure that the performance is only evaluated at the PCC.

4. Only the worst voltage sag (lowest rms voltage) in a 15-min interval at each location can qualify. The 15-min interval begins when the first sag in a chronological list of sags is detected and ends when either the last sag in the interval is detected or at a point 15 min after the first. Voltage sags that occur after that 15-min interval are considered part of the next interval and are assessed separately. This type of processing is called 15-min temporal aggregation with spatial aggregation by location.

5. If a voltage interruption is measured during a 15-min interval, then any voltage sags that are also measured at the location will not qualify.

According to Detloff and Sabin,[14] approximately 20 percent of all the voltage sags measured by the Detroit Edison power quality monitoring system (nonaggregated) fall below 0.75 pu. Only about 8 percent of the nonaggregated sags qualify after applying the five sag rules and for which sag scores are computed.

Sag Score Definition

The sag score is the average per-unit voltage lost by each of the three phase voltages for the lowest qualifying voltage sag within a 15-min interval. It is defined by Eq. (8.7):

$$\text{Sag score} = 1 - \frac{V_A + V_B + V_C}{3} \qquad (8.7)$$

For interruptions, the sag score is defined to be zero to prevent overlap with the administration of voltage interruptions. If any of the phase voltages are greater than 1 pu because of a neutral shift during a voltage sag, then each is set to 1 pu before the computation of the sag score. These two policies confine scores to the range 0.0833 to 1. The minimum sag score of 0.0833 corresponds to a condition where the voltage on one phase is 0.75 pu, which is the threshold for a qualifying voltage sag, and the other two phase voltages are set to 1 pu.

Detloff and Sabin reported that the average sag score from January 1, 1998, until their paper was prepared in 2000 was 0.31, with 87 percent of the sag scores being less than or equal to 0.50.

Sag Score Targets

A sag score target is the maximum sum of sag score values allowable for a group of locations before compensation is due. Two of the automakers have only one group score target, while the third has six. The sag scores for all qualifying sags in a group are summed and compared to the group sag score target. If the sag score total exceeds the target, compensation is computed.

The SMC agreement allows sag score targets to be recomputed for the eight groups at the start of each calendar year. The group sag score targets are determined by computing the average group sag score totals for the voltage sag data collected from 1995 and up to the present year. The targets are expected to gradually stabilize as data from more years are included.

Voltage Sag Payment

The payment due to a location is computed by determining the sag score sum in excess of the sag score target multiplied by the SGPA subject to an annual payment cap. For example, assume that a location has an SGPA of $100,000 and a sag score target of 3.0. If the sag score is 3.28, then a payment of $0.28 \times \$100,000 = \$28,000$ is due.

An alternative to the sag score method would be to categorize sags into bins based on their magnitudes. Each bin would have a certain level of payout relative to the SGPA. However, this could produce large swings in payments and could disappoint customers. One significant benefit of the sag score method is the lack of abrupt changes in sag scores and payment amounts.

8.6 Power Quality Insurance*

This section proposes a brief overview of a pricing strategy for premium power services that is founded on an insurance policy model.[15,16]

8.6.1 Overview of Power Quality Insurance Concept

Offering premium power services requires the provider, either a distribution company or an energy services provider (referred to hereafter as the utility), to price the services in such a way as to provide benefits to both customers and to the utility. Using an insurance model in which customers subscribe to their desired level of improved power quality (PQ) ensures that no customer will pay more than its own perception of the value or benefit associated with the PQ services. Customer benefits are unique in that they reflect each individual customer's damage function, including the customer's risk aversion. Utility benefits must reflect the risk associated with offering insurance and include returns commensurate with operating in a new competitive environment.

The premium PQ service program uses a business model involving premiums and claims. The utility offers PQ services under an insurance plan. Customers pay premiums for a defined level of service, and the utility pays the customer directly for events exceeding the terms of that service. Customers are motivated to pay a premium

---

*Contributed by Snuller Price, Greg Ball, and C. K. Woo of Energy and Environmental Economics, Inc.

to reduce the uncertainty and/or the expected value of their damage costs. Utilities assume the financial risk associated with the claims in exchange for a return on the aggregate premiums.

The utility's insurance service can make use of a purely financial policy or a policy that incorporates investments in PQ equipment or service. In both cases, the critical advantage of the insurance approach over a cost-of-service approach is that it allows customers to self-select an appropriate solution from policies that are designed without use of customer damage cost data.

Insurance as a Financial Product

The utility can create a purely financial insurance product in which it offers to pay customers for reliability events covered by the policy, and customers pay premiums. Utilities will use customer location to estimate an expected frequency of claims. Because the expected frequency of claims is different for various customer groupings, insurance premiums will likewise be different. Premiums are calculated using principles of fair insurance with a margin that incorporates an appropriate level of risk mitigation and return for the utility. The utility makes no additional investments in reliability equipment or services with the purely financial product. Customers who are risk averse, or who have a higher expectation of their claims than the utility, will subscribe.

Incorporating PQ Investments into Insurance Products

A utility can greatly increase the types of PQ insurance products it can offer if it considers the role that PQ investments can play. By making such investments, the utility can offer insurance products with higher payout ratios and improve customer service quality. To develop such products, the utility must investigate the types of PQ solutions that could be implemented for a group of customers with a similar event rate and location. For example, for a given section of a feeder, there may be four different types of PQ investments a utility could make, ranging from a tree-trimming program to installation of UPS systems at customer sites. For each possible solution, the utility estimates the cost of the solution and the improvement in power quality that would result. Based on these costs, and the expected claims after the investment is made, the utility can design insurance products that will cover their combined costs and provide the customers with an ensured level of improvement in their power quality. Power quality insurance can be provided for a number of distribution system events. Table 8.3 shows five general categories of PQ problems and the associated claim payment structure.

8.6.2   Designing an Insurance Policy

The goals of a PQ insurance scheme are to recover the cost of providing the plan, treat all customers within a group equally at

PQ category	Claim payment structure
RMS variations	$/event categorized by amount of variation as necessary. Incorporate any impact of duration into an event total.
Sustained interruptions	$/event $/hour
Voltage regulation	$/hour categorized as necessary by magnitude indices.
Harmonics	$/hour categorized as necessary by component and magnitude indices.
Transients	$/event categorized as necessary by magnitude indices.

TABLE 8.3 Proposed Claim Payment Structure for PQ Insurance in Five General Categories

cost-based premiums, improve efficient use of resources, and be comprehensible and acceptable.

Fairness

An insurance scheme is considered fair if the expected cost of claims equals the premiums paid. For example, assume a customer's value of service (net of the energy rate) to be $x/kW, which is unobservable. Suppose the probability of an outage is r and the expected benefit to the customer of electricity consumption is $(1 - r)x$.

Now consider an insurance scheme in which a premium of $p/kW results in an insurance payment of $x/kW in the event of an outage. This means that the customer by buying insurance will obtain $[rx + (1 - r)x - p] = x - p$ with certainty. The customer will buy the insurance if

$$x - p > rx \qquad (8.8)$$

When $p = (1 - r)x$, the insurance scheme is fair and cost-based.

Implementation

Designing a basic area financial insurance option involves the following steps:

Step 1. Compute the area-specific probability of outage using historic outage data. For example, the probability of an outage with a duration of more than 1 h is

$$r(>1\text{h}) = \frac{\text{annual unserved hours for such outages}}{8760\,\text{h}}$$

Step 2. Compute the fair insurance premium for a given payoff. For example, if payoff = $1/kW unserved and r (>1 h) = 0.0002, the fair premium is $0.0002/kW unserved.

Step 3. Adjust the premium to collect margin. Suppose the adder is \$0.0001/kW unserved; then the posted premium is \$0.0003/kW unserved.

Step 4. Design service conditions. Here is an example.

- *Subscription.* A participating customer subscribes to the example option in step 3. The customer must specify the amount of kilowatts unserved to be insured over a 1-year period. The annual premium is computed as: kW unserved subscribed × premium/kW. The premium is an up-front payment, irrespective of whether outages actually occur.

- *Eligibility for payoffs.* A participating customer will be paid according to its subscription level, subject to the utility's receipt of the premium from the customer. In this example, payoffs will only be for outages lasting more than 1 h.

8.6.3   Adjusting for PQ Investment Costs

The derivation in Sec. 8.6.2 applies to an insurance scheme in which the utility does not make any investments in PQ improvement technologies. In a case where the utility does make such investments, the cost of these investments is added into the premium:

$$p = (r')\, y + \text{investment cost} + \text{margin} \tag{8.9}$$

where r' is the probability of an outage after the investment is made and y is the payout per outage.

The fact that investments in PQ solutions will often be financed over a period of several years, and that many of these investments are permanent in nature, raises the issue of whether or not the customer will hold the insurance policy long enough to pay for the investment cost. In such cases, the insurance policy should have a clause stating that the policy duration is the full length of the financial life of the PQ solution, and that the customer is responsible for paying for its share of the remaining cost of the solution should it cancel the policy in advance.

Likewise, the utility's ability to offer certain insurance products may depend upon attaining a minimum level of customer subscription (i.e., a minimum total revenue requirement from customer premiums). In such a case, a utility's insurance policy offering can be predicated upon the level of customer subscription.

8.7   Power Quality State Estimation

Power quality monitoring is a relatively new idea in the electric power industry. While there are now many PQ monitors installed and operable, there are still considerably large gaps in coverage of the distribution feeders in the United States. As part of the EPRI RBM project,[4] investigators explored the idea of estimating the voltages at locations without monitors given the data at only one monitor or a few monitors. This resulted in the development of the EPRI power quality state estimator (PQSE), which uses feeder models and recorded data to estimate what would have been recorded on the customer side of the service transformer. In most cases, this can be done with sufficient accuracy to estimate the indices such as SARFI.

The following describes, in very general terms, the approach taken in the PQSE to determine the best match between the measured results and the computed results. Most of the effort was focused on rms variations, since this is the disturbance classification that has the most widespread impact on the power quality. An overview of the algorithm is given here. Additional details are available from EPRI to eligible parties. Many of the algorithms can also be implemented in the open-source EPRI Open DSS Program available on the Internet.

8.7.1 General Approach

State estimation techniques have been used for many years on transmission systems to determine steady-state voltage and load quantities. The most common mathematical methods are based on the weighted least squares (WLS) approach to minimizing the square of the error between the measurements and the computed estimates of the load quantities. By iterating through some kind of optimization solution procedure, an estimate of the state of the circuit variables can be obtained.

Baran and Kelley[17] describe many of the issues related to performing state estimation on distribution feeders. As with nearly all papers on state estimation, their primary concern is the load-related, steady-state voltages and currents throughout the feeders. Many of the same principles apply to estimating the various PQ variation phenomena. Of course, there are some additional challenges in crafting specific algorithms.

The WLS method is typically summarized by first writing the equation relating the measurements and the state variables in the system:

$$z = h(x) + v$$

where z = measurements
x = state variables
$h(x)$ = function relating state variables to measurements
v = measurement errors

Then the WLS method boils down to minimizing the WLS errors over all the measurements:

$$\min_{x} \sum_{i} w_i [z_i - h_i(x)]^2$$

That is, this method finds vector x of state variables that minimize the WLS error between the measurements and what they are computed to be. The weights w_i are assigned to denote the relative accuracy of the measurements. The higher the accuracy of the measurement, the greater the weight.

Many of the papers dealing with the state estimation for load quantities have elaborate mathematical representations of the system based on common formulations of the power flow problem and the numerical methods for solving the WLS minimization. It is not easy, and perhaps impossible, to accomplish a closed-form solution for all kinds of PQ phenomena. Iterative simulation methods are employed instead. Basically, a number of disturbance events that could have caused what was observed are simulated and the result is compared with measurements. Estimated values are stored from the simulation matching the measurements most closely.

In many cases on the distribution feeder, we are constrained by the lack of measurements in critical areas. We are forced to make some intelligent guesses founded on simple rules and to take the best of these results. Fortunately, a high degree of accuracy is generally not required to achieve a reasonable estimate of state variables suitable for computing power quality indices. For example, the voltage sag indices have relatively coarse intervals (90, 70, 50, and 10 percent). A voltage sag of 55 percent falls in the same bin as one of 65 percent, so the estimate can have a significant error while still yielding meaningful indices.

The circuit model for PQ state estimation divides the system into a number of segments across which the power quality is assumed to be constant. The segmentation is determined by the locations of the available monitors and switches. Figure 8.11 shows the basic concept. This system is simulated for several candidate events that might have caused the disturbance. The simulations yielding the best fit to the measurements are assumed to provide good estimates of the voltages at the locations without monitors.

8.7.2   Number of Monitors

The EPRI PQSE was designed to function acceptably for only one monitor at the substation. Of course, it would function better with more monitoring data. While some feeders have two or three monitors, most have only one in the substation. This forces one to assume that all customers experience the same voltage disturbance.

Despite the lack of confirming data along the feeder, useful results can be produced with only one monitor. For example, there is generally more concern with how the secondary-side voltages appear at important three-phase customers than with voltages appearing along the feeder. Often these customers are served with transformer connections that distort the primary voltages during disturbances. This problem can be addressed simply by representing a portion of the feeder and an appropriately connected transformer.

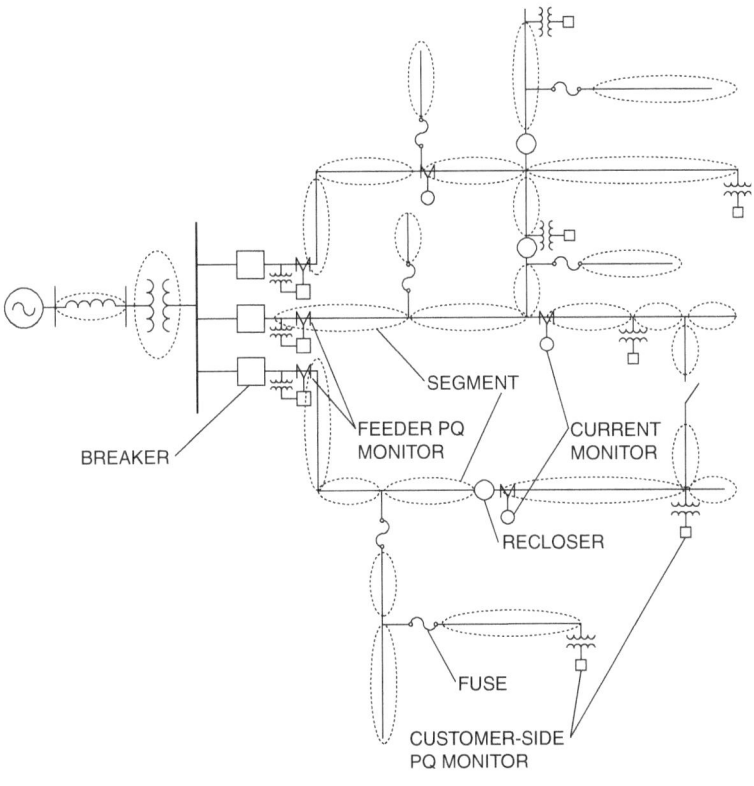

FIGURE 8.11   Circuit with PQ monitors at various points.

This will yield results similar to having an actual monitor on the secondary side, particularly if the customer is near the substation.

The accuracy in predicting the voltages along the feeder improves dramatically with the addition of just one more monitor on the feeder. This monitor should be placed approximately one-half to two-thirds of the way down the feeder so that there is considerable separation from the substation monitor. In staged tests with the rms variation estimation algorithm, the estimator never failed to find the correct fault location (to the accuracy of the feeder model), whereas with only one monitor, the estimator produced somewhat random predictions of where the fault was located. Of course, in practice the models will be imprecise and the measurements will also contain errors, so we cannot expect that the fault position will always be predicted with such accuracy. However, it should be expected that the estimates will improve considerably with just one more monitor—even a customer-side monitor. Using three well-spaced monitors as in the EPRI DPQ project gives good coverage of the main feeder.

Ideally, power quality state estimation would work best with fully capable PQ monitors near the substation and on all the major branches of a three-phase feeder. The limitations are three levels of monitoring listed in the table:

Monitor configuration level	Capabilities
1. Substation only	Adequate for cases in which it can be assumed all customers on the feeder see the same voltage.
2. Substation + customer-side monitors	Accuracy of prediction of voltages along the feeder is considerably enhanced if customer sites are significantly downline from the substation. However, it is still difficult to predict fault locations accurately since the fault current path is not known.
3. Substation + PQ monitors on main three-phase feeder branches	Should yield the most accurate results.
	Improves on capabilities gained by adding customer-side monitors by providing information on the feeder current flows.

8.7.3   Estimating RMS Variations

Voltage sags and interruptions are almost always due to short-circuit faults. Therefore, the procedure for estimating voltages due to rms variations is basically to find a fault on the feeder that produces voltages and currents that most closely match what was measured on the few existing monitors. Then the voltages are computed at all other buses of interest and assumed to what would have been measured if a monitor had been present. Of particular interest are the voltages on the load side of the service transformers.

One can always employ a brute force approach for searching out a fault location on the feeder. There may be no alternative if there is only one monitor on the system, and today's fast computers actually make this a practical approach in many cases. If there is more than one monitor, one can program some intelligence into the search algorithm to limit the search area. This is illustrated in Fig. 8.12. By observing the currents measured by the monitors, it is determined that the fault is downline from the first two monitors on the feeder. Therefore, the search area is restricted to the shaded area.

Some new types of PQ monitoring equipment can provide an estimate of the distance to the fault. This can be quite helpful in providing a starting point for the simulation.

Customer-owned monitors can be quite helpful in identifying the probable fault location. The load current data are probably not

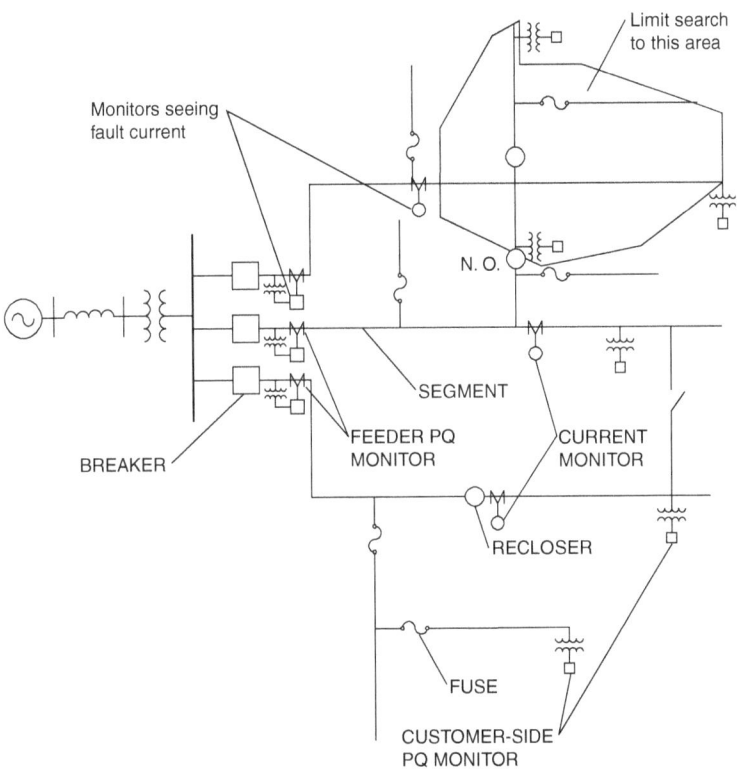

FIGURE 8.12 Intelligently limiting the search area for possible fault locations.

useful, but the measured voltages can provide intelligence to improve the estimation.

While the customer-side monitors can be very helpful, there are a number of potential problems with respect to rms variations:

1. The transformer connection, if not grounded wye-wye, may alter the perception of the voltages seen on the primary feeder and, therefore, by other customers. Thus, the monitor values do not represent the primary system voltage. An effective rms variation estimator would accommodate various transformer connections.

2. The phasing may not agree with the utility-side monitoring. This is a constant bookkeeping problem. End users frequently alter their installations, and any type of automatic estimating system for computing PQ indices must have a facility for periodically correcting the phasing.

3. The transformer impedance will alter the voltage measured for certain events.

4. The current measured at the end-user site represents only the current into that individual load. This will be of limited use in determining the state of the feeder for PQ indices.

5. There is no direct communications link to get the data back to a central site for timely processing.

8.7.4   Simulation Engine Requirements

The simulation engine for performing this type of PQ state estimation has requirements similar to that of the fault simulator used to estimate voltage sags for PQ planning (see Sec. 8.8). One of the most critical pieces of information is what the voltage looks like on the end-user side of the service transformer. In the United States, this transformer can have one of several different connections. Therefore, the ability to model a three-phase feeder in complete detail during fault conditions is imperative. Some representation of load is also called for. However, traditional power flow models will be inadequate once the voltage drops below 90 percent or so.

Another feature is the ability to scan through a range of possible fault locations and magnitudes quickly. There may be hundreds of fault events per year to be evaluated. Although modern computers are becoming very fast, combining this need with the detailed circuit modeling capability creates an application that is quite taxing computationally. Nevertheless, useful calculations can be performed within a reasonable time on modern personal computers.

One simplication that eases the computation and modeling burden is to model only a few service transformers explicitly. Sometimes it is sufficient to model only one example of each type of transformer connection found on the system. This will give a reasonably good accounting of the phase-shifting effect of the connection without having to model each individual transformer.

Likewise, classes of loads may be lumped. Loads with large amounts of rotating machines will often have a tendency to counter the effects of unbalanced voltage sags, while resistive loads are relatively passive. Modeling at least one of each class will often provide a good picture of the voltages experienced by end users. Of course, the solution engine must be capable of providing suitable models for these loads.

8.8   Including Power Quality in Distribution Planning

The traditional approach to distribution system planning calls for the most economical system upgrades, timed to meet projected increases in peak load. The main driver is reliability. Unless the utility has entered into PQ contracts with severe penalties, power quality is not explicitly included in planning decisions.

However, PQ considerations may justify modifying the investment plan to provide better quality of service as well as sufficient capacity and reliability.[18-20] If the costs for lost production, cleanup, and equipment damage were included to some degree, it is more likely that the investment plan chosen would benefit both the utility and the customers. The indices described previously provide a means of quantifying power quality and should prove useful for evaluating planning options where the service quality is of high value. The challenge is to compute them from predictive models rather than historical data.

8.8.1 Planning Process

Figure 8.13 is one way to compare the general utility investment planning criteria with both traditional and competitive criteria. The traditional criteria assume the customer's cost of unserved load is very high, so the best decision is what can serve the expected peak load at minimum cost to the utility. This is often called *least-cost planning*. The risk that the peak load will differ from the forecast has been assumed by the utility ratepayers, so planning errors tend to be on the conservative side. The utility invests in more capacity than actually needed at a given time to keep ahead of the load growth.

In a competitive business environment, the utility must assume more of the risk of uneconomical investments, while both customers and the utility share the risks of unserved load. This leads to consideration of customer damage costs in the planning decision shown by the dashed curve in Fig. 8.13. As the capacity of the system increases, there is likely to be less customer cost due to outages. The

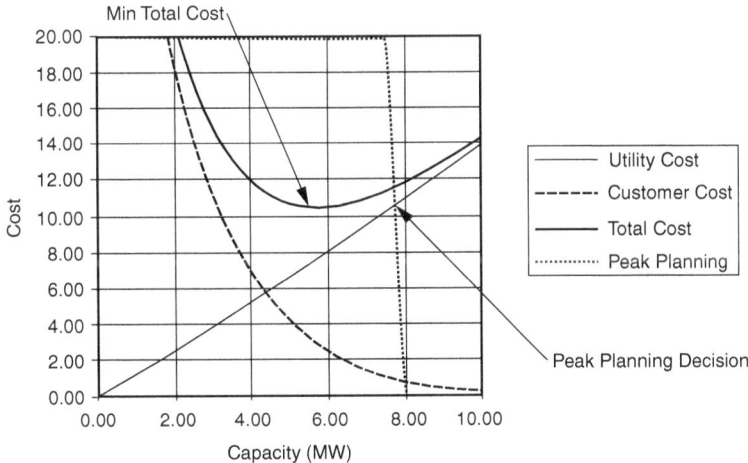

FIGURE 8.13   Costs impacting the planning decision.

optimal planning decision would theoretically minimize the U-shaped total-cost curve (exaggerated for clarity) in Fig. 8.13. If customer costs were higher, the optimal system capacity point would be shifted to the right. While this new decision may be more economical overall, the utility may find it difficult to accurately estimate the customer damage costs to justify building less capacity. This might be hard to justify even if customer costs are known to be lower. In this section, we are mainly interested in looking at more costly investments that might be justified based on costs to end users with sensitive loads.

The planning decision in Fig. 8.13 is mainly driven by system capacity, which indirectly relates to reliability. Power quality impacts are an important, but secondary, factor. To incorporate capacity and PQ costs in the same planning process, only the investments that improve power quality over the whole system, affecting more than one customer, are considered. If there are localized PQ problems, or just one high-value customer, it becomes more appropriate to consider point solutions or even dedicated feeders.

Figure 8.14 shows a planning framework that can support planning with PQ and other additional considerations. The base case

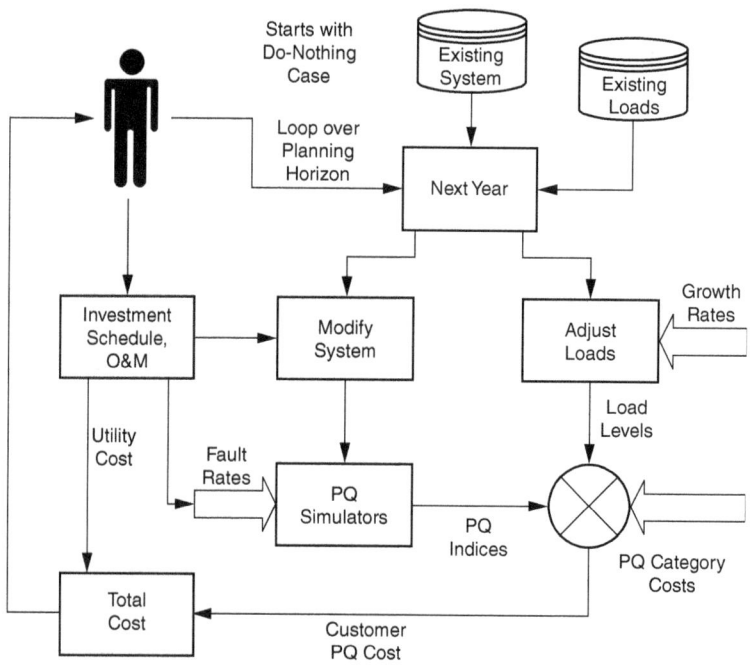

Figure 8.14   Power quality planning process—thick arrows show inputs with significant uncertainty.

is the existing system with no planned investments; the so-called do-nothing case. System performance is then simulated over the planning horizon, typically 5 to 10 years to reflect the riskier business environment for utilities. Utility incremental costs might be set at zero in the base case, or they might include vegetation control and fault restoration costs if those will be under study later on. In each year, the system PQ performance is calculated and the customer damage cost is accumulated over the study system. The damage cost is a product of the system performance indices at each load bus, and the customer costs corresponding to each index. To simplify the study, these costs might only include high-value customer loads for which reasonable cost estimates can be determined. At the end of the planning horizon simulation, the net present value of the total utility-plus-customer cost is evaluated. This becomes the cost of the base case, or do-nothing case. Any attractive planning option must have a net present value of total cost less than the base case.

To consider design alternatives for improved power quality, the planner schedules new investments, changes protective device settings, changes the vegetation control practice, etc. Some changes affect the fault rates, some affect restoration times, some alter the base system, and some may result in scheduled system upgrade investments. The system is simulated again over the planning horizon with these changes, and a new present worth of total cost is calculated. Note that the customer load growth and cost data should not be changed during this simulation, but they may be altered while assessing the risk of different planning scenarios.

This process may also be used to evaluate distribution generation, demand-side management (DSM), and distribution automation (DA) by incorporating the appropriate simulators.

8.8.2 Risk versus Expected Value

The basic planning process in Fig. 8.14 is tailored to minimize the expected value of total cost. This is the most common practice today, but it should be adjusted to better consider risk.[21]

The planning decision is based on uncertain forecasts, to which probabilities can be assigned. After the passage of time, the actual planning scenario will become known, and the optimal solution will also be known. It is not likely that the optimal solution will be chosen, nor is it likely that errors in the planning decision can be completely corrected. Therefore, there will be a deviation between the actual cost and the optimal cost as inductively determined. The typical cost minimization is equivalent to minimizing the linear norm of actual deviations from the optimal solution. An extreme risk-based approach would minimize the infinity norm of actual deviations from the optimal solution; in other words, it would minimize the maximum regret that might occur. An intermediate approach would use the popular euclidean norm to minimize the square of deviations from the optimal. Both the expected value and the risk-based approaches require probability estimates for the various planning forecasts, but the quantity minimized is different.

Expected value minimization works well when repeated trials can be performed to mask undesirable results. For large investments in a competitive business environment, however, a risk-based minimization is probably more appropriate.

8.8.3   System Simulation Tools

The data requirements and analytical complexity for simulation software in this planning method range from simple but uncertain, for long-term capacity planning, to more detailed and complex, for design and operational studies. Power quality data and analysis requirements tend toward more detail, even for planning studies. For an example, readers are referred to the modeling capabilities of the EPRI Open DSS Program.

Capacity Planning

A positive-sequence load flow program, with balanced load models, can adequately support traditional capacity planning studies. The result of these studies would be a schedule of investments in new capacity to serve expected load growth. The major investments traditionally include substations, transformers, and feeders. The schedule might call for substantial upgrades to existing facilities or for brand new construction. Distributed generation has recently been considered as an alternative to investments in substations, transformers, and feeders. This may require a new generator model, but in other respects the existing software would already be adequate.

Fault Analysis

To include rms variation impacts in the planning process, it will be necessary to construct a multi-phase system model. The rms voltage variation magnitudes computed from simulations will be more accurate when full multiphase models are used. For steady-state voltage unbalance, harmonic distortion, and transient overvoltages, the accurate phase impedances are even more important. For all types of variations, the loads must be modeled per phase.

The customer service transformers must also be modeled, with explicit winding connections, in order to calculate rms variations. This may be a new data requirement for the utility. While European systems tend to have the same MV and LV connection throughout, others, particularly North American systems, use many different connections. The customer transformer data may already be available for high-value customers, for use in utility-customer device coordination studies. Smaller customers can be aggregated by type of service transformer and phase connection.

Harmonics and Transients

Harmonics and transients represent an additional level of complexity in PQ simulation. One area where there is an interesting planning issue is in the relationship between capacitors and harmonics. Utilities routinely add capacitors to increase system capacity. Others require end users to correct the power factor of their loads to reduce demand on the system. Either can result in serious harmonic resonance problems. Of course, simulating these can be time-consuming. It is not clear when it will be possible to have sufficient tools and data to routinely include these phenomena in the planning cycle explicitly. Although there are tools for scanning proposed plans for harmonic problems, most utilities will choose to address these issue after they arise.

8.8.4 Fault Incidence Rates

One of the key inputs to rms voltage variation simulation is the incidence rate for different kinds of faults. The simplest approach is to use an average fault rate per unit length of line, with an average percentage of permanent faults and a nominal fault resistance based on system design characteristics. In the United States, 0.06 fault/km (0.1 fault/mi) per year is commonly assumed, with about 20 percent of the faults being permanent. The fault resistance is typically from 1 to 5 Ω, although some utilities may use as high as 20 Ω.

For planning studies, it is important to segregate faults caused by lightning from faults caused by vegetation. This can be done by outage data analysis, by lightning performance simulation, or by using the results of work in progress on vegetation-induced faults. With separate fault categories, it becomes possible to simulate the effect of investing in improved lightning protection such as line arresters, more vegetation control, and better fault-locating equipment on the system's PQ performance. Without separate fault categories, some of the important PQ improvement methods cannot be evaluated for planning purposes.

The model should also include a separate category for faults on the subtransmission or transmission system (HV system), because these faults are not mitigated by some of the countermeasures that work for faults on distribution feeders.

8.8.5 Overcurrent Device Response

A key component of rms voltage variation simulation is the response of the overcurrent protective device system. Each feeder segment is assumed to experience faults at some annual rate. The faults can be subdivided according to phases involved, percentage of temporary faults, fault resistance, etc. This can result in different sequences of events and patterns of rms variations.

The behavior of utility fault-clearing devices can be quite complicated. Figure 8.15 shows one algorithm for performing the simulation for a specific fault location and type. It uses a priority queue to establish the overcurrent device operating time. A detailed, multiphase short-circuit algorithm calculates the load voltages and protective device currents while the fault is applied. Each protective device in the system uses the calculated currents and voltages to predict the next state change for that device and pushes the time of that state change onto the priority queue of device events. After all

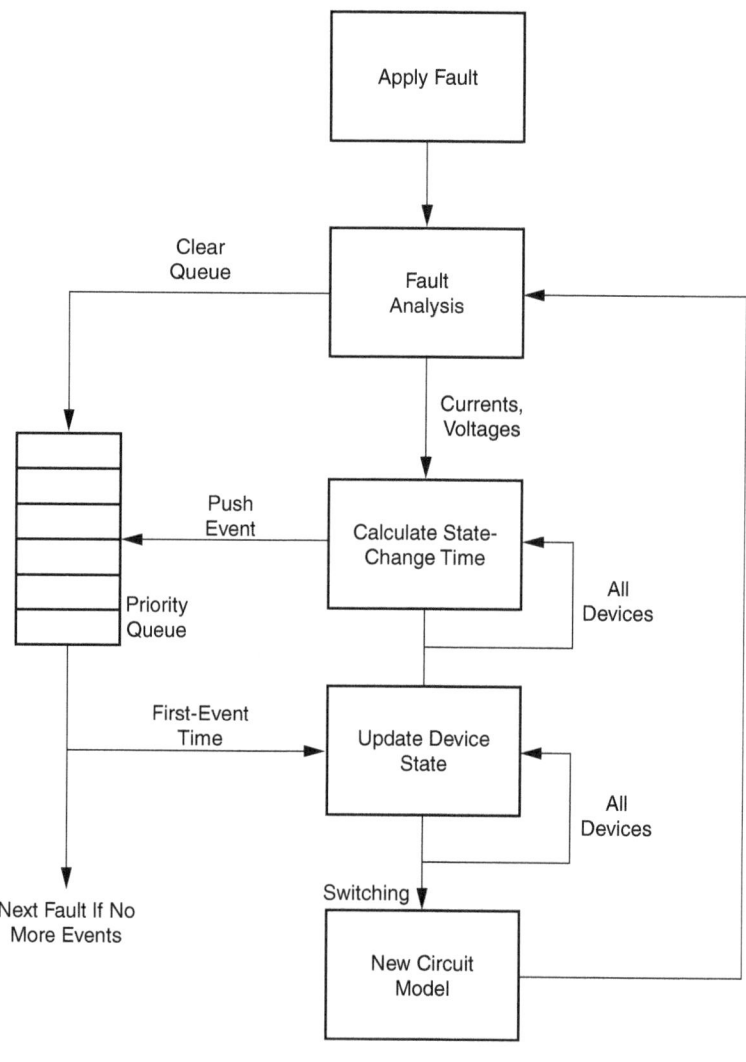

FIGURE 8.15   Event-based fault simulation using a priority queue.

protective devices have been analyzed, the event with highest priority (lowest time) is executed. That device changes state between open and closed, changing the circuit model. All the other devices use the actual time of the state change to update their internal states using the calculated currents and voltages. The priority queue of device events is cleared, and the fault currents and voltages are recalculated for the new circuit topology. The simulation of the fault ends when there are no events pushed onto the priority queue by any of the devices. The fault itself is a "device" and may push a clearing time for temporary faults, whenever the fault has been deenergized by a device opening.

The calculated load voltages and event durations are used to determine interruptions and rms variations according to the definition described in this chapter. When the fault simulation ends, all the load costs are calculated and the next fault type and location are considered.

The following briefly describes how the protective devices are handled. All devices are assumed to operate correctly, except for relay overtravel and sympathetic tripping, since this can circumvent measures taken specifically to improve power quality.

Circuit Breaker with Relay

A circuit breaker will schedule an opening event if its currents, adjusted by the associated current transformer ratio, exceed the associated relay pickup setting. If the relay has an instantaneous setting and the current exceeds that level, the event time will be the relay instantaneous pickup time plus the breaker clearing time. Otherwise, the event time will depend on the relay's time-current characteristic. If the relay is of the definite-time type, this will be a constant relay setting plus breaker clearing time. If the relay is of the inverse type, this will be a current-dependent time plus the breaker clearing time. We use approximate time-current curves for both relays and reclosers.

If the fault current is removed before the breaker opens, an internal relay travel state variable is updated. This may produce a sympathetic trip due to relay inertia. If no sympathetic trip is predicted, an event for full reset is then pushed onto the priority queue.

The circuit breaker may have one or two reclosure settings. If the breaker has opened, it will schedule a closing operation at the appropriate time. In case there are subsequent events from other devices, the breaker model must manage an internal state variable of time accumulated toward the reclose operation. The time between opening and reclosing is a constant. Once the breaker recloses, it follows the defined fault-clearing behavior. There may be two reclosings, at different time settings, before the breaker locks out and pushes no more events.

Fault

A permanent fault will not schedule any events for the priority queue, but will have an associated repair time. Any customers without power at the end of the fault simulation will experience a sustained interruption, of duration equal to the repair time.

A temporary fault will schedule a clearing event whenever its voltage is zero. Whenever the fault is reenergized before clearing, any accumulated clearing time is reset to zero. Upon clearing, the fault switch state changes from closed to open, and then the fault simulation must continue to account for subsequent device reclosures.

Fuse

A fuse will open when the fault current and time applied penetrate the minimum melting curve, or when the I^2t product reaches the minimum melting I^2t. We use *minimum melt* rather than *total clearing time* in order to be conservative in studies of fuse saving; this would not be appropriate for device coordination studies. Expulsion fuses are modeled with a spline fit to the manufacturer's time-current curve, while current-limiting fuses are modeled with I^2t. In both cases, if the fault is interrupted before the fuse melts, an internal preheating state variable is updated in case the fault is reapplied. However, we do not specifically track possible fuse damage during the simulation.

If the fuse currents will penetrate the time-current curve or minimum melting I^2t, then a fuse melting time is pushed onto the priority queue. If the fuse currents are too low to melt the fuse, no event is pushed. Once the fuse opens, downstream customers will experience a sustained interruption equal to the fuse repair time.

Recloser

The recloser model is very similar to the circuit breaker with relay model previously discussed. The main differences are that the recloser can have up to four trips during the fault sequence, and two different time-current curves can be used.

Sectionalizer

A sectionalizer will count the number of times the current drops to zero and will open after this count reaches a number that can vary from 1 to 3. The device will not open under either load or fault current.

8.8.6   Customer Damage Costs

Customer damage costs are determined by survey, PQ contract amounts, or actual spending on mitigation. In terms of kilowatthours unserved, estimates range from $2/kWh to more than $50/kWh. A typical cost for an average feeder with some industrial and commercial load is $4 to $6/kWh. For approximating purposes, weighting factors can be used to extend these costs to momentary interruptions and rms variations assuming that the event has caused an equivalent amount of unserved energy. Alternatively, one can use a model similar to the example in Sec. 8.5, which basically is based on event count. Average costs per event for a wide range of customer classes are typically stated in the range of $3000 to $10,000.

With such high cost values, customer damage costs will drive the planning decisions. However, these costs are very uncertain. Surveys have been relatively consistent, but the costs are seldom "verified" with customer payments to improve reliability or power quality. For example, aggregating the effect on a large number of residential customers may indicate a significant damage cost, but there is no evidence that residential customers will pay any additional amount for improved power quality, in spite of the surveys. There may be a loss of goodwill, but this is a soft cost. Planning should focus on high-value customers for which the damage costs are more verifiable.

Costs for other types of PQ disturbances are less defined. For example, the economic effect of long-term steady-state voltage unbalance on motors is not well known, although it likely causes premature failures. Likewise, the costs are not well established for harmonic distortion and transients that do not cause load tripping.

The costs may be specified per number of customers (residential, small commercial), by energy served, or by peak demand. If the cost is specified by peak demand, it should be weighted using a load duration curve. For steady-state voltage, harmonic distortion, and transients, the load variation should be included in the electrical simulations, but this is not necessary for sustained interruptions and rms variations.

Several examples and algorithm descriptions are provided in the EPRI *Power Quality for Distribution Planning* report[19] showing how the planning method can be used for making decisions about various investments for improving the power quality. We've addressed only the tip of the iceberg here but hopefully have provided some inspiration for readers.

8.9 References

1. EPRI TR-106294-V2, *An Assessment of Distribution System Power Quality.* Vol. 2: *Statistical Summary Report,* Electric Power Research Institute, Palo Alto, Calif., May 1996.
2. M. McGranaghan, A. Mansoor, A. Sundaram, R. Gilleskie, "Economic Evaluation Procedure for Assessing Power Quality Improvement Alternatives," *Proceedings of PQA North America,* Columbus, Ohio, 1997.
3. Daniel Brooks and Bill Howe, *Establishing PQ Benchmarks,* E Source, Boulder, Colo., May 2000.
4. EPRI TR-107938, *EPRI Reliability Benchmarking Methodology,* EPRI, Palo Alto, Calif., 1997.
5. IEEE Standard 1366–1998, *IEEE Guide for Electric Power Distribution Reliability Indices.*

6. D. D. Sabin, T. E. Grebe, M. F. McGranaghan, A. Sundaram, "Statistical Analysis of Voltage Dips and Interruptions—Final Results from the EPRI Distribution System Power Quality Monitoring Survey," *Proceedings 15th International Conference on Electricity Distribution (CIRED '99),* Nice, France, June 1999.

7. IEEE Standard 1159–1995, *IEEE Recommended Practice on Monitoring Electric Power.*

8. Dan Sabin, "Indices Used to Assess RMS Variations," presentation at the Summer Power Meeting of IEEE PES and IAS Task Force on Standard P1546, Voltage Sag Indices, Edmonton, Alberta, Canada, 1999.

9. D. L. Brooks, R. C. Dugan, M. Waclawiak, A. Sundaram, "Indices for Assessing Utility Distribution System RMS Variation Performance," *IEEE Transactions on Power Delivery,* PE-920-PWRD-1-04-1997.

10. IEEE Standard 519–1992, *IEEE Recommended Practices and Requirements for Harmonic Control in Electrical Power Systems.*

11. A. E. Emanuel, J. Janczak, D. J. Pileggi, E. M. Gulachenski, "Distribution Feeders with Nonlinear Loads in the NE USA: Part I. Voltage Distortion Forecast," *IEEE Transactions on Power Delivery,* Vol. 10, No. 1, January 1995, pp. 340–347.

12. Barry W. Kennedy, *Power Quality Primer,* McGraw-Hill, New York, 2000.

13. M. F. McGranaghan, B. W. Kennedy, et. al., *Power Quality Standards and Specifications Workbook,* Bonneville Power Administration, Portland, Oreg., 1994.

14. Andy Detloff and Daniel Sabin, "Power Quality Performance Component of the Special Manufacturing Contracts between Power Provider and Customer," *Proceedings of the ICHPQ Conference,* Orlando, Fla., 2000.

15. Shmuel S. Oren and Joseph A. Doucet, "Interruption Insurance for Generation and Distribution of Power Generation," *Journal of Regulatory Economics,* Vol. 2, 1990, pp. 5–19.

16. Joseph A. Doucet and Shmuel S. Oren, "Onsite Backup Generation and Interruption Insurance for Electricity Distribution," *The Energy Journal,* Vol. 12, No. 4, 1991, pp. 79–93.

17. Mesut E. Baran and Arthur W. Kelley, "State Estimation for Real-Time Monitoring of Distribution Systems," *IEEE Transactions on Power Systems,* Vol. 9, No. 3, August 1994, pp. 1601–1609.

18. T. E. McDermott, R. C. Dugan, G. J. Ball, "A Methodology for Including Power Quality Concerns in Distribution Planning," *EPQU '99,* Krakow, Poland, 1999.

19. EPRI TR-110346, *Power Quality for Distribution Planning,* EPRI, Palo Alto, CA, April 1998.

20. M. T. Bishop, C. A. McCarthy, V. G. Rose, E. K. Stanek, "Considering Momentary and Sustained Reliability Indices in the Design of Distribution Feeder Overcurrent Protection," *Proceedings of 1999 IEEE T&D Conference,* New Orleans, La., 1999, pp. 206–211.

21. V. Miranda and L. M. Proenca, "Probabilistic Choice vs. Risk Analysis— Conflicts and Synthesis in Power System Planning," *IEEE Transactions on Power Systems,* Vol. 13, No. 3, August 1998, pp. 1038–1043.

8.10 Bibliography

Sabin, D. D., Brooks, D. L., Sundaram, A., "Indices for Assessing Harmonic Distortion from Power Quality Measurements: Definitions and Benchmark Data." *IEEE Transactions on Power Delivery,* Vol. 14, No. 2, April 1999, pp. 489–496.

EPRI Reliability Benchmarking Application Guide for Utility/Customer PQ Indices, EPRI, Palo Alto, Calif., 1999.

CHAPTER 9

Distributed Generation and Power Quality

Many involved in power quality have also become involved in distributed generation (DG) because there is considerable overlap in the two technologies. Since the second edition there has been a significant increase in interest in the impacts of renewable DG, which has a variable output that can affect voltages on distribution feeders. Therefore, it is very appropriate to include a chapter on this topic.

As the name implies, DG uses smaller-sized generators than does the typical central station plant. They are distributed throughout the power system closer to the loads. The term *smaller-sized* can apply to a wide range of generator sizes. Because this book is primarily concerned with power quality of the primary and secondary distribution system, the discussion of DG will be confined to generator sizes less than 10 MW. Generators larger than this are typically interconnected at transmission voltages where the system is designed to accommodate many generators.

The typical distribution system delivers electric energy through wires from a single source of power to a multitude of loads. Thus, several power quality issues arise when there are multiple sources. Will DG improve the power quality or will it degrade the service end users have come to expect? There are arguments supporting each side of this question, and several of the issues that arise are examined here.

9.1 Resurgence of DG

For more than seven decades, the norm for the electric power industry in developed nations has been to generate power in large, centralized generating stations and to distribute the power to end users through transformers, transmission lines, and distribution lines. This is often collectively referred to as the "wires" system in DG literature. In essence, this book describes what can go wrong with delivery of power by wires.

The original electrical power systems, consisting of relatively small generators configured in isolated islands, used DG. That model gave way to the present centralized system largely because of economies of scale. Also, there was the desire to sequester electricity generation facilities away from population centers for environmental reasons and to locate them closer to the source of fuel and water.

The passage of the Public Utilities Regulatory Act of 1978 (PURPA) in the United States in 1978 was intended to foster energy independence. Tax credits were given, and power was purchased at avoided-cost rates to spur development of renewable and energy-efficient, low-emission technologies. This led to a spurt in the development of wind, solar, and geothermal generation as well as gas-fired cogeneration (combined heat and power) facilities. In the mid-1990s, interest in DG once again peaked with the development of improved DG technologies and the deregulation of the power industry allowing more power producers to participate in the market. Also, the appearance of critical high-technology loads requiring much greater reliability than can be achieved by wire delivery alone has created a demand for local generation and storage to fill the gap.

Some futurists see a return to a high-tech version of the original power system model. New technologies would allow the generation to be as widely dispersed as the load and interconnected power grids could be small (i.e., microgrids). The generation would be powered by renewable resources or clean-burning, high-efficiency technologies. Energy distribution might be shifted from wires to pipes containing some type of fuel, which many think will ultimately be hydrogen. As this is being written there is an acceleration of the installation of renewable generation based on solar photovoltaics (PVs) and wind turbines. How the industry moves from its present state to this future, if it can at all, is open to question. Recent efforts to deregulate electric power have been aimed not only at achieving better prices for power but at enabling new technologies. However, it is by no means certain that the power industry will evolve into DG sources. Despite the difficulties in wire-based delivery described in this book, wires are very robust compared to generation technologies. Once installed, they remain silently in service for decades with remarkably little maintenance.

9.1.1 Perspectives on DG Benefits

One key to understanding the DG issue is to recognize that there are multiple perspectives on every relevant issue. To illustrate, we discuss the benefits of DG from three different perspectives.

1. *End-user perspective.* This is where most of the value for DG is found today. End users who place a high value on electric power can

generally benefit greatly by having backup generation to provide improved reliability. Others will find substantial benefit in high-efficiency applications, such as combined heat and power, where the total energy bill is reduced. End users may also be able to receive compensation for making their generation capacity available to the power system in areas where there are potential power shortages. Homeowners in regions with net metering laws or other government subsidies may find it beneficial to install rooftop solar PV generation to offset consumption.

2. *Distribution utility perspective.* The distribution utility is interested in selling power to end users through its existing network of lines and substations. DG can be used for transmission and distribution (T&D) capacity relief. In most cases, this application has a limited life until the load grows sufficiently to justify building new T&D facilities. Thus, DG serves as a hedge against uncertain load growth. It also can serve as a hedge against high price spikes on the power market (if permitted by regulatory agencies).

3. *Commercial power producer perspective.* Those looking at DG from this perspective are mainly interested in selling power or ancillary services into the area power market. In the sense that DG is discussed here, most units are too small to bid individually in the power markets. Commercial aggregators will bid the capacities of several units. The DG may be directly interconnected into the grid or simply serve the load off-grid. The latter avoids many of the problems associated with interconnection but does not allow the full capacity of the DG to be utilized.

Disadvantages of DG

There are also different perspectives on the disadvantages of DG. Utilities are concerned with power quality issues, and a great deal of the remainder of this chapter is devoted to that concern. End users should be mainly concerned about costs and maintenance. Do end users really want to operate generators? Will electricity actually cost less and be more reliable? Will power markets and government incentives continue to be favorable toward DG? There are many unanswered questions. However, it seems likely that the amount of DG interconnected with the utility system will continue to increase for the foreseeable future.

9.1.2   Perspectives on Interconnection

There are also opposing perspectives on the issue of interconnecting DG to the utility system. This is the source of much controversy in efforts to establish industry standards for interconnection. Figures 9.1 and 9.2 illustrate the views of the two key opposing positions.

Figure 9.1 depicts the viewpoint of end users and DG owners who want to interconnect to extract one or more of the benefits

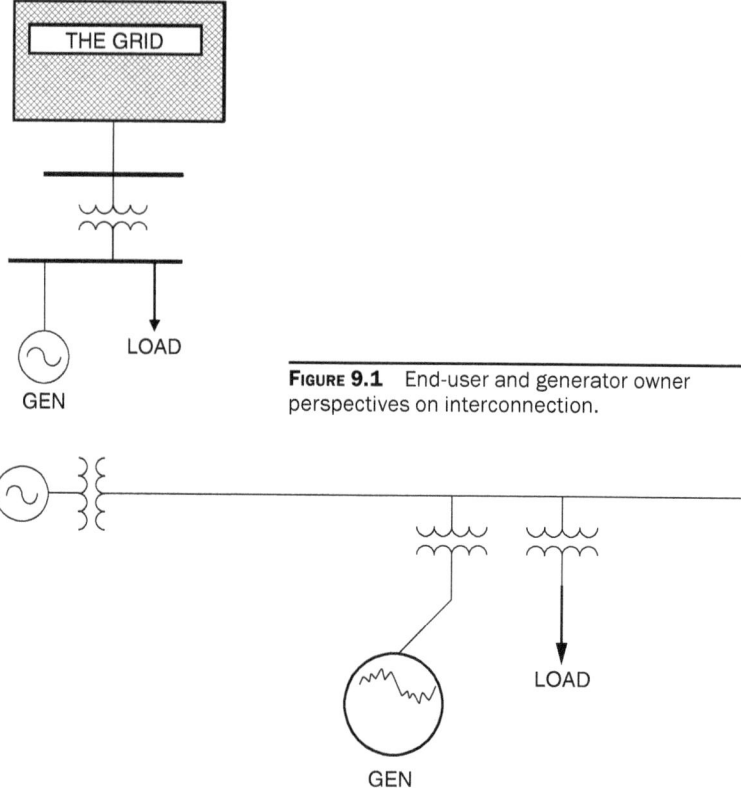

FIGURE 9.1   End-user and generator owner perspectives on interconnection.

FIGURE 9.2   Distribution planner perspective on interconnection.

previously mentioned. Drawings like this can be found in many different publications promoting the use of DG. The implied message related to power quality is that the DG is small compared to the grid. This group often has the view that the grid is a massive entity too large to be affected by their relatively small generator. For this reason, many have a difficult time understanding why utilities balk at interconnecting and view the utility requirements simply as obstructionist and designed to avoid competition.

Another aspect of the end-user viewpoint that is not captured in this drawing is that despite the large mass of the grid, it is viewed as unreliable and providing "dirty" power. DG proponent literature often portrays DG as improving the reliability of the system (including the grid) and providing better-quality power.

The perspective on interconnected DG of typical utility distribution engineers, most of whom are very conservative in their approach to planning and operations, is captured in Fig. 9.2. The size of customer-owned DG is magnified to appear much larger than its actual size, and it produces dirty power. It is also a little off-center in its design, suggesting that it is not built and maintained as well as utility equipment.

There are elements of truth to each of these positions. The intent in this book is not to take sides in this debate but to present the issues as fairly as possible while pointing out how to solve problems related to power quality.

9.2 DG Technologies

The emphasis of this chapter is on the power aspects of DG, and only a cursory description of the relevant issue with the technologies will be given. Readers are referred to Refs. 1 and 2 for more details. Also, the Internet contains a multitude of resources on DG. A word of caution: As with all things on the Internet, it is good to maintain a healthy skepticism of any material found there. Proponents and marketers for particular technologies have a way of making things seem very attractive while neglecting to inform the reader of major pitfalls.

9.2.1 Reciprocating Engine Genset

A commonly applied DG technology is the reciprocating engine-generator set. A typical unit is shown in Fig. 9.3. This technology is generally the least expensive DG technology, often by a factor of 2. Reciprocating gas or diesel engines are mature technologies and are readily available.

Utilities currently favor mobile gensets mounted on trailers so that they can be moved to sites where they are needed. A common application is to provide support for the transmission and distribution system in emergencies. The units are placed in substations and interconnected to the grid through transformers that typically step up the voltage from the 480 V produced by the generators. Manufacturers of these units have geared up production in recent years to meet demands to relieve severe grid constraints that have occurred in some areas. One side effect of this is that the cost of the units has dropped, widening the cost gap between this technology and the next least costly option, which is generally some sort of combustion turbine.

Diesel gensets are quite popular with end users for backup power. One of the disadvantages of this technology is high NOx and SOx emissions. This severely limits the number of hours the units, particularly diesels, may operate each year to perhaps as few as 150. Thus, the main applications will be for peaking generation and emergency backup.

Natural gas-fired engines produce fewer emissions and can generally be operated several thousand hours each year. Thus, they are popular in combined heat and power cogeneration applications

FIGURE 9.3   Diesel reciprocating engine genset. (*Courtesy of Cummins Inc.*)

in schools, government, and commercial buildings where they operate at least for the business day.

The unit shown in Fig. 9.3 has a synchronous alternator, which would be the most common configuration for standby and utility grid support applications. However, it is also common to find reciprocating engines with induction generators. This is particularly true for cogeneration applications of less than 300 kW because it is often simpler to meet interconnection requirements with induction machines that are not likely to support islands.

Reciprocating engine gensets have consistent performance characteristics over a wide range of environmental conditions with efficiencies in the range of 35 to 40 percent. They are less sensitive to ambient conditions than combustion turbines whose power efficiency declines considerably as the outside air temperature rises. However, the waste heat from a combustion turbine is at a much higher temperature than that from a reciprocating engine. Thus, turbines are generally the choice for combined heat and power applications that require process steam.

9.2.2   Combustion (Gas) Turbines

Combustion turbines commonly used in cogeneration applications interconnected to the distribution system generally range in size from 1 to 10 MW. The turbines commonly turn at speeds of 8000 to 12,000 rpm and are geared down to the speed required by the synchronous alternator (typically 1800 or 3600 rpm for 60-Hz systems). Units of 10 MW or larger in size, in either simple- or combined-cycle configurations, are commonly found connected to the transmission grid. Natural gas is a common fuel, although various liquid fuels may also be used.

The microturbine is another combustion turbine technology used in DG applications. Figure 9.4 shows a microturbine being employed in a combined heat and power application with the heat exchanger shown on the left. One of the major advantages of this technology is that installations are clean and compact. This allows deployment near living and working areas, although there may be some issues with the high-pitched turbine noise in some environments.

FIGURE 9.4 Microturbine in a combined heat and power installation. (*Courtesy of Capstone Turbine Corporation.*)

The only moving part in a microturbine is a one-piece turbine with a permanent-magnet rotor. The assembly spins at speeds typically ranging from 10,000 to 100,000 rpm. The alternator output is rectified to direct current immediately and fed into an inverter that interfaces with the ac electric power system. Thus, the characteristic of the microturbine that is of interest to power quality engineers is the response of the inverter to system disturbances.

Microturbines are produced in sizes of 30 to 75 kW, which are most commonly matched to small commercial loads. They may be paralleled in packs to achieve higher ratings. Larger sizes of approximately 300 to 400 kW are also becoming available and are sometimes called miniturbines.

Microturbine electricity generation efficiency is often claimed to be as high as 30 percent, but 25 percent is a more likely value. Because of its low efficiency, it is not generally cost competitive for electricity generation alone. However, when teamed with an appropriate thermal load, net energy efficiencies exceeding 60 percent can be achieved. This technology is best suited for combined heat and power applications in small- to medium-sized commercial and industrial facilities.

There are niche applications where microturbines are used strictly for electricity generation. Because microturbines have compact packaging and low emissions, they make convenient and environmentally friendly standby and peaking generators. They are also used in some base load applications; have the ability to accept a wide variety and quality of fuels; and are a convenient means to extract energy from biomass gas, flare gas, or natural gas that is not economical to transport to pipelines.

9.2.3   Fuel Cells

Another exciting DG technology is the fuel cell (Fig. 9.5). This technology also occupies a relatively small footprint, is very quiet, and has virtually no harmful emissions during operation. Fuel cells are efficient electricity generators and may be employed in combined heat and power applications to achieve among the very best possible energy-conversion efficiencies. Those who see the future energy economy based on hydrogen see the fuel cell as the dominant energy-conversion technology.

A fuel cell is basically a battery powered by an electrochemical process based on the conversion of hydrogen. It produces dc voltage, and an inverter is required for interfacing to the ac power system.

The chief drawback to fuel cells at present is cost. Fuel cell technologies are on the order of 10 times more expensive than reciprocating gensets. This will limit the implementation of fuel cells for electricity production to niche applications until there is a price

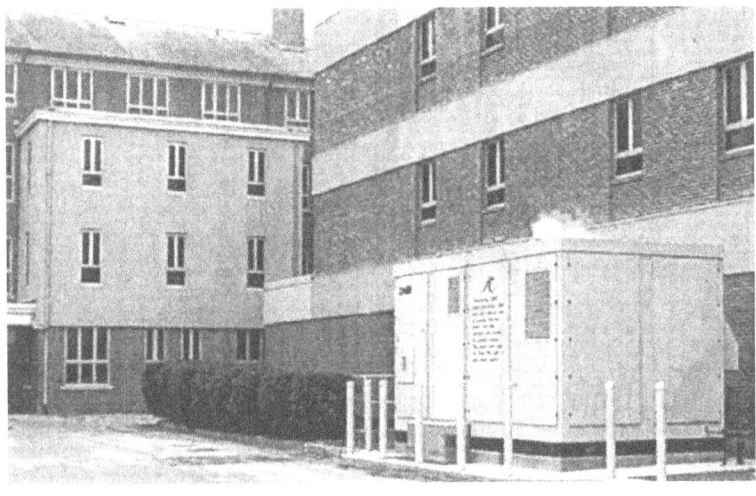

FIGURE 9.5   A fuel cell producing electricity and heat for a hospital. (*Courtesy of International Fuel Cells, LLC.*)

breakthrough. Many expect this breakthrough to occur if the fuel cell is widely adopted for transportation applications.

9.2.4   Wind Turbines

Wind generation capacity has been increasing rapidly and has become cost competitive with other means of generation in some regions. A common implementation in the United States is to group a number of wind turbines ranging in size from 700 to 2500 kW each into a "wind farm" having a total maximum capacity range of 200 to 500 MW. One example is shown in Fig. 9.6. Such large farms are interconnected to the transmission system rather than the distribution system. However, smaller farms of a few MW have been proposed for connection directly to distribution feeders.

The chief power quality issue associated with wind generation is voltage regulation. Wind generation tends to be located in sparsely populated areas where the electrical system is weak relative to the generation capacity. This results in voltage variations that are difficult to manage. Thus, it is sometimes impossible to serve loads from the same feeder that serves a wind farm.

There are three main classes of generator technologies used for the electrical system interface for wind turbines:

1. Conventional squirrel-cage induction machines or wound-rotor induction machines. These frequently are supplemented by switched capacitors to compensate for reactive power needs.

FIGURE **9.6** Wind farm in the western United States.

2. Doubly fed wound-rotor induction machines that employ power converters to control the rotor current to provide reactive power control.

3. Non–power frequency generation that requires an inverter interface.

9.2.5 Photovoltaic Systems

The passage of net metering legislation and various government incentives has spurred the installation of rooftop PV solar systems. Figure 9.7 shows a large system on a commercial building in California. A typical size for a residential unit would be between 2 and 6 kW. Once installed, the incremental cost of electricity is very low with the source of energy being essentially free while it is available. However, the first cost is very substantial even with buy-down incentives from government programs. Installed costs currently range from $3,500–4,500 to $20,000/kW. Despite this high cost, PV solar technology is favored by many environmentalists and installed capacity can be expected to continue growing.

Solar PV systems generate dc power while the sun is shining on them and are interfaced to the utility system through inverters. Some systems do not have the capability to operate stand-alone—the inverters operate only in the utility-interactive mode and require the presence of the grid.

FIGURE 9.7   Rooftop PV solar system. (*Courtesy of Sunpower Corporation.*)

9.3   Interface to the Utility System

The primary concern here is the impact of DG on the distribution system power quality. While the energy-conversion technology may play some role in the power quality, most power quality issues relate to the type of electrical system interface.

Some notable exceptions include:

1. The power variation from renewable sources such as wind and solar can cause voltage fluctuations.

2. Some fuel cells and microturbines do not follow step changes in load well and must be supplemented with battery or flywheel storage to achieve the improved reliability expected from standby power applications.

3. Misfiring of reciprocating engines can lead to a persistent and irritating type of flicker, particularly if it is magnified by the response of the power system.

The main types of electrical system interfaces are

1. Synchronous machines

2. Asynchronous (induction) machines

3. Electronic power inverters

The key power quality issues for each type of interface are described in Secs. 9.3.1 to 9.3.3.

9.3.1 Synchronous Machines

Even though synchronous machines use old technology, are common on power systems, and are well understood, there are some concerns when they are applied in grid parallel DG applications. They are the primary type of electric machine used in backup generation applications. With proper field and governor control, the machine can follow any load within its design capability. The inherent inertia allows it to be tolerant of step-load changes. While this is good for backup power, it is the source of much concern to utility distribution engineers because this technology can easily sustain inadvertent islands that could occur when the utility feeder breaker opens. It also can feed faults and possibly interfere with utility overcurrent protection.

Unless the machines are large relative to system capacity, interconnected synchronous generators on distribution systems are usually operated with a constant power factor or constant var exciter control. For one thing, small DG does not have sufficient capacity to regulate the voltage while interconnected. Attempting to do so would generally result in the exciter going to either of the two extremes. Secondly, this avoids having the voltage controls of several small machines competing with each other and the utility voltage regulation scheme. A third reason this is done is to reduce the chances that an inadvertent island will be sustained. A nearly exact match of the load at the time of separation would have to exist for the island to escape detection.

It is possible for a synchronous machine that is large relative to the capacity of the system at the PCC to regulate the utility system voltage. This can be a power quality advantage in certain weak systems. However, this type of system should be carefully studied and coordinated with the utility system protection and voltage regulation equipment. It would be possible to permit only one generator on each substation bus to operate in this fashion without adding elaborate controls. The generation will likely take over voltage regulation and can drive voltage regulators to undesirable tap positions. Conversely, utility voltage regulators can drive the generator exciter to undesirable set points. To ensure detection of utility-side faults when the interconnected generator is being operated under automatic voltage control, many utilities will require a direct transfer trip between the utility breaker and the generation interconnection breaker.

One aspect of synchronous generators that is often overlooked is their impedance. Compared to the utility electrical power system, generators sized for typical backup power purposes have high impedances. The subtransient reactance $X_{d'}$, which is seen by harmonics, is often about 15 percent of the machine's rating. The transient reactance, $X_{d'}$, which governs much of the fault contribution, might be around 25 percent. The synchronous reactance X_d is generally over 100 percent. In contrast, the impedance of the power system seen from the main load bus is generally only 5 to 6 percent of the service transformer rating, which is normally larger than the machine rating. Thus, end users expecting a relatively seamless transfer from interconnected operation to isolated backup operation are often disappointed. Some actual examples of unexpected consequences are

1. The harmonic voltage distortion increases to intolerable levels when the generator is attempting to supply adjustable-speed-drive loads.

2. There is not enough fault current to trip breakers or blow fuses that were sized based on the power system contribution.

3. The voltage sag when elevator motors are being started causes fluorescent lamps to extinguish.

Generators must be sized considerably larger than the load to achieve satisfactory power quality in isolated operation.

Another aspect that is often overlooked is that the voltage waveform produced by a synchronous machine is not perfect. In certain designs, there is significant third-harmonic distortion in the voltage. Utility central station generation may also have this imperfection, but the delta winding of the generator step-up transformer blocks the flow of this harmonic. The service transformer connection for many potential end-user DG locations is not configured to do this and will result in high third-harmonic currents flowing in the generator and, possibly, onto the utility system. This is discussed is greater detail in Sec. 9.5. The net result is that synchronous generators for grid parallel DG applications should generally be designed with a 2/3 winding pitch or some other winding design that minimizes the third-harmonic component. Otherwise, special attention must be given to the interface transformer connection, or additional equipment such as a neutral reactor and shorting switch must be installed.

9.3.2 Asynchronous (Induction) Machines

In many ways, it is simple to interface induction machines to the utility system. Induction generators are induction motors that are driven slightly faster than synchronous speed. They require another source to provide excitation, which greatly reduces the chances of inadvertent islanding. No special synchronizing equipment is necessary. In fact, if the capacity of the electrical power system permits, induction generators can be started across the line. For weaker systems, the prime mover is started and brought to near-synchronous speed before the machine is interconnected. There will be an inrush transient upon closure, but this would be relatively minor in comparison to starting from a standstill across the line.

The requirements for operating an induction generator are essentially the same as for operating an induction motor of the same size. The chief issue is that a simple induction generator requires reactive power (vars) to excite the machine from the power system to which it is connected. Sometimes, this is an advantage when the DG results in overvoltages, but there can also be low-voltage problems in induction generator applications. The usual fix is to add power factor correction capacitors to supply the reactive power locally. While this works well most of the time, it can bring about another set of power quality problems.

One of the problems is that the capacitor bank will yield resonances that coincide with harmonics produced in the same facility. This can bring about the problems described in Chaps. 5 and 6.

Another issue is self-excitation. An induction generator that is suddenly isolated on a capacitor bank can continue to generate for some period of time. This is an unregulated voltage and will likely deviate outside the normal range quickly and be detected. However, this situation can often result in a ferroresonant condition with damaging voltages.[3] Induction generators that can become isolated on capacitor banks and load that is less than 3 times rated power are usually required to have instantaneous overvoltage relaying.

One myth surrounding induction generators is that they do not feed into utility-side faults. Textbook examples typically show the current contribution into a fault from an induction machine dying out in 1.5 cycles. While this is true for three-phase faults near the machine terminals that collapse the terminal phase voltages, there are not many faults like this on a utility distribution system. Most are single-line-to-ground (SLG) faults, and the voltage on the faulted phase does not collapse to zero (see the examples in Chap. 3). In fact, generators served by delta-wye transformers may detect very little disturbance in the voltage. There are many complex dynamics occurring within the machine during unbalanced faults, and a detailed electromagnetic transients analysis is needed to compute them precisely. A common rule of thumb is that if the voltage supplying the induction machine remains higher than 60 percent, assume that it will continue to feed into the fault as if it were a synchronous machine. This voltage level is sufficient to maintain excitation levels within the machine.

9.3.3 Electronic Power Inverters

All DG technologies that generate either dc or non–power frequency ac must use an electronic power inverter to interface with the electrical power system.

The early thyristor-based, line-commutated inverters quickly developed a reputation for being undesirable on the power system. In fact, the development of much of the harmonics analysis technology described in Chaps. 5 and 6 was triggered by proposals to install hundreds of rooftop PV solar arrays with line-commutated inverters.[4] These inverters produced harmonic currents in similar proportion to loads with traditional thyristor-based converters. Besides contributing to the distortion on the feeders, one fear was that this type of DG would produce a significant amount of power at the harmonic frequencies. Such power does little more than heat up wires.

To achieve better control and to avoid harmonics problems, the inverter technology has changed to switched, pulse-width modulated technologies. This has resulted in a more friendly interface to the electrical power system.

Figure 9.8 shows the basic components of a utility interactive inverter that meets the requirements of IEEE Standard 929–2000.[5] Direct current is supplied on the left side of the diagram either from a conversion technology that produces direct current directly or from the rectification of ac generator output. Variations of this type of inverter are commonly employed on fuel cells, microturbines, solar PV systems, and some wind turbines.

The dc voltage is switched at a very high rate with an insulated gate bipolar transistor (IGBT) switch to create a sinusoid voltage or current of power frequency. The switching frequency is typically on the order of 50 to 100 times the power frequency. The filter on the output attenuates these high-frequency components to a degree that they are usually negligible. However, resonant conditions on the power system can sometimes make these high frequencies noticeable. The largest low-order harmonic (usually, the fifth) is generally less than 3 percent, and the others are often negligible. The total harmonic

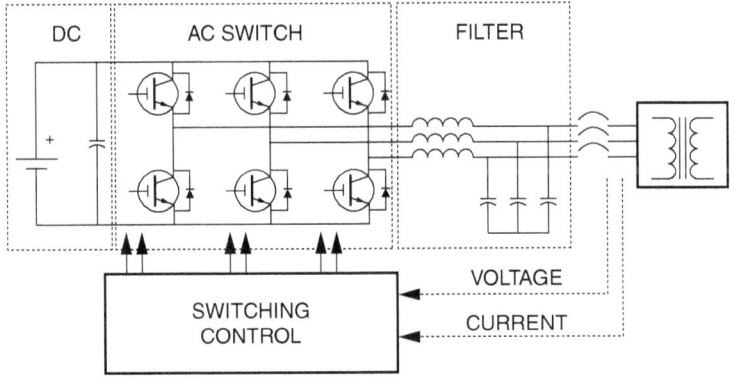

FIGURE 9.8 Simplified schematic diagram of a modern switching inverter.

distortion limit is 5 percent, based on the requirements of IEEE Standard 519–1992. Occasionally, some inverters will exceed these limits under specific conditions. Manufacturers may skimp on filtering, or there may be a flaw in the switch control algorithm. Nevertheless, the power system supply harmonic issue with modern inverters is certainly much less of a concern than those based on older technologies. Switching inverters can produce significant electromagnetic interference (EMI) in some circumstances. Harmonics of the switching frequency or other frequencies associated with the application can be coupled with the distribution system and propagate long distances. While this is rarely damaging to power distribution system components, the EMI may interfere with the operation of loads in other utility customer facilities.

While interconnected to the utility, commonly applied inverters for small DG basically attempt to generate a sine-wave current that follows the voltage waveform. Thus, they would produce power at unity power factor. It is possible to program other strategies into the switching control, but the unity power factor strategy is the simplest and most common. Also, it allows the full current-carrying capability of the switch to be used for delivering active power (watts). If the inverter has stand-alone capability, the control objective would change to producing a sinusoidal voltage waveform at power frequency and the current would follow the load.

One of the advantages of such an inverter for DG applications is that it can be switched off very quickly when trouble is detected. There may be some lag in determining that something has gone wrong, particularly if there are synchronous machines with substantial inertia maintaining the voltage on the system. When a disturbance requiring disconnection is detected, the switching simply ceases. Inverters typically exhibit very little inertia and changes can take place in milliseconds. Rotating machines may require several cycles to respond. It may be possible to reclose out of phase on inverters without damage provided current surge limits in the semiconductor switches are not exceeded. Thus, reconnection and resynchronization are less of an issue than with synchronous machines.

The ability of inverters to feed utility-side faults is usually limited by the maximum current capability of the IGBT switches. Analysts commonly assume that the current will be limited to 2 times the rated output of the inverter. Of course, once the current reaches these values, the inverter will likely assume a fault and cease operation for a predetermined time. This can be an advantage for utility interactive operation but can also be a disadvantage for applications requiring a certain amount of fault current to trip relays.

Utility interactive inverters compliant with IEEE Standard 929–2000 also have a destabilizing signal that is constantly trying to change the frequency of the control. The purpose is to help ensure that inadvertent islands are promptly detected. While interconnected with the utility, the strength of the electrical power system overpowers this tendency toward destabilization. If the inverter system is suddenly isolated on load, the frequency will quickly deviate, allowing it to be detected both within the control and by external relays.

9.4   Power Quality Issues

The main power quality issues affected by DG are

1. *Sustained interruptions.* This is the traditional reliability area. Many generators are designed to provide backup power to the load in case of power interruption. However, DG has the potential to increase the number of interruptions in some cases.

2. *Voltage regulation.* This is often the most limiting factor for how much DG can be hosted on a distribution feeder without making changes. The voltage may be either too high or too low.

3. *Harmonics.* There are harmonics concerns with both rotating machines and inverters, although concern with inverters is less with modern technologies.

4. *Voltage sags.* This a special case because DG may or may not help.

Each of these issues is discussed in turn.

9.4.1   Sustained Interruptions

Much of the DG that is already in place was installed as backup generation. The most common technology used for backup generation is diesel gensets. The bulk of the capacity of this form of DG can be realized simply by transferring the load to the backup system. However, there will be additional power that can be extracted by paralleling with the power system. Many DG installations will operate with better power quality while paralleled with the utility system because of its large capacity. However, not all backup DG can be paralleled without great expense.

Not all DG technologies are capable of significant improvements in reliability. To achieve improvement, the DG must be capable of serving the load when the utility system cannot.

For example, a homeowner may install a rooftop solar PV system with the expectation of being able to ride through rotating blackouts. Unfortunately, the less costly systems do not have the proper inverter and storage capacity to operate stand-alone. Therefore, there can be no improvement in reliability.

Utilities may achieve improved reliability by employing DG to cover contingencies when part of the delivery system is out of service. In this case, the DG does not serve all the load, but only enough to cover for the capacity that is out of service. This can allow deferral of major construction expenses for a few years. The downside is that reliance on this scheme for too many years can ultimately lead to worse reliability. The load growth will overtake the base capacity of the system, requiring load shedding during peak load conditions or resulting in the inability to operate the system at acceptable voltage after a fault.

9.4.2 Voltage Regulation

It may initially seem that DG should be able to improve the voltage regulation on a feeder. Generator controls are much faster and smoother than conventional tap-changing transformers and switched capacitor banks. With careful engineering, this can be accomplished with sufficiently large DG capable of voltage regulation. However, there are many problems associated with voltage regulation. In cases where the DG is located relatively far from the substation for the size of DG, voltage regulation issues are often the most limiting for being able to accommodate the DG without changes to the utility system.

It should first be recognized that some technologies are unsuitable for regulating voltages. This is the case for simple induction machines and for most of the smaller utility interactive inverters that produce no reactive power. Secondly, most utilities do not want the DG to attempt to regulate the voltage because that would interfere with utility voltage regulation equipment and increase the chances of supporting an unwanted island. Multiple DG would interfere with each other. Finally, small DG is simply not powerful enough to regulate the voltage and will be dominated by the daily voltage changes on the utility system. Small DG is almost universally required to interconnect with a fixed power factor or fixed reactive power control.

Large DG greater than 30 percent of the feeder capacity that is set to regulate the voltage will often require special communications and control to work properly with the utility voltage-regulating equipment. One common occurrence is that the DG will take over the voltage-regulating duties and drive the substation load tap changer (LTC) into a significant bucking position as the load cycles up and down. This results in a problem when the DG suddenly disconnects, as it would for a fault. The voltage is then too low to support the load and takes a minute or more to recover. One solution is to establish a control scheme that locks the LTC at a preselected tap when the generator is operating and interconnected. One common strategy for line regulators is to take the regulator to the neutral tap when the reverse power is sensed by its control.

Large voltage changes are also possible if there were a significant penetration of dispersed, smaller DG producing reactive power at a constant power factor. Suddenly connecting or disconnecting such generation can result in a relatively large voltage change that will persist until recognized by the utility voltage-regulating system. This could be a few minutes, so the change should be no more than about 5 percent. One condition that might give rise to this would be fault clearing on the utility system. All the generation would disconnect when the fault occurs, wait 5 min, and then reconnect. Customers would first see low voltage for a minute, or so, until the voltage regulators react. This is followed 5 min later by high voltages because the voltage regulator taps are too high. Options for dealing with this include faster tap-changing voltage regulators and requiring the load to be disconnected whenever the DG is forced off due to a disturbance. There is less voltage excursion when the DG is operating near unity power factor. In fact, the voltage excursion can be reduced by having the DG absorb a small amount of reactive power. However, there may be some instances where it will be advantageous in normal operation to have the DG produce reactive power.

9.4.3 Harmonics

There are many who still associate DG with bad experiences with harmonics from electronic power converters. If thyristor-based, line-commutated inverters were still the norm, this would be a large problem. Fortunately, the technologies requiring inverters have adopted the switching inverters like the one described previously in this chapter. This has eliminated the bulk of the power system harmonics problems from these technologies.

One problem that occurs infrequently arises when a switching inverter is installed in a system that is resonant at frequencies produced by the switching process. The symptom is usually high-frequency hash appearing on the voltage waveform. The usual power quality complaint, if any, is that clocks supplied by this voltage run fast at times. This problem is generally solved by adding a capacitor to the bus that is of sufficient size to shunt off the high-frequency components without causing additional resonances.

Harmonics from rotating machines are not always negligible, particularly in grid parallel operation. The utility power system acts as a short circuit to zero-sequence triplen harmonics in the voltage, which can result in surprisingly high currents. For grounded wye-wye or delta-wye service transformers, only synchronous machines with 2/3 pitch can be paralleled without special provisions to limit neutral current. For service transformer connections with a delta-connected winding on the DG side, nearly any type of three-phase alternator can be paralleled without this harmonic problem.

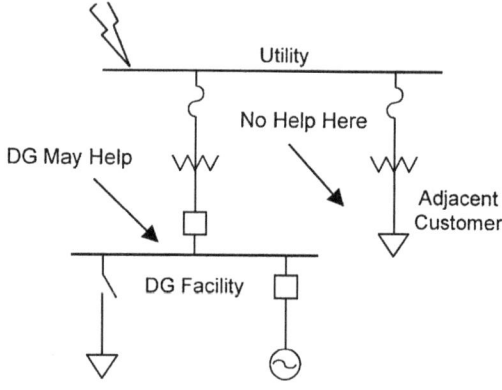

FIGURE 9.9   DG may help reduce voltage sags on local facility bus, but impedance of interconnection transformer inhibits any impact on adjacent utility customers.

9.4.4   Voltage Sags

The most common power quality problem is a voltage sag, but the ability of DG to help alleviate sags is very dependent on the type of generation technology and the interconnection location. Figure 9.9 illustrates a case in which DG is interconnected on the load side of the service transformer. During a voltage sag, DG might act to counter the sag. Large rotating machines can help support the voltage magnitudes and phase relationships. Although not a normal feature, it is conceivable to control an inverter to counteract voltage excursions.

The DG influence on sags at its own load bus is aided by the impedance of the service transformer, which provides some isolation from the source of the sag on the utility system. However, this impedance hinders the ability of the DG to provide any relief to other loads on the same feeder. DG larger than 1 MW will often be required to have its own service transformer. The point of common coupling with any load is the primary distribution system. Therefore, it is not likely that DG connected in this manner will have any impact on the voltage sag characteristic seen by other loads served from the feeder.

9.5   Operating Conflicts

Deploying generation along utility distribution systems naturally creates some conflicts because the design of the system assumes only one source of power.[6] A certain amount of generation can be accommodated without making any changes. At some point, the conflicts will be too great and changes must be made.

In this section, several of the operating conflicts that can result in power quality problems are described.

9.5.1   Utility Fault-Clearing Requirements

Figure 9.10 shows the key components of the overcurrent protection system of a radial feeder.[7] The lowest-level component is the lateral fuse, and the other devices (reclosers and breakers) are designed to conform to the fuse characteristic. There will frequently be two to four feeders off the same substation bus. This design is based mostly on economic concerns. This is the least costly protection scheme that is able to achieve acceptable reliability for distributing the power. One essential characteristic is that only one device has to operate to clear and isolate a short circuit, and local intelligence can accomplish the task satisfactorily. In contrast, faults on the transmission system, which easily handles generation, usually require at least two breakers to operate and local intelligence is insufficient in some cases.

In essence, this design is the source of most of the conflicts for interconnecting DG with the utility distribution system. Because there is too much infrastructure in place to consider a totally different distribution system design to better accommodate DG, the DG must adapt to the way the utility system works. With only one utility device operating to clear a fault, all other DG devices must independently detect the fault and separate to allow the utility protection system to complete the clearing and isolation process. This is not always simple to do from the information that can be sensed at the generator. The remainder of this section describes some of the difficulties that occur. Refer to Chap. 3 for more details on the fault-clearing process on radial distribution systems.

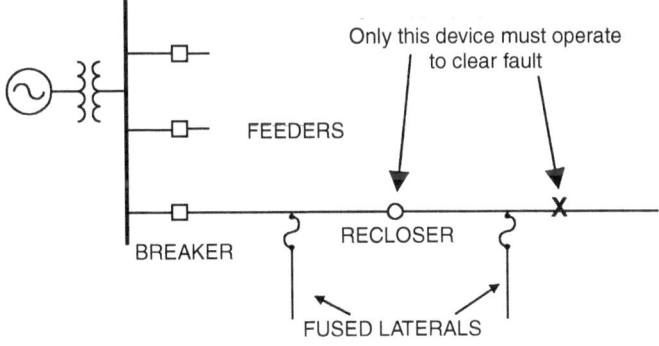

Figure 9.10   Typical overcurrent protection of a utility distribution feeder.

9.5.2   Reclosing

Reclosing utility breakers after a fault is a very common practice, particularly throughout North America. Most of the distribution lines are overhead, and it is common to have temporary faults. Once the current is interrupted and the arc dispersed, the line insulation is restored. Reclosing enables the power to be restored to most of the customers within seconds.

Reclosing presents two special problems with respect to DG:

1. DG must disconnect early in the reclose interval to allow time for the arc to dissipate so that the reclose will be successful.

2. Reclosing on DG, particularly those systems using rotating machine technologies, can cause damage to the generator or prime mover.

Figure 9.11 illustrates the reclose interval between the first two operations of the utility breaker (this represents an unsuccessful reclose because the fault is still present). The DG relaying must be able to detect the presence of the fault followed by the opening of the utility fault interrupter so that it can disconnect early in the reclose interval as shown.

Normally, this detection and disconnection process should be straightforward. However, some transformer connections make it difficult to detect certain faults, which could delay disconnection.

A greater complicating factor is the use of instantaneous reclosing by many utilities. This is used for the first reclose interval for the purpose of improving power quality to sensitive customers. The blinking clock problem can be largely averted, and many other

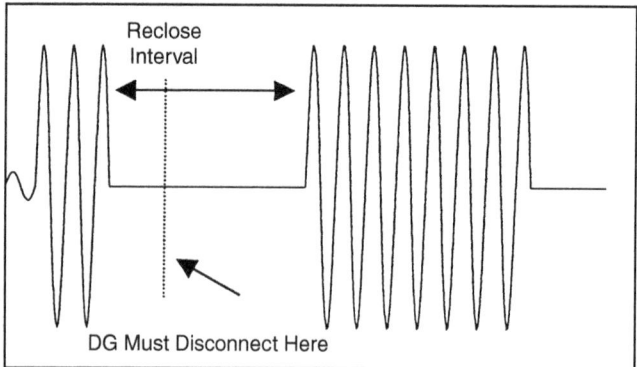

Figure 9.11   DG must disconnect early in the first reclose interval to allow the fault-clearing process to proceed.

types of loads can ride through this brief dead time. The interval for instantaneous reclose is nominally 0.5 s, but can be as fast as 0.2 s. This is in the range of relaying and opening times for some DG breakers. Thus, instantaneous reclose is very likely to be incompatible with DG. It greatly increases the probability that some DG will still be connected when the reclose occurs or that the fault did not have enough time to clear, resulting in an unsuccessful reclose.

A reclose interval of at least 1.0 s is safer when there is DG on the feeder. Many utilities use 2.0 or 5.0 s for the first reclose interval when DG is installed. This minimizes the risk that the DG will not disconnect in time. If it is deemed necessary to maintain the instantaneous reclose, it is generally necessary to employ direct transfer trip so that the DG breaker is tripped simultaneously with the utility breaker. This can be a very expensive proposition for smaller DG installations. Thus, for some distribution systems it will be necessary to compromise one aspect of power quality to better accommodate significant amounts of DG.

9.5.3   Interference with Relaying

Three of the more common cases where DG can interfere with the overcurrent protection relaying on distribution feeders will be examined here:

1. Reduction of reach
2. Sympathetic tripping of feeder breakers
3. Defeat of fuse saving

Figure 9.12 illustrates the reduction-of-reach concept. Each overcurrent relay device has an assigned zone of protection that is determined by its minimum pickup value. Some refer to this generically as the "reach" of the relay. DG infeed can reduce the

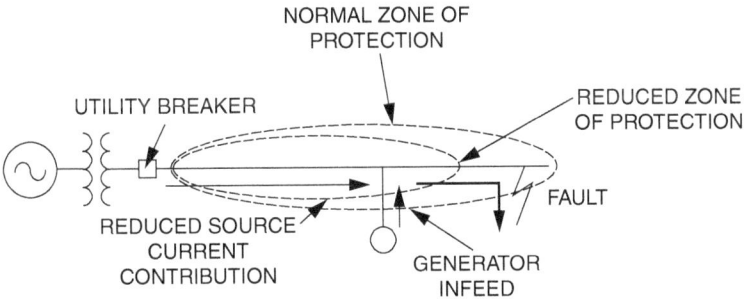

FIGURE 9.12   Infeed from DG can reduce the reach of the relay.

current that the relay sees, thereby shortening its reach. When the total DG capacity increases to a certain amount, the infeed into faults can desensitize the relays and leave remote sections of the feeder unprotected. A low-current (high-impedance) fault near the end of the feeder is more likely to go undetected until it does sufficient damage to develop into a major fault. The power quality consequences of this are that voltage sags will be prolonged for some customers and the additional fault damage will eventually lead to more sustained interruptions.

This issue can be a particular problem for peaking generation located near the end of the feeder. This generation is on at peak load level where the overcurrent relaying would normally be very sensitive to a high-impedance fault. The DG infeed has the potential to mask many faults that would otherwise be detected.

Solutions include:

1. Decrease the relay minimum pickup current to increase the zone. This may not be practical for ground relays that are already set to a very sensitive level.

2. Add a line recloser to create another protection zone that extends far past the end of the feeder.

3. Use a transformer connection that minimizes DG contribution to ground faults, since high-impedance faults are likely to be ground faults.

Sympathetic tripping describes a condition where a breaker that does not see fault current trips "in sympathy" with the breaker that did. The most common circuit condition on utility distribution feeders is backfeed into a ground fault. For the situation shown in Fig. 9.13, the source of the backfeed current is the DG. Most utility feeder breakers do not have directional sensing. Therefore, the ground relay sees the DG contribution as a fault and trips the breaker needlessly. This situation is exacerbated if the service transformer for the DG has a grounded wye-delta connection.

The main solution to this problem is to use directional overcurrent relaying. If appropriate potential transformers are not already

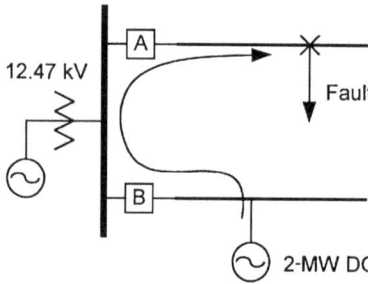

Figure 9.13 Sympathetic tripping of feeder breaker (B) for DG infeed into faults on other feeders.

present, this could end up being an expensive alteration. Since the DG contribution in breaker B is likely to be much lower than the fault current through breaker A, it may be possible to achieve coordination with the appropriate time-delay characteristic or by raising the instantaneous (or fast) trip pickup above the amount of DG infeed.

The power quality impact of the sympathetic tripping is that many customers are interrupted needlessly. The DG is also forced off-line, which could be a problem for the DG owner. There could be impacts from the solutions as well. By slowing the ground trip, there will be more arcing damage to lines and through-fault duty on transformers. This could eventually lead to increased failures.

Fuse saving is commonly practiced in utility overcurrent protection schemes, particularly in more rural regions. The desired sequence for the situation depicted in Fig. 9.14 is for the recloser R to operate before the lateral fuse has a chance to blow. If the fault is temporary, the arc will extinguish and service will be restored upon the subsequent reclose, which normally takes place within 1 or 2 s. This saves the cost of sending a line crew to change the fuse and improves the reliability of customers served on the fused lateral.

Fuse-saving action is a "horse race" in the best of circumstances. It is a challenge for the mechanical recloser to detect the fault and operate fast enough to prevent damage to the fuse element. DG infeed adds to the current in the fuse and makes this race even tighter. At some amount of DG capacity that is capable of feeding the fault, it will no longer be possible to save the fuse.

This phenomenon limits the amount of synchronous machine DG that can be accommodated without making changes to the system. Fuse-saving coordination fails for about the same level of generation that causes voltage regulation problems.

Solutions include:

1. Increase the size of the lateral fuses. All fused cutouts in the zone would have to be changed, which could be quite expensive.

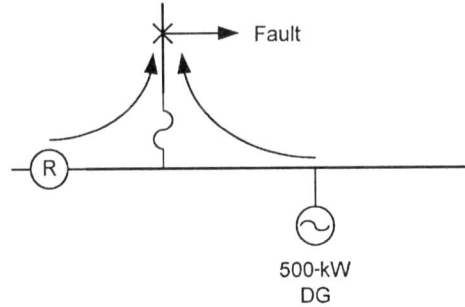

FIGURE 9.14   Infeed from DG can defeat fuse saving.

2. Choose to simply abandon fuse saving, particularly if the DG is only connected intermittently.

3. Require DG to have transformer connections that do not feed single-line-to-ground faults.

The power quality impacts of this are mixed. While the utility generally views fuse saving as an improvement in power quality, customers tend to view the short blink as poor service. Therefore, many utilities have already abandoned fuse saving in many areas.

9.5.4 Voltage Regulation Issues

While there is great concern for various dynamics and transients issues that are difficult to analyze, voltage regulation issues are more likely to occur and cause interconnection problems. Figure 9.15 illustrates one voltage regulation problem that can arise when the total DG capacity on a feeder becomes significant. This problem is a consequence of the requirement to disconnect all DG when a fault occurs.

Figure 9.15*a* shows the voltage profile along the feeder prior to the fault occurring. The intent of the voltage regulation scheme is to

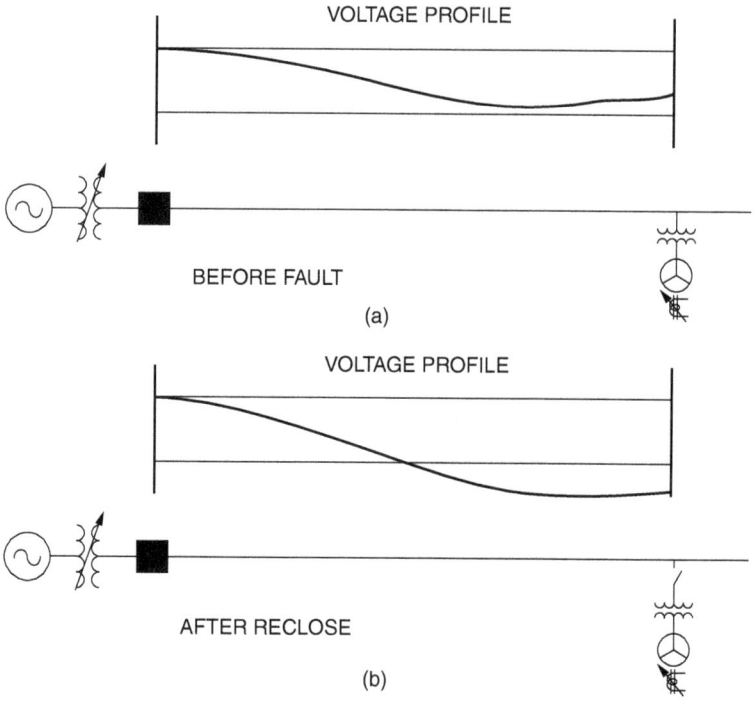

FIGURE 9.15   Voltage profile change when DG is forced off to clear faults.

keep the voltage magnitude between the two limits shown. In this case, the DG helps keep the voltage above the minimum and, in fact, is large enough to give a slight voltage rise toward the end of the feeder.

When the fault occurs, the DG disconnects and may remain disconnected for up to 5 min. The breaker recloses within a few seconds, resulting in the condition shown in Fig. 9.15b. The load is now too great for the feeder and the present settings of the voltage regulation devices. Therefore, the voltage at the end of the feeder sags below the minimum and will remain low until voltage regulation equipment can react. This can be the better part of a minute or longer, which increases the risk of damage to load equipment due to excessively low voltages.

Of course, this assumes that the voltage regulation devices are not already at the maximum tap position. Utility planners will often point out that this is one of the dangers of relying on DG to meet capacity. It masks the true load growth on the system, and there is insufficient base capacity in the wires to deliver the power.

This issue can be one of the more limiting with respect to how much DG can be accommodated on a feeder. It is particularly an issue for lengthy feeders on which the DG is located some considerable distance from the substation. This may be an attractive application of DG because it defers the construction of major wire facilities to serve the remote area. However, it can come at the cost of having to modify long-established operating practices and sacrificing some reliability of the system.

It also suggests one test an analyst can perform to determine if a proposed DG application will likely require changes on the utility system. The test would be to compute the voltage change that occurs at peak load if the DG suddenly disconnects. This change should be less than about 5 percent unless there is fast-acting voltage regulation equipment that can compensate for a larger change. A larger change will require at least some special studies and possible changes to the voltage regulation control on the feeder.

The amount of generation that can be accommodated by this test will obviously vary with position on the feeder. One useful analysis is to determine how much DG capacity can be accommodated (without change) at various distances along the feeder. For example, if one were to establish a 5-percent change criterion for the limit, there would be a curve of generation limit versus distance similar to that shown in Fig. 9.16. If a proposed DG application falls to the left of the curve, it is likely to be acceptable. If it falls to the right, more engineering study is needed to determine how to accommodate the DG.

This simple analysis will work for one DG site per feeder. This may be adequate if penetrations of DG are low, but the problem can get complicated quickly as more sites are added. One approach is to study many random distributions of small DG at peak load. This will

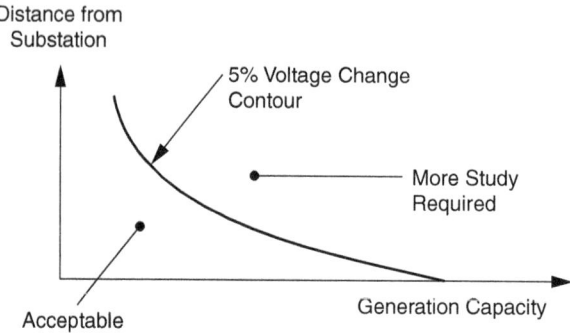

FIGURE 9.16 Simple voltage change test for screening DG applications.

result in a more conservative screening curve that is shifted downward and to the left.

When something must be done, solutions include:

1. Requiring customer load to disconnect with the DG. This may not be practical for widespread residential and small commercial loads. Also, it is difficult to make this transition seamlessly and the load may suffer downtime anyway, negating positive reliability benefits of DG.

2. Installing more voltage regulators, each with the ability to bypass the normal time delay of 30 to 45 s and begin changing taps immediately. This will minimize the inconvenience to other customers.

3. Allow DG to reconnect more quickly than the standard 5-min disconnect time. This would be done more safely by using direct communications between the DG and utility system control.

4. Limit the amount of DG on the feeder.

Another voltage regulation issue involving step-voltage regulators is illustrated in Fig. 9.17. Utility voltage regulators commonly come with a reverse-power feature that allows the regulators to be used when a feeder is supplied from its alternate source. The logic is that when the net power through the regulator is in the reverse direction, the regulator control switches direction and regulates the original source terminal so that the regulator can work properly. Otherwise, the control will attempt to regulate the alternate source side, which would not be possible. The tap position would generally move to one extreme or the other and stay there.

Assume, for example, that several cogeneration sites have been added to a feeder and there is excess generation when the load is

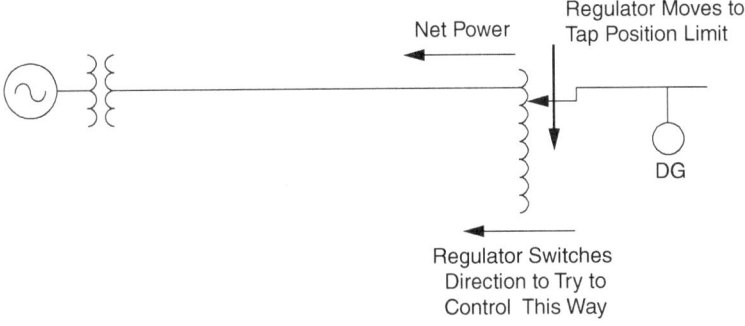

low. The regulator now senses reverse power and attempts to regulate the utility source. However, the DG is not nearly as strong as the utility source and the regulator will not succeed. Similar to the case where the controls fail to switch direction on the alternate feed, the tap will be run to an extreme position, often in the worst possible direction.

To prevent this, regulator vendors have come up with cogeneration features on the controls that can detect this condition. The desired result is to keep the regulator looking in the forward direction. The line-drop compensator R and X settings may also be changed while the reverse-power condition exists.

Generation technologies whose output varies rapidly can be difficult to handle on a distribution feeder. Wind-turbine generation is the most difficult because there is seldom a substation near the proposed site. The generation is typically sited several miles from the nearest substation on a feeder that already may have several switched capacitors and a voltage regulator. One example based on a proposed wind farm at a ski resort is shown in Fig. 9.18. The line is a typical untransposed, horizontal crossarm geometry that leads to special issues. As the power output of the generator varies, one outside phase will tend to rise in voltage while the other tends to drop. Not only is there a magnitude issue but a balance issue.

The results of simulating the system voltages for approximately 40 min with the unchanged control settings are shown in Fig. 9.19. Interestingly, the bank of single-phase voltage regulators keeps the three-phase voltages reasonably well balanced. There were a few tap changes during this period, but the greater problem here is the number of capacitor-switching operations. After changing the capacitor control setting to avoid excessive operations, the regulator tap changes increase significantly, which is typical of this kind of generation. Normally, capacitors will switch once or twice a day and there may be a dozen or so regulator tap changes. Therefore,

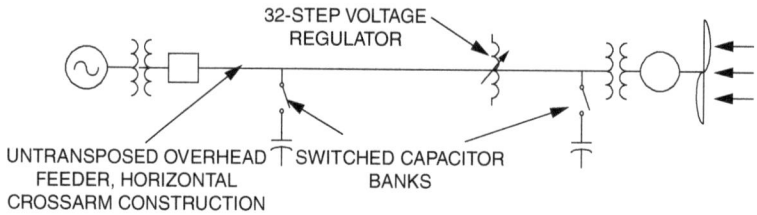

FIGURE **9.18**   Varying generation can cause excessive duty on utility voltage regulation equipment.

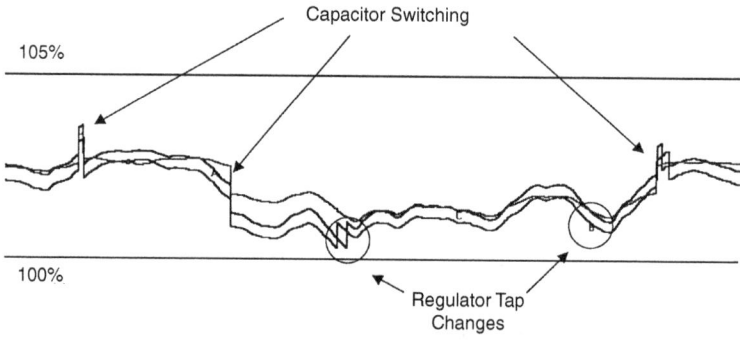

FIGURE **9.19**   Simulated capacitor switching and regulator tap changes for a wind farm connected as shown in Fig. 9.18.

subjecting capacitors and regulators to this kind of duty will almost certainly result in premature failure.

One solution is to increase the bandwidth on the controls so that there are fewer actions. This will result in more voltage variation, but it may be tolerable. While the magnitude variation in this case exceeded 2 percent, it is not changing fast enough to likely cause voltage fluctuation complaints.

When the fluctuation is too great, the main recourse is to build a separate feeder for the wind generation. Also, some wind generators employ doubly fed wound-rotor induction machines that can control reactive power very quickly. With proper control, this can help tame the voltage fluctuations.

The accelerating increase in the amount of solar PV generation interconnected with the utility distribution system has introduced new concerns for voltage regulation issues. The chief concerns are:

1. Voltage rise that exceeds the upper limit of the standard service voltage range.

2. Fluctuations due to *cloud transients,* particularly on feeders with line voltage regulators.

The basic problem with voltage rise is that most utility distribution feeders have been designed assuming that the voltage is going to drop as one travels from the substation to the ends of the feeder. Therefore, utilities tend to regulate feeder voltages toward the upper end of the allowable voltage range so that all voltages are within range at maximum load demand. This leaves very little "headroom" for voltage rise from any type of DG, but especially for solar PV generation. In addition to the possible rise on the primary feeder, there also may be a voltage rise on the secondary behind the meter that drives the voltage on the customer premises above the limits. There are numerous reports worldwide of the solar PV inverters not turning on because the voltage is too high.

As clouds cover a PV array there is a sudden drop in power output. The rate of drop is frequently on the order of 10 percent per second. A similar rate of increase is observed as the power output recovers after the cloud no longer shades the sun. Figure 9.20 illustrates one type of voltage fluctuation problem that can occur when the voltage excursion is sufficient to cause regulator action. The regulator responds to the voltage decline due to decreasing power output by tapping up. This leaves the regulator tap position too high when the full power output resumes and the voltage overshoots to over 106 percent until the regulators act once again to bring it back down.

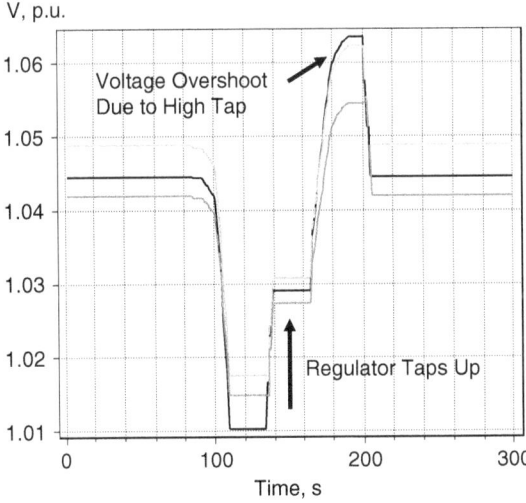

FIGURE 9.20   Solar PV ramping resulting in voltage overshoot.

Figure 9.21 shows the voltage profile on the feeder with the DG at the peak voltage instant. There is simply not enough "headroom" in the voltage profile to host any more DG that causes a voltage rise.

If this were to happen but once in a while, it would probably not be noticeable. However, clouds tend to come in large numbers resulting in several cloud transients in sequence. Figure 9.22 shows a simulation using an actual PV output during a series of cloud transients. Although the voltage did not exceed limits in the case shown, there are numerous regulator tap changes within a 45-m period. This would likely cause excessive wear on utility voltage regulator tap changers.

Note that to observe these effects via simulation one must use a small time step of approximately 1 s.

As mentioned previously, the voltage change due to DG is one of the key indicators of whether a distribution system is able to host a given amount of DG. While planners might typically allow up to a 5-percent change in voltage due to DG if it occurs infrequently—once or twice per day—the criterion to trigger studies for renewable DG has become tighter at 1 to 2 percent change.[13] On many distribution systems, without changes to the voltage regulation scheme this relatively small change can drive the voltage over the upper limit many times during the day. Studies would be required to determine how the voltage regulation scheme might be adjusted to increase the hosting capacity and still meet voltage requirements at all expected demand levels.

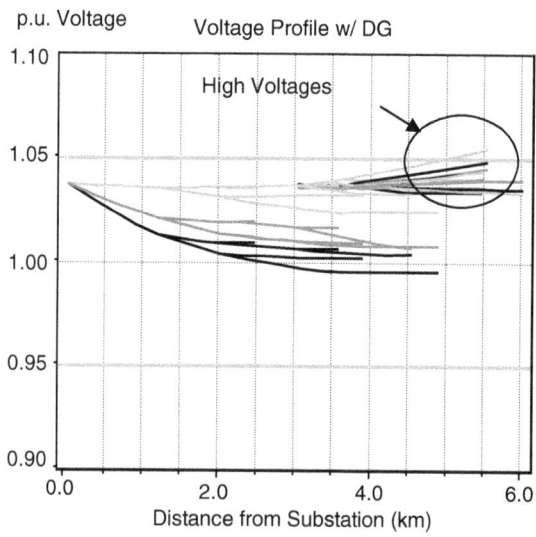

Figure 9.21 Feeder voltage profile illustrating voltage rise at peak DG output.

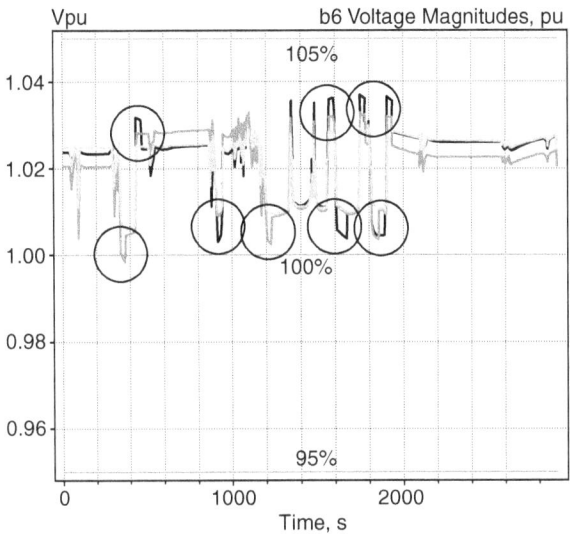

Figure 9.22 Solar PV generation simulation of cloud transients on a 1-s interval using actual PV array power output. Regulator operations circled.

9.5.5 Harmonics

Harmonics from DG come from inverters and some synchronous machines. In the earlier discussion on inverters in this chapter, the measures to eliminate the larger, low-order harmonics were described. The PWM switching inverters produce a much lower harmonic current content than earlier line-commutated, thyristor-based inverters.

In IEEE Standard 519–1992, generators are limited to the most restrictive values in the tables on the allowable amount of harmonic current injection. While generator inverters are not necessarily any worse than power converters used in loads, the developers of the IEEE standard allocated all the capacity in the system to loads, leaving very little for generators. Fortunately, the shift to PWM switching technology has made it relatively easy for inverters to meet the standard.

One new distortion problem that arises with the modern inverters is that the switching frequencies will occasionally excite resonances in the primary distribution system. This creates nonharmonic frequency signals typically at the 35th harmonic and higher riding on the voltage waveform. This has an impact on clocks and other circuitry that depend on a clean voltage zero crossing. A typical situation in which this might occur is an industrial park fed by its own substation and containing a few thousand feet of cable. A quick fix is to add more capacitance in the form of power factor correction capacitors, being careful not to cause additional harmful resonances.

As mentioned in the discussion of the characteristics of synchronous machines, there can be harmonics problems related to zero-sequence triplen harmonics. Figure 9.23 shows a typical situation where this occurs. The facility where the generator is located is served at 480 V by a common delta-wye transformer. When the generator is paralleled to the utility system through this transformer, the operator is frequently surprised to find a large amount of current circulating in the neutral. In the example shown, the current is 26 percent of the machine's rated current and is entirely third-harmonic current. This can adversely affect the operation and efficiency of the machine and may result in the failure of some circuit element. In this case, the problem is confined to the generator side of the transformer and does not affect the primary distribution system because the triplen harmonics are trapped by the delta winding. The same thing can happen with a grounded wye-wye transformer, except that the harmonic currents do reach the primary distribution system.

This problem is well known among vendors of standby generation equipment. If known beforehand, most will recommend a machine with a 2/3 winding pitch that can be paralleled without this difficulty. If it is necessary to parallel a design that does produce significant triplen harmonics, a reactor can be added in the neutral to limit the current flow (as in Fig. 9.28). A shorting switch is closed when the generator is used for backup power to maintain solid grounding.

9.5.6 Islanding

DG protective relays will generally perform their function independently of any outside knowledge of the system to which they

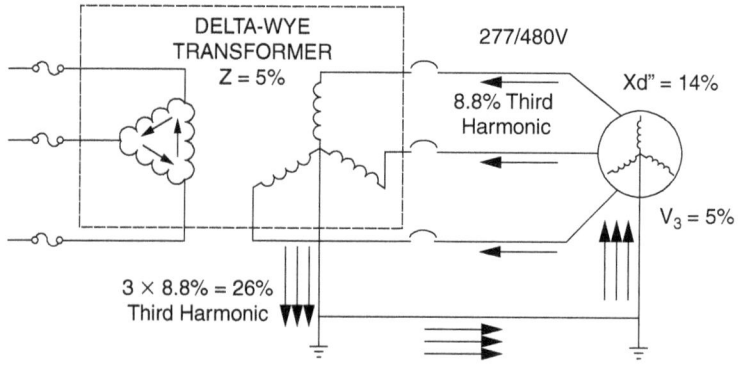

FIGURE 9.23 Generators with significant third-harmonic voltage distortion can produce large circulating third-harmonic currents when paralleled with the utility system.

are connected. Perhaps the greatest fear of the utility protection engineer is that DG relaying will fail to detect the fact that the utility breaker has opened and will continue to energize a portion of the feeder. Therefore, much attention has been paid to detecting islands or forcing islands to become unstable so they can be detected. The reliability concern is that other customers will be subjected to such poor-quality voltage that damage will be sustained. The utility is fearful it will be held liable for the damage. There is also the safety concern of a generator accidentally energizing the line resulting in injuries to the public and utility personnel.

Another concern is the DG itself. Since reclosing is common, it is essential that the DG detect the island promptly and disconnect. If it is still connected when the utility breaker recloses, damage can occur to prime movers, shafts, and components of machines due to the shock from out-of-phase reclosing. This highlights one area of potential conflict with utility practices: Those utilities using instantaneous reclose may have to extend the reclose intervals to ensure that there is sufficient time for DG to detect the island and disconnect.

Relaying is one way to address the issue. The main keys are to detect the deviations in voltage and frequency that are outside the values normally expected while interconnected.

Another approach to anti-islanding is to make requirements for the operating mode for the DG while interconnected that significantly reduce the chances that the generation will match the load when an inadvertent island forms:

- Inverters operating in parallel are less likely to form an island if they are acting as current sources and have a destabilizing signal that is constantly trying to shift the frequency reference out of band.[5] Islanding would require another source to provide a voltage for the inverter to follow. Of course, this source could be provided by any synchronous machine DG that remains on the island.

- Interconnected DG should operate in a mode that does not attempt to regulate voltage. This usually means a constant power factor or constant reactive power mode. For many inverter-based devices, this will be unity power factor, producing watts only. Automatic voltage control should be avoided while DG is interconnected to the distribution system unless the generator is directly connected to a control center to receive dispatching and transfer trip signals.

Without the ability to regulate voltage, the match between load and generation would have to be almost perfect to escape detection by the protective relaying.

While these measures will work in the vast majority of cases, there will be some cases where islanding detection by local intelligence at the DG site is too uncertain. One example would be large generation that is permitted to operate with automatic voltage control. In such cases, direct transfer trip is usually required.

9.5.7 Ferroresonance

This section describes an interesting dilemma that illustrates the conflicting requirements that arise when trying to fit DG into a system that is not designed for it. Ferroresonance is a special kind of resonance in which the inductive element is the nonlinear characteristic of an iron-core device.[8] Most commonly, ferroresonance occurs when the magnetizing reactance of a transformer inadvertently is in series with cable or power factor capacitance.

One interesting case occurs for DG served by cable-fed transformers. It is common practice for the larger DG installations to have their own transformer. Also, it is nearly universal to require DG to disconnect at the first sign of trouble on the utility system. This combination of requirements can lead to a common ferroresonant condition.

The circuit is shown in Fig. 9.24. Underground cable runs are normally fused at the point where they are tapped off the overhead feeder line. This is variously called the riser pole or dip pole. Should something happen that causes one or two fuses to blow, the relaying on the DG will detect an unbalanced condition and trip the generator breaker. This leaves the transformer isolated on the cable with one or two open phases and no load. Either condition is conducive to ferroresonance because the cable capacitance in an open phase, or phases, now appears in series with the transformer's magnetizing impedance (Fig. 9.25).

There are several reasons why the riser-pole fuse may blow or become open. Normally, they are designed to blow for faults in the

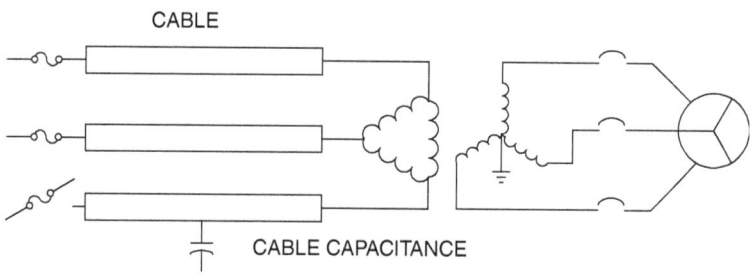

CABLE

CABLE CAPACITANCE

FIGURE **9.24**   DG breaker is required to open when riser-pole fuse blows, leading to ferroresonance.

cable, but there are other reasons. Squirrels or snakes may climb the pole and get in contact with the line. Fuse elements may also fatigue due to frequent inrush currents or lightning surge currents. Fused cutouts may open due to corrosion or improper installation. Finally, they may be operated by a line crew for maintenance purposes. Whether a blown fuse results in damage is dependent on many variables and the specific design of the equipment. A number of ferroresonance modes are possible, depending on the connection of the transformer, its size, and the length of cable. The most susceptible transformer connections are the ungrounded ones. The delta configuration with one phase open is shown in Fig. 9.24. The overvoltages for this condition that can occur can easily reach a value of 3 to 4 pu unless limited by arresters.

Figure 9.26 shows the voltages computed for a 300-kVA delta-connected transformer fed by cable that has 30 nF of capacitance.

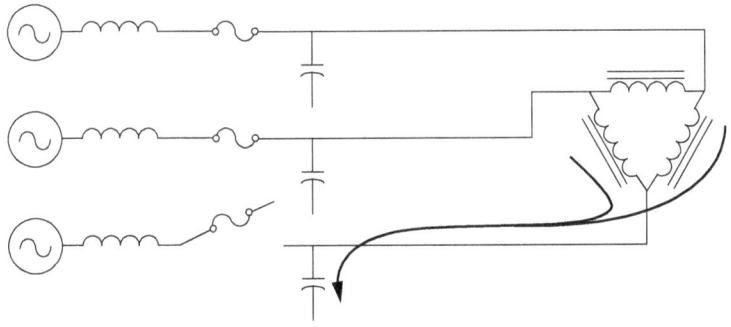

Figure 9.25   Schematic showing magnetizing impedance of a delta-connected transformer in series with cable capacitance when fused cutout is opened.

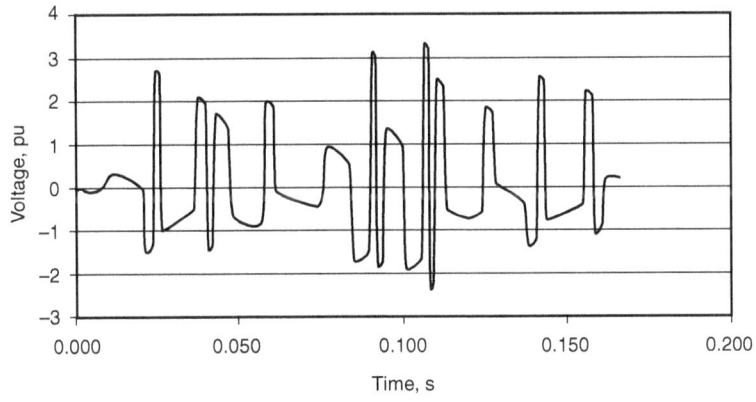

Figure 9.26   Example ferroresonant overvoltages for delta primary.

This models a case in which there is no load on the transformer and no arresters. Arresters would clamp the voltage to a lower value unless they had thermally failed from prolonged exposure to this waveform. The high voltages and the chaotic waveshape are due to the transformer slamming in and out of saturation. The magnetic forces associated with this change cause the core to emit very loud noises that are sometimes described as the sound of a large bucket of bolts being shaken or a chorus of hammers on anvils.

This can cause failures of both the arresters protecting the transformers and the transformers themselves. Arresters fail thermally, leaving the transformers unprotected. Then the transformers may fail either from thermal effects or from dielectric failure. It is common for low-voltage arresters and transient voltage surge suppressors to suffer failures during this type of ferroresonance.

At one time, it was believed that grounded-wye connections were impervious to ferroresonance. However, this theory was shown to be false in a landmark paper.[9] While grounded-wye transformers made up of three single-phase transformers, or a three-phase shell core design, are immune to this type of ferroresonance, the majority of pad-mounted transformers used in commercial installations are of three-legged or five-legged core design. Both are susceptible to ferroresonance due to phase coupling through the magnetic core. Although not immune, the overvoltages are lower than with ungrounded connections, typically ranging from 120 to 200 percent. Sometimes, the voltages are not high enough to cause failure of the transformer. The utility line crew responding to a trouble call encounters the transformer making a lot of noise, but it is still functional with no detectable damage. In other cases, there could be a burned spot on the paint on the top of the tank where the high fluxes in the core have caused heating in the tank. Primary arresters should be tested and the secondary system inspected for failed equipment before reenergizing.

This situation is not necessarily unique to DG installations. Many modern commercial facilities are supplied with cable-fed service transformers that are disconnected from the mains when there is a problem on the utility system. The purpose would be to switch to UPS systems or backup generation. Unfortunately, it leaves the transformer isolated with little or no load.

As a general rule there should be no line fuses or single-phase reclosers between the generator and the utility substation. This is to prevent single-phasing the generator, which could not only result in ferroresonance but could thermally damage rotating machines. This rule is particularly appropriate for DG with cable-fed service transformers. The riser-pole fuses may be replaced with solid blades (no fuses) or three-phase switchgear such as a recloser or sectionalizer. Replacing the fuses with solid blades will reduce the reliability of the feeder section somewhat. Each time there is a fault on the cable system, the entire feeder or feeder section will be out of service. If the cable is short and dig-ins unlikely, this may very well be the lowest-cost option. If protection is required, the three-phase switchgear option is preferred.

The type of ferroresonance shown in Fig. 9.26 is very sensitive to the amount of load. If the system can be arranged so that there is always a *resistive* load attached to the secondary bus, the resonance can be damped out. The load need not always be large, but must be significant. In the example cited, a 2-percent load (6 kW) was sufficient. However, in other cases, more than a 10-percent load may be required.

9.5.8   Shunt Capacitor Interaction

Utilities use switched capacitors to help support the voltage during high-load periods. These banks are mostly controlled by local intelligence, switching at predetermined times or at loading levels as measured by either voltage, current, or kvar. Some types of DG can also produce reactive power (vars), and this can create control hunting and other difficulties. There can be several capacitor banks on the feeder as illustrated in Fig. 9.27. The capacitors switch independently of the generator control unless special communications and control have been added to coordinate dispatch. A 2- to 3-percent increase in the voltage is common when a typical capacitor bank is energized.

Generators in parallel operation are generally maintaining a constant power and power factor. The reactive power of the machine is controlled by the exciter field, which will have certain minimum and maximum voltage or field excitation limits as indicated on the diagram. The generator control attempts to maintain a constant reactive power output until it bumps up against one of these limits.

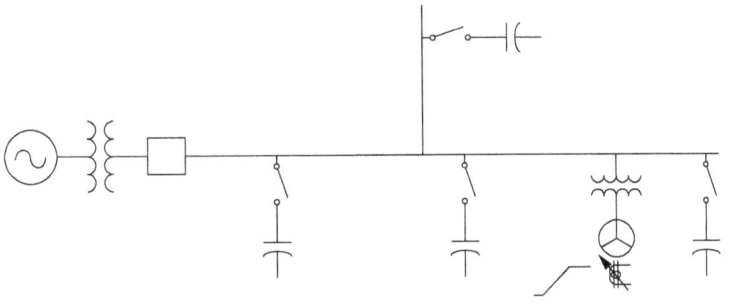

FIGURE 9.27   A typical distribution feeder may employ numerous switched capacitor banks that may interact with generator excitation control and cause nuisance tripping.

There can easily be conditions in which the total reactive power output of the generators and capacitors is too great, resulting in high voltages. This is particularly likely when capacitors are switched by time clock or by current magnitude (without voltage override). There are at least three things that can happen at this point to trip the generator:

1. The generator control senses overvoltage at its terminals and attempts to back down the field to compensate. However, the utility system overpowers the generator and the field reduces to a level deemed to be too low for safe operation of the machine.

2. When the generator reaches its voltage limit, reactive power flows back into the machine. When it reaches a certain level, the generator protection interprets this as a malfunction.

3. DG that does not produce reactive power simply trips on overvoltage.

9.5.9   Transformer Connections

The service transformer used for interconnection can have a great influence on the impact DG will have on the power quality. The advantages and disadvantages of the common three-phase transformer connections are discussed in this section.

Grounded Wye-Wye Connection

This is the most common connection applied in North America for three-phase loads. It is favored because of its reduced susceptibility to ferroresonance on cable-fed loads and fewer operating restrictions when being switched for maintenance. It is also generally well behaved with respect to DG interconnection, but there are a couple of issues.

Advantages include:

- No phase-shifting of utility-side voltages. This makes detection of utility faults by DG protection relays more certain.

- Less concern for ferroresonance, but it is not immune to ferroresonance.

Disadvantages include:

- Allows DG to feed all types of faults on the utility system.

- Does not inhibit the flow of zero-sequence harmonic currents that might be produced from certain kinds of generators.

Because of these two concerns, it may be difficult to parallel some generators using this transformer connection. If the DG is a synchronous machine, it may produce a small amount of third-harmonic voltage distortion, depending on the winding pitch of the machine. If a synchronous generator does not have a 2/3 winding pitch, paralleling to the utility system provides a very low impedance path for the third harmonics and the resulting neutral currents may damage generator equipment or simply add unwanted harmonic currents to the utility system. A neutral reactor may be necessary for some wye-connected machines while they are paralleled to the utility system to

- Limit the flow of zero-sequence harmonics (principally, the third)
- Limit the contribution of the generator to ground faults

The reactor would be shorted when operating the generator stand-alone to provide emergency backup power so that a stable neutral is presented to the load.

Delta-wye Connection

This is the second most common connection for three-phase loads in North America, and the most common in Europe. It would probably be favored for serving loads in nearly all cases if it were not for the susceptibility of the connection to ferroresonance in cable-fed systems.

Advantages include:

- There is less infeed into utility-side ground faults.
- Third harmonics from the DG do not reach the utility system.
- Some isolation from voltage sags due to utility-side SLG faults is provided.

Disadvantages include:

- It is difficult to detect some SLG faults from the secondary side by voltage relaying alone.
- It is susceptible to ferroresonance in cable-fed installations.[8]
- Third harmonics in the DG may cause excessive current in the secondary-side neutral.
- If islanded on an SLG fault, utility arresters can be subjected to overvoltages (see Sec. 9.6).
- If arresters are islanded on an SLG fault and there is little load, resonant overvoltages can result.

The last two items are common to all transformers with an ungrounded primary connection.

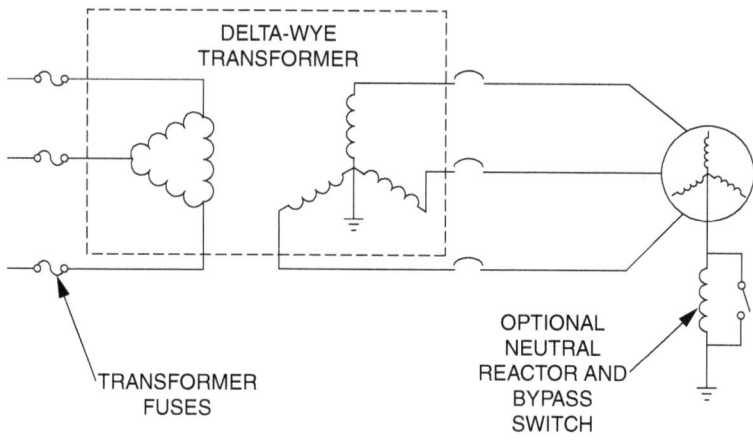

FIGURE 9.28   Delta-wye transformer connection.

Note that while this connection prevents third harmonics from the generator from reaching the utility system, it does not prevent their flow on the DG side (see Fig. 9.23). As with the grounded wye-wye connection, it is generally not advisable to directly connect synchronous alternators that are not 2/3 pitch without inserting an impedance in the neutral to limit the third-harmonic current flow (Fig. 9.28).

While the phase shift can be beneficial to the load in reducing the impact of voltage sags due to SLG faults, it also makes some SLG faults on the utility system more difficult to detect. This increases the chances of islanding at least briefly because it delays fault detection until the utility breaker operates.

Therefore, it is common to add other relaying functions to aid in the early detection of utility-side faults. A negative-sequence relay can make the detection more reliable. While the voltage magnitudes seen on the secondary may not change much during a fault, they will be unbalanced, resulting in detectable negative-sequence voltages and currents.

Another approach is to add relaying on the primary side of the transformer, such as a type 59G ground overvoltage relay that can detect the presence of the SLG fault. This is frequently implemented as an overvoltage relay placed in the corner of a broken delta potential transformer that measures zero-sequence voltage. Some multifunction microprocessor-based DG interconnection relays can provide this function directly from the Y-connected potential transformers.

Delta-Delta or Ungrounded Wye-Delta Connection

While not in the majority, these connections are still common for commercial and industrial loads. Both have similar behavior with respect to serving DG. Neither would be the preferred connection for serving most new DG installations, but could be encountered in legacy systems where a customer wishes to parallel DG.

Some inverter-based systems (fuel cells, PVs, microturbines, etc.) require an ungrounded connection on the DG side because the dc side of the inverter is grounded. This is often accomplished by use of a separate isolation transformer rather than the main service transformer. However, either of these connections would also suffice.

The delta secondary is sometimes a four-wire connection with one of the delta legs center-tapped and grounded to serve single-phase 120-V loads. This is common in smaller commercial facilities that have three-phase HVAC equipment along with typical office load. If this is the case, no part of a three-phase DG can be grounded while paralleled with the grid.

Advantages include:

- More economical transformer installation for smaller three-phase service with some single-phase loads is possible.
- The load is isolated from ground faults on the utility side.
- DG would not typically feed utility-side ground faults except when resonance occurs.
- Ungrounded interconnection can be provided for inverter-based systems requiring it.

Disadvantages include:

- Utility-side SLG faults are difficult to detect.
- Utility arresters are subjected to high steady-state overvoltages if islanded on an SLG fault (see Fig. 9.29). This is true for delta-wye connections as well.

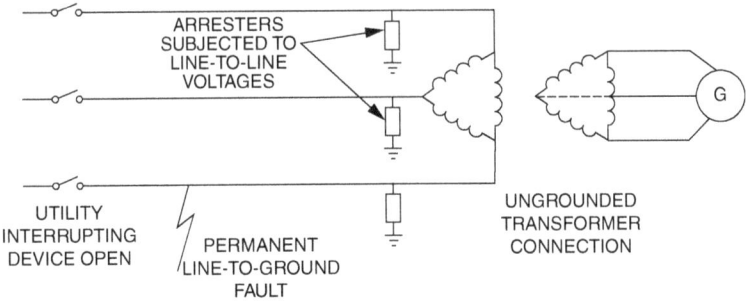

FIGURE 9.29 Isolating DG with ungrounded service transformer connection on an SLG fault can lead to arrester failure.

- These connections are highly susceptible to ferroresonance in cable-fed installations.

- There are more restrictions on switching for utility maintenance. Three-phase switchgear may be required on the primary because there are several problems that can occur if one attempts to perform single-phase switching. This will increase the cost of the interconnection.

The prompt detection of SLG utility faults using voltage relaying is a problem with these connections. This will delay fault detection until after the utility breaker has opened, resulting in at least a brief island. This can result in overvoltages and a resonant condition common to all ungrounded primary connections. Supplementing voltage relaying with negative-sequence relaying on the DG side can make the detection more certain. Also, it is common to add a ground overvoltage relay (59G) on the primary side to detect the continuing presence of a ground fault.

Grounded Wye-Delta Connection

This is an interesting connection because of the conflicting application considerations. Many utility engineers believe this is the best winding connection for interconnecting generation to the utility system. This is the connection used for nearly all central station generation. There are many advantages, including:

1. Utility-side faults are easily detected partly because the transformer itself actively participates in ground faults.

2. Triplen harmonic voltages produced by the generator do not cause any current to flow because it is blocked by the delta winding. Therefore, nearly any generator can be paralleled with this connection.

3. Protection schemes are well understood based on many years of experience with utility generation.

Despite these benefits, one may be surprised to learn that this connection is not permitted on many utility distribution systems without a great deal of study and special considerations that may result in costly modifications to the system. In fact, it may not be possible to accommodate the connection on some distribution systems because of the inconvenience to other customers.

The connection is often referred to as a "ground source" because it contributes to ground faults and will generally disrupt the ground fault relaying coordination on the feeder. Other feeders connected to the same substation bus may be disrupted also. Figure 9.30 shows how the connection contributes to an SLG fault on a four-wire, multigrounded neutral distribution system, the most common in the United States. The thicker arrows show the

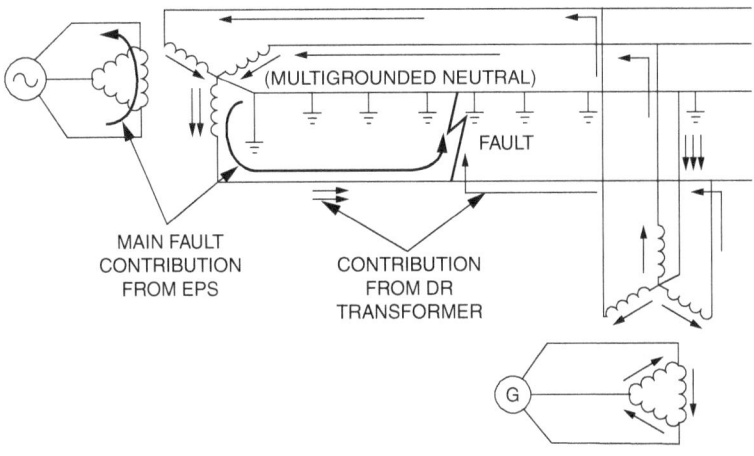

Figure 9.30 Grounded-wye-delta connection acts as a "ground source" feeding ground faults.

normal contribution expected from the main utility source. Only one phase is involved on the distribution side, and the fault appears to be a line-to-line fault from the transmission side. The thinner arrows show the paths of the current from the grounded wye-delta interconnection transformer. The currents flow back through the substation and contribute additional current to the fault. The amount contributed would depend on the size and impedance of the transformer.

The generator contribution is not shown in Fig. 9.30 for clarity. This contribution will be dependent on the capability of the DG to feed a short circuit. In some cases, the contribution due to the transformer alone will be larger.

This characteristic has a number of possible adverse side effects when present on the distribution system:

- Increased fault current means increased damage at the fault site, which will eventually lead to more sustained interruptions and reduced reliability.

- The connection is likely to cause sympathetic tripping of the feeder breaker for faults on other feeders. The transformer supplies ground current to other feeders connected to the same substation bus. Many customers who would normally see only a sag would be subjected to interruptions.

- Ground trip pickup levels must be increased, and more delay must be used to maintain coordination, which results in less sensitive fault protection. (An alternative is to use directional overcurrent relaying.)

- Sags for ground faults will generally be somewhat deeper (the transformer makes the system appear more solidly grounded).

- If fuse saving is being attempted, the fault infeed, which is likely to be larger than from the DG itself, makes this much more difficult to achieve.

- The transformer itself is subject to short-circuit failure when a ground fault occurs. This is particularly true for smaller transformer banks with impedances less than 4 percent. A special transformer must generally be ordered.

- The transformer is also subject to failure thermally because the feeder load is rarely balanced. Thus, the transformer will act as a sink for zero-sequence load currents.

The distribution system can almost always be engineered to work with grounded wye-delta connections. This makes the DG interconnection protection more certain and straightforward. However, this may require costly modifications that could become an insurmountable barrier for small- and medium-sized DG. The utility must also be willing to accept special transformers and operating procedures that different from the rest of the system. Some utilities are unwilling to do this. One danger is that if the transformer were to fail at some time in the future, those replacing it might not be aware that it requires a certain minimum impedance to prevent failure. Replacing it with a conventional transformer may result in catastrophic failure of the transformer. This is a particular issue when line crews from other companies have to be brought in during disasters. Thus, many utility companies understandably resist use of special options. One common approach to limiting SLG fault contributions and the amount of unbalanced load current the transformer has to handle is to add a neutral reactor on the Y side. There are many considerations to sizing this reactor.[14] The one related to DG is that there may be more than one DG on a feeder or fed from the same bus. The reactors must be sized considering all such installations and possible future DG installations.

9.6   DG on Low-Voltage Distribution Networks*

The discussions in this chapter focus primarily on DG interconnected to radial primary or secondary distribution circuits, which is the most common distribution configuration. However, in large cities a

---

*The text in this section is derived from the lecture material originally developed by D. C. Dawson and W. E. Feero for the University of Wisconsin—Madison course "Interconnecting Distributed Generation to Utility Distribution Systems" and is used here with permission.

number of utilities use a low-voltage network method of distribution.

These low-voltage network systems are of two major subtypes, the *secondary network* (also referred to as an area network, grid network, or street network) and the *spot network*. Secondary networks serve numerous sites, usually several city blocks, from a grid of low-voltage mains at 120/208 V, three-phase. Spot networks serve a single site, usually a large building or even a portion of a large building. The secondary voltage of spot networks is often 277/480 V, three-phase, but 120/208-V spot networks are also used.

Street networks and spot networks are supplied from two or more primary distribution feeders through integrated transformer/breaker/protection combinations called *network units*. These network units are often located in transformer vaults within the building or in underground vaults in the street.

Figure 9.31 shows a spot network arrangement with three primary feeders. The primary feeders may be dedicated to the spot network, may serve other network units at different sites, or may serve ordinary radial distribution loads as well.

The objective of the network distribution design is to achieve high service reliability with high power quality. To accomplish this, the primary feeders are often chosen so that they originate at different substations or, at least at different bus sections of the same substation. As will be explained in Sec. 9.6.1, the high power quality is achieved by having full service capability with any feeder out of service and

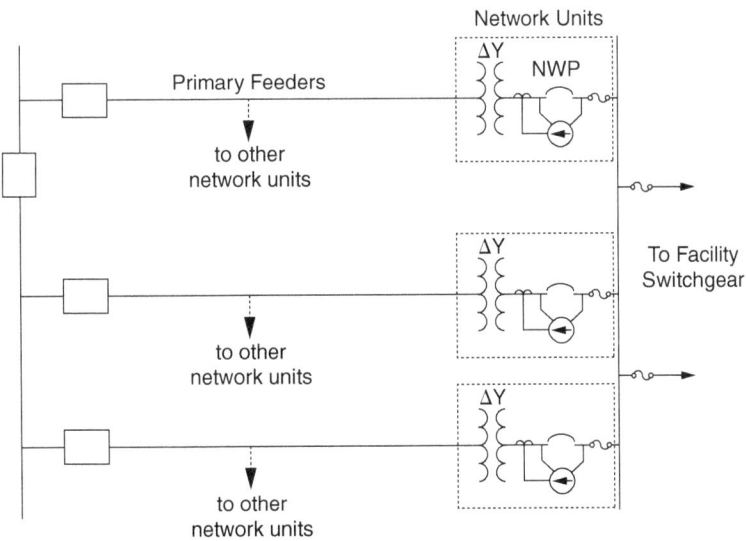

Figure 9.31 Spot network arrangement.

rapidly removing any faulted feeder from connection to the low-voltage network.

9.6.1 Fundamentals of Network Operation

To gain an understanding of how the operation of networks differs from radial service, we will concentrate on the spot network. In normal operation, the spot network is supplied simultaneously from all the primary feeders, by paralleling the low-voltage secondaries of the network transformers on the spot network bus. In order that the spot network can continue to operate if a primary feeder becomes faulted, the network units are each equipped with a low-voltage circuit breaker, called the *network protector,* and a directional-power relay called the *network relay* or *master relay.*

When a primary feeder is faulted, the network relay senses reverse power flow (from the network toward the primary feeder) and opens the network protector, thereby isolating the network bus from the faulted feeder and allowing service on the network to continue without interruption. This function is the reason for the name network protector and the reason why DG interconnection to networks becomes a complex issue.

Figure 9.32 shows the operation of a network protector in isolating a faulted primary feeder. Later, when the faulted primary feeder is repaired and returned to service, the network relay senses voltage at

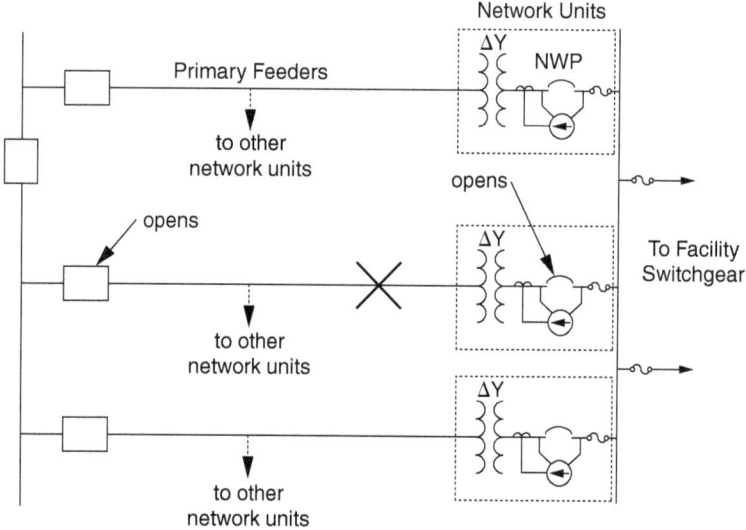

FIGURE 9.32 Network primary feeder fault.

the transformer side of the open network protector. If this voltage is such that power will flow from the network unit to the bus when the protector is closed, the network relay commands the protector switch to reclose. Determining when this reclose will take place may become an interconnection issue.

The network relay is a very sensitive reverse-power relay, with a pickup level on the order of 0.1 percent of the rated power of the network transformer. These settings can be as low as 1 to 2 kW. It is the mission of the reverse-power relay to be capable of sensing reverse power flow with no other feeder loads than the core losses of its own network transformer. This great sensitivity is necessary because the network protector must operate for all types of faults on the primary feeder, including ground faults. Since a delta-connected primary is commonly used on network transformers, no ground fault current will flow from the network toward the primary feeder once the source substation feeder circuit breaker has opened. The primary feeder may have no other loads, or the only other loads may be other network units that have already disconnected from the faulty feeder. Thus the network relay must be capable of sensing reverse power using only its own transformer's losses.

This sensitive reverse-power function means that no DG can be connected to the network with the intent to export power to the utility system. It further means that even momentary power reversals under abnormal conditions must be considered in the interconnection design.

The traditional network relay is an electromechanical device and has no intentional time delay. The typical operating time is about 0.05 s (3 cycles) at normal voltage levels, thus the reason that even momentary power reversals caused by the DG are of concern. Microprocessor-based network relays have replaced the electromechanical types in new network units, and these relays can be retrofitted into many types of existing network units. The basic performance of the microprocessor types is similar to the electromechanicals, but they have more flexibility and new features.

The network protector is an air circuit breaker specifically designed for the fault current conditions encountered on low-voltage network systems. It operates only under the control of the network relay. The network protector has no overcurrent protection and does not open for faults on the low-voltage secondary system. Low-voltage faults are cleared by fuses or by circuit breakers within the served facility. The most critical design characteristic of almost all network protectors in service is that they are not intended to separate two operating electrical systems. Therefore, a DG can never be allowed to island on a network bus.

9.6.2   Summary of Network Interconnection Issues

From the discussion in this section it is clear that installing DG in facilities served by a spot network has a number of special application problems that do not arise in the usual radial service arrangement.

1. Exporting power from a spot network, or even serving the entire facility load from a DG, is not practical because of the reverse-power method of protection used on the network units. If DG exceeds the on-site load, even momentarily, power flows from the network toward the primary feeders and the network relays will open their network protectors, isolating the network from its utility supply. Minimum site loads, such as late-night or weekend loads, may severely limit the size or operating hours of a DG. Even if a DG is sized to the site's minimum load, consideration has to be given to the possibility of sudden loss of a large load, which might cause reverse power flow through the network units.

2. Network protectors built in accordance with ANSI/IEEE Standard C57.12.44–1994 are not required to withstand the 180° out-of-phase voltages that could exist across an open switch with DG on the network. They also are not required to interrupt fault currents with higher X/R ratios than those usually encountered in low-voltage network systems. A serious failure of a network protector on a network in New Mexico during installation test of DG demonstrated the reality of this problem.

3. The fault current delivery from synchronous DG to external faults can cause network protectors to open, potentially isolating the network. Figure 9.33 shows an example of how this can happen. As noted earlier in this chapter, it cannot be determined how induction generators will contribute to unbalanced and high-impedance faults at such locations without detailed studies. If such studies are absent, induction generators should be treated as if they have synchronous generation capability in selecting the appropriate interconnection response to this remote fault issue.

4. If the network protectors open, isolating the network and the DG from the utility source, the network relay may repeatedly attempt to reclose the network protector, leading to destruction of the protector and the possibility of catastrophic failure of the network unit.

5. The network relays are part of an integrated assembly in a submersible enclosure, often mounted in vaults in the street, and are not as easily modified as a typical relay control scheme.

6. Referring back to Fig. 9.31, if the bus tie breaker is operated open, or a second substation is used to supply the network, then there is a possibility that protector cycling would occur under light-load conditions. The addition of DG to the network bus will worsen this condition. Making the determination of when and where the cycling

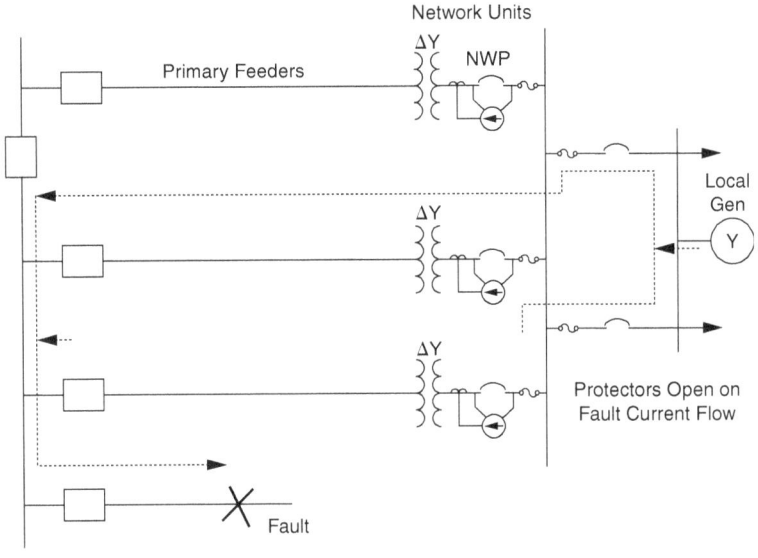

FIGURE 9.33 Fault current from synchronous local generation.

problem might emerge is particularly difficult on street networks without the aid of sophisticated load flow simulations.

9.6.3 Integration Techniques for DG on Networks

Despite this daunting list of problems, there are some ways that DG may be accommodated on spot networks, by taking advantage of new technology. Inverter-based DG has the advantages that fault current is very limited, to about 100 to 200 percent of the normal inverter load current, and that an inverter can respond very rapidly to signals controlling its power output level. Figure 9.34 shows a network DG installation that uses this approach.

Because the most critical aspect of DG on networks is that the local generation must never exceed the local load, this example uses a tie-line load control scheme that senses the total incoming power to the facility and adjusts the DG power output to ensure that power flow is always inward.

Other approaches that might be used are

- Size the DG to be less than the minimum load that ever exists on the network, with a margin for sudden loss of a large load.

- Total the incoming power to the network and trip the DG whenever the inward power flow falls below a safe value.

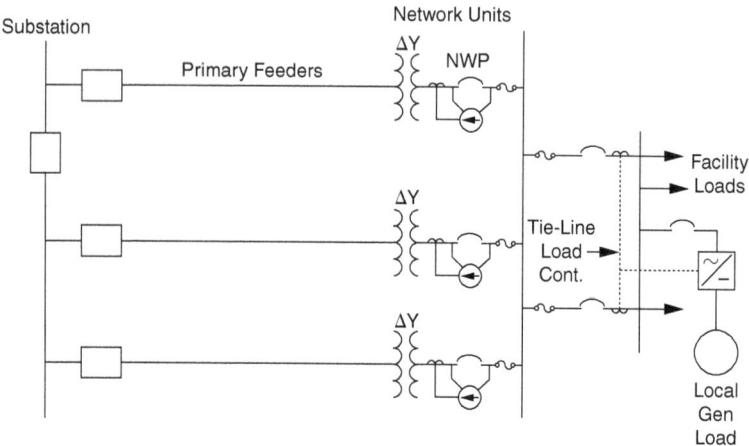

FIGURE 9.34   Inverter-based DG on a spot network.

All three of these approaches seek to avoid opening the network protectors by limiting the DG to less than the on-site load. Note that if cycling is a problem, then these techniques can be used to establish minimum load where cycling will not be experienced. Because of the high-speed response of the network relays, there may not be much time available to measure power flow and make a control decision. This problem can be eased by arranging the network relays to have a time delay at low reverse-power levels, e.g., reverse flow less than the rated capacity of the network transformer. For the high levels of reverse power flow that occur during multiphase faults on the primary feeders, the network relays still operate instantaneously. Figure 9.35 shows the adjustable features of the time-delayed network relays.

Time delay on low reverse power is a technique that has been used for many years to deal with regenerative loads such as elevators. Modern microprocessor-based network relays generally have this capability built in. The advantage of the microprocessor is that this time delay can be as short as 6 cycles. If the instantaneous trip threshold can be kept at or below the rating of the transformer, then such a brief time delay may be acceptable from a power quality standpoint. A supplementary overcurrent and time-delay relay is needed to add the capability to electromechanical network relays. In the past, these supplemental relays had minimum time delays of 1 s or more since their mission was to wait for the elevator to descend. However, not all utilities endorse this low-current, time-delay technique. Some feel that any time delay in opening the network protectors degrades the high service quality that the network system is intended to provide.

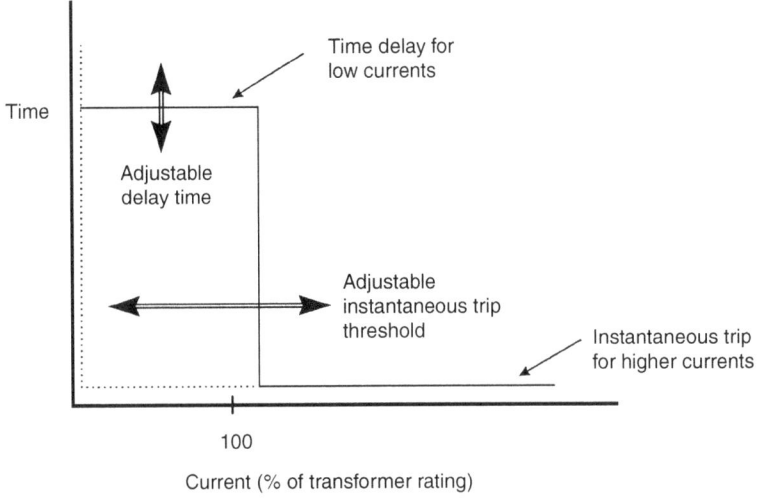

Current (% of transformer rating)

FIGURE 9.35 Adjustable reverse-power characteristic.

The load-generation control and DG tripping schemes mentioned above are intended to ensure that the network protectors are never opened by exported power. As long as the schemes work properly, the network protectors are never exposed to the out-of-phase voltage conditions that may exceed the switch capability. However, because of the potentially catastrophic consequences of causing a network protector failure, it is prudent to provide a backup. An interlocking scheme that trips the DG instantaneously when a certain number of network protectors have opened ensures that the network protectors will not be exposed to out-of-phase voltages for more than a few cycles. The decision as to how many protectors must open before the DG is tripped (one, two, or all) is a tradeoff between security of the protectors and nuisance tripping of the DG. Note that this scheme does not relieve the DG installer from the responsibility of providing stuck-breaker backup protection for the DG's switching device.

An even more secure approach to avoiding overstressing the network protectors is to replace existing protectors with new designs that are capable of interrupting fault currents from sources with higher X/R ratios and of withstanding out-of-phase voltages across the open switch. One major U.S. manufacturer of network protector units has recently introduced such high-capacity protectors in 800- to 2250-A ratings and plans to introduce them in ratings up to 6000 A. These protectors are designed to be retrofitted in many existing types of network units.

A possible DG interconnection problem exists that would involve network protectors without a network bus interconnection. If a DG is interconnected on a feeder that also supplies a network unit, then if its feeder breaker is tripped and the DG is not rapidly isolated, it may impact one or more of the network units as if it were isolated on the network bus. For this type of event to occur, the DG output does not have to be matched to the feeder load. For the excess generation case, it only has to be momentarily greater than the load on the network bus. Under this condition the power continues to flow to the network bus from the feeder with the interconnected DG, which keeps that protector closed. However, the excess power flows through the network back to the other feeders, resulting in the opening of the protectors connected to those feeders. Once open, these protectors will be separating two independent systems. For the case of less generation than load, the protector connecting to the feeder with the generation may trip. Again, such a condition would have a protector separating two independent systems. Therefore, such DG applications should be avoided unless the DG breaker is interlocked with the feeder breaker with a direct transfer trip scheme.

9.7   Siting DG

The value of DG to the power delivery system is very much dependent on time and location. It must be available when needed and must be where it is needed. This is an often neglected or misunderstood concept in discussions about DG. Many publications on DG assume that if 1 MW of DG is added to the system, 1 MW of additional load can be served. This is not always true.

Utility distribution engineers generally feel more comfortable with DG installed on facilities they maintain and control. The obvious choice for a location is a substation where there is sufficient space and communications to control centers. This is an appropriate location if the needs are capacity relief on the transmission system or the substation transformer. It is also adequate for basic power supply issues, and one will find many peaking units in substations. However, to provide support for distribution feeders, the DG must be sited out on the feeder away from the substation. Such generation will also relieve capacity constraints on transmission and power supply. In fact, it is more effective than the same amount of DG installed in the substation. Unfortunately, this generation is usually customer-owned and distribution planners are reluctant to rely on it for capacity.

The application of DG to relieve feeder capacity constraints is illustrated in Fig. 9.36. The feeder load has grown to where it exceeds a limit on the feeder. This limit could be imposed by either current ratings on lines or switchgear. It could also be imposed by bus voltage limits. There is DG on the feeder at a location where it can actually relieve the constraint and is dispatched near the daily peak to help serve the load. The straightforward message of the figure is that the load that would otherwise have to be curtailed can now be served. Therefore, the reliability has been improved.

This application is becoming more common as a means to defer expansion of the wire-based power delivery infrastructure. The generation might be leased for a peak load period. However, it is more common to offer capacity credits to customers located in appropriate areas to use their backup generation for the benefit of the utility system. If there are no customers with DG in the area, utilities may lease space to connect generation or, depending on regulatory rules, may provide some incentives for customers to add backup generation.

There is by no means universal agreement that this is a permanent solution to the reliability problem. When utility planners are shown (Fig. 9.36), most will concede the obvious, but not necessarily agree that this situation represents an improvement in reliability. Three of the stronger arguments are

1. If the feeder goes out, only the customer with the DG sees an improvement in reliability. There is no noticeable change in the service reliability indices.

2. Customer generation cannot be relied upon to start when needed. Thus, the reliability cannot be expected to improve.

3. Using customer-owned generation in this fashion masks the true load growth. Investment in wire facilities lags behind demand, increasing the risk that the distribution system will eventually not be able to serve the load.

It should also be noted that the capacity relief benefit is nullified when the distribution system is upgraded and no longer has a

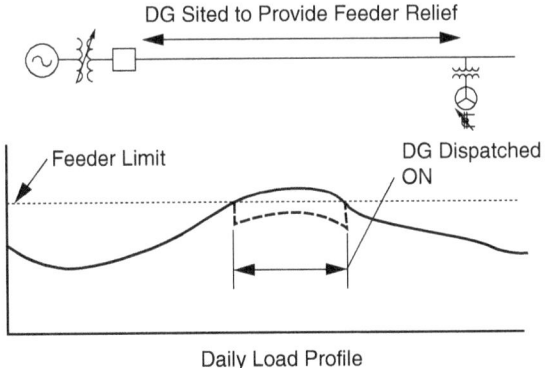

FIGURE **9.36**   DG sited to relieve feeder overload constraint.

constraint. Thus, capacity credits offered for this application generally have a short term ranging from 6 months to 1 year.

If one had to choose a location on the distribution feeder, where should the DG be located? The optimal DG siting problem is similar to the optimal siting problem for shunt capacitor banks. Many of the same algorithms can be used with the chief difference being that the object being added produces watts in addition to vars. Some of the same rules of thumb also apply. For example, if the load is uniformly distributed along the feeder, the optimal point for loss reduction and capacity relief is approximately two-thirds of the way down the main feeder. When there are more generators to consider, the problem requires computer programs for analysis.

The utility does not generally have a choice in the location of feeder-connected DG. The location is given for customer-owned generation, and the problem is to determine if the location has any capacity-related value to the power delivery system. Optimal siting algorithms can be employed to evaluate the relative value of alternative sites.

One measure of the value of DG in a location is the additional amount of load that can be served relative to the size of the DG. Transmission networks are very complex systems that are sometimes constrained by one small area that affects a large geographical area. A relatively small amount of load reduction in the constrained area allows several times that amount of load to be served by the system.

This effect can also be seen on distribution feeders. Because of the simple, radial structure of most feeders, there is generally not a constraint so severe that DG application will allow the serving of additional load several times greater than the size of the generator. However, there can be a multiplying effect as illustrated in Fig. 9.37.

This example assumes that the constraint is on the feeder rather than on the substation. If 1 MW of generation were placed in the substation, no additional load could be served on the feeder because no feeder relief has been achieved. However, if there is a good site on the feeder, the total feeder load often can grow by as much as 1.4 MW. This is a typical maximum value for this measure of DG benefit on radial distribution feeders.

Another application that is becoming common is the use of DG to cover contingencies. Traditionally, utilities have built sufficient wire-based delivery capacity to serve the peak load assuming one major failure (the so-called N-1 contingency design criterion). At the distribution feeder level, this involves adding sufficient ties to other feeders so that the load can be conveniently switched to an alternate feeder when a failure occurs. There must also be sufficient substation capacity to serve the normal load and the

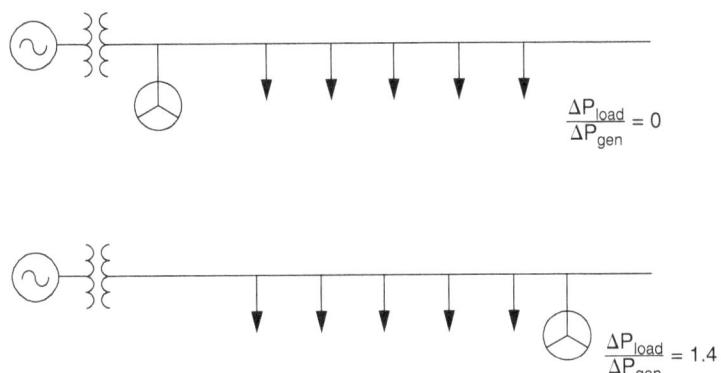

Figure 9.37 Ability of DG to increase the capacity of a distribution feeder is dependent on DG location.

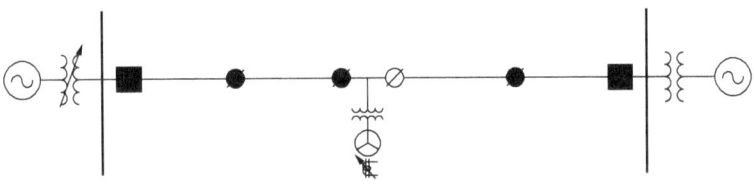

Figure 9.38 DG sited near the tie-point between two feeders to help support contingencies.

additional load expected to be switched over during a failure. This results in substantial overcapacity when the system is in its normal state with no failures.

One potentially good economic application of DG is to provide support for feeders when it is necessary to switch them to an alternate source while repairs are made. Figure 9.38 depicts the use of DG located on the feeder for this purpose. This will be substantially less costly than building a new feeder or upgrading a substation to cover this contingency.

The DG in this case is located near the tie-point between two feeders. It is not necessarily used for feeder support during normal conditions although there would often be some benefits to be gained by operating the DG at peak load. When a failure occurs on either side of the tie, the open tie switch is closed to pick up load from the opposite side. The DG is dispatched on and connected to help support the backup feeder.

Locating the DG in this manner gives the utility additional flexibility and more reconfiguration options. Currently, the most common DG technology used for this application is currently diesel gensets. The gensets may be mounted on portable trailers and leased only for the peak load season when a particular contingency leaves the system vulnerable. One or more units may be interconnected through a pad-mounted transformer and may also employ a recloser with a DG protection relay. This makes a compact and safe interconnection package using equipment familiar to utility personnel.

9.8 Interconnection Standards

Standards for interconnection of DG to distribution systems are examined in this section. Two examples illustrating the range of requirements for interconnection protection are presented.

9.8.1 Industry Standards Efforts

There have been two main DG interconnection standards efforts in the United States. IEEE Standard 929–2000[5] was developed to address requirements for inverters used in PV systems interconnected with utility systems. The standard has been generally applied to all technologies requiring an inverter interface. One of the main issues this standard addresses is the anti-islanding scheme. The basic idea is to introduce a destabilizing signal into the switching control so that it will quickly drift in frequency if allowed to run isolated while the control thinks it is still interconnected. Amid fears that vendors would independently choose schemes that might cancel out each other, agreement was reached on a uniform direction to drive the frequency.

Another effort has been the development of IEEE Standard 1547–2003.[10] The intent is to develop a national standard that will apply to the interconnection of all types of DG to both the radial and network distribution systems. Vendors, utilities, and end users have joined in this effort, which is continuing with several additions to the standard. This standard addresses many of the issues described in this chapter, and the approach taken here is largely consistent with the approach described here.

9.8.2 Interconnection Requirements

The basic requirements for interconnecting DG to the utility distribution system are listed here.

Voltage Regulation

DG shall not attempt to regulate voltage while interconnected unless special agreement is reached with the utility. As pointed out previously, this generally means that the DG will operate at a constant power factor or constant reactive power output acceptable to the operation of the system. Inverters in utility-interactive mode would typically operate by producing a current in phase with the voltage to achieve a particular power output level.

Anti-Islanding

DG shall have relaying that is capable of detecting when it is operating as an island and disconnect from the power system. Inverters should be compliant with IEEE Standard 929–2000 such that they would naturally drift in frequency when isolated from the utility source. Relaying to detect resonant conditions that might occur should be applied in susceptible DG applications.

Fault Detection

DG shall have relaying capable of detecting faults on the utility system and disconnecting after a time delay of typically 0.16 to 2.0 s, depending on the amount of deviation from normal. DG should disconnect sufficiently early in the first reclose interval to allow temporary faults to clear. (The utility may have to extend the first reclose interval to ensure that this can be accomplished.) However, to prevent nuisance tripping of the DG, the tripping should not be too fast. The 0.16-s (10 cycles at 60 Hz) delay is to allow time for faults on the transmission system or adjacent feeder to clear before tripping the DG needlessly.

Settings proposed for voltage and frequency relays for this application are given in Table 9.1.[10] The cutoff voltages are nominal guidelines and may have to be modified for some applications. A common adjustment is to decrease the voltage trip levels to avoid nuisance tripping for faults on parallel feeders. For example, faults on parallel feeders will sometimes give voltages less than 50 percent, requiring the setting on the 10-cycle trip to be reduced to perhaps 40 percent. The frequency trip settings may be adjusted according to local standards. Some utilities may want larger DG to remain connected to a much lower frequency (e.g., 57 Hz) to help with system stability issues following loss of a major generating plant or a tie-line.

Condition	Clearing time, s
$V \leq 50\%$	0.16
$50\% < V \leq 88\%$	2.0
$110\% < V \leq 120\%$	1.0
$V > 120\%$	0.16
$f > 59.3$ Hz	0.16
$f < 60.5$ Hz	0.16

TABLE 9.1 Typical Voltage and Frequency Relay Settings for DG Interconnection for a 60-Hz System

Direct Transfer Trip (Optional)

For applications where it is difficult to detect islands and utility-side faults, or where it is not possible to coordinate with utility fault-clearing devices, direct transfer trip should be applied such that the DG interconnect breaker is tripped simultaneously with the utility breaker. Transfer trip is usually advisable when DG is permitted to operate with automatic voltage control because this situation is much more likely to support an inadvertent island. Transfer trip is relatively costly and is generally applied only on large DG systems. Two relaying schemes for meeting these requirements are presented in Secs. 9.8.3 and 9.8.4.

9.8.3 A Simple Interconnection

The protection scheme shown in Fig. 9.39 applies to small systems that are not expected to be able to support islands by themselves. There is not universal agreement on what constitutes a "small" DG system. Some utilities draw the line at 30 kW, while others might restrict this to less than 10 kW. Some may allow this kind of interface protection for sizes up to 100 kW, or more. The two relaying functions shown are expected to do most of the work even for large DG systems. Large systems have additional relaying to provide a greater margin of safety.

Small DG systems would commonly be connected to the load bus at secondary voltage levels. There would not be a separate transformer, although there may be separate metering. Overcurrent protection is

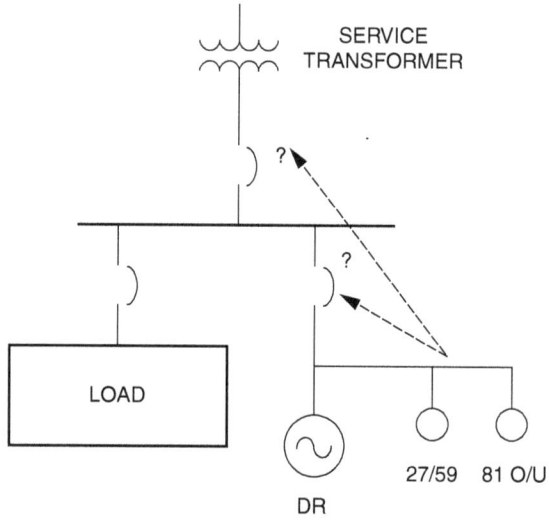

FIGURE 9.39 Simple interconnection protection scheme for smaller generators.

provided by molded case circuit breakers. The main DG interface protection functions are

1. Over/under (O/U) voltage (27/59 relay)
2. Over/under frequency (81 O/U relay)

These relays can be used to trip either the generator breaker or the main service breaker, depending on the desired mode of operation. Tripping only the generator leaves the load connected, and this is probably the desired operation for most loads employing small cogeneration or peaking generators. However, the utility may require the main breaker to be tripped if the DG system is running when a disturbance occurs.

The main service breaker would also be tripped if the DG system is to be used for backup power so that the DG system can continue to supply the load off-line. It should be noted that special controls (not shown in Fig. 9.39) may be required for this transfer to occur seamlessly. It is not always easy to accomplish.

The over/under voltage relay has the primary responsibility to detect utility-side disturbances. There should be no frequency deviation until the utility fault interrupter opens. If the fault is very close to the generator interconnection point and the voltage sag is deep, the overcurrent relaying may also see the fault. This will depend on the capability of the DG system to supply fault current. The overcurrent breakers are necessary for protecting the DG system in case of an internal fault.

Once the distribution feeder is separated from the utility bulk power system, an island forms. The voltage and frequency relays then work in concert to detect the island. One would normally expect the voltage to collapse very quickly and be detected by the undervoltage relay. If this does not happen for some reason, the frequency should quickly drift outside the narrow band expected while interconnected so that the 81 O/U relay would detect it.

9.8.4 A Complex Interconnection

The second protection scheme described here represents the other extreme from the simple scheme presented in Sec. 9.8.3. Figure 9.40 shows the key functions in an actual distribution-connected DG installation that employs a primary-side recloser. This is a relatively complex interconnection protection scheme for a large synchronous generator. There are many other variant schemes that may also be applied, and the reader is referred to vendors of DG packages whose literature describes these in great detail.

A large DG installation on the distribution system would typically correspond to generators in the 1- to 10-MW range. Most generators larger than this will be interconnected at the

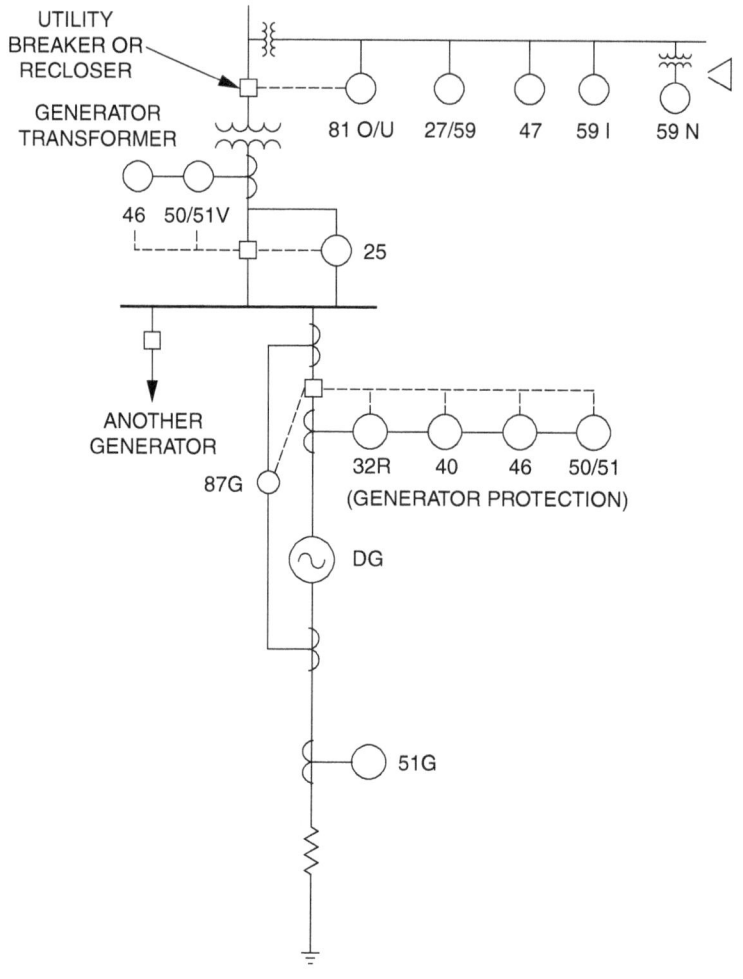

UTILITY
BREAKER OR
RECLOSER

GENERATOR
TRANSFORMER

81 O/U   27/59   47   59 I   59 N

46 50/51V

25

ANOTHER
GENERATOR

87G

32R   40   46   50/51

(GENERATOR PROTECTION)

DG

51G

FIGURE 9.40   Protection scheme for a large synchronous generator with high-side recloser.

transmission level and have relaying similar to utility central station generation.

Figure 9.40 shows the relays necessary for interface protection as well as some of the relays necessary for generator protection. Not all the functions that might be necessary for proper control of the generator, interlocking of breakers, etc., are shown. This installation is comprised of multiple generators connected identically.

In this example, there is a primary-side utility breaker for which utilities will typically use a common three-phase recloser. This is a convenient switchgear package for utilities to install and probably the least costly as well. The recloser comes with overcurrent relaying (not shown), and a separate DG relay package has been added that operates off a separate potential transformer. This is the main breaker used to achieve or ensure separation of the generator(s) from the utility.

The relaying elements in the system and their function are as follows.

Primary Side

- *27/59: standard under/over voltage relay.* This serves as the primary means of fault and island detection. This can be used to block closing of the breaker until there is voltage present on the utility system, or there may be a separate relay for that purpose.
- *81 O/U: standard over/under frequency relay for islanding detection.*
- *47: negative-sequence voltage relay (optional).* This is a backup means for detecting utility-side faults that can be more sensitive than voltage magnitudes in some cases. Also, it helps prevent generator damage due to unbalance, although there is another relay for that here.
- *59I: instantaneous (peak) overvoltage.* This is a supplemental islanding detection function. This would be employed in cases where ferroresonance or other resonance phenomena are likely. This would occur when utility-side capacitors interact with the generator reactance. Since such overvoltages can cause damage quickly, the time delay is much shorter than for the other relays—but not so short that it trips on utility capacitor-switching transients.
- *59N (or 59G): neutral or ground overvoltage.* This relay is installed in the corner of a broken delta connection on the potential transformer. It is a supplemental fault and islanding detection relay function that measures the zero-sequence voltage. This would detect conditions in which the generator is islanded on an SLG fault. It is more necessary when the primary connection of the transformer is delta or ungrounded-wye.

These relaying functions may be moved to the secondary side of the service transformer if there is no high-side breaker. The relays would then trip the main breaker on the secondary side.

No reverse power (32) function is used at this interface because net export is expected.

Generator Side

- *50/51: overcurrent relay.* Responsible for tripping the main breaker for faults within the generator system. May also trip for faults on the utility system that the generator feeds. Therefore, the time delay must be coordinated with the other relays so that it does not trip inadvertently.

- *46 relay at transformer: negative-sequence current.* Assists in the detection of faults on the utility system, particularly open-phase conditions, and trips the main breaker. (Generators have a separate 46 relay.)

- *25: synchronizing relay.* Controls closing of the main breaker when the generators are being interconnected to the utility. (This scheme would also require synchronous check relays on the individual generators if they are to be interconnected separately.)

Generator Protection

- *87G: differential ground relay.* For fast detection of ground faults within the generator.

- *51G: ground overcurrent.* Trips the generator for high neutral currents, an indicative of a ground fault on the secondary system.

- *32R: reverse-power relay.* This relay detects power going into the generator, which would indicate a fault. Can be set very sensitive.

- *40: loss of field relay.*

- *46: negative-sequence current.* Protects the machine against excessive unbalanced currents, which may result from an internal fault but may also be due to unbalance on the utility system.

- *50/51: overcurrent relays.* Protects the generator against excessive loads and faults on either side of the generator breaker.

9.9   Summary

Readers might easily get the impression from the material in this chapter that interconnecting a DG installation to the distribution system is fraught with Gordian knot–like entanglement power quality problems. However, few problems can be expected for most DG applications in the near future while the total penetration is relatively low. There is a significant amount of DG that can be accommodated without affecting the operation of the distribution system, but there is a limit. The grid is not infinite in capacity.

As a general rule, problems begin to appear when the total interconnected DG capacity approaches 15 percent of the feeder capacity.[11,12] This might drop to as little as 5 percent of capacity on more rural feeders or be as high as 30 percent if the DG is clustered near the substation. Voltage regulation problems are often the first to appear, followed by interference with the utility fault-clearing process, which includes concerns for islanding.

Changes can be made to accommodate nearly any amount of DG. As the amount of DG increases, the simple, low-cost distribution system design must be abandoned in favor of a more capable design. It will almost certainly be more costly, but engineers can make it work. Deciding who pays for it is another matter.

In a future of massively distributed generation, as some see it, communications and control will be key. Today, most of the control of distribution systems is accomplished by local intelligence operating autonomously. Systems with high penetrations of DG would benefit greatly from fast, interconnected communications networks. This is one technology shift that must accompany the spread of DG if it is to be successful in contributing to reliable, high-quality electric power.

9.10 References

1. H. L. Willis and W. G. Scott, *Distributed Power Generation Planning and Evaluation*, Marcel Dekker, New York, 2000.
2. N. Jenkins, R. Allan, P. Crossley, D. Kirschen, G. Strbac, *Embedded Generation*, The Institute of Electrical Engineers, London, U.K., 2000.
3. W. E. Feero and W. B. Gish, "Overvoltages Caused by DSG Operation: Synchronous and Induction Generators," *IEEE Transactions on Power Delivery*, January 1986, pp. 258–264.
4. R. C. Dugan and D. T. Rizy, *Harmonic Considerations for Electric Distribution Feeders*, ORNL/Sub/81–95011/4, Oak Ridge National Laboratory, U.S. DOE, March 1988.
5. IEEE Standard 929–2000, *Recommended Practice for Utility Interface of Photovoltaic Systems*.
6. R. C. Dugan, and T. E. McDermott, "Operating Conflicts for Distributed Generation on Distribution Systems," *IEEE IAS 2001 Rural Electric Power Conference Record*, IEEE Catalog No. 01CH37214, Little Rock, Ark., May 2001, Paper No. 01-A3.
7. *Electrical Distribution-System Protection*, 3d ed., Cooper Power Systems, Franksville, Wis., 1990.
8. R. H. Hopkinson, "Ferroresonance Overvoltage Control Based on TNA Tests of Three-Phase Delta-Wye Transformer Banks," *IEEE Transactions on Power Apparatus and Systems*, Vol. 86, No. 10, October 1967, pp. 1258–1265.
9. D. R. Smith, S. R. Swanson, J. D. Borst, "Overvoltages with Remotely-Switched Cable-Fed Grounded Wye-Wye Transformers," *IEEE Transactions on Power Apparatus and Systems*, Vol. PAS-94, No. 5, September/October 1975, pp. 1843–1853.
10. IEEE Standard 1547–2003, *Interconnecting Distributed Resources with Electric Power Systems*, Draft 8, P1547 Working Group of IEEE SCC 21, T. Basso, Secretary.
11. *Protection of Electric Distribution Systems with Dispersed Storage and Generation (DSG) Devices*, Oak Ridge National Laboratory, Report ORNL/CON-123, September 1983.

12. R. C. Dugan, T. E. McDermott, D. T. Rizy, S. Steffel, "Interconnecting Single-Phase Backup Generation to the Utility Distribution System," *Transmission and Distribution Conference and Exposition,* 2001 IEEE/PES, Vol. 1, 2001, pp. 486–491.

13. R. Dugan and J. Smith, "Determining Practical Planning Limits for DG on Distribution Circuits, CIRED 2011, Paper No. 1277, Frankfort, June, 2011. *Planning Methodology to Determine Practical Circuit Limits for Distributed Generation: With Emphasis on Solar PV and Other Renewable Generation,* EPRI, Palo Alto, CA: 2010. 1020157.

14. R. F. Arritt and R. C. Dugan, *Distributed Generation Interconnection Transformer and Grounding Selection,* 2008 IEEE PES General Meeting Proceedings, Pittsburgh, PA, July 20–24, 2008.

9.11 Bibliography

Dugan, R. C., McDermott, T. E., Ball, G. J., "Distribution Planning for Distributed Generation," *IEEE IAS Rural Electric Power Conference Record,* IEEE Catalog No. 00CH37071, Louisville, Ky., May 7–9, 2000, pp. C4–1–C4–7.

Engineering Handbook for Dispersed Energy Systems on Utility Distribution Systems, EPRI Final Report, TR-105589, November 1995.

Integration of Distributed Resources in Electric Utility Systems: Current Interconnection Practice and Unified Approach, EPRI Final Report, TR-111489, November 1998.

"Interconnecting Distributed Generation to Utility Distribution Systems," Short Course, The Department of Engineering Professional Development, University of Wisconsin—Madison, 2001.

Dugan, Roger C., "Computing Incremental Capacity Provided By Distributed Resources For Distribution Planning," *IEEE PES 2007 General Meeting Conference Proceedings,* Tampa, FL, 24–28 June, 2007.

Smith, Jeff, Brooks, Daniel, Taylor, Jason, and Dugan, Roger, "Interconnection Studies for Wind Generation," *2004 IEEE Rural Electric Power Conference Proceedings,* IEEE Catalog No. 04CH37528, Scottsdale, AZ, May 2004, Paper No. C3.

Dugan, Roger C., Key, Thomas S., and Ball, Greg J., "On Standards for Interconnecting Distributed Resources ", *2005 IEEE Rural Electric Power Conference Proceedings,* San Antonio, TX, May8–10, 2005.

CHAPTER 10

Wiring and Grounding

Many power quality variations that occur within customer facilities are related to wiring and grounding problems. It is commonly stated at power quality conferences and in journals that 80 percent of all the power quality problems reported by customers are related to wiring and grounding problems within a facility. While this may be an exaggeration, many power quality problems are solved by simply tightening a loose connection or replacing a corroded conductor. Therefore, an evaluation of wiring and grounding practices is a necessary first step when evaluating power quality problems in general.

The *National Electrical Code®* (*NEC®*)* and other important standards provide the minimum standards for wiring and grounding. It is often necessary to go beyond the requirements of these standards to achieve a system that also minimizes the impact of power quality variations (harmonics, transients, noise) on connected equipment. While the intent of this book is to concentrate on subjects that are more amenable to engineering analysis, the basic principles of wiring and grounding are presented in this chapter to provide the reader with at least a fundamental understanding of why things are done. References are provided throughout the text for readers interested in further details.

10.1 Resources

Selected definitions are presented here from the *IEEE Dictionary* (Standard 100), the *IEEE Green Book* (IEEE Standard 142), and the *NEC*. These are the fundamental resources on wiring and grounding. The *IEEE Green Book* and the *NEC* provide extensive information on proper grounding practices for safety considerations and proper system operation. However, these documents do not address concerns for power quality.

National Electrical Code® and *NEC®* are registered trademarks of the National Fire Protection Association, Inc., Quincy, Mass. 02269.

Power quality considerations associated with wiring and grounding practices are covered in Federal Information Processing Standard (FIPS) 94, *Guideline on Electrical Power for ADP Installations* (1983). This is the original source of much of the information interpreted and summarized here.

The *IEEE Emerald Book* (ANSI/IEEE Standard 1100–1992, *IEEE Recommended Practice for Powering and Grounding Sensitive Electronic Equipment*) updates the information presented in FIPS 94. This is an excellent resource for wiring and grounding with respect to power quality issues and is highly recommended.

Grounding guidelines to minimize noise in electronic circuits are also covered in IEEE Standard 518, *IEEE Guide for the Installation of Electrical Equipment to Minimize Electrical Noise Inputs to Controllers from External Sources.* EPRI's *Wiring and Grounding for Power Quality* (Publication CU.2026.3.90) provides an excellent summary of typical wiring and grounding problems along with recommended solutions. Additional resources are provided in the Bibliography at the end of this chapter.

10.2   Definitions

Some of the key definitions of wiring and grounding terms from these documents are included here.

IEEE Dictionary (Standard 100) Definition*

Grounding

A conducting connection, whether intentional or accidental, by which an electric circuit or equipment is connected to the earth, or to some conducting body of relatively large extent that serves in place of the earth. It is used for establishing and maintaining the potential of the earth (or of the conducting body) or approximately that potential, on conductors connected to it; and for conducting ground current to and from the earth (or the conducting body).

IEEE Green Book (IEEE Standard 142) Definitions†

ungrounded system   A system, circuit, or apparatus without an intentional connection to ground, except through potential indicating or measuring devices or other very high impedance devices.

---

*Reprinted from IEEE Standard 100–1992, *IEEE Standard Dictionary of Electrical and Electronic Terms*, copyright © 1993 by the Institute of Electrical and Electronics Engineers, Inc. The IEEE disclaims any responsibility or liability resulting from the placement and use in this publication. Information is reprinted with the permission of the IEEE.

†Reprinted from IEEE Standard 142–1991, *IEEE Recommended Practice for Grounding of Industrial and Commerical Power Systems*, copyright © 1991 by the Institute of Electrical and Electronics Engineers, Inc. The IEEE disclaims any responsibility or liability resulting from the placement and use in this publication. Information is reprinted with the permission of the IEEE.

grounded system A system of conductors in which at least one conductor or point (usually the middle wire or neutral point of transformer or generator windings) is intentionally grounded, either solidly or through an impedance.

grounded solidly Connected directly through an adequate ground connection in which no impedance has been intentionally inserted.

grounded effectively Grounded through a sufficiently low impedance such that for all system conditions the ratio of zero sequence reactance to positive sequence reactance (X0/X1) is positive and less than 3, and the ratio of zero sequence resistance to positive sequence reactance (R0/X1) is positive and less than 1.

resistance grounded Grounded through impedance, the principal element of which is resistance.

inductance grounded Grounded through impedance, the principal element of which is inductance.

NEC definitions.* Refer to Fig. 10.1.

grounding electrode The grounding electrode shall be as near as practicable to and preferably in the same area as the grounding conductor connection to the system. The grounding electrode shall be: (1) the nearest available effectively grounded structural metal

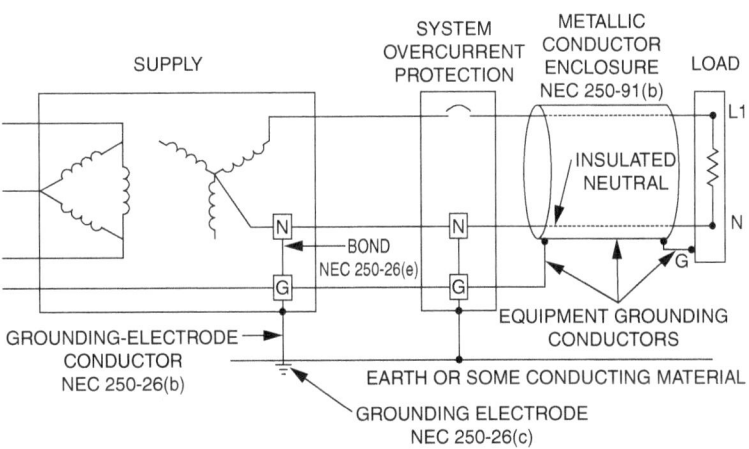

FIGURE **10.1** Terminology used in *NEC* definitions.

member of the structure; or (2) the nearest available effectively grounded metal water pipe; or (3) other electrodes (Section 250–81 & 250–83) where electrodes specified in (1) and (2) are not available.

grounded   Connected to earth or to some conducting body that serves in place of the earth.

grounded conductor   A system or circuit conductor that is intentionally grounded (the neutral is normally referred to as the grounded conductor).

grounding conductor   A conductor used to connect equipment or the grounded circuit of a wiring system to a grounding electrode or electrodes.

grounding conductor, equipment   The conductor used to connect the noncurrent-carrying metal parts of equipment, raceways, and other enclosures to the system grounded conductor and/or the grounding electrode conductor at the service equipment or at the source of a separately derived system.

grounding electrode conductor   The conductor used to connect the grounding electrode to the equipment grounding conductor and/or to the grounded conductor of the circuit at the service equipment or at the source of a separately derived system.

grounding electrode system   Defined in *NEC* Section 250–81 as including: (a) metal underground water pipe; (b) metal frame of the building; (c) concrete-encased electrode; and (d) ground ring. When these elements are available, they are required to be bonded together to form the grounding electrode system. Where a metal underground water pipe is the only grounding electrode available, it must be supplemented by one of the grounding electrodes specified in Section 250–81 or 250–83.

bonding jumper, main   The connector between the grounded circuit conductor (neutral) and the equipment grounding conductor at the service entrance.

branch circuit   The circuit conductors between the final overcurrent device protecting the circuit and the outlets.

conduit enclosure bond   (bonding definition) The permanent joining of metallic parts to form an electrically conductive path, which will assure electrical continuity and the capacity to conduct safely any current likely to be imposed.

feeder   All circuit conductors between the service equipment of the source of a separately derived system and the final branch circuit overcurrent device.

outlet   A point on the wiring system at which current is taken to supply utilization equipment.

overcurrent   Any current in excess of the rated current of equipment or the capacity of a conductor. It may result from overload, short circuit, or ground fault.

panel board   A single panel or group of panel units designed for assembly in the form of a single panel; including buses, automatic overcurrent devices, and with or without switches for the control of light, heat, or power circuits; designed to be placed in a cabinet or cutout box placed in or against a wall or partition and accessible only from the front.

separately derived systems   A premises wiring system whose power is derived from a generator, a transformer, or converter windings and has no direct electrical connection, including a solidly connected grounded circuit conductor, to supply conductors originating in another system.

service equipment   The necessary equipment, usually consisting of a circuit breaker switch and fuses, and their accessories, located near the point of entrance of supply conductors to a building or other structure, or an otherwise defined area, and intended to constitute the main control and means of cutoff of the supply.

ufer ground   A method of grounding or connection to the earth in which the reinforcing steel (rebar) of the building, especially at the ground floor, serves as a grounding electrode.

10.3   Reasons for Grounding

The most important reason for grounding is safety. Two important aspects to grounding requirements with respect to safety and one with respect to power quality are

1. *Personnel safety.* Personnel safety is the primary reason that all equipment must have a safety equipment ground. This is designed to prevent the possibility of high touch voltages when there is a fault in a piece of equipment (Fig. 10.2). The touch voltage is the voltage between any two conducting surfaces that can be simultaneously touched by an individual. The earth may be one of these surfaces.

There should be no "floating" panels or enclosures in the vicinity of electric circuits. In the event of insulation failure or inadvertent application of moisture, any electric charge which appears on a panel, enclosure, or raceway must be drained to "ground" or to an object which is reliably grounded.

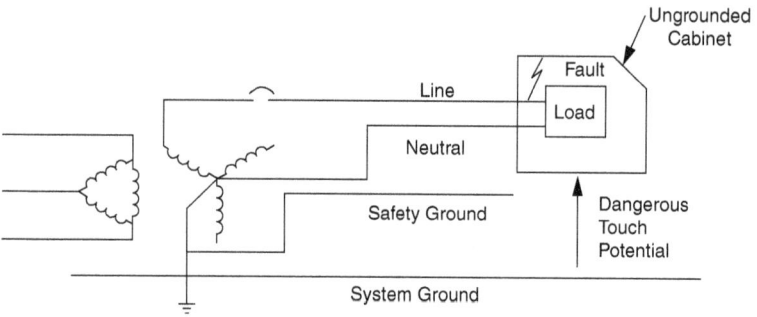

FIGURE 10.2 High touch voltage created by improper grounding.

2. *Grounding to assure protective device operation.* A ground fault return path to the point where the power source neutral conductor is grounded is an essential safety feature. The *NEC* and some local wiring codes permit electrically continuous conduit and wiring device enclosures to serve as this ground return path. Some codes require the conduit to be supplemented with a bare or insulated conductor included with the other power conductors.

An insulation failure or other fault that allows a phase wire to make contact with an enclosure will find a low-impedance path back to the power source neutral. The resulting overcurrent will cause the circuit breaker or fuse to disconnect the faulted circuit promptly.

NEC Article 250–51 states that an effective grounding path (the path to ground from circuits, equipment, and conductor enclosures) shall

a. Be permanent and continuous

b. Have the capacity to conduct safely any fault current likely to be imposed on it

c. Have sufficiently low impedance to limit the voltage to ground and to facilitate the operation of the circuit protective devices in the circuit

d. Not have the earth as the sole equipment ground conductor

3. *Noise control.* Noise control includes transients from all sources. This is where grounding relates to power quality. Grounding for safety reasons defines the minimum requirements for a grounding system. Anything that is done to the grounding system to improve the noise performance must be done in addition to the minimum requirements defined in the *NEC* and local codes.

The primary objective of grounding for noise control is to create an equipotential ground system. Potential differences between different ground locations can stress insulation, create circulating ground currents in low-voltage cables, and interfere with sensitive equipment that may be grounded in multiple locations.

Ground voltage equalization of voltage differences between parts of an automated data processing (ADP) grounding system is accomplished in part when the equipment grounding conductors are connected to the grounding point of a single power source. However, if the equipment grounding conductors are long, it is difficult to achieve a constant potential throughout the grounding system, particularly for high-frequency noise. Supplemental conductors, ground grids, low-inductance ground plates, etc., may be needed for improving the power quality. These must be used in addition to the equipment ground conductors, which are required for safety, and not as a replacement for them.

10.4   Typical Wiring and Grounding Problems

Sections 10.4.1 to 10.4.7 describe some typical power quality problems that are due to inadequacies in the wiring and grounding of electrical systems. It is useful to be aware of these typical problems when performing site surveys because many of the problems can be detected through simple observations. Other problems require measurements of voltages, currents, or impedances in the circuits.

10.4.1   Problems with Conductors and Connectors

One of the first things to be done during a site survey is to inspect the service entrance, main panel, and major subpanels for problems with conductors or connections. A bad connection (faulty, loose, or resistive) will result in heating, possible arcing, and burning of insulation. Table 10.1 summarizes some of the wiring problems that can be uncovered during a site survey.

10.4.2   Missing Safety Ground

If the safety ground is missing, a fault in the equipment from the phase conductor to the enclosure results in line potential on the exposed surfaces of the equipment. No breakers will trip, and a hazardous situation results (see Fig. 10.2).

10.4.3   Multiple Neutral-to-Ground Connections

Unless there is a separately derived system, the only neutral-to-ground bond should be at the service entrance. The neutral and ground should be kept separate at all panel boards and junction boxes. Downline neutral-to-ground bonds result in parallel paths for the load return current where one of the paths becomes the ground circuit. This can cause misoperation of protective devices. Also,

Problem observed	Possible cause
Burnt smell at the panel, junction box, or load equipment	Faulted conductor, bad connection, arcing, or overloaded wiring
Panel or junction box is warm to the touch	Faulty circuit breaker or bad connection
Buzzing (corona effect)	Arcing
Scorched insulation	Overloaded wiring, faulted conductor, or bad connection
No voltage at load equipment	Tripped breaker, bad connection, or faulted conductor
Intermittent voltage at load equipment	Bad connection or arcing
Scorched panel or junction box	Bad connection or faulted conductor

TABLE 10.1   Problems with Conductors and Connectors

during a fault condition, the fault current will split between the ground and the neutral, which could prevent proper operation of protective devices (a serious safety concern). This is a direct violation of the *NEC*.

10.4.4   Ungrounded Equipment
Isolated grounds are sometimes used due to the perceived notion of obtaining a "clean" ground. The proper procedure for using an isolated ground must be followed (see Sec. 10.5.5). Procedures that involve having an illegal insulating bushing in the power source conduit and replacing the prescribed equipment grounding conductor with one to an "isolated dedicated computer ground" are dangerous, violate code, and are unlikely to solve noise problems.

10.4.5   Additional Ground Rods
Ground rods should be part of a facility grounding system and connected where all the building grounding electrodes (building steel, metal water pipe, etc.) are bonded together. Multiple ground rods can be bused together at the service entrance to reduce the overall ground resistance. Isolated grounds can be used for sensitive equipment, as described previously. However, these should not include isolated ground rods to establish a new ground reference for the equipment. One very important power quality problem with additional ground rods is that they create additional

paths for lightning stroke currents to flow. With the ground rod at the service entrance, any lightning stroke current reaching the facility goes to ground at the service entrance and the ground potential of the whole facility rises together. With additional ground rods, a portion of the lightning stroke current will flow on the building wiring (green ground conductor and/or conduit) to reach the additional ground rods. This creates a possible transient voltage problem for equipment and a possible overload problem for the conductors.

10.4.6   Ground Loops

Ground loops are one of the most important grounding problems in many commercial and industrial environments that include data processing and communication equipment. If two devices are grounded via different paths and a communication cable between the devices provides another ground connection between them, a ground loop results. Slightly different potentials in the two power system grounds can cause circulating currents in this ground loop if there is indeed a complete path. Even if there is not a complete path, the insulation that is preventing current flow may flash over because the communication circuit insulation levels are generally quite low.

Likewise, very low magnitudes of circulating current can cause serious noise problems. The best solution to this problem in many cases is to use optical couplers in the communication lines, thereby eliminating the ground loop and providing adequate insulation to withstand transient overvoltages. When this is not practical, the grounded conductors in the signal cable may have to be supplemented with heavier conductors or better shielding. Equipment on both ends of the cable should be protected with arresters in addition to the improved grounding because of the coupling that can still occur into signal circuits.

10.4.7   Insufficient Neutral Conductor

Switch-mode power supplies and fluorescent lighting with electronic ballasts are widely used in commercial environments. The high third-harmonic content present in these load currents can have a very important impact on the required neutral conductor rating for the supply circuits.

Third-harmonic currents in a balanced system appear in the zero-sequence circuit. This means that third-harmonic currents from three single-phase loads will add in the neutral, rather than cancel as is the case for the 60-Hz current. In typical commercial buildings with a diversity of switched-mode power supply loads, the neutral current is typically in the range 140 to 170 percent of the fundamental frequency phase current magnitude.

The possible solutions to neutral conductor overloading include the following:

- Run a separate neutral conductor for each phase in a three-phase circuit that serves single-phase nonlinear loads.

- When a shared neutral must be used in a three-phase circuit with single-phase nonlinear loads, the neutral conductor capacity should be approximately double the phase conductor capacity.

- Delta-wye transformers (see Sec. 10.5.6) designed for nonlinear loads can be used to limit the penetration of high neutral currents. These transformers should be placed as close as possible to the nonlinear loads (e.g., in the computer room). The neutral conductors on the secondary of each separately derived system must be rated based on the expected neutral current magnitudes.

- Filters to control the third-harmonic current that can be placed at the individual loads are becoming available. These will be an alternative in existing installations where changing the wiring may be an expensive proposition.

- Zigzag transformers provide a low impedance for zero-sequence harmonic currents and, like filters, can be placed at various places along the three-phase circuit to shorten the path of third-harmonic currents and better disperse them.

10.5   Solutions to Wiring and Grounding Problems

10.5.1   Proper Grounding Practices

Figure 10.3 illustrates the basic elements of a properly grounded electrical system. The important elements of the electrical system grounding are described in Secs. 10.5.2 to 10.5.5.

10.5.2   Ground Electrode (rod)

The ground rod provides the electrical connection from the power system ground to earth. The item of primary interest in evaluating the adequacy of the ground rod is the resistance of this connection. There are three basic components of resistance in a ground rod:

- *Electrode resistance.* Resistance due to the physical connection of the grounding wire to the grounding rod.

- *Rod-earth contact resistance.* Resistance due to the interface between the soil and the rod. This resistance is inversely proportional to the surface area of the grounding rod (i.e., more area of contact means lower resistance).

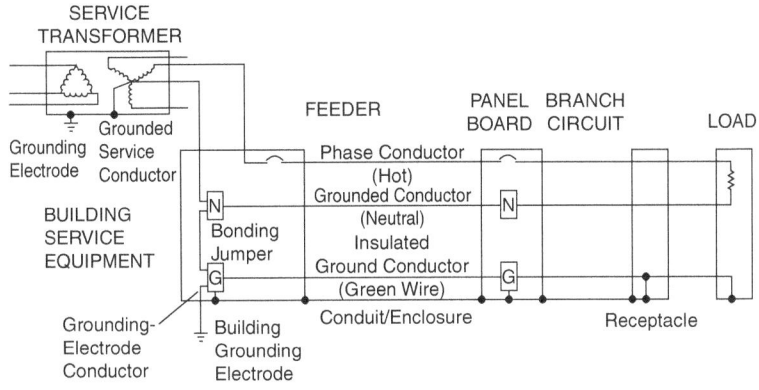

FIGURE **10.3** Basic elements of a properly grounded electrical system.

- *Ground resistance.* Resistance due to the resistivity of the soil in the vicinity of the grounding rod. The soil resistivity varies over a wide range, depending on the soil type and moisture content.

The resistance of the ground-rod connection is important because it influences transient voltage levels during switching events and lightning transients. High-magnitude currents during lightning strokes result in a voltage across the resistance, raising the ground reference for the entire facility. The difference in voltage between the ground reference and true earth ground will appear at grounded equipment within the facility, and this can result in dangerous touch potentials.

10.5.3 Service Entrance Connections

The primary components of a properly grounded system are found at the service entrance. The neutral point of the supply power system is connected to the grounded conductor (neutral wire) at this point. This is also the one location in the system (except in the case of a separately derived system) where the grounded conductor is connected to the ground conductor (green wire) via the bonding jumper. The ground conductor is also connected to the building grounding electrode via the grounding-electrode conductor at the service entrance. For most effective grounding, the grounding-electrode conductor should be exothermically welded at both ends.

The grounding-electrode conductor is sized based on guidelines in the *NEC* (Section 250–94). *NEC* table 250–94 (reproduced in Table 10.2) provides the basic guidelines.

There are a number of options for the building grounding electrode. It is important that all of the different grounding electrodes

Size of largest service entrance conductor or equivalent area for parallel conductors		Size of grounding-electrode conductor	
Copper	Aluminum or copper-clad aluminum	Copper	Aluminum or copper-clad aluminum
2 or smaller	0 or smaller	8	6
1 or 0	2/0 or 3/0	6	4
2/0 or 3/0	4/0 or 250 MCM*	4	2
Over 3/0–350 MCM	Over 250 MCM–500 MCM	2	0
Over 350 MCM–600 MCM	Over 500 MCM–900 MCM	0	3/0
Over 600 MCM–1100 MCM	Over 900 MCM–1750 MCM	2/0	4/0
Over 1100 MCM	Over 1750 MCM	3/0	250 MCM

*MCM = thousand circular mil (unit of wire size).

TABLE **10.2**   Grounding-Electrode Conductor for AC Systems

used in a building are connected together at the service entrance. The following are permissible for use as grounding electrodes:

- *Underground water pipe.* (See *NEC* table 250–94 for grounding-electrode conductor requirements for connection to the neutral bus.)

- *Building steel.* (See *NEC* table 250–94 for grounding-electrode conductor requirements for connection to the neutral bus or the underground water pipe.)

- *Ground ring.* A ground ring can be used in addition to building steel to provide a better equipotential ground for the grounding electrode. It is connected to the main grounding electrode with a conductor that is not larger than the ground ring conductor.

- *Concrete encased electrode.* This can serve a similar purpose to a ground ring and is connected to the main grounding electrode with a conductor that has a minimum size of #4 AWG.

- *Ground rod.* The ground rod is connected to the main building grounding electrode with a conductor that has a minimum size of #6 AWG.

Throughout the system, a safety ground must be maintained to ensure that all exposed conductors that may be touched are kept at an equal potential. This safety ground also provides a ground fault return path to the point where the power source neutral conductor is grounded. The safety ground can consist of the conduit itself or the conduit and a separate conductor (ground conductor or green wire) in the conduit. This safety ground originates at the service entrance and is carried throughout the building.

10.5.4   Panel Board

The panel board is the point in the system where the various branch circuits are supplied by a feeder from the service entrance. The panel board provides breakers in series with the phase conductors; connects the grounded conductor (neutral) of the branch circuit to that of the feeder circuit; and connects the ground conductor (green wire) to the feeder ground conductor, conduit, and enclosure. It is important to note that there should not be a neutral-to-ground connection at the panel board. This neutral-to-ground connection is prohibited in the *NEC* as it would result in load return currents flowing in the ground path between the panel board and the service entrance. In order to maintain an equipotential grounding system, the ground path should not contain any load return current. Also, fault currents would split between the neutral conductor and the ground return path. Protection is based on the fault current flowing in the ground path.

10.5.5   Isolated Ground

The noise performance of the supply to sensitive loads can sometimes be improved by providing an isolated ground to the load. This is done using isolated ground receptacles, which are orange in color. If an isolated ground receptacle is being used downline from the panel board, the isolated ground conductor is not connected to the conduit or enclosure in the panel board, but only to the ground conductor of the supply feeder (Fig. 10.4). The conduit is the safety ground in this case and is connected to the enclosure. A separate conductor can also be used for the safety ground in addition to the conduit. This technique is described in the *NEC,* Article 274, Exception 4 on receptacles. It is not described as a grounding technique.

The isolated ground receptacle is orange in color for identification purposes. This receptacle does not have the ground conductor connected to the receptacle enclosure or conduit. The isolated ground conductor may pass back through several panel boards without being connected to local ground until grounded at the service entrance or other separately derived ground. The use of isolated ground receptacles requires careful wiring practices to avoid unintentional connections between the isolated ground and the safety ground. In general, dedicated branch circuits accomplish the same objective as isolated ground receptacles without the concern for complicated wiring.

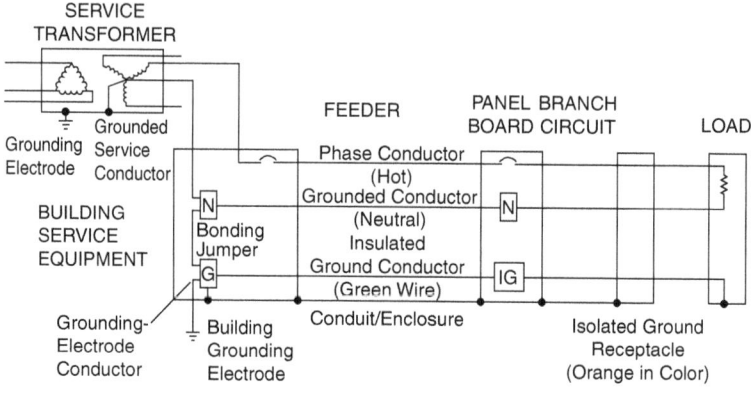

FIGURE 10.4   Grounding configuration for an isolated ground.

A special case of isolated grounds is used for grounding some hospital equipment. These procedures are described in the *NEC* and in the *White Book* (IEEE Standard 602).

10.5.6   Separately Derived Systems

A separately derived system has a ground reference that is independent from other systems. A common example of this is a delta-wye isolation transformer (Fig. 10.5). The wye-connected secondary neutral is connected to local building ground (not a separate ground rod) to provide a new ground reference independent from the rest of the system. The point in the system where this new ground reference is defined is like a service entrance in that the system neutral is connected to the grounded conductor (neutral wire), which is connected to the ground conductor with a bonding jumper.

Separately derived systems are used to provide a local ground reference for sensitive loads. The local ground reference can have significantly reduced noise levels as compared to the system ground if an isolation transformer is used to supply the separately derived system. An additional benefit is that neutral currents are localized to the load side of the separately derived system. This can help reduce neutral current magnitudes in the overall system when there are large numbers of single-phase nonlinear loads.

10.5.7   Grounding Techniques for Signal Reference

Most of the grounding requirements previously described deal with the concerns for safety and proper operation of protective devices. Grounding is also used to provide a signal reference point for equipment exchanging signals over communication or control circuits within a facility. The requirements for a signal reference

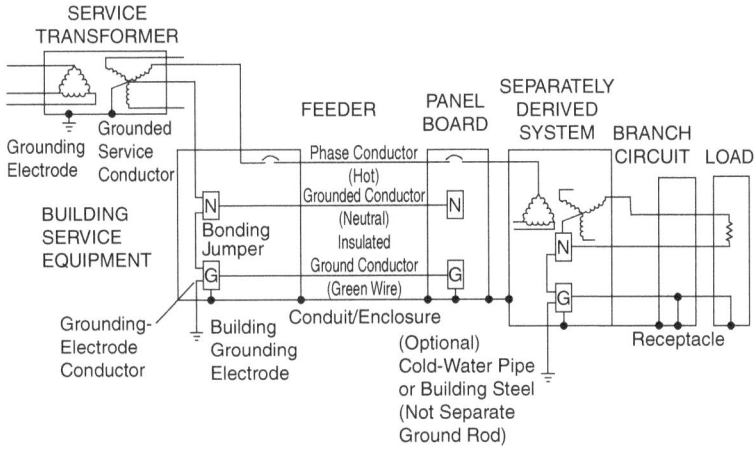

FIGURE 10.5 Configuration for a separately derived system.

ground are often significantly different from the requirements for a safety ground. However, the safety ground requirements must always be considered first whenever designing a grounding scheme.

The most important characteristic of a signal reference ground is that it must have a low impedance over a wide range of frequencies. One way to accomplish this (at least for low frequencies) is to use an adequately sized ground conductor. Conduit is particularly bad for a signal reference ground because it relies on continuity of connections and the impedance is high relative to the phase and neutral conductors. Undersized ground conductors have the same problem of high impedance. For reducing power quality problems, the ground conductor should be at least the same size as the phase conductors and the neutral conductor (the neutral conductor may need to be larger than the phase conductors in some special cases involving nonlinear single-phase loads).

As frequency increases, the wavelength becomes short enough to cause resonances for relatively short lengths of wire. A good rule of thumb is that when the length of the ground conductor is greater than one-twentieth of the signal wavelength, the ground conductor is no longer effective at that frequency. Since the grounding system is more complicated than a simple conductor, there is actually a complicated impedance versus frequency characteristic involved (Fig. 10.6).

One way to provide a signal reference ground to sensitive equipment that is effective over a wide range of frequencies (0 to 30 MHz) is to use a signal reference grid or zero reference grid (Fig. 10.7). This technique uses a rectangular mesh of copper wire with

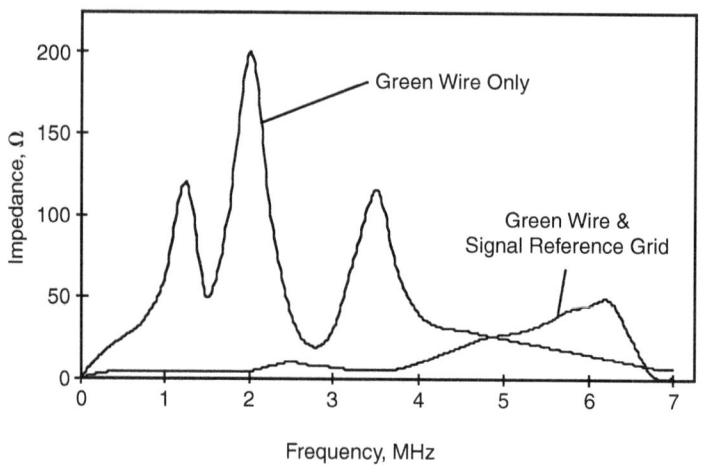

Figure 10.6   Effect of signal reference grid on ground impedance.

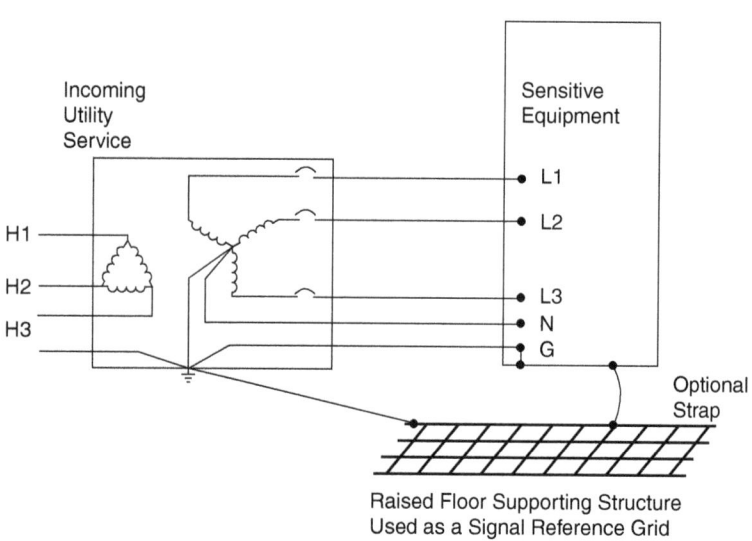

Figure 10.7   Use of a signal reference grid

about 2-ft spacing. It is commonly applied in large data-processing equipment rooms. Even if a portion of the conductor system is in resonance at a particular frequency, there will always be other paths of the grid that are not in resonance due to the multiple paths available for current to flow. When using a signal reference grid, the enclosure of each piece of equipment must still be connected to a single common ground via the ground conductor (*NEC* requirement). The enclosures may also be connected to the closest interconnection of the grid to provide a high-frequency, low-impedance signal reference. Figure 10.6 illustrates the effect of the signal reference grid on the overall ground impedance versus frequency characteristic.

10.5.8   More on Grounding for Sensitive Equipment

The following practices are appropriate for any installation with equipment that may be sensitive to noise or disturbances introduced due to coupling in the ground system:

- Whenever possible, use individual branch circuits to power sensitive equipment. Individual branch circuits provide good isolation for high-frequency transients and noise.

- Conduit should never be the sole source of grounding for sensitive equipment (even though it may be legal). Currents flowing on the conduit can cause interference with communications and electronics.

- Green-wire grounds should be the same size as the current-carrying conductors, and the individual circuit conduit should be bonded at both ends.

- Use building steel as a ground reference, whenever available. The building steel usually provides an excellent, low-impedance ground reference for a building. Additional ground electrodes (water pipes, etc.) can be used as supplemental to the building steel.

These practices are often applied in computer rooms, where the frequency response of the grounding system is even more important due to communication requirements between different parts of a computer system:

- Either install a signal reference grid under a raised floor or use the raised floor as a signal reference grid. This is not a replacement for the safety ground, but augments the safety ground for noise reduction.

- Add a transient suppression plate at or near the power entry point (with the power cabling laid on top of it) to provide a controlled capacitive and magnetic coupling noise bypass between building reinforced steel and the electrical ground conductors.

10.5.9   Summary of Wiring and Grounding Solutions

The grounding system should be designed to accomplish these minimum objectives:

1. There should never be load currents flowing in the grounding system under normal operating conditions. One can likely measure very small currents in the grounding system due to inductive coupling, capacitive coupling, and the connection of surge suppressors and the like. In fact, if the ground current is exactly zero, there is probably an open ground connection. However, these currents should be only a tiny fraction of the load currents.

2. There should be, as near as possible, an equipotential reference for all devices and locations in the system.

3. To avoid excessive touch potential safety risks, the housings of all equipment and enclosures should be connected to the equipotential grounding system.

The most important implications resulting from these objectives are:

1. There can only be one neutral-to-ground bond for any subsystem. A separately derived system may be created with a transformer, which establishes a new neutral-to-ground bond.

2. There must be sufficient interconnections in the equipotential plane to achieve a low impedance over a wide frequency range.

3. All equipment and enclosures should be grounded.

10.6   Bibliography

In addition to the standard reference books cited earlier, the following books are recommended for further reading on this subject:

Hasse, P., *Overvoltage Protection of Low Voltage Systems*, IEE Power Series 12, Peter Peregrinus, Ltd., London, 1992.

Mardiguian, Michael, *A Handbook Series on Electromagnetic Interference and Compatibility*, Vol. 2, *Grounding and Bonding*, Interference Control Technologies, Inc., Gainesville, Va., 1988.

Morrison, Ralph, and Lewis, Warren H., *Grounding and Shielding for Facilities*, John Wiley & Sons, New York, 1990.

Ott, Henry W., *Noise Reduction Techniques in Electronic Systems*, 2d ed., John Wiley & Sons, New York, 1988.

CHAPTER 11

Power Quality Monitoring

P ower quality monitoring is the process of gathering, analyzing, and interpreting raw measurement data into useful information. The process of gathering data is usually carried out by continuous measurement of voltage and current over an extended period. The process of analysis and interpretation has been traditionally performed manually, but recent advances in signal processing and artificial intelligence fields have made it possible to design and implement intelligent systems to automatically analyze and interpret raw data into useful information with minimum human intervention.

Power quality monitoring programs are often driven by the demand for improving the systemwide power quality performance. Many industrial and commercial customers have equipment that is sensitive to power disturbances, and, therefore, it is more important to understand the quality of power being provided. Examples of these facilities include computer networking and telecommunication facilities, semiconductor and electronics manufacturing facilities, biotechnology and pharmaceutical laboratories, and financial data-processing centers. Hence, in the last decade many utility companies have implemented extensive power quality monitoring programs.

In this chapter, various issues relating to power quality monitoring are described. Section 11.1 details the objectives and procedures for performing monitoring. Section 11.2 provides historical perspective on various monitoring instruments. Section 11.3 provides a description of various power quality monitoring instruments and their typical functions. Section 11.4 describes methods of assessment of power quality data. Section 11.5 details the applications of intelligent systems in automating analysis and interpretation of raw power quality measurement data. Section 11.6 reviews standards dealing with power quality monitoring.

11.1   Monitoring Considerations

Before embarking on any power quality monitoring effort, one should clearly define the monitoring objectives. The monitoring objectives often determine the choice of monitoring equipment, triggering thresholds, methods for data acquisition and storage, and analysis and interpretation requirements. Several common objectives of power quality monitoring are summarized here.

Monitoring to Characterize System Performance

This is the most general requirement. A power producer may find this objective important if it has the need to understand its system performance and then match that system performance with the needs of customers. System characterization is a **proactive** approach to power quality monitoring. By understanding the normal power quality performance of a system, a provider can quickly identify problems and can offer information to its customers to help them match their sensitive equipment's characteristics with realistic power quality characteristics.

Monitoring to Characterize Specific Problems

Many power quality service departments or plant managers solve problems by performing short-term monitoring at specific customer sites or at difficult loads. This is a **reactive** mode of power quality monitoring, but it frequently identifies the cause of equipment incompatibility, which is the first step to a solution.

Monitoring as Part of an Enhanced Power Quality Service

Many power producers are currently considering additional services to offer customers. One of these services would be to offer differentiated levels of power quality to match the needs of specific customers. A provider and customer can together achieve this goal by modifying the power system or by installing equipment within the customer's premises. In either case, monitoring becomes essential to establish the benchmarks for the differentiated service and to verify that the utility achieves contracted levels of power quality.

Monitoring as Part of Predictive or Just-in-Time Maintenance

Power quality data gathered over time can be analyzed to provide information relating to specific equipment performance. For example, a repetitive arcing fault from an underground cable may signify impending cable failure, or repetitive capacitor-switching restrikes may signify impending failure on the capacitor-switching device. Equipment maintenance can be quickly ordered to avoid catastrophic failure, thus preventing major power quality disturbances which ultimately will impact overall power quality performance.

The monitoring program must be designed based on the appropriate objectives, and it must make the information available in a convenient form and in a timely manner (i.e., immediately). The most comprehensive monitoring approach will be a permanently installed monitoring system with automatic collection of information about steady-state power quality conditions and energy use as well as disturbances.

11.1.1   Monitoring as Part of a Facility Site Survey

Site surveys are performed to evaluate concerns for power quality and equipment performance throughout a facility. The survey will include inspection of wiring and grounding concerns, equipment connections, and the voltage and current characteristics throughout the facility. Power quality monitoring, along with infrared scans and visual inspections, is an important part of the overall survey.

The initial site survey should be designed to obtain as much information as possible about the customer facility. This information is especially important when the monitoring objective is intended to address specific power quality problems. This information is summarized here.

1. Nature of the problems (data loss, nuisance trips, component failures, control system malfunctions, etc.)

2. Characteristics of the sensitive equipment experiencing problems (equipment design information or at least application guide information)

3. The times at which problems occur

4. Coincident problems or known operations (e.g., capacitor switching) that occur at the same time

5. Possible sources of power quality variations within the facility (motor starting, capacitor switching, power electronic equipment operation, arcing equipment, etc.)

6. Existing power conditioning equipment being used

7. Electrical system data (one-line diagrams, transformer sizes and impedances, load information, capacitor information, cable data, etc.)

Once these basic data have been obtained through discussions with the customer, a site survey should be performed to verify the one-line diagrams, electrical system data, wiring and grounding integrity, load levels, and basic power quality characteristics. Data forms that can be used for this initial verification of the power distribution system are provided in Figs. 11.1 to 11.4. They can be

Supply Transformer Data:

    Manufacturer:

    Connection:

    kVA Rating:

    Primary Voltage:

    Secondary Voltage:

        Taps:

    Tap Position:

Test Data:

Primary Voltages:		Primary Currents:	
A-B		A	
B-C		B	
C-A		C	
A-N		Neutral	
B-N		Ground	
C-N			

Secondary Voltages:		Secondary Currents:	
A-B		A	
B-C		B	
C-A		C	
A-N		Neutral	
B-N		Ground	
C-N			

    N-G Bond?

FIGURE 11.1   Form for recording supply transformer test data.

used to organize the power quality monitoring results from throughout the facility.

11.1.2   Determining What to Monitor

Power quality encompasses a wide variety of conditions on the power system. Important disturbances can range from very high frequency impulses caused by lightning strokes or current chopping during circuit interruptions to long-term overvoltages caused by a regulator tap switching problem. The range of conditions that must be

Panel Identification: _____

Location: _____

Voltages: Feeder Currents: _____

    A-B _____   A _____

    B-C _____   B _____

    C-A _____   C _____

    A-N _____   Neutral _____

    B-N _____   Ground _____

    C-N _____

                             Feeder Wire Sizes: _____

   N-G Bond? _____   Phase _____

                              Neutral _____

                              Ground _____

Comments: _____

_____

_____

_____

FIGURE 11.2   Form for recording feeder circuit test data (from panel).

Panel Identification: _____

Location: _____

Circuit Identifier	Breaker	Phase A	Phase B	Phase C	Neutral	Ground	Loads Served

FIGURE 11.3   Form for recording branch circuit test data (from panel).

| Branch Circuit Identification: | | | | | | |
| Location: | | | | | | |

Equipment/Location	Volts Ph-Ph	Volts Ph-N	Volts N-G	Load Current	Ground Z	Neutral Z

FIGURE 11.4   Form for recording test data at individual loads.

characterized creates challenges both in terms of the monitoring equipment performance specifications and in the data-collection requirements. Chapter 2 details various categories of power quality variations along with methods for characterizing the variations and the typical causes of the disturbances.

The methods for characterizing the quality of ac power are important for the monitoring requirements. For instance, characterizing most transients requires high-frequency sampling of the actual waveform. Voltage sags can be characterized with a plot of the rms voltage versus time. Outages can be defined simply by a time duration. Monitoring to characterize harmonic distortion levels and normal voltage variations requires steady-state sampling with results analysis of trends over time.

Extensive monitoring of all the different types of power quality variations at many locations may be rather costly in terms of hardware, communications charges, data management, and report preparation. Hence, the priorities for monitoring should be determined based on the objectives of the effort. Projects to benchmark system performance should involve a reasonably complete monitoring effort. Projects designed to evaluate compliance with IEEE Standard 519–1992 for harmonic distortion levels may only require steady-state monitoring of harmonic levels. Other projects focused on specific industrial problems may only require monitoring of rms variations, such as voltage sags.

11.1.3   Choosing Monitoring Locations

Obviously, we would like to monitor conditions at virtually all locations throughout the system to completely understand the overall power quality. However, such monitoring may be prohibitively expensive and there are challenges in data management, analysis, and interpretation. Fortunately, taking measurements from all possible locations is usually not necessary since measurements taken from several strategic locations can be used to determine characteristics of the overall system. Thus, it is very important that the monitoring locations be selected carefully based on the monitoring objectives. We now present examples of how to choose a monitoring location.

The monitoring experience gained from the EPRI DPQ project[1] provides an excellent example of how to choose monitoring locations. The primary objective of the DPQ project was to characterize power quality on the U.S. electric utility distribution feeders. Actual feeder monitoring began in June 1992 and was completed in September 1995. Twenty-four different utilities participated in the data-collection effort with almost 300 measurement sites. Monitoring for the project was designed to provide a statistically valid set of data of the various phenomena related to power quality.

Since the primary objective was to characterize power quality on primary distribution feeders, monitoring was done on the actual feeder circuits. One monitor was located near the substation, and two additional sites were selected randomly (see Fig. 11.5). By randomly choosing the remote sites, the overall project results represented power quality on distribution feeders in general. It may not be realistic, however, to assume that the three selected sites completely characterized power quality on the individual feeders involved.

When a monitoring project involves characterizing specific power quality problems that are actually being experienced by customers on the distribution system, the monitoring locations should be at actual customer service entrance locations because it includes the effect of step-down transformers supplying the customer. Data collected at the service entrance can also characterize the customer load current variations and harmonic distortion levels. Monitoring at customer service entrance locations has the additional advantage of reduced transducer costs. In addition, it provides indications of the origin of the disturbances, i.e., the utility or the customer side of the meter.

Another important aspect of the monitoring location when characterizing specific power quality problems is to locate the monitors as close as possible to the equipment affected by power quality variations. It is important that the monitor sees the same variations that the sensitive equipment sees. High-frequency transients, in particular,

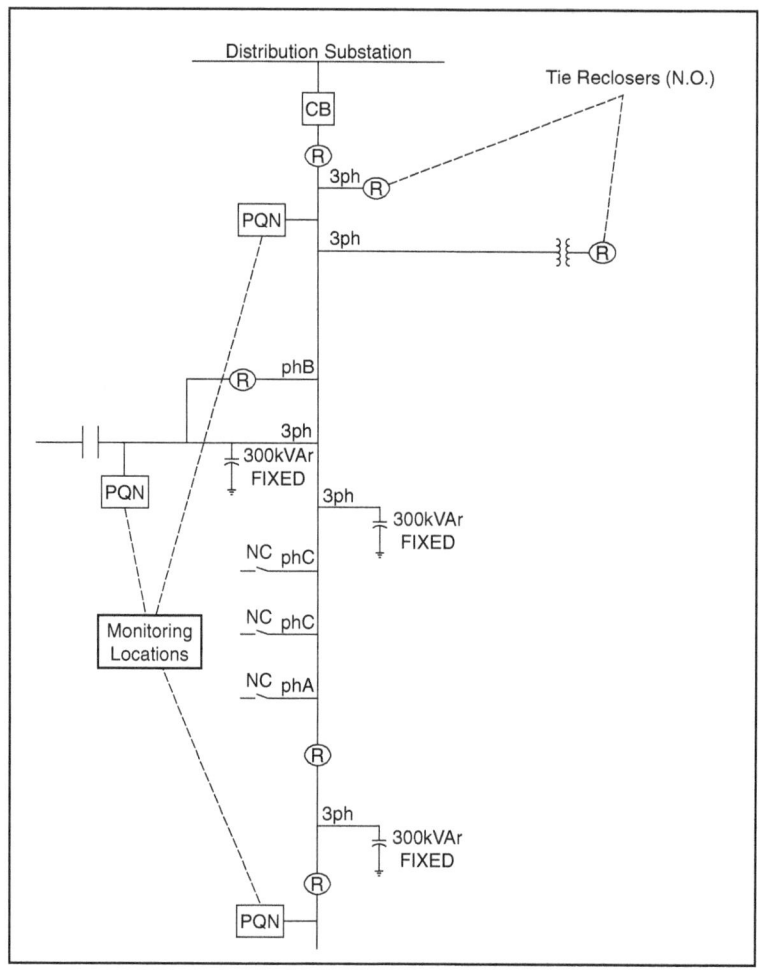

Figure 11.5 Typical distribution feeder monitoring scheme.

can be significantly different if there is significant separation between the monitor and the affected equipment.

A good compromise approach is to monitor at the substation and at selected customer service entrance locations. The substation is important because it is the PCC for most rms voltage variations. The voltage sag experienced at the substation during a feeder fault is experienced by all the customers on other feeders supplied from the same substation bus. Customer equipment sensitivity and location on a feeder together determine the service entrance locations for monitoring. For instance, it is valuable to have a location immediately downline from each protective device on the feeder. Figure 11.6 illustrates the concept of a monitoring system based on monitoring at substations and customer sites. It also illustrates how the Internet can be used to provide access to the information for all interested parties (discussed in Sec. 11.1.4).

11.1.4   Options for Permanent Power Quality Monitoring Equipment

Permanent power quality monitoring systems, such as the system illustrated in Fig. 11.6, should take advantage of the wide variety of equipment that may have the capability to record power quality information. Some of the categories of equipment that can be incorporated into an overall monitoring system include the following:

1. *Digital fault recorders (DFRs).* These may already be in place at many substations. DFR manufacturers do not design the devices specifically for power quality monitoring. However, a DFR will typically trigger on fault events and record the voltage and current waveforms that characterize the event. This makes them valuable for characterizing rms disturbances, such as voltage sags, during power

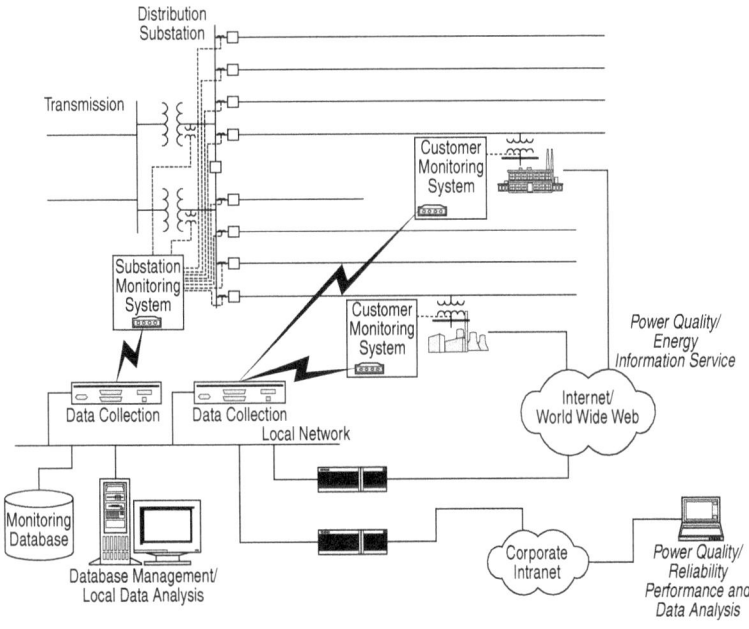

Figure 11.6   Illustration of system power quality monitoring concept with monitoring at the substation and selected customer locations.

system faults. DFRs also offer periodic waveform capture for calculating harmonic distortion levels.

2. *Smart relays and other IEDs.* Many types of substation equipment may have the capability to be an intelligent electronic device (IED) with monitoring capability. Manufacturers of devices like relays and reclosers that monitor the current anyway are adding on the capability to record disturbances and make the information available to an overall monitoring system controller. These devices can be located on the feeder circuits as well as at the substation.

3. *Voltage recorders.* Power providers use a variety of voltage recorders to monitor steady-state voltage variations on distribution systems. We are encountering more and more sophisticated models fully capable of characterizing momentary voltage sags and even harmonic distortion levels. Typically, the voltage recorder provides a trend that gives the maximum, minimum, and average voltage within a specified sampling window (for example, 2 s). With this type of sampling, the recorder can characterize a voltage sag magnitude adequately. However, it will not provide the duration with a resolution less than 2 s.

4 *In-plant power monitors.* It is now common for monitoring systems in industrial facilities to have some power quality capabilities. These monitors, particularly those located at the service entrance, can be used as part of a utility monitoring program. Capabilities usually include waveshape capture for evaluation of harmonic distortion levels, voltage profiles for steady-state rms variations, and triggered waveshape captures for voltage sag conditions. It is not common for these instruments to have transient monitoring capabilities.

5. *Special-purpose power quality monitors.* The monitoring instrument developed for the EPRI DPQ project was specifically designed to measure the full range of power quality variations. This instrument features monitoring of voltage and current on all three phases plus the neutral. A 14-bit analog-to-digital (A/D) board provides a sampling rate of 256 points per cycle for voltage and 128 points per cycle for current. This high sampling rate allowed detection of voltage harmonics as high as the 100th and current harmonics as high as the 50th. Most power quality instruments can record both triggered and sampled data. Triggering should be based upon rms thresholds for rms variations and on waveshape for transient variations. Simultaneous voltage and current monitoring with triggering of all channels during a disturbance is an important capability for these instruments. Power quality monitors have proven suitable for substations, feeder locations, and customer service entrance locations.

6. *Revenue meters.* Revenue meters monitor the voltage and current anyway, so it seems logical to offer alternatives for more advanced

monitoring that could include recording of power quality information. Virtually all the revenue meter manufacturers are moving in this direction, and the information from these meters can then be incorporated into an overall power quality monitoring system.

11.1.5 Disturbance Monitor Connections

The recommended practice is to provide input power to the monitor from a circuit other than the circuit to be monitored. Some manufacturers include input filters and/or surge suppressors on their power supplies that can alter disturbance data if the monitor is powered from the same circuit that is being monitored.

The grounding of the power disturbance monitor is an important consideration. The disturbance monitor will have a ground connection for the signal to be monitored and a ground connection for the power supply of the instrument. Both of these grounds will be connected to the instrument chassis. For safety reasons, both of these ground terminals should be connected to earth ground. However, this has the potential of creating ground loops if different circuits are involved.

Safety comes first. Therefore, both grounds should be connected whenever there is a doubt about what to do. If ground loops can be a significant problem such that transient currents might damage the instruments or invalidate the measurements, it may be possible to power the instrument from the same line that is being monitored (check to make sure there is no signal conditioning in the power supply). Alternatively, it may be possible to connect just one ground (signal to be monitored) and place the instrument on an insulating mat. Appropriate safety practices such as using insulated gloves when operating the instrument must be employed if it is possible for the instrument to rise in potential with respect to other apparatus and ground references with which the operator can come into contact.

11.1.6 Setting Monitor Thresholds

Disturbance monitors are designed to detect conditions that are abnormal. Therefore, it is necessary to define the range of conditions that can be considered normal. Some disturbance monitors have preselected (default) thresholds that can be used as a starting point.

The best approach for selecting thresholds is to match them with the specifications of the equipment that is affected. This may not always be possible due to a lack of specifications or application guidelines. An alternative approach is to set the thresholds fairly tight for a period of time (collect a lot of disturbance data) and then use the data collected to select appropriate thresholds for longer-duration monitoring.

Some monitoring systems advertise the advantage of no setups and no thresholds to set. Of course, there have to be thresholds because no monitor (so far) has enough storage capacity to save every single cycle of the voltages and currents being monitored. The thresholds in these cases are essentially fixed in the instruments, and algorithms may be adjusted internally based on the disturbances being recorded. This type of system is convenient for the user since the setup is simple, but it is still a compromise since you lose the capability to change the thresholds based on the specific circumstances at a particular location.

11.1.7   Quantities and Duration to Measure

Sometimes, when characterizing system disturbances, it is sufficient to monitor only the voltage signals. For instance, the voltages provide information about the quality of power being delivered to a facility and characterize the transients and voltage sags that may affect customer equipment. However, there is a tremendous amount of information in the currents associated with these disturbances that can help determine the cause and whether or not equipment was impacted. Also, current measurements are required if harmonics are a concern since the currents characterize the harmonic injection from the customer onto the power system.

Current measurements are used to characterize the generation of harmonics by nonlinear loads on the system. Current measurements at individual loads are valuable for determining these harmonic generation characteristics. Current measurements on feeder circuits or at the service entrance characterize a group of loads or the entire facility as a source of harmonics. Current measurements on the distribution system can be used to characterize groups of customers or an entire feeder.

Voltage measurements help characterize the system response to the generated harmonic currents. Resonance conditions will be indicated by high harmonic voltage distortion at specific frequencies. In order to determine system frequency response characteristics from measurements, voltages and currents must be measured simultaneously. In order to measure harmonic power flows, all three phases must be sampled simultaneously.

The duration of monitoring depends on the monitoring objectives. For instance, if the objective is to solve problems that are caused by voltage sags during remote faults on the utility system, monitoring may be required for a significant length of time because system faults are probably rare. If the problem involves capacitor switching, it may be possible to characterize the conditions over the period of a couple days. Harmonic distortion problems and flicker problems should be characterized over a period of at least 1 week to get a picture of how the load changes and how system variations may affect these levels.

The duration of monitoring is becoming less of an issue as the general trend is to use permanent power quality monitoring systems, taking advantage of the wide variety of equipment that can provide information as part of the system.

11.1.8   Finding the Source of a Disturbance

The first step in identifying the source of a disturbance is to correlate the disturbance waveform with possible causes, as outlined in Chap. 2. Once a category for the cause has been determined (e.g., load switching, capacitor switching, remote fault condition, recloser operation), the identification becomes more straightforward. The following general guidelines can help:

- High-frequency voltage variations will be limited to locations close to the source of the disturbance. Low-voltage (600 V and below) wiring often damps out high-frequency components very quickly due to circuit resistance, so these frequency components will only appear when the monitor is located close to the source of the disturbance.

- Power interruptions close to the monitoring location will cause a very abrupt change in the voltage. Power interruptions remote from the monitoring location will result in a decaying voltage due to stored energy in rotating equipment and capacitors.

- The highest harmonic voltage distortion levels will occur close to capacitors that are causing resonance problems. In these cases, a single frequency will usually dominate the voltage harmonic spectrum.

11.2   Historical Perspective of Power Quality Measuring Instruments

Early monitoring devices were bulky, heavy boxes that required a screwdriver to make selections. Data collected were recorded on strip-chart paper. One of the earliest power quality monitoring instruments is a lightning strike recorder developed by General Electric in the 1920s[2] (see Fig. 11.7). The instrument makes an impulse-like mark on strip-chart paper to record a lightning strike event along with its time and date of occurrence. The data were more qualitative then quantitative, making the data interpretation rather difficult. The principal component of the device consisted of a windup clockwork motor that moved the strip of paper from one spool to another, and a pair of electrodes that struck an arc across the paper.

Figure 11.7   Early power quality instrument developed by GE in the 1920s to record lightning strikes. *(Courtesy of Alex McEachern, from his personal collection of historical power instruments; www.Alex.McEachern.com.)*

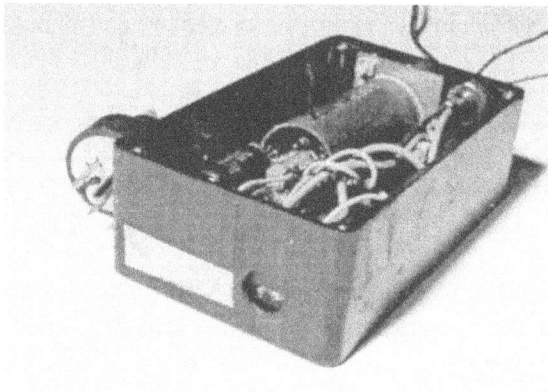

Figure 11.8   Martzloff's 1967 surge counter. Recorded data are quantitative. *(Courtesy of Alex McEachern, from his personal collection of historical power instruments; www.Alex.McEachern.com.)*

Significant development on power quality devices was not made until the 1960s when Martzloff developed a surge counter that could capture a voltage waveform of lightning strikes.[2] The device consisted of a high-persistence analog oscilloscope with a logarithmic sweep rate (see Fig. 11.8). The improvement of this device over its predecessor was that the recorded data were quantitative (voltage waveforms) as opposed to qualitative (marks on strip charts). By the mid-1960s, limitations of power quality devices relating to the trigger mechanism and the preset frequency response were well understood.

Many engineers consider that the first generation of power quality monitors began in the mid-1970s when Dranetz Engineering Laboratories (now Dranetz-BMI) introduced the Series 606 power line disturbance analyzer shown in Fig. 11.9.[3,4] This was a microprocessor-based monitor-analyzer first manufactured in 1975, and many units are still in service. The output of these monitors was text-based, printed on a paper tape. The printout described a disturbance by the event type (sag, interruption, etc.) and voltage magnitude. These monitors had limited functionalities compared to modern monitors, but the triggering mechanics were already well developed.

Second-generation power quality instruments debuted in the mid-1980s. This generation of power quality monitors generally featured full graphic display and digital memory to view and store

FIGURE 11.9 A first-generation power line disturbance monitor-analyzer, the Dranetz 606. (*Courtesy of Dranetz-BMI.*)

captured power quality events, including both transients and steady-state events. Some instruments had a capability of transmitting data from a remote monitoring site to a central location for further analysis. Second-generation power quality instruments virtually had perfected the basic requirements of the triggering mechanism. Since the occurrence of a power quality disturbance is highly unpredictable, data must be continuously recorded and processed without any dead time. Complex triggering engines determine what data and how much data should be saved to the digital memory. Trigger methods include fixed and floating limits and sensitivities, waveshape changes, and specific event characteristic parameters. These methods optimize the probability that what is important to the user will be captured and stored.

By the mid-1990s, the third-generation power quality instruments emerged. The development of the third-generation power monitors was inspired in part by the EPRI DPQ project. This generation of monitors was more appropriate as part of a complete power quality monitoring system, and the software systems to collect and manage the data were also developed. Since the conclusion of the project, substantial field experience gained revealed some of the difficulties in managing a large system of power quality monitors:[5]

1. Managing the large volume of raw measurement data that must be collected, analyzed, and archived becomes a serious challenge as the number of monitoring points grows.

2. The data volume collected at each monitoring point can strain communication mechanisms employed to move that data from monitor to analysis point.

3. As understanding of system performance grows through the feedback provided by the monitoring data, detailed views of certain events, such as normal capacitor switching, become less valuable and would be of more use in a summary or condensed form.

4. The real value of any monitoring system lies in its ability to generate information rather than in collecting and storing volumes of detailed raw data.

Based on the experience gained from the EPRI DPQ project, it was realized that the information system aspect of a power monitoring program plays a very important role in tracking power quality performance. Thus, the development of the most recent generation of power quality monitors was geared toward meeting the new information system demand, i.e., to be able to discover knowledge or information from the collected data as they are captured and to disseminate the information rapidly. This type of instrument employs expert system and advanced communication technologies.

11.3 Power Quality Measurement Equipment

From Chap. 2, it is clear that power quality phenomena cover a wide range of frequencies. They include everything from very fast transient overvoltages (microsecond time frame) to long-duration outages (hours or days time frame). Power quality problems also include steady-state phenomena, such as harmonic distortion, and intermittent phenomena, such as voltage flicker. Definitions for the different categories were presented in Chap. 2. This wide variety of conditions that make up power quality makes the development of standard measurement procedures and equipment very difficult.

11.3.1 Types of Instruments

Although instruments have been developed that measure a wide variety of disturbances, a number of different instruments may be used, depending on the phenomena being investigated. Basic categories of instruments that may be applicable include

- Wiring and grounding test devices
- Multimeters
- Oscilloscopes
- Disturbance analyzers
- Harmonic analyzers and spectrum analyzers
- Combination disturbance and harmonic analyzers
- Flicker meters
- Energy monitors

This section and Secs. 11.3.2 to 11.3.12 discuss the application and limitations of these different instruments. Besides these instruments, which measure steady-state signals or disturbances on the power system directly, there are other instruments that can be used to help solve power quality problems by measuring ambient conditions:

- Infrared meters can be very valuable in detecting loose connections and overheating conductors. An annual procedure of checking the system in this manner can help prevent power quality problems due to arcing, bad connections, and overloaded conductors.

- Noise problems related to electromagnetic radiation may require measurement of field strengths in the vicinity of affected equipment. Magnetic gauss meters are used to measure magnetic field strengths for inductive coupling concerns. Electric field meters can measure the strength of electric fields for electrostatic coupling concerns.

- Static electricity meters are special-purpose devices used to measure static electricity in the vicinity of sensitive equipment. Electrostatic discharge (ESD) can be an important cause of power quality problems in some types of electronic equipment.

Regardless of the type of instrumentation needed for a particular test, there are a number of important factors that should be considered when selecting the instrument. Some of the more important factors include

- Number of channels (voltage and/or current)
- Temperature specifications of the instrument
- Ruggedness of the instrument
- Input voltage range (e.g., 0 to 600 V)
- Power requirements
- Ability to measure three-phase voltages
- Input isolation (isolation between input channels and from each input to ground)
- Ability to measure currents
- Housing of the instrument (portable, rack-mount, etc.)
- Ease of use (user interface, graphics capability, etc.)
- Documentation
- Communication capability (modem, network interface)
- Analysis software

The flexibility (comprehensiveness) of the instrument is also important. The more functions that can be performed with a single instrument, the fewer the number of instruments required. Recognizing that there is some crossover between the different instrument categories, we discuss the basic categories of instruments for direct measurement of power signals in Secs. 11.3.2 to 11.3.12.

11.3.2 Wiring and Grounding Testers

Many power quality problems reported by end users are caused by problems with wiring and/or grounding within the facility. These problems can be identified by visual inspection of wiring, connections, and panel boxes and also with special test devices for detecting wiring and grounding problems.

Important capabilities for a wiring and grounding test device include

- Detection of isolated ground shorts and neutral-ground bonds

- Ground impedance and neutral impedance measurement or indication

- Detection of open grounds, open neutrals, or open hot wires

- Detection of hot/neutral reversals or neutral/ground reversals

Three-phase wiring testers should also test for phase rotation and phase-to-phase voltages. These test devices can be quite simple and provide an excellent initial test for circuit integrity. Many problems can be detected without the requirement for detailed monitoring using expensive instrumentation.

11.3.3 Multimeters

After initial tests of wiring integrity, it may also be necessary to make quick checks of the voltage and/or current levels within a facility. Overloading of circuits, undervoltage and overvoltage problems, and unbalances between circuits can be detected in this manner. These measurements just require a simple multimeter. Signals used to check for these include

- Phase-to-ground voltages

- Phase-to-neutral voltages

- Neutral-to-ground voltages

- Phase-to-phase voltages (three-phase system)

- Phase currents

- Neutral currents

The most important factor to consider when selecting and using a multimeter is the method of calculation used in the meter. All the commonly used meters are calibrated to give an rms indication for the measured signal. However, a number of different methods are used to calculate the rms value. The three most common methods are

1. *Peak method.* Assuming the signal to be a sinusoid, the meter reads the peak of the signal and divides the result by 1.414 (square root of 2) to obtain the rms.

2. *Averaging method.* The meter determines the average value of a rectified signal. For a clean sinusoidal signal (signal containing only one frequency), this average value is related to the rms value by a constant.

3. *True rms.* The rms value of a signal is a measure of the heating that will result if the voltage is impressed across a resistive load. One method of detecting the true rms value is to actually use a thermal detector to measure a heating value. More modern digital meters use

a digital calculation of the rms value by squaring the signal on a sample-by-sample basis, averaging over the period, and then taking the square root of the result.

These different methods all give the same result for a clean, sinusoidal signal but can give significantly different answers for distorted signals. This is very important because significant distortion levels are quite common, especially for the phase and neutral currents within the facility. Table 11.1 can be used to better illustrate this point.

Each waveform in Table 11.1 has an rms value of 1.0 pu (100.0 percent). The corresponding measured values for each type of meter are displayed under the associated waveforms, normalized to the true rms value.

11.3.4 Digital Cameras

Photographs are extremely useful for documentation purposes. Those conducting the measurements often get distracted trying to get instruments to function properly and tests coordinated. They are rushed and fail to write down certain key data that later turn out to be important. Unfortunately, human memory is unreliable when there are dozens of measurement details to remember. The

		Meter Type		
		True RMS	Peak Method	Average Responding
		Circuit Type		
		RMS Converter	Peak / 1.414	Sine Avg. X 1.11
Sine Wave		100 %	100 %	100 %
Square Wave		100 %	82 %	110 %
Triangle Wave		100 %	121 %	96 %
ASD Current		100 %	127 %	86 %
PC Current		100 %	184 %	60 %
Light Dimmer		100 %	113 %	84 %

TABLE 11.1 Comparison of Methods for Measuring Voltages and Currents with Multimeters

modern digital camera has become an indispendable tool when taking field measurements. It is a simple matter to take photographs to document the tests. The photographer can immediately tell if the shot failed and retake it with a different exposure. Typical items to record photographically during field measurements include

1. Nameplates of transformers, motors, etc.

2. Instrumentation setups

3. Transducer and probe connections

4. Key waveform displays from instruments

5. Substations, switchgear arrangements, arrester positions, etc.

6. Dimensions of key electrical components such as cable lengths

Video cameras are similarly useful when there is moving action or random events. For example, they may be used to help identify the locations of flashovers. Many industrial facilities will require special permission to take photographs and may place stringent limitations on the distribution of any photographs.

11.3.5 Oscilloscopes

An oscilloscope is valuable when performing real-time tests. Looking at the voltage and current waveforms can provide much information about what is happening, even without performing detailed harmonic analysis on the waveforms. One can get the magnitudes of the voltages and currents, look for obvious distortion, and detect any major variations in the signals.

There are numerous makes and models of oscilloscopes to choose from. A digital oscilloscope with data storage is valuable because the waveform can be saved and analyzed. Oscilloscopes in this category often also have waveform analysis capability (energy calculation, spectrum analysis). In addition, the digital oscilloscopes can usually be obtained with communications so that waveform data can be uploaded to a personal computer for additional analysis with a software package.

The latest developments in oscilloscopes are hand-held instruments with the capability to display waveforms as well as performing some signal processing. These are quite useful for power quality investigations because they are very portable and can be operated like a volt-ohm meter (VOM), but yield much more information. These are ideal for initial plant surveys. A typical device is shown in Figs. 11.10 and 11.11. This particular instrument also has the capability to analyze harmonics and permits connection with

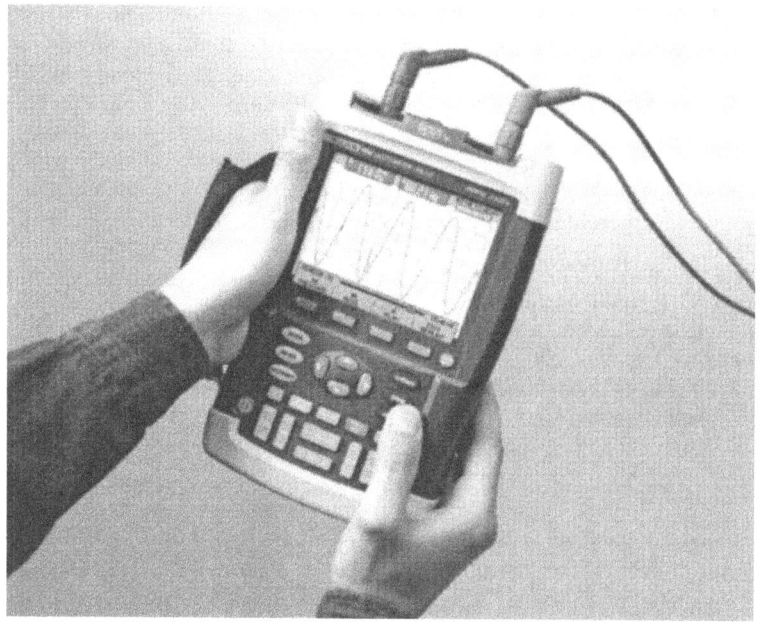

Figure 11.10   A hand-held oscillographic monitoring instrument. (*Courtesy of* Fluke Corporation.)

personal computers for further data analysis and inclusion into reports as illustrated.

11.3.6   Disturbance Analyzers

Disturbance analyzers and disturbance monitors form a category of instruments that have been developed specifically for power quality measurements. They typically can measure a wide variety of system disturbances from very short duration transient voltages to long-duration outages or undervoltages. Thresholds can be set and the instruments left unattended to record disturbances over a period of time. The information is most commonly recorded on a paper tape, but many devices have attachments so that it can be recorded on disk as well.

There are basically two categories of these devices:

1. *Conventional analyzers* that summarize events with specific information such as overvoltage and undervoltage magnitudes, sags and surge magnitude and duration, transient magnitude and duration, etc.

2. *Graphics-based analyzers* that save and print the actual waveform along with the descriptive information which would be generated by one of the conventional analyzers

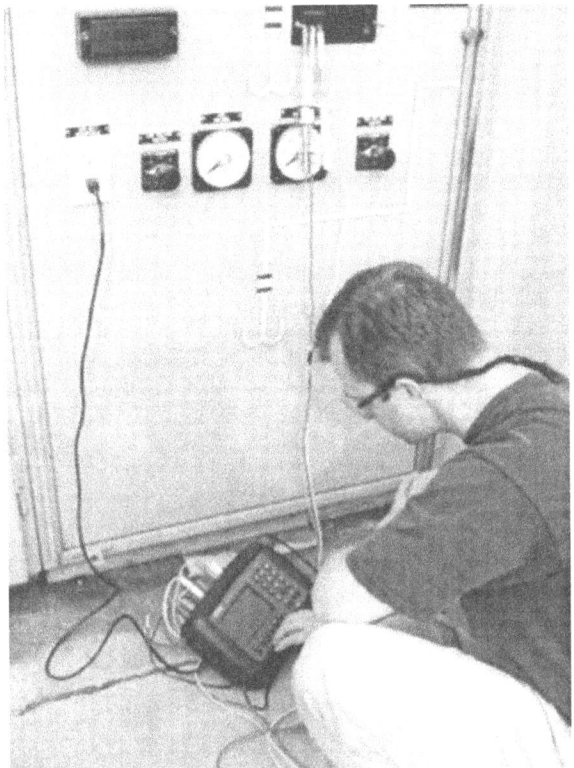

FIGURE 11.11 Demonstrating the use of a hand-held, three-phase power quality monitoring instrument to quickly evaluate voltages at the mains.

It is often difficult to determine the characteristics of a disturbance or a transient from the summary information available from conventional disturbance analyzers. For instance, an oscillatory transient cannot be effectively described by a peak and a duration. Therefore, it is almost imperative to have the waveform capture capability of a graphics-based disturbance analyzer for detailed analysis of a power quality problem (Fig. 11.12). However, a simple conventional disturbance monitor can be valuable for initial checks at a problem location.

11.3.7 Spectrum Analyzers and Harmonic Analyzers

Instruments in the disturbance analyzer category have very limited harmonic analysis capabilities. Some of the more powerful analyzers have add-on modules that can be used for computing fast Fourier transform (FFT) calculations to determine the lower-order harmonics. However, any significant harmonic measurement requirements will

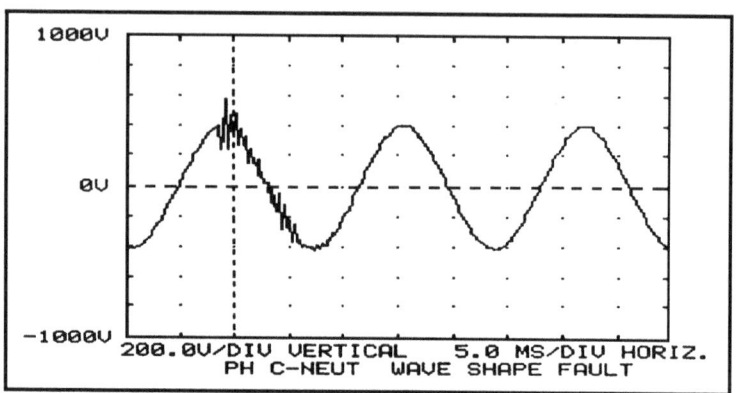

FIGURE **11.12**   Graphics-based analyzer output.

demand an instrument that is designed for spectral analysis or harmonic analysis. Important capabilities for useful harmonic measurements include

- Capability to measure both voltage and current simultaneously so that harmonic power flow information can be obtained.

- Capability to measure both magnitude and phase angle of individual harmonic components (also needed for power flow calculations).

- Synchronization and a sampling rate fast enough to obtain accurate measurement of harmonic components up to at least the 37th harmonic (this requirement is a combination of a high sampling rate and a sampling interval based on the 60-Hz fundamental).

- Capability to characterize the statistical nature of harmonic distortion levels (harmonics levels change with changing load conditions and changing system conditions).

There are basically three categories of instruments to consider for harmonic analysis:

1. *Simple meters.* It may sometimes be necessary to make a quick check of harmonic levels at a problem location. A simple, portable meter for this purpose is ideal. There are now several hand-held instruments of this type on the market. Each instrument has advantages and disadvantages in its operation and design. These devices generally use microprocessor-based circuitry to perform the necessary calculations to determine individual harmonics up to the

50th harmonic, as well as the rms, the THD, and the telephone influence factor (TIF). Some of these devices can calculate harmonic powers (magnitudes and angles) and can upload stored waveforms and calculated data to a personal computer.

2. *General-purpose spectrum analyzers.* Instruments in this category are designed to perform spectrum analysis on waveforms for a wide variety of applications. They are general signal analysis instruments. The advantage of these instruments is that they have very powerful capabilities for a reasonable price since they are designed for a broader market than just power system applications. The disadvantage is that they are not designed specifically for sampling power frequency waveforms and, therefore, must be used carefully to assure accurate harmonic analysis. There are a wide variety of instruments in this category.

3. *Special-purpose power system harmonic analyzers.* Besides the general-purpose spectrum analyzers just described, there are also a number of instruments and devices that have been designed specifically for power system harmonic analysis. These are based on the FFT with sampling rates specifically designed for determining harmonic components in power signals. They can generally be left in the field and include communications capability for remote monitoring.

11.3.8   Combination Disturbance and Harmonic Analyzers

The most recent instruments combine harmonic sampling and energy monitoring functions with complete disturbance monitoring functions as well. The output is graphically based, and the data are remotely gathered over phone lines into a central database. Statistical analysis can then be performed on the data. The data are also available for input and manipulation into other programs such as spreadsheets and other graphical output processors.

One example of such an instrument is shown in Fig. 11.13. This instrument is designed for both utility and end-user applications, being mounted in a suitable enclosure for installation outdoors on utility poles. It monitors three-phase voltages and currents (plus neutrals) simultaneously, which is very important for diagnosing power quality problems. The instrument captures the raw data and saves the data in internal storage for remote downloading. Off-line analysis is performed with powerful software that can produce a variety of outputs such as that shown in Fig. 11.14. The top chart shows a typical result for a voltage sag. Both the rms variation for the first 0.8 s and the actual waveform for the first 175 ms are shown. The middle chart shows a typical wave fault capture from a capacitor-switching operation. The bottom chart demonstrates the capability to report harmonics of a distorted waveform. Both the actual waveform and the harmonic spectrum can be obtained.

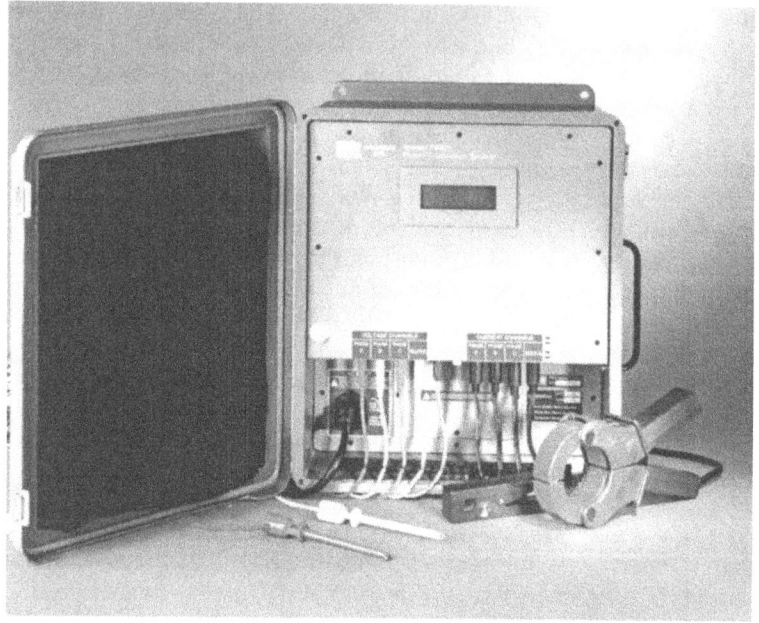

FIGURE 11.13   A power quality monitoring instrument capable of monitoring disturbances, harmonics, and other steady-state phenomena on both utility systems and end-user systems. (*Courtesy of Dranetz-BMI.*)

Another device is shown in Fig. 11.15. This is a power quality monitoring system designed for key utility accounts. It monitors three-phase voltages and has the capability to capture disturbances and page power quality engineers. The engineers can then call in and hear a voice message describing the event. It has memory for more than 30 events.

Thus, while only a few short years ago power quality monitoring was a rare feature to be found in instruments, it is becoming much more commonplace in commercially available equipment.

11.3.9   Flicker Meters*

Over the years, many different methods for measuring flicker have been developed. These methods range from using very simple rms meters with flicker curves to elaborate flicker meters that use exactly tuned filters and statistical analysis to evaluate the level of voltage flicker. This section discusses various methods available for measuring flicker.

---

*This subsection was contributed by Jeff W. Smith and Erich W. Gunther.

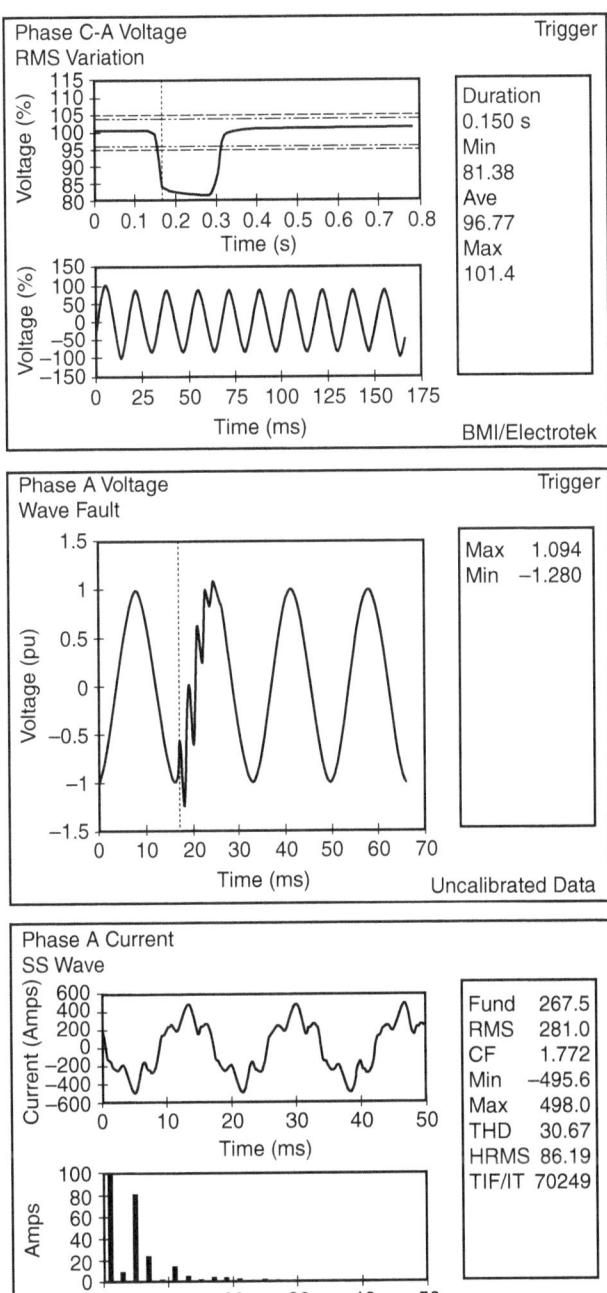

Figure 11.14   Output from combination disturbance and harmonic analyzer.

FIGURE **11.15**   A low-cost power quality monitor that can page power quality engineers when disturbances occur.

Flicker Standards

Although the United States does not currently have a standard for flicker measurement, there are IEEE standards that address flicker. IEEE Standards 141–1993[6] and 519–1992[7] both contain flicker curves that have been used as guides for utilities to evaluate the severity of flicker within their system. Both flicker curves, from Standards 141 and 519, are shown in Fig. 11.16.

In other countries, a standard methodology for measuring flicker has been established. The IEC flicker meter is the standard for measuring flicker in Europe and other countries currently adopting IEC standards. The IEC method for flicker measurement, defined in IEC Standard 61000–4–15[8] (formerly IEC 868), is a very comprehensive approach to flicker measurement and is further described in "Flicker Measurement Techniques" below. More recently, the IEEE has been working toward adoption of the IEC flicker monitoring standards with an additional curve to account for the differences between 230-V and 120-V systems.

Flicker Measurement Techniques

RMS Strip Charts.   Historically, flicker has been measured using rms meters, load duty cycle, and a flicker curve. If sudden rms voltage deviations occurred with specified frequencies exceeding values found in flicker curves, such as one shown in Fig. 11.16, the

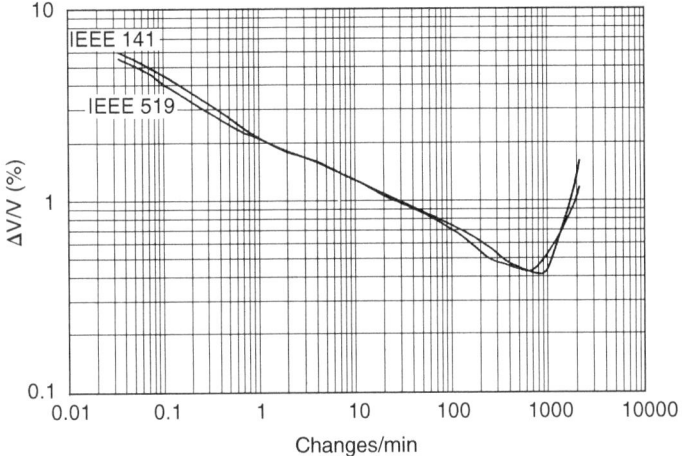

FIGURE 11.16   Flicker curves from IEEE Standards 141 and 519.

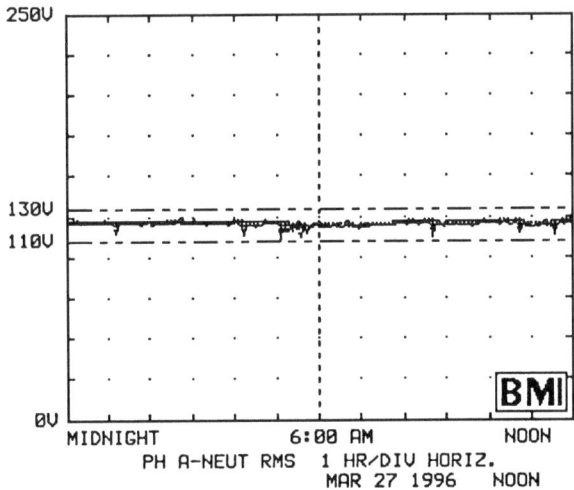

FIGURE 11.17   RMS voltage variations.

system was said to have experienced flicker. A sample graph of rms voltage variations is shown in Fig. 11.17 where large voltage deviations up to 9.0 V rms ($\Delta V/V = \pm 8.0$ percent on a 120-V base) are found. Upon comparing this to the flicker curve in Fig. 11.16, the feeder would be experiencing flicker, regardless of the duty cycle of the load producing the flicker, because any sudden total change in voltage greater than 7.0 V rms results in objectionable flicker, regardless of the frequency. The advantage to such a method is that it is quite simple in nature and the rms data required are rather easy to acquire. The apparent disadvantage to such a method would be the lack of accuracy and inability to obtain the exact frequency content of the flicker.

Fast Fourier Transform.   Another method that has been used to measure flicker is to take raw samples of the actual voltage waveforms and implement a fast Fourier transform on the demodulated signal (flicker signal only) to extract the various frequencies and magnitudes found in the data. These data would then be compared to a flicker curve. Although similar to using the rms strip charts, this method more accurately quantifies the data measured due to the magnitude and frequency of the flicker being known. The downside to implementing this method is associated with quantifying flicker levels when the flicker-producing load contains multiple flicker signals. Some instruments compensate for this by reporting only the dominant frequency and discarding the rest.

Flicker Meters.   Because of the complexity of quantifying flicker levels that are based upon human perception, the most comprehensive approach to measuring flicker is to use flicker meters. A flicker meter is essentially a device that demodulates the flicker signal, weights it according to established "flicker curves," and performs statistical analysis on the processed data.

Generally, these meters can be divided up into three sections. In the first section the input waveform is demodulated, thus removing the carrier signal. As a result of the demodulator, a dc offset and higher-frequency terms (sidebands) are produced. The second section removes these unwanted terms using filters, thus leaving only the modulating (flicker) signal remaining. The second section also consists of filters that weight the modulating signal according to the particular meter specifications. The last section usually consists of a statistical analysis of the measured flicker.

The most established method for doing this is described in IEC Standard 61000–4–15.[8] The IEC flicker meter consists of five blocks, which are shown in Fig. 11.18.

Block 1 is an input voltage adapter that scales the input half-cycle rms value to an internal reference level. This allows flicker measurements to be made based upon a percent ratio rather than be dependent upon the input carrier voltage level.

Block 2 is simply a squaring demodulator that squares the input to separate the voltage fluctuation (modulating signal) from the main voltage signal (carrier signal), thus simulating the behavior of the incandescent lamp.

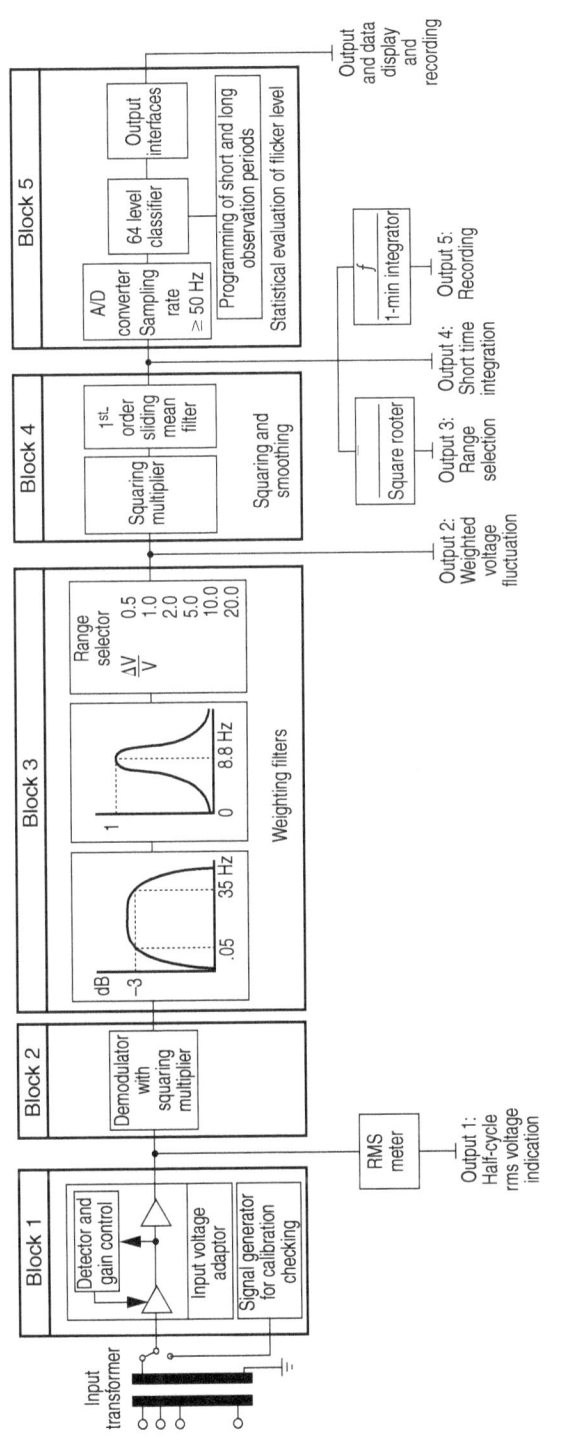

FIGURE 11.18 Diagram of the IEC flicker meter.

517

Block 3 consists of multiple filters that serve to filter out unwanted frequencies produced from the demodulator and also to weight the input signal according to the incandescent lamp eye-brain response. The basic transfer function for the weighting filter is

$$H(s) = \frac{k\omega_1 s}{s^2 + 2\lambda s + \omega_1^2} \cdot \frac{1 + s/\omega_2}{(1 + s/\omega_s)(1 + s/\omega_4)}$$

(See IEC Standard 61000–4–15 for a description of the variables used above.)

Block 4 consists of a squaring multiplier and sliding mean filter. The voltage signal is squared to simulate the nonlinear eye-brain response, while the sliding mean filter averages the signal to simulate the short-term storage effect of the brain. The output of this block is considered to be the instantaneous flicker level. A level of 1 on the output of this block corresponds to perceptible flicker.

Block 5 consists of a statistical analysis of the instantaneous flicker level. The output of block 4 is divided into suitable classes, thus creating a histogram. A probability density function is created based upon each class, and from this a cumulative distribution function can be formed.

Flicker level evaluation can be divided into two categories, short-term and long-term. Short-term evaluation of flicker severity P_{ST} is based upon an observation period of 10 min. This period is based upon assessing disturbances with a short duty cycle or those that produce continuous fluctuations. P_{ST} can be found using the equation

$$P_{ST} = \sqrt{0.0314P_{0.1} + 0.0525P_{1s} + 0.0657P_{3s} + 0.28P_{10s} + 0.08P_{50s}}$$

where the percentages $P_{0.1}$, P_{1s}, P_{3s}, P_{10s}, and P_{50s} are the flicker levels that are exceeded 0.1, 1.0, 3.0, 10.0, and 50.0 percent of the time, respectively. These values are taken from the cumulative distribution curve discussed previously. A P_{ST} of 1.0 on the output of block 5 represents the objectionable (or irritable) limit of flicker.

For cases where the duty cycle is long or variable, such as in arc furnaces, or disturbances on the system that are caused by multiple loads operating simultaneously, the need for the long-term assessment of flicker severity arises. Therefore, the long-term flicker severity P_{LT} is derived from P_{ST} using the equation

$$P_{LT} = \sqrt{\frac{\sum_{i=1}^{N} P_{STi}^3}{N}}$$

where N is the number of P_{ST} readings and is determined by the duty cycle of the flicker-producing load. The purpose is to capture one duty cycle of the fluctuating load. If the duty cycle is unknown, the recommended number of P_{ST} readings is 12 (2-h measurement window).

The advantage of using a single quantity, like Pst, to characterize flicker is that it provides a basis for implementing contracts and describing flicker levels in a much simpler manner. Figure 11.19 illustrates the Pst levels measured at the PCC with an arc furnace over a 24-h period. The melt cycles when the furnace was operating can be clearly identified by the high Pst levels. Note that Pst levels greater than 1.0 are usually considered to be levels that might result in customers being aware of lights flickering.

11.3.10 Smart Power Quality Monitors

All power quality measurement instruments previously described are designed to collect power quality data. Some instruments can send the data over a telecommunication line to a central processing location for analysis and interpretation. However, one common feature among these instruments is that they do not possess the capability to locally analyze, interpret, and determine what is happening in the power system. They simply record and transmit data for postprocessing.

Since the conclusion of the EPRI DPQ project in Fall 1995, it was realized that these monitors, along with the monitoring practice previously described, were inadequate. An emerging trend in power

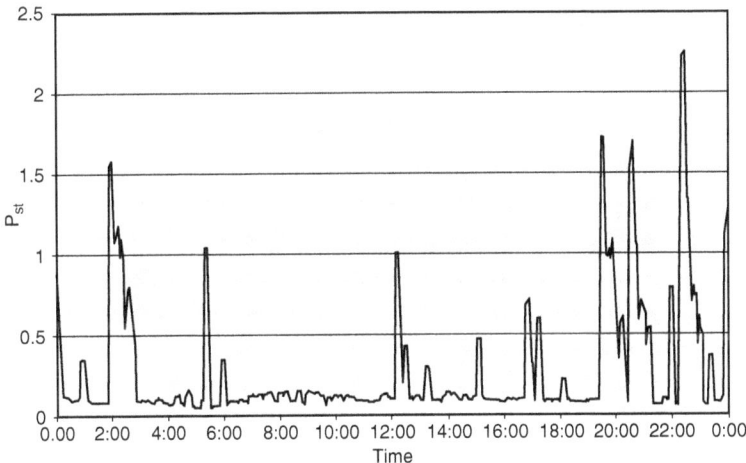

FIGURE 11.19 Flicker variations at the PCC with an arc furnace characterized by the P_{st} levels for a 24-h period (March 1, 2001) (note that there is one P_{st} value every 10 min).

quality monitoring practice is to collect the data, turn them into useful information, and disseminate it to users. All these processes take place within the instrument itself. Thus, a new breed of power quality monitor was developed with integrated intelligent systems to meet this new challenge. This type of power quality monitor is an intelligent power quality monitor where information is directly created within the instrument and immediately available to the users. A smart power quality monitor allows engineers to take necessary or appropriate actions in a timely manner. Thus, instead of acting in a reactive fashion, engineers will act in a proactive fashion.

One such smart power quality monitor was developed by Electrotek Concepts, Dranetz-BMI, EPRI, and the Tennessee Valley Authority (TVA) (Fig. 11.20). The system features on-the-spot data analysis with rapid information dissemination via Internet technology, e-mails, pagers, and faxes. The system consists of data acquisition, data aggregation, communication, Web-based visualization, and enterprise management components. The data acquisition component (DataNode) is designed to measure the actual power system voltages, currents, and other quantities. The data aggregation, communication, Web-based visualization, and enterprise management components are performed by a mission-specific computer system called the InfoNode. The communication between the data acquisition device and the InfoNode is accomplished through serial RS-232/485/422 or Ethernet communications using industry standard protocols (UCA MMS and Modbus). One or more data acquisition devices, or DataNodes, can be connected to an InfoNode.

The InfoNode has its own firmware that governs the overall functionality of the monitoring system. It acts as a special-purpose data-base manager and Web server. Various special-purpose intelligent

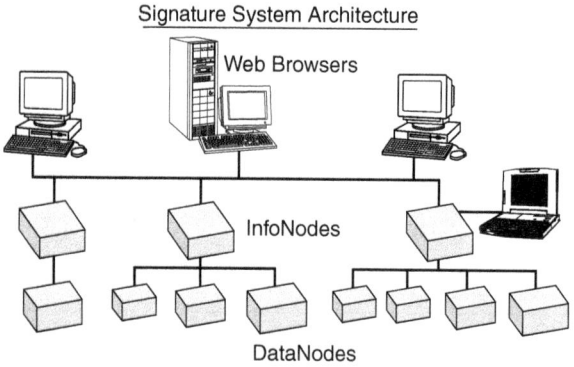

Signature System Architecture

Web Browsers

InfoNodes

DataNodes

FIGURE 11.20 A smart power quality monitoring system—it turns data into information on the spot and makes it available over the Internet. (*Courtesy of Dranetz-BMI.*)

systems are implemented within this computer system. Since it is a Web server, any user with Internet connectivity can access the data and its analysis results stored in its memory system. The monitoring system supports the standard file transfer protocol (FTP). Therefore, a database can be manually archived via FTP by simply copying the database to any personal computer with connectivity to the mission-specific computer system via network or modem. Proprietary software can be used to archive data from a group of InfoNodes.

11.3.11 Transducer Requirements

Monitoring of power quality on power systems often requires transducers to obtain acceptable voltage and current signal levels. Voltage monitoring on secondary systems can usually be performed with direct connections, but even these locations require current transformers (CTs) for the current signal.

Many power quality monitoring instruments are designed for input voltages up to 600 V rms and current inputs up to 5 A rms. Voltage and current transducers must be selected to provide these signal levels. Two important concerns must be addressed in selecting transducers:

1. *Signal levels.* Signal levels should use the full scale of the instrument without distorting or clipping the desired signal.

2. *Frequency response.* This is particularly important for transient and harmonic distortion monitoring, where high-frequency signals are particularly important.

These concerns and transducer installation considerations will now be discussed.

Signal Levels

Careful consideration to sizing of voltage transducers (VTs) and CTs is required to take advantage of the full resolution of the instrument without clipping the measured signal. Improper sizing can result in damage to the transducer or monitoring instrument.

Digital monitoring instruments incorporate the use of analog-to-digital (A/D) converters. These A/D boards convert the analog signal received by the instrument from the transducers into a digital signal for processing. To obtain the most accurate representation of the signal being monitored, it is important to use as much of the full range of the A/D board as possible. The noise level of a typical A/D board is approximately 33 percent of the full-scale bit value (5 bits for a 16-bit A/D board). Therefore, as a general rule, the signal that is input to the instrument should never be less than one-eighth of the full-scale value so that it is well above the noise level of the A/D board. This can be accomplished by selecting the proper transducers.

Voltage Transducers.   VTs should be sized to prevent measured disturbances from inducing saturation in the VT. For transients, this generally requires that the knee point of the transducer saturation curve be at least 200 percent of nominal system voltage.

Example 1.   When monitoring on a 12.47-kV distribution feeder and measuring line-to-ground, the nominal voltage across the primary of the voltage transducer will be 7200 V rms.

A VT ratio of 60:1 will produce an output voltage on the VT of 120 V rms (170 V peak) for a 7200-V rms input. Therefore, if the full-range value of the instrument is 600 V rms and the instrument incorporates a 16-bit A/D board, 131 bits of the A/D board will be used.

It is always good practice to incorporate some allowance in the calculations for overvoltage conditions. The steady-state voltage should not be right at the full-scale value of the monitoring instrument. If an overvoltage occurred, the signal would be clipped by the A/D board, and the measurement would be useless. Allowing for a 200 percent overvoltage is suggested. This can be accomplished by changing the input scale on the instrument, or sizing the VT accordingly.

Current Transducers.   Selecting the proper transducer for currents is more difficult. The current in any system changes more often and with greater magnitude than the voltage. Most power quality instrument manufacturers supply CTs with their equipment. These CTs come in a wide range of sizes to accommodate different load levels. The CTs are usually rated for maximum continuous load current.

The proper CT current rating and turns ratio depend on the measurement objective. If fault or inrush currents are of concern, the CT must be sized in the range of 20 to 30 times normal load current. This will result in low resolution of the load currents and an inability to accurately characterize load current harmonics.

If harmonics and load characterization are important, CTs should be selected to accurately characterize load currents. This permits evaluation of load response to system voltage variations and accurate calculation of load current harmonics.

Example 2.   The desired current signal to the monitoring instrument is 1 to 2 A rms. Assuming a 1-A value, the optimum CT ratio for an average feeder current of 120 A rms is 120:1. Manufacturer's data commonly list a secondary current base of 5 A to describe CT turns ratios rather than 1 A. The primary rating for a CT with a 5-A secondary rating is calculated as follows:

$$CT_{PRI} = \frac{I_{PRI}CT_{SEC}}{I_{SEC}} = \frac{120 \cdot 5}{1} = 600$$

Thus, a 600:5 CT would be specified.

Frequency Response

Transducer frequency response characteristics can be illustrated by plotting the ratio correction factor (RCF), which is the ratio of the expected output signal (input scaled by turns ratio) to the actual output signal, as a function of frequency.

Voltage Transducers. The frequency response of a standard metering class VT depends on the type and burden. In general, the burden should be a very high impedance (see Figs. 11.21 and 11.22). This is generally not a problem with most monitoring equipment available today. Power quality monitoring instruments, digital multimeters (DMMs), oscilloscopes, and other instruments all present a very high impedance to the transducer. With a high impedance burden, the response is usually adequate to at least 5 kHz. A typical RCF is plotted in Figs. 11.21 and 11.22 for two VT burdens.[9]

Some substations use capacitively coupled voltage transformers (CCVTs) for voltage transducers. These *should not* be used for general power quality monitoring. There is a low-voltage transformer in parallel with the lower capacitor in the capacitive divider. This configuration results in a circuit that is tuned to 60 Hz and will not provide accurate representation of any higher-frequency components.

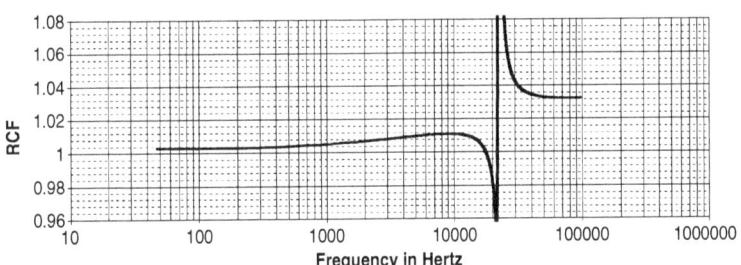

FIGURE 11.21 Frequency response of a standard VT with 1-MV burden.

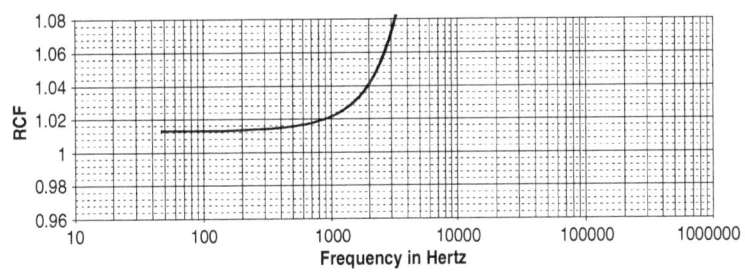

FIGURE 11.22 Frequency response of a standard VT with 100-V burden.

Measuring very high frequency components in the voltage requires a capacitive divider or pure resistive divider. Figure 11.23 illustrates the difference between a CCVT and a capacitive divider. Special-purpose capacitor dividers can be obtained for measurements requiring accurate characterization of transients up to at least 1 MHz.

Current Transducers. Standard metering class CTs are generally adequate for frequencies up to 2 kHz (phase error may start to become significant before this).[10] For higher frequencies, *window-type* CTs with a high turns ratio (doughnut, split-core, bar-type, and clamp-on) should be used.

Additional desirable attributes for CTs include

1. Large turns ratio, e.g., 2000:5 or greater.

2. Window-type CTs are preferred. *Primary wound* CTs (i.e., CTs in which system current flows through a winding) may be used, provided that the number of turns is less than five.

3. Small remnant flux, e.g., ±10 percent of the core saturation value.

4. Large core area. The more steel used in the core, the better the frequency response of the CT.

5. Secondary winding resistance and leakage impedance as small as possible. As shown in Fig. 11.24, this allows more of the output signal to flow into the burden, rather than the stray capacitance and core exciting impedance.

Installation Considerations

Monitoring on the distribution primary requires both voltage and current transducers. Selection of the best combination of these transducers depends on a number of factors:

- Monitoring location (substation, overhead, underground, etc.)
- Space limitations
- Ability to interrupt circuit for transducer installation
- Need for current monitoring

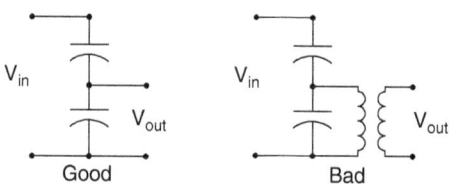

FIGURE 11.23 Capacitively coupled voltage dividers.

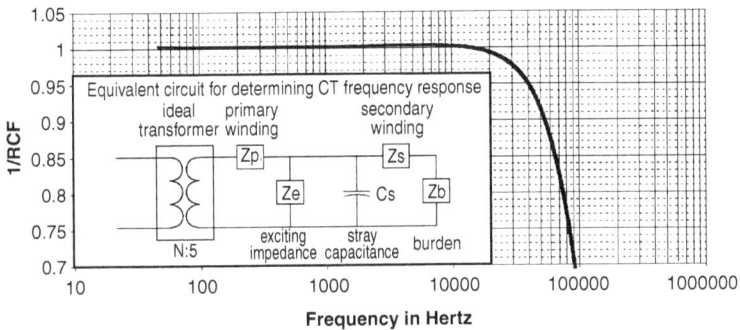

Substation Transducers.   Usually, existing substation CTs and VTs (except CCVTs) can be used for power quality monitoring.

Utility Overhead Line Locations.   For power quality monitoring on distribution primary circuits, it is often desirable to use a transducer that could be installed without taking the circuit out of service. Recently, transducers for monitoring both voltage and current have been developed that can be installed on a live line.

These devices incorporate a resistive divider-type VT and window-type CT in a single unit. A split-core choke is clamped around the phase conductor and is used to shunt the line current through the CT in the insulator. This method allows the device to be installed on the crossarm in place of the original insulator. By using the split-core choke, the phase conductor does not have to be broken, and thus, the transducers can be installed on a live line.

Initial tests indicated adequate frequency response for these transducers. However, field experience with these units has shown that the frequency response, even at 60 Hz, is dependent on current magnitude, temperature, and secondary cable length. This makes this type of device difficult to use for accurate power quality monitoring. Care must be exercised in matching these transducers to the instruments.

In general, all primary sites should be monitored with metering class VTs and CTs to obtain accurate results over the required frequency spectrum. Installation will require a circuit outage, but convenient designs can be developed for pole-top installations to minimize the outage.

Another option for monitoring primary sites involves monitoring at the secondary of an unloaded distribution transformer. This will give accurate results up to at least 3 kHz. This option does not help with the current transducers, but it is possible to get by without the currents at some circuit locations (e.g., end of the feeder). This option may be particularly attractive for underground circuits where the monitor can be installed on the secondary of a pad-mounted transformer.

Primary wound CTs are available from a variety of CT manufacturers. Reference 2 concludes that any primary wound CT with a single turn, or very few turns, should have a frequency response up to 10 kHz.

End-User (Secondary) Sites.   Transducer requirements at secondary sites are much simpler. Direct connection for the voltage is possible for 120/208- or 277/480-V rms systems. This permits full utilization of the instrument's frequency-response capability.

Currents can be monitored with either metering CTs (at the service entrance, for example) or with clamp-on CTs (at locations within the facility). Clamp-on CTs are available in a wide range of turns ratios. The frequency range is usually published by the manufacturer.

Summary of Transducer Recommendations
Table 11.2 describes different monitoring locations and the different types of transducers that are adequate for monitoring at these locations.

Table 11.3 describes the different power quality phenomena and the proper transducers to measure that type of power quality problem. Tables 11.2 and 11.3 should be used in conjunction with each other to determine the best transducer for a given application.

Summary of Monitoring Equipment Capabilities
Figure 11.25 summarizes the capabilities of the previously described metering instruments as they relate to the various categories of power quality variations.

Location	VT	CT
Substation	Metering VTs Special-purpose capacitive or resistive dividers Calibrated bushing taps	Metering CTs Relaying CTs
Overhead lines	Metering VTs	Metering CTs
Underground locations	Metering VTs Pad-mounted transformer Special-purpose dividers	Metering CTs
Secondary sites' service entrance	Direct connection	Metering CTs Clamp-on CTs
In facility	Direct connection	Clamp-on CTs

TABLE 11.2   VT and CT Options for Different Locations

Concern	VTs*	CTs
Voltage variations	Standard metering	Standard metering
Harmonic levels	Standard metering	Window-type
Low-frequency transients (switching)	Standard metering with high-kneepoint saturation	Window-type
High-frequency transients (lightning)	Capacitive or resistive dividers	Window-type

*VTs are usually not required at locations below 600 V rms nominal.

TABLE 11.3   VT and CT Requirements for Different Power Quality Variations

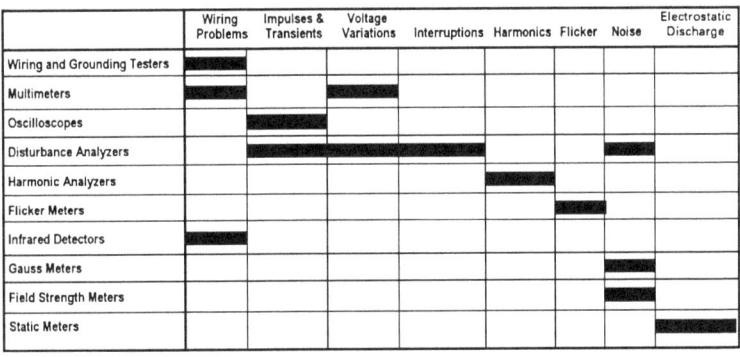

FIGURE 11.25   Power quality measurement equipment capabilities.

11.4   Assessment of Power Quality Measurement Data

As utilities and industrial customers have expanded their power quality monitoring systems, the data management, analysis, and interpretation functions have become the most significant challenges in the overall power quality monitoring effort. In addition, the shift in the use of power quality monitoring from off-line benchmarking to on-line operation with automatic identification of problems and concerns has made the task of data management and analysis even more critical.

There are two streams of power quality data analysis, i.e., off-line and on-line analyses. The off-line power quality data analysis, as the term suggests, is performed off-line at the central processing locations. On the other hand, the on-line data analysis is performed within the instrument itself for immediate information dissemination. Both types of power quality data assessment are described in Secs. 11.4.1 and 11.4.2.

11.4.1   Off-Line Power Quality Data Assessment

Off-line power quality data assessment is carried out separately from the monitoring instruments. Dedicated computer software is used for this purpose. Large-scale monitoring projects with large volumes of data to analyze often present a challenging set of requirements for software designers and application engineers. First, the software must integrate well with monitoring equipment and the large number of productivity tools that are currently available. The storage of vast quantities of both disturbance and steady-state measurement data requires an efficient and well-suited database. Data management tools that can quickly characterize and load power quality data must be devised, and analysis tools must be integrated with the database. Automation of data management and report generation tasks must be supported, and the design must allow for future expansion and customizing.

The new standard format for interchanging power quality data—the Power Quality Data Interchange Format (PQDIF)—makes sharing of data between different types of monitoring systems much more feasible. This means that applications for data management and data analysis can be written by third parties and measurement data from a wide variety of monitoring systems can be accessible to these systems. PQView (www.pqview.com) is an example of this type of third-party application. The PQDIF standard is described in Sec. 11.6.

The off-line power quality data assessment software usually performs the following functions:

- Viewing of individual disturbance events.

- RMS variation analysis which includes tabulations of voltage sags and swells, magnitude-duration scatter plots based on CBEMA, ITI, or user-specified magnitude-duration curves, and computations of a wide range of rms indices such as SARFI, SIARFI, and CAIDI.

- Steady-state analysis which includes trends of rms voltages, rms currents, and negative- and zero-sequence unbalances. In addition, many software systems provide statistical analysis of various minimum, average, maximum, standard deviation, count, and cumulative probability levels. Statistics can be temporally aggregated and dynamically filtered. Figures 11.26 and 11.27 show the time trend of phase A rms voltage along with its histogram representation.

- Harmonic analysis where users can perform voltage and current harmonic spectra, statistical analysis of various harmonic indices, and trending overtime.

- Transient analysis which includes statistical analysis of maximum voltage, transient durations, and transient frequency.

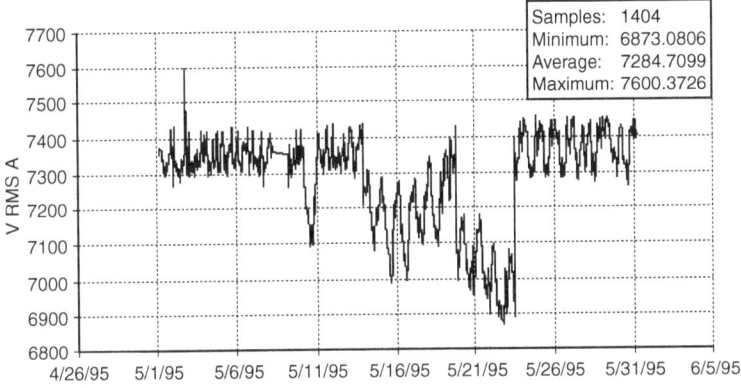

FIGURE 11.26 Time trend of an rms voltage is a standard feature in many power quality analysis software programs.

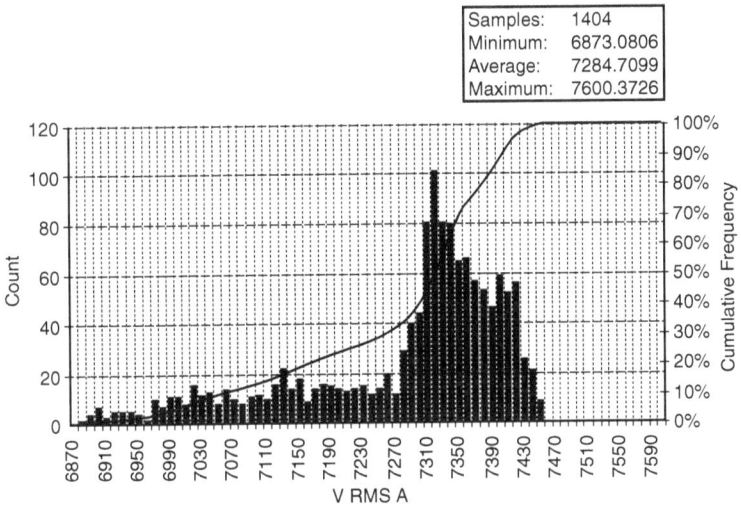

FIGURE 11.27 Histogram representation of rms voltage indicates the statistical distribution of the rms voltage magnitude.

- Standardized power quality reports (e.g. daily reports, monthly reports, statistical performance reports, executive summaries, customer power quality summaries).
- Analysis of protective device operation (identify problems).
- Analysis of energy use.

- Correlation of power quality levels or energy use with important parameters (e.g., voltage sag performance versus lightning flash density).

- Equipment performance as a function of power quality levels (equipment sensitivity reports).

11.4.2   On-Line Power Quality Data Assessment

On-line power quality data assessment analyzes data as they are captured. The analysis results are available immediately for rapid dissemination. Complexity in the software design requirement for on-line assessment is usually higher than that of off-line. Most features available in off-line analysis software can also be made available in an on-line system. One of the primary advantages of on-line data analysis is that it can provide instant message delivery to notify users of specific events of interest. Users can then take immediate actions upon receiving the notifications. Figure 11.28 illustrates a simple message delivered to a user reporting that a capacitor bank located upstream from a data acquisition node called "DataNode H09_5530" was energized at 05–15–2002 at 04:56:11 a.m. The message also details the transient characteristics such as the magnitude, frequency, and duration along with the relative location of the capacitor bank from the data acquisition node.

Figure 11.29 shows another example of the on-line power quality assessment. It shows the time trend of a fifth-harmonic current magnitude along with its statistical distribution. The data and its analysis are displayed on a standard Web browser. Here a

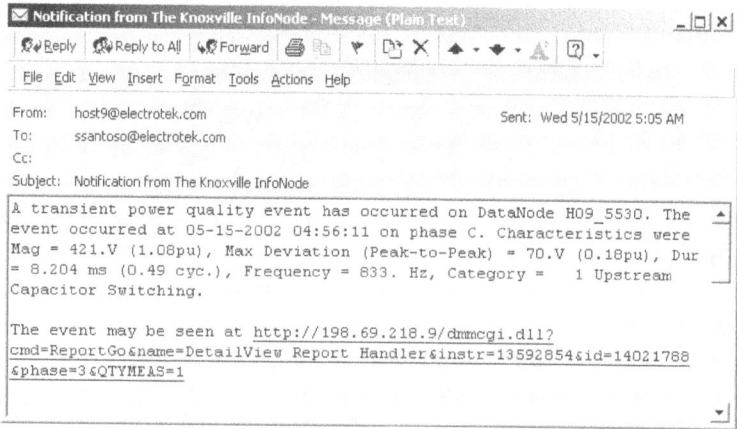

Figure 11.28 On-line data analysis can send e-mail notifications to users about the occurrence of specific events.

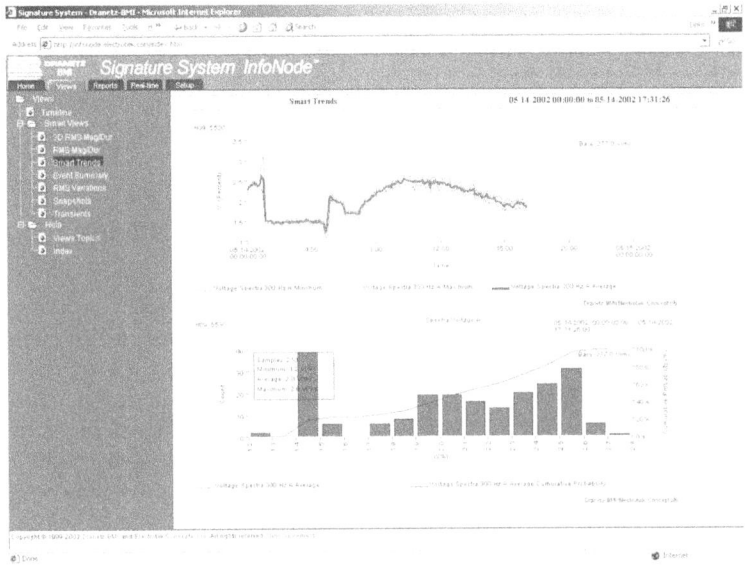

FIGURE **11.29** On-line data analysis displayed on a standard Web browser. The analysis includes the trend of minimum, maximum, and average values of the fifth-harmonic voltage distortion along with a statistical distribution of the average values.

user can analyze data up to the current time. This on-line system has the capability of performing a full range of transient, harmonic, and steady-state characterization along with their statistical distribution analysis comparable to that in off-line assessment analysis.

11.5 Application of Intelligent Systems

Many advanced power quality monitoring systems are equipped with either off-line or on-line intelligent systems to evaluate disturbances and system conditions so as to make conclusions about the cause of the problem or even predict problems before they occur. The applications of intelligent systems or autonomous expert systems in monitoring instruments help engineers determine the system condition rapidly. This is especially important when restoring service following major disturbances.

The implementation of intelligent systems within a monitoring instrument can significantly increase the value of a monitoring application since it can generate information rather than just collect data.[11] The intelligent systems are packaged as individual autonomous expert system modules, where each module performs specific functions. Examples include an expert system module that analyzes capacitor-switching transients and determines the relative location of the capacitor bank, and an expert system module to determine the relative location of the fault causing a voltage sag. Sections 11.5.1 and 11.5.2 describe the approach in designing an autonomous expert system for power quality data assessment, and give application examples.

11.5.1   Basic Design of an Expert System for Monitoring Applications

The development of an autonomous expert system calls for many approaches such as signal processing and rule-based techniques along with the knowledge-discovery approach commonly known as data mining. Before the expert system module is designed, the functionalities or objectives of the module must be clearly defined. In other words, the designers or developers of the expert system module must have a clear understanding about what knowledge they are trying to discover from volumes of raw measurement data. This is very important since they will ultimately determine the overall design of the expert system module.

The process of turning raw measurement data into knowledge involves data selection and preparation, information extraction from selected data, information assimilation, and report presentation. These steps (illustrated in Fig. 11.30) are commonly known as knowledge discovery or data mining.[12]

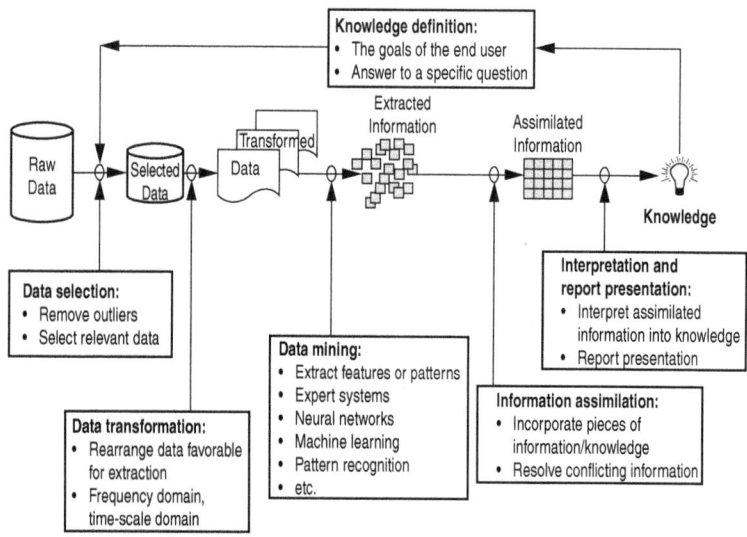

FIGURE 11.30   Process of turning raw data into answers or knowledge.

The first step in the knowledge discovery is to select appropriate measurement quantities and disregard other types of measurements that do not provide relevant information. In addition, during the data-selection process preliminary analyses are usually carried out to ensure the quality of the measurement. For example, an expert system module is developed to retrieve a specific answer, and it requires measurements of instantaneous three-phase voltage and current waveforms to be available. The data-selection task is responsible for ensuring that all required phase voltage and current waveform data are available before proceeding to the next step. In some instances, it might be necessary to interpolate or extrapolate data in this step. Other preliminary examinations include checking any outlier magnitudes, missing data sequences, corrupted data, etc. Examination on data quality is important as the accuracy of the knowledge discovered is determined by the quality of data.

The second step attempts to represent the data and project them onto domains in which a solution is more favorable to discover. Signal-processing techniques and power system analysis are applied. An example of this step is to transform data into another domain where the information might be located. The Fourier transform is performed to uncover frequency information for steady-state signals, the wavelet transform is performed to find the temporal and frequency information for transient signals, and other transforms may be performed as well.

Now that the data are already projected onto other spaces or domains, we are ready to extract the desired information. Techniques to extract the information vary from sophisticated ones, such as pattern recognition, neural networks, and machine learning, to simple ones, such as finding the maximum value in the transformed signal or counting the number of points in which the magnitude of a voltage waveform is above a predetermined threshold value. One example is looking for harmonic frequencies of a distorted waveform. In the second step the waveform is transformed using the Fourier transform, resulting in a frequency-domain signal. A simple harmonic frequency extraction process might be accomplished by first computing the noise level in the frequency-domain signal, and subsequently setting a threshold number to several-fold that of the noise level. Any magnitude higher than the threshold number may indicate the presence of harmonic frequencies.

The data mining step usually results in scattered pieces of information. These pieces of information are assimilated to form knowledge. In some instances assimilation of information is not readily possible since some pieces of information conflict with each other. If the conflicting information cannot be resolved, the quality of the answer provided might have limited use. The last step in the chain is interpretation of knowledge and report presentation.

11.5.2   Example Applications of Expert Systems

One or more autonomous expert system modules can be implemented within an advanced power quality monitoring system. When a power quality event is captured, all modules will be invoked. Each module will attempt to discover the unique knowledge it is designed to look for. Once the unique knowledge is discovered, the knowledge will be available for users to inspect. The knowledge can be viewed on a standard browser, or sent as an e-mail, pager, or fax message. We present a few examples of autonomous expert systems.

Voltage Sag Direction Module

Voltage sags are some of the most important disturbances on utility systems. They are usually caused by a remote fault somewhere on the power system; however, they can also be caused by a fault inside end-user facilities. Determining the location of the fault causing the voltage sag can be an important step toward preventing voltage sags in the future and assigning responsibility for addressing the problem. For instance, understanding the fault location is necessary for implementing contracts that include voltage sag performance specifications. The supplier would not be responsible for sags that are caused by faults within the customer facility. This is also important when trying to assess performance of the distribution system in comparison to the transmission system as the cause of voltage sag events that can impact customer operations. The fault locations can help identify future problems or locations where maintenance or system changes are required. An expert system to identify the fault location (at least upstream or downstream from the monitoring location) can help in all these cases.

An autonomous expert system module called the voltage sag direction module is designed just for that purpose, i.e., to detect and identify a voltage sag event and subsequently determine the origin (upstream or downstream from the monitoring location) of the voltage sag event. If a data acquisition node is installed at a customer PCC, the source of the voltage sag will be either on the utility or the customer side of the meter. If the monitoring point is at a distribution substation transformer, the source of the voltage sag will be either the distribution system or the transmission system.

The voltage sag direction module works by comparing current and voltage rms magnitudes both before and after the sag event. It tracks phase angle changes from prefault to postfault. By assembling information from the rms magnitude comparison and the phase angle behavior, the origin of the voltage sag event can be accurately determined. In addition, the voltage sag direction module is equipped with algorithms to assess the quality of the knowledge or answer discovered. If the answer is deemed accurate, it will be sent as an output; otherwise, it will be neglected and no answer will be provided. In this way, inaccurate or false knowledge can be minimized. Inaccurate knowledge can be due to a number of factors, primarily to missing data and unresolved conflicting characteristics.

Outputs of the voltage sag direction module can be displayed on a computer screen using Web browser software, displayed in printed paper format, sent to a pager, or sent as an e-mail. Figure 11.31 shows an output of a voltage sag direction expert system module. The first column indicates the event time, the second column indicates the monitor identification, the third column indicates event types, the fourth column indicates the triggered channel, and finally the fifth column indicates the characteristics of the event and outputs of the answer module.

Figure 11.31 shows an event table with several voltage sag events that occurred at 11:16:55 a.m. on April 24, 2002. A tree branch that fell across a 13-kV overhead line caused the sag events. A total of five automatic reclosure operations were performed before the breaker finally tripped and locked out. There were two data acquisition nodes available to capture this disturbance: one at the substation, i.e., at the secondary of 161/13-kV transformer (LCUBSub), where the affected overhead line was served, and one at the service entrance of a Electrotek office complex (H09_5530) located about 0.5 mi from the substation. (See Fig. 11.38 for the geographical locations of these data acquisition nodes.) Obviously, the LCUBSub and H09_5530 data acquisition nodes should report that the directions or the relative origin of voltage sags are downstream and upstream, respectively. Analysis provided by the voltage sag direction module reports the direction of the voltage sag

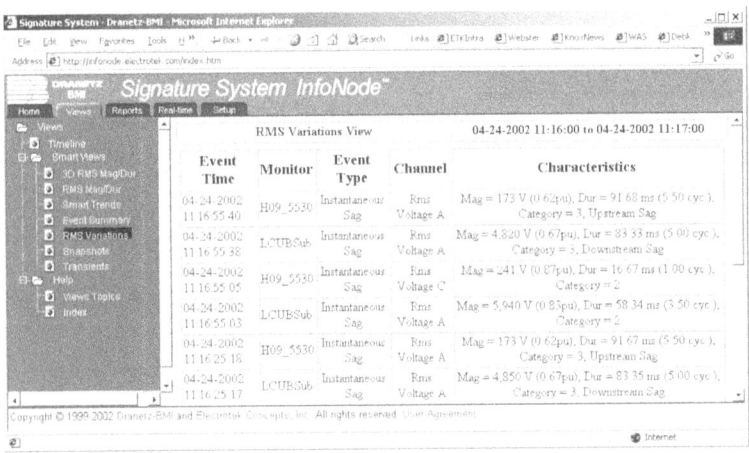

Figure 11.31 A standard Web browser is the interface between the monitoring system and users. Outputs of the voltage sag direction module are shown in the last column of the table.

correctly. Note that there are two voltage sag events where the module does not provide any knowledge about the origin of the sag event. This happens since the algorithms were unable to resolve conflicting characteristics extracted from the data. Figure 11.32 shows the table of the sag events associated with this fault.

Radial Fault Locator Module

Radial distribution feeders are susceptible to various short-circuit events such as symmetrical faults (three-phase) and unsymmetrical faults, including single-line-to-ground, double-line-to-ground, and line-to-line faults. These system faults arise from various conditions ranging from natural causes such as severe weather conditions and animal contacts to human intervention and errors, including equipment failure. Quickly identifying the source and location of faults is the key to cost-efficient system restoration. The current practice to locate the faults is to send a lineperson to patrol the suspected feeders. While this is a proven method, it is certainly not a cost-effective way to restore power.

RMS Variations View				04-24-2002 11:16:00 to 04-24-2002 11:17:00
Event Time	**Monitor**	**Event Type**	**Channel**	**Characteristics**
04-24-2002 11:16:55.40	H09_5530	Instantaneous Sag	Rms Voltage A	Mag = 173.V (0.62pu), Dur = 91.68 ms (5.50 cyc.), Category = 3, Upstream Sag
04-24-2002 11:16:55.38	LCUBSub	Instantaneous Sag	Rms Voltage A	Mag = 4,820.V (0.67pu), Dur = 83.33 ms (5.00 cyc.), Category = 3, Downstream Sag
04-24-2002 11:16:55.05	H09_5530	Instantaneous Sag	Rms Voltage C	Mag = 241.V (0.87pu), Dur = 16.67 ms (1.00 cyc.), Category = 2
04-24-2002 11:16:55.03	LCUBSub	Instantaneous Sag	Rms Voltage A	Mag = 5,940.V (0.83pu), Dur = 58.34 ms (3.50 cyc.), Category = 2
04-24-2002 11:16:25.18	H09_5530	Instantaneous Sag	Rms Voltage A	Mag = 173.V (0.62pu), Dur = 91.67 ms (5.50 cyc.), Category = 3, Upstream Sag
04-24-2002 11:16:25.17	LCUBSub	Instantaneous Sag	Rms Voltage A	Mag = 4,850.V (0.67pu), Dur = 83.35 ms (5.00 cyc.), Category = 3, Downstream Sag
04-24-2002 11:16:07.42	H09_5530	Instantaneous Sag	Rms Voltage A	Mag = 173.V (0.62pu), Dur = 91.66 ms (5.50 cyc.), Category = 3, Upstream Sag
04-24-2002 11:16:07.40	LCUBSub	Instantaneous Sag	Rms Voltage A	Mag = 4,850.V (0.67pu), Dur = 83.33 ms (5.00 cyc.), Category = 3, Downstream Sag
04-24-2002 11:16:06.46	H09_5530	Instantaneous Sag	Rms Voltage A	Mag = 174.V (0.63pu), Dur = 91.68 ms (5.50 cyc.), Category = 3, Upstream Sag
04-24-2002 11:16:06.44	LCUBSub	Instantaneous Sag	Rms Voltage A	Mag = 4,840.V (0.67pu), Dur = 83.33 ms (5.00 cyc.), Category = 3, Downstream Sag

Figure 11.32 An event summary report detailing time of occurrence and event characteristics. There are five voltage sag events associated with the autoreclosure operation following a fault. The voltage sag direction module identifies the origin of the sag correctly.

An expert system module called the radial fault locator is developed to estimate the distance to a fault location from the location where the measurements were made. The unique feature of this module is that it only requires a set of three-phase voltages and currents from a single measurement location with the sequence impedance data of the primary distribution feeder. The module works by first identifying a permanent fault event based on the ground fault and phase fault pickup current threshold. Users can enter these values in the answer module setup window shown in Fig. 11.33. Once a permanent fault event is identified, the distance to fault estimation is carried out based on the apparent impedance approach.[13] Estimates of the distance to the fault are then displayed in a computer screen with the Web browser illustrated in Fig. 11.34 or sent to a lineperson via a pager. The lineperson can quickly pinpoint the fault location. This example illustrates the emerging trend in smart power quality monitoring, i.e., collect power quality data and extract and formulate information for users to perform necessary actions.

Capacitor-Switching Direction Module

Capacitor-switching operations are the most common cause of transient events on the power system. When a capacitor bank is

Properties	Values
Activate AnswerModule	✓
Ground fault pickup current threshold (amperes):	150.0000
Phase fault pickup current threshold (amperes):	800.0000
Ratio of fault peak current to pre-fault peak current:	2.0000
Sequence impedance unit:	Ohms per 1000 ft
Length of primary feeder (unit is based on the unit length in sequence impedance unit)	62.0000
Positive-sequence impedance of the primary feeder (real):	0.0570
Positive-sequence impedance of the primary feeder (imaginary):	0.1225
Zero-sequence impedance of the primary feeder (real):	0.1790
Zero-sequence impedance of the primary feeder (imaginary):	0.4150

FIGURE 11.33   Setup window for the radial fault location answer module.

DRANETZ
BMI

CDRWest
All time
MaryRosenbalt

Radial Fault Report

Event Time	Monitor	Characteristics (Mag/Dur)	Fault Type	Lower Distance Estimate	Upper Distance Estimate
12/28/1999 13.49.11	CDRWest	Mag=19797.24V ,Duration= 0.17 secs	3 Phase Fault	20.44 (in 1000 ft)	20.44 (in 1000 ft)
02/28/2000 11.23.13	CDRWest	Mag=19797.24V ,Duration= 0.17 secs	3 Phase Fault	22.18 (in 1000 ft)	22.18 (in 1000 ft)
03/18/2000 03.09.37	CDRWest	Mag=19725.27V ,Duration= 0.17 secs	LLF at Phase BC	34.07 (in 1000 ft)	35.59 (in 1000 ft)
03/21/2000 16.41.32	CDRWest	Mag=14387.99V ,Duration= 0.17 secs	SLG at Phase A	20.45 (in 1000 ft)	21.43 (in 1000 ft)

FIGURE 11.34   The distance estimates presented in tabular form.

energized, it interacts with the system inductance, yielding oscillatory transients. The transient overvoltage in an uncontrolled switching is between 1.0 to 2.0 pu with typical overvoltages of 1.3 to 1.4 pu and frequencies of 250 to 1000 Hz. Transients due to energizing utility capacitor banks can propagate into customer facilities. Common problems associated with the switching transients include tripping off sensitive equipment such adjustable-speed drives and other electronically controlled loads. Some larger end-user facilities may also have capacitor banks to provide reactive power and voltage support as well.

When a sensitive load trips off due to capacitor-switching transients, it is important to know where the capacitor bank is, whether it is on the utility side or in the customer facility. A capacitor-switching direction expert system module is designed to detect and identify a capacitor-switching event and determine the relative location of the capacitor bank from the point where measurements were collected. It only requires a set of three-phase voltages and currents to perform the tasks mentioned. This module is useful to determine the responsible parties, i.e., the utility or customer, and help engineers pinpoint the problematic capacitor bank.

The capacitor-switching transient direction module works as follows. When an event is captured, the module will extract the information and represent it in domains where detection and identification are more favorable. The domains where the information is represented are in the time-, frequency-, and time-scale (wavelet) domains. If the root cause of the event is due to a capacitor bank energization, the answer module will proceed to determine the most probable location of the capacitor bank.

There are only two possible locations with respect to the monitoring location, i.e., upstream or downstream. The expert system module works well with grounded, ungrounded, delta-configured, and wye- (or star-) configured capacitor banks. It also works well for back-to-back capacitor banks. The capacitor-switching transient direction module is equipped with algorithms to determine the quality of the information it discovers. Thus, the module may provide an undetermined answer. This answer is certainly better than an incorrect one.

An example application of the answer module to analyze data capture from a data acquisition node installed at an office complex service entrance is shown in Fig. 11.35. The analysis results are shown in Fig. 11.36, which is a screen capture from a standard Web browser. Since the office complex has no capacitor banks, any capacitor-switching transients must originate from the utility side located upstream from the data acquisition node. The module correctly determines the relative location of the capacitor bank. Note that there are some instances where the expert system was not able to determine the relative location of the capacitor bank. From the time stamp of the

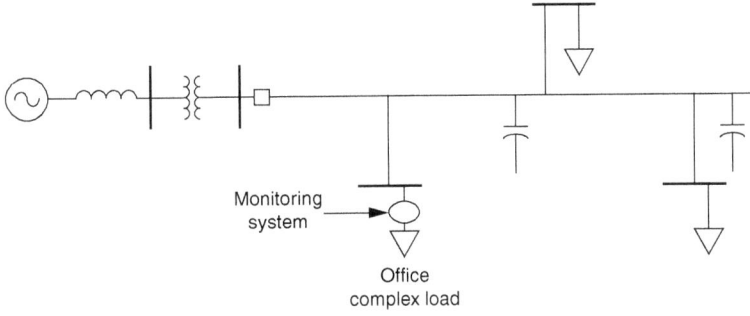

FIGURE 11.35 All capacitors are upstream from the monitoring location. Therefore, the answer module should report upstream capacitor switching when such an event is captured.

Event Time	Monitor	Event Type	Channel	Characteristics
04-19-2002 19:01:17.36	H09_5530	Transient	Instantaneous Voltage C	Mag = 448 V (1.14pu), Max Deviation (Peak-to-Peak) = 174 V (0.44pu), Dur = 1.692 ms (0.10 cyc.), Frequency = 420 Hz, Category = 2, Upstream Capacitor Switching
04-19-2002 05:03:48.96	H09_5530	Transient	Instantaneous Voltage A	Mag = 406 V (1.04pu), Max Deviation (Peak-to-Peak) = 92 V (0.24pu), Dur = 3.646 ms (0.22 cyc.), Upstream Capacitor Switching
04-19-2002 04:57:05.33	H09_5530	Transient	Instantaneous Voltage A	Mag = 474 V (1.21pu), Max Deviation (Peak-to-Peak) = 310 V (0.79pu), Dur = 3.516 ms (0.21 cyc.), Frequency = 3,218 Hz, Category = 3, Upstream Capacitor Switching
04-19-2002 04:57:04.84	H09_5530	Transient	Instantaneous Voltage B	Mag = 485 V (1.24pu), Max Deviation (Peak-to-Peak) = 270 V (0.69pu), Dur = 1.693 ms (0.10 cyc.), Frequency = 2,453 Hz, Category = 3, Upstream Capacitor Switching
04-19-2002 04:56:31.69	H09_5530	Transient	Instantaneous Voltage A	Mag = 412 V (1.05pu), Max Deviation (Peak-to-Peak) = 106 V (0.27pu), Dur = 5.079 ms (0.30 cyc.), Frequency = 1,013 Hz, Category = 1, Direction Unknown Capacitor Switching
04-18-2002 19:01:19.69	H09_5530	Transient	Instantaneous Voltage C	Mag = 408 V (1.04pu), Max Deviation (Peak-to-Peak) = 169 V (0.43pu), Dur = 2.343 ms (0.14 cyc.), Frequency = 795 Hz, Upstream Capacitor Switching
04-18-2002 04:57:06.26	H09_5530	Transient	Instantaneous Voltage A	Mag = 405 V (1.03pu), Max Deviation (Peak-to-Peak) = 121 V (0.31pu), Dur = 2.865 ms (0.17 cyc.), Direction Unknown Capacitor Switching
04-18-2002 04:57:05.81	H09_5530	Transient	Instantaneous Voltage A	Mag = 404 V (1.03pu), Max Deviation (Peak-to-Peak) = 168 V (0.43pu), Dur = 45.19 ms (2.71 cyc.), Upstream Capacitor Switching
04-17-2002 04:56:53.15	H09_5530	Transient	Instantaneous Voltage A	Mag = 422 V (1.08pu), Max Deviation (Peak-to-Peak) = 220 V (0.56pu), Dur = 2.735 ms (0.16 cyc.), Frequency = 3,180 Hz, Category = 1, Upstream Capacitor Switching
04-17-2002 04:56:52.73	H09_5530	Transient	Instantaneous Voltage B	Mag = 479 V (1.22pu), Max Deviation (Peak-to-Peak) = 125 V (0.32pu), Dur = 2.734 ms (0.16 cyc.), Category = 3, Direction Unknown Capacitor Switching
04-17-2002 04:56:52.65	H09_5530	Transient	Instantaneous Voltage B	Mag = 497 V (1.27pu), Max Deviation (Peak-to-Peak) = 230 V (0.59pu), Dur = 2.344 ms (0.14 cyc.), Frequency = 2,468 Hz, Category = 3, Direction Unknown Capacitor Switching

Transients View — 04-17-2002 00:00:00 to 04-19-2002 23:23:55

FIGURE 11.36 The output of the capacitor-switching answer module for the one-line diagram presented in Fig. 11.35.

events, it is clear that capacitor bank energizations occur at about 5:00 a.m. and 7:00 p.m. each day.

Capacitor-Switching Operation Inspection Module

As described, capacitor-switching transients are the most common cause of transient events on the power system and are results of capacitor bank energization operation. One common thing that can go wrong with a capacitor bank is for a fuse to blow. Some capacitor banks may not be operating properly for months before utility personnel notice the problem. Routine maintenance is usually performed by driving along the line and visually inspecting the capacitor bank.

An autonomous expert system was developed for substation applications to analyze downstream transient data and determine if a capacitor-switching operation is performed successfully and display a warning message if the operation was not successful.[14] With the large number of capacitor banks on most power systems, this expert system module can be a significant benefit to power systems engineers in identifying problems and correlating them with capacitor-switching events.

Successful capacitor bank energization is characterized by a uniform increase of kvar on each phase whose total corresponds to the capacitor kvar size. For example, when a 1200-kvar capacitor bank is energized, reactive power of approximately 400 kvar should appear on each phase. The total kvar increase can be determined by computing kvar changes in individual phases from the current and voltage waveforms before and after the switching operation. This total computed kvar change is then compared to the actual or physical capacitor bank kvar supplied by a user. If the expected kvar was not realized, the capacitor bank or its switching device may be having some problems.

Figure 11.37 shows the application of the capacitor-switching operation inspector expert system in a commercial monitoring system. The monitoring location is at the substation; thus, all capacitor banks along the feeders are downstream from the monitoring

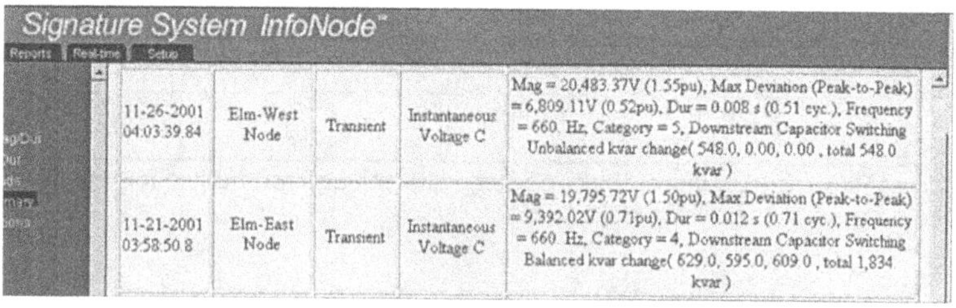

11-26-2001 04:03:39.84	Elm-West Node	Transient	Instantaneous Voltage C	Mag = 20,483.37V (1.55pu), Max Deviation (Peak-to-Peak) = 6,809.11V (0.52pu), Dur = 0.008 s (0.51 cyc.), Frequency = 660. Hz, Category = 5, Downstream Capacitor Switching Unbalanced kvar change(548.0, 0.00, 0.00 , total 548.0 kvar)
11-21-2001 03:58:50.8	Elm-East Node	Transient	Instantaneous Voltage C	Mag = 19,795.72V (1.50pu), Max Deviation (Peak-to-Peak) = 9,392.02V (0.71pu), Dur = 0.012 s (0.71 cyc.), Frequency = 660. Hz, Category = 4, Downstream Capacitor Switching Balanced kvar change(629.0, 595.0, 609.0 , total 1,834 kvar)

Figure 11.37 Analysis results of the capacitor-switching inspector expert system.

location. The first capacitor-switching event indicates that two phases of the capacitor are out of service. Either the fuses have blown or the switch is malfunctioning. The second event shows a successful capacitor-switching operation.

Lightning Correlation Module

The majority of voltage sags and outages in the United States are attributed to weather-related conditions such as thunderstorms. For example, TVA has approximately 17,000 mi of transmission lines where lightning accounts for as much as 45 percent of the faults on their system. The lightning correlation expert system module is designed to correlate lightning strikes with measured power quality events and make that information available in real time directly at the point of measurement. Armed with the correlation results, engineers can evaluate the cause and impact of voltage sags for a specific customer at a specific monitoring point as well as evaluate the impact on all customers for a given event.

When the lightning correlation module detects a voltage sag or transient event, it queries a lightning database via the Internet. The lightning data are provided by the U.S. National Lightning Detection Network operated by Global Atmospherics, Inc. If the query returns a result set, the lightning correlation module will store this information in the monitoring system database along with the disturbance data for information dissemination. The lightning data include the event time of the strike, the latitude and longitude of strike location, the current magnitude, and number of strokes.

The following example illustrates how the module performs its function. On Easter Sunday, March 31, 2002, in Knoxville, Tennessee, the location of the Electrotek Concepts primary engineering office, thunderstorms moved through the area around 11:00 a.m. Figure 11.38 shows the four lightning strikes in the Cedar Bluff area of Knoxville. The office complex, marked as "Electrotek," has a data acquisition node at the service entrance. Another data acquisition node marked as "Cedar Bluff Substation" is connected at the 161/13-kV transformer on the 13-kV side at the substation that feeds the Electrotek office about 1/2 mi away.

During the time of the storm, the power monitoring system captured a number of events. The strikes shown in Fig. 11.38 were all within several kilometers of the substation and are located directly on known distribution system right of way. Specifically, there were two events that correlated within 100 ms of lightning strikes during the storm as shown in the output of the lightning correlation module in Fig. 11.39. Lightning strike data are summarized in Table 11.4.

The output of the lightning correlation module clearly shows the correlation between power quality events and lightning strikes. Note that the power monitoring system is equipped with GPS-based time synchronization; thus, it is capable of making precise multisource

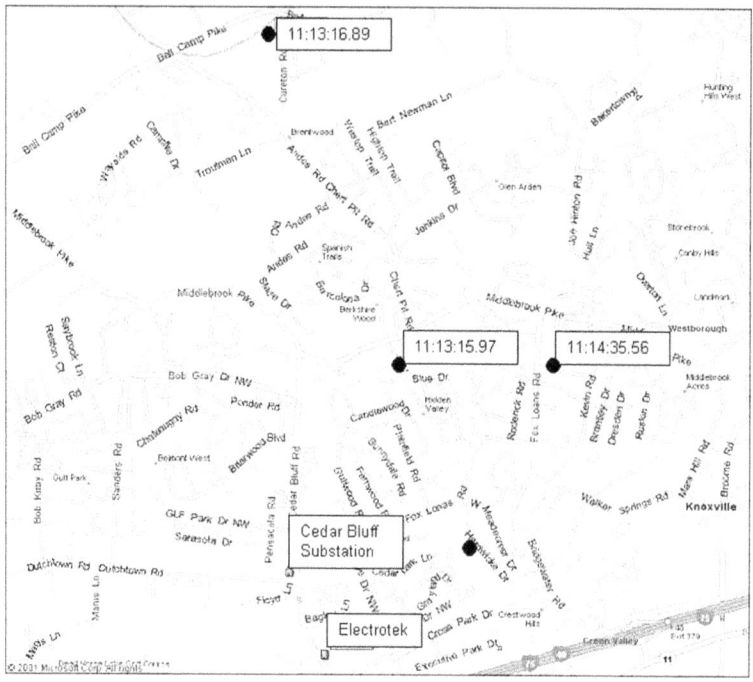

FIGURE 11.38   Lightning strikes near a substation serving the Electrotek office on Easter Sunday morning in 2002.

03-31-2002 11:14:34.56	LCUBSub	Transient	Instantaneous Voltage C	Mag = 9,800.V (0.96pu), Max Deviation (Peak-to-Peak) = 2,160.V (0.21pu), Dur = 8.075 ms (0.48 cyc.), Frequency = 390. Hz, 2 lightning strokes 3 km away 62 kA
03-31-2002 11:14:34.55	H09_5530	Transient	Instantaneous Voltage B	Mag = 420.V (1.07pu), Max Deviation (Peak-to-Peak) = 128.V (0.33pu), Dur = 7.945 ms (0.48 cyc.), Frequency = 398. Hz, Category = 1, 2 lightning strokes 3 km away 62 kA
03-31-2002 11:13:15.97	LCUBSub	Momentary Sag	Rms Voltage C	Mag = 5,730.V (0.80pu), Dur = 691.7 ms (41.50 cyc.), Category = 2, Downstream Sag, 1 lightning stroke 2 km away 12 kA
03-31-2002 11:13:15.97	H09_5530	Momentary Sag	Rms Voltage B	Mag = 237.V (0.86pu), Dur = 683.3 ms (41.00 cyc.), Category = 2, Upstream Sag, 1 lightning stroke 2 km away 12 kA

FIGURE 11.39   Outputs of the lightning correlation expert system module.

event correlation. Two lightning strikes, at 11:13:15 a.m. and 11:14:34 a.m., shown in Table 11.4 are captured in both data acquisition nodes installed at the Electrotek office and the substation.

11.5.3   Future Applications

There are many applications for the intelligent power quality monitoring concept. Some of the more important applications are listed in this section. The examples described in Sec. 11.5.2 are also included in this listing.

Time stamp		Latitude	Longitude	Mag (kA)	Distance (km)	Strokes	Type
3/31/2002	11:15:21.700	35.962	84.048	58	6	10	CG*
3/31/2002	11:14:42.700	31.335	88.921	132	678	1	CG
3/31/2002	11:14:34.500	35.941	84.073	62	3	2	CG
3/31/2002	11:14:31.800	31.696	88.760	42	638	1	CG
3/31/2002	11:14:11.100	31.549	88.745	32	649	2	CG
3/31/2002	11:14:09.800	31.885	88.406	10	600	1	CG
3/31/2002	11:14:08.900	31.705	88.347	10	612	1	CG
3/31/2002	11:13:36.300	31.739	88.577	14	623	1	CG
3/31/2002	11:13:36.300	31.887	88.849	14	627	1	CG
3/31/2002	11:13:36.200	32.150	88.553	42	587	1	CG
3/31/2002	11:13:36.000	31.725	88.720	22	633	1	CG
3/31/2002	11:13:16.000	35.963	84.097	8	5	2	CG
3/31/2002	11:13:15.900	35.941	84.086	12	2	1	CG
3/31/2002	11:13:03.800	31.769	88.768	18	632	3	CG

*CG = cloud to ground.

TABLE 11.4   Lightning Stroke Data

Industrial Power Quality Monitoring Applications

- Energy and demand profiling with identification of opportunities for energy savings and demand reduction
- Harmonics evaluations to identify transformer loading concerns, sources of harmonics, problems indicating mis-operation of equipment (such as converters), and resonance concerns associated with power factor correction
- Voltage sag impacts evaluation to identify sensitive equipment and possible opportunities for process ride-through improvement
- Power factor correction evaluation to identify proper operation of capacitor banks, switching concerns, resonance concerns, and optimizing performance to minimize electric bills
- Motor starting evaluation to identify switching problems, inrush current concerns, and protection device operation
- Short-circuit protection evaluation to evaluate proper operation of protective devices based on short-circuit current characteristics, time-current curves, etc.

Power System Performance Assessment and Benchmarking

- Trending and analysis of steady-state power quality parameters (voltage regulation, unbalance, flicker, harmonics) for performance trends, correlation with system conditions (capacitor banks, generation, loading, etc.), and identification of conditions that need attention

- Voltage sag characterizing and assessment to identify the cause of the voltage sags (transmission or distribution) and to characterize the events for classification and analysis (including aggregation of multiple events and identification of subevents for analysis with respect to protective device operations)

- Capacitor-switching characterization to identify the source of the transient (upline or downline), locate the capacitor bank, and characterize the events for database management and analysis

- Performance index calculations and reporting for system benchmarking purposes and for prioritizing of system maintenance and improvement investments

Applications for System Maintenance, Operations, and Reliability

- Locating faults. This is one of the most important benefits of the monitoring systems. It can improve response time for repairing circuits dramatically and also identify problem conditions related to multiple faults over time in the same location.

- Capacitor bank performance assessment. Smart applications can identify fuse blowing, failures, switch problems (restrikes, reignitions), and resonance concerns.

- Voltage regulator performance assessment to identify unusual operations, arcing problems, regulation problems, etc.

- Distributed generator performance assessment. Smart systems should identify interconnection issues, such as protective device coordination problems, harmonic injection concerns, islanding problems, etc.

- Incipient fault identifier. Research has shown that cable faults and arrester faults are often preceded by current discharges that occur weeks before the actual failure. This is an ideal expert system application for the monitoring system.

- Transformer loading assessment can evaluate transformer loss of life issues related to loading and can also include harmonic loading impacts in the calculations.

- Feeder breaker performance assessment can identify coordination problems, proper operation for short-circuit conditions, nuisance tripping, etc.

11.5.4 Power Quality Monitoring and the Internet

Many utilities have adopted power quality monitoring systems to continuously assess system performance and provide faster response to system problems. It is clear that intranet and Internet access to the information has been key to the success of these systems. An example of a completely Web based power quality monitoring system is the result of research initiated by TVA and EPRI (Fig. 11.40). Specifications for the system were developed with the help of all the members of the EPRI Power Quality Target group to support the variety of applications which must be supported by such a system. The result was a modular system with a completely open architecture so that it can be interfaced with a wide variety of platforms.

After helping with the development of the system, TVA is deploying the Web-based monitoring systems at important customers and substations throughout their system. TVA distributors are also taking advantage of the system. It already had an extensive power quality monitoring system in place, and the new system is integrated with the existing monitoring system infrastructure at the central data management level (enterprise level), as illustrated in Fig. 11.41. This provides the capability to provide systemwide analysis of the power quality information.

The future of these systems involves integration with other data-collection devices in the substation and the facility. Standard

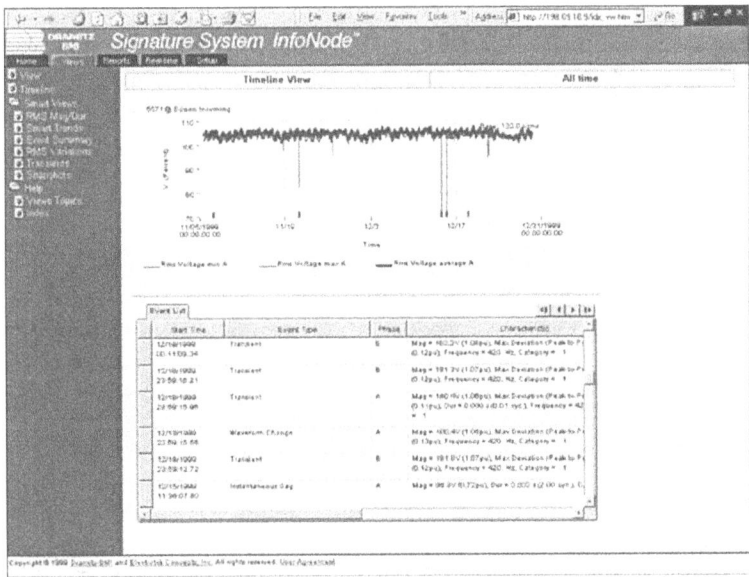

FIGURE 11.40 Example of Web-based interface to the power quality monitoring system for easy evaluation of system performance and individual disturbances.

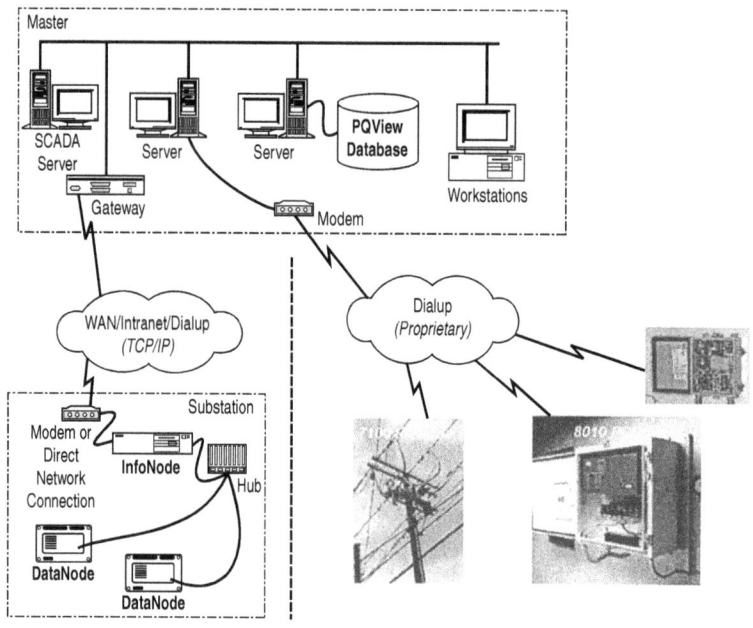

Figure 11.41   Enterprise-level integration of power quality information from a Web-based power quality monitoring system with the overall monitoring system database at TVA.

interfaces like the Power Quality Data Interchange Format (PQDIF) and COMTRADE are used to share the information, and standard protocols like UCA are used for the communications. The intelligent applications described will be applied at both the substation level and at the enterprise level, as appropriate.

11.5.5   Summary and Future Direction

Power quality monitoring is fast becoming an integral component of general distribution system monitoring, as well as an important customer service. Power producers are integrating power quality monitoring with monitoring for energy management, evaluation of protective device operation, and distribution automation functions. The power quality information should be available throughout the company via the intranet and should be made available to customers for evaluation of facility power conditioning requirements.

The power quality information should be analyzed and summarized in a form that can be used to prioritize system expenditures and to help customers understand the system performance. Therefore, power quality indices should be based on customer equipment sensitivity. The SARFI index for voltage sags is an excellent example of this concept.

Power quality encompasses a wide range of conditions and disturbances. Therefore, the requirements for the monitoring system can be quite substantial, as described in this chapter. Table 11.5 summarizes the basic requirements as a function of the different types of power quality variations.

The information from power quality monitoring systems can help improve the efficiency of operating the system and the reliability of customer operations. These are benefits that cannot be ignored. The capabilities and applications for power quality monitors are continually evolving. Ongoing development and new applications are described on various Internet sites. One such useful site is www.powermonitoring.com.

11.6 Power Quality Monitoring Standards

Standards are very important in the area of power quality monitoring. Power quality levels must be defined consistently and characterized using the same methods if they are going to be compared from one site to another and from one system to another. IEEE 1159 is the IEEE Working Group that coordinates the development of power quality monitoring standards. The existing IEEE 1159 provides general guidelines and definitions for monitoring power quality, and there are three separate task forces that are working on more specific guidelines and requirements. Much of this work is being coordinated with IEC activities so that the monitoring requirements can be more consistent internationally. This section describes the most important IEEE and IEC standards. These standards are a moving target as there are many groups working on new developments.

11.6.1 IEEE 1159: Guide for Power Quality Monitoring

IEEE Standard 1159[15] was developed to provide general guidelines for power quality measurements and to provide standard definitions for the different categories of power quality problems. These definitions were provided previously in Chap. 2 and provide the basis for a common language in describing power quality phenomena. Power quality monitoring equipment can use this language to correctly differentiate between different power quality variations and disturbances.

After publication of the basic monitoring guidelines, working groups were established for development of more advanced guides for power quality monitoring. Three working groups were established. Progress can be tracked at the IEEE 1159 website: http://grouper.ieee.org/groups/1159/.

The IEEE 1159.1 Working Group is developing guidelines for instrumentation requirements associated with different types of power quality phenomena. These requirements address issues like

Type of power quality variation		Requirements for monitoring	Analysis and display requirements
Voltage regulation and unbalance		Three-phase voltages RMS magnitudes Continuous monitoring with periodic maximum, minimum, and average samples Currents for response of equipment	Trending Statistical evaluation of voltage levels and unbalance levels
Harmonic distortion		Three-phase voltages and currents Waveform characteristics 128 samples per cycle minimum Synchronized sampling of all voltages and currents Configurable sampling characteristics	Individual waveforms and FFTs Trends of harmonic levels (THD and individual harmonics) Statistical characteristics of harmonic levels Evaluation of neutral conductor loading issues Evaluation with respect to standards (e.g., IEEE 519, EN 50160) Evaluation of trends to indicate equipment problems
Voltage sags, swells, and short-duration interruptions		Three-phase voltages and currents for each event that is captured Configurable thresholds for triggering events	Waveform plots and rms vs. time plots with pre- and postevent information included Evaluation of cause of each event (fault upline or downline from the monitoring) Voltages and currents to evaluate load interaction issues

	Characteristics of events with actual voltage and current waveforms, as well as rms vs time plots RMS resolution of 1 cycle or better during the rms vs. time events and for triggering	Magnitude duration plots superimposed with equipment ride-through characteristics (e.g., ITI curve or SEMI curve) Statistical summary of performance (e.g., bar charts) for benchmarking Evaluation of power conditioning equipment performance during events
Transients	Three-phase voltages and currents with complete waveforms Minimum of 128 samples per cycle for events from the power supply system (e.g., capacitor switching) Configurable thresholds for triggering Triggering based on waveform variations, not just peak voltage	Waveform plots Evaluation of event causes (e.g., capacitor switching upline or downline from monitor) Correlation of events with switching operations Statistical summaries of transient performance for benchmarking

TABLE 11.5 Summary of Monitoring Requirements for Different Types of Power Quality Variations

sampling rate requirements, synchronization, A/D sampling accuracy, and number of cycles to sample.

The IEEE 1159.2 Working Group is developing guidelines for characterizing different power quality phenomena. This includes definition of important characteristics that may relate to the impacts of the power quality variations (such as minimum magnitude, duration, phase shift, and number of phases for voltage sags). Example waveforms have been collected illustrating the importance of different characteristics of the power quality variations.

Recently, the work of the IEEE 1159.1 and 1159.2 Working Groups has been combined into a single task force and is being coordinated with the development on an international standard for characterizing power quality variations with monitoring equipment—IEC 61000–4–30[16] (see Sec. 11.6.2).

The IEEE 1159.3 Working Group is defining an interchange format that can be used to exchange power quality monitoring information between different applications. IEEE developed the COMTRADE format for exchanging waveform data between fault recorders and other applications, such as relay testing equipment. A more complete data interchange format is needed for power quality data, which can include harmonic spectra, rms envelopes, characterized power quality data, and statistical power quality data, as well as steady-state and disturbance waveforms. The new Power Quality Data Interchange Format (PQDIF) has been defined, and the standard is being balloted at the time of this writing.[17] The common data interchange format will allow software developers to develop applications for analyzing power quality events and problems independently from the manufacturers of the actual power quality monitoring equipment.

11.6.2   IEC 61000–4–30: *Testing and Measurement Techniques—Power Quality Measurement Methods*

IEC standards for monitoring power quality phenomena are provided in a series of documents with the numbers 61000–4-xx. The individual standards in this series cover specific requirements for each type of power quality variation or concern. For instance, IEC 61000–4–7 provides the specifications for monitoring harmonic distortion levels. IEC 61000–4–15 provides the specifications for monitoring flicker, as previously described. The overall requirements for characterizing power quality phenomena are summarized in a new standard that is just being completed within IEC (61000–4–30). This new standard refers to the appropriate individual standards (like 61000–4–7 and 61000–4–15) for detailed specifications where appropriate.

This standard provides detailed requirements for the measurement procedures and the accuracy requirements of the measurements. Not all monitoring equipment will be able to meet the exact requirements of this standard. As a result, two classes of measurement equipment have been defined which can both be considered compliant with the procedures of IEC 61000–4–30:

- Class A performance is for measurements where very precise accuracy is required. Two instruments that comply with the requirements of class A should give the same results (within the specified levels of accuracy) for any of the types of power quality variations considered. These instruments could be appropriate for laboratories or for special applications where highly precise results are required.

- Class B performance still indicates that the recommended procedures for characterizing power quality variations are used but that the exact accuracy requirements may not be met. These instruments are appropriate for most system power quality monitoring (surveys, troubleshooting, characterizing performance, etc.).

The concept of aggregation is also introduced in this standard. Aggregation is used so that multiple measurements that are associated with essentially the same event are not counted multiple times. For example, multiple voltage sags caused by reclosing operations should only be counted as a single event for evaluating the impact on customers and the number of problem events on the system. Three different aggregation intervals are defined in IEC 61000–4–30: 3 s, 10 min, and 2 h.

As mentioned, the work in IEC 61000–4–30 is also becoming the basis for updates and enhancements to the IEEE power quality monitoring standards (IEEE 1159 series). This is part of the general trend toward internationalizing power quality standards.

These aggregation periods are also very important intervals for characterizing steady-state power quality variations like voltage magnitude, unbalance, harmonics, and flicker. All these quantities are described statistically using 10-min values as the most important quantity.

Note that the basic measurement period for the steady-state power quality parameters is 200 ms. This permits characterization of interharmonics in 5-Hz bins, and it provides some smoothing of very fast changes that should not be considered part of steady-state power quality performance. Of course, voltage sags and transients are characterized with actual waveforms and rms versus time plots.

11.7 References

1. EPRI-RP3098–01, *An Assessment of Distribution System Power Quality.*
2. A. McEachern, "Roles of Intelligent Systems in Power Quality Monitoring: Past, Present, and Future," *Conference Record,* Power Engineering Society Summer Meeting, 2001, Vol. 2, pp. 1103–1105.
3. R. P. Bingham, "Recent Advancements in Monitoring the Quality of the Supply," Power Engineering Society Summer Meeting, 2001, Vol. 2, pp. 1106–1109.
4. Dranetz Engineering Laboratories, "Series 606 Power-Line Disturbance Analyzer," December 1975.
5. E. W. Gunther and J. Rossman, "Application of Advanced Characterization Algorithms, UCA and Internet Communications Technology at the Point of Power Quantity and Quality Measurement," Conference Proceedings of EPRI PQA 1999.
6. IEEE Standard 141–1993: *Recommended Practice for Power Distribution in Industrial Plants.*
7. IEEE Standard 519–1992: *Recommended Practices and Requirements for Harmonic Control in Electrical Power Systems.*
8. IEC 61000–4–15, *Electromagnetic Compatibility (EMC).* Part 4: Testing and Measuring Techniques. Section 15: Flickermeter—Functional and Design Specifications.
9. D. A. Douglass, *Potential Transformer Accuracy at 60-Hz Voltages above and below Rating and at Frequencies above 60 Hz.* Presented at the IEEE Power Engineering Society Summer Meeting, Minneapolis, Minn., July 13–18, 1980.
10. D. A. Douglass, "Current Transformer Accuracy with Asymmetric and High Frequency Fault Currents," *IEEE Transactions on Power Apparatus on Systems,* Vol. PAS-100, No. 3, March 1981.
11. S. Santoso and J. D. Lamoree, "Answer Module: A Custom-Built Module to Meet Specific Power Monitoring Tasks," Conference Proceedings of EPRI PQA 2001, Pittsburgh, Pa.
12. U. Fayyad, G. Piatetsky-Shapiro, P. Smyth, "From Data Mining to Knowledge Discovery: An Overview," in U. Fayyad, G. Piatetsky-Shapiro, P. Smyth, R. Uthurusamy, eds., *Advances in Knowledge Discovery and Data Mining,* MIT Press, 1996, pp. 471–494.
13. S. Santoso, R. C. Dugan, J. D. Lamoree, A. Sundaram, "Distance Estimation Technique for Single Line-to-Ground Faults in a Radial Distribution System," *Conference Record,* Power Engineering Society Winter Meeting, 2000 IEEE, Vol. 4, pp. 2551–2555.
14. S. Santoso, J. D. Lamoree, M. F. McGranaghan, "Signature Analysis to Track Capacitor Switching Performance," *Conference Record,* Transmission and Distribution Conference and Exposition, 2001 IEEE/PES, Vol. 1, pp. 259–263.
15. IEEE Standard 1159–1995, *Recommended Practice on Monitoring Electric Power.*
16. Draft Standard IEC 61000–4–30 77A/356/CDV, *Power Quality Measurement Methods.*
17. IEEE Draft Standard P1159.3, *Recommended Practice for a Power Quality Data Interchange Format—An Extensible File Format for the Exchange of Power Quality Measurement and Simulation Data,* SCC 22.

11.8 Bibliography

Cokkinides, C. J., Banta, L. E., Meliopoulos, A. P., "Transducer Performances for Power System Harmonic Measurements," *Proceedings of the International Conference on Harmonics,* Worcester, Mass., October 1984.

"Computation of Current Transformer Transient Performance," *IEEE Transactions on Power Delivery,* Vol. PWRD-3, No. 4, October 1988.

Greenwood, A. N., *Electrical Transients in Power Systems,* 2d ed., John Wiley & Sons, New York, 1991, chap. 18.

McShane, E. L., and Colbaugh, M. E., *Advance Current and Voltage Transformers for Power Distribution Systems,* PRI Report EL-6289, 1989.

Index